INTERNATIONAL UNION OF PURE AND APPLIED CHEMISTRY

ENVIRONMENTAL ANALYTICAL AND PHYSICAL CHEMISTRY SERIES

Environmental Particles

Volume 2

ENVIRONMENTAL ANALYTICAL AND PHYSICAL CHEMISTRY SERIES

Environmental Particles

Volume 2

Volume Editors

Jacques Buffle
Department of Inorganic,
Analytical, and Applied Chemistry
University of Geneva
Geneva, Switzerland

Herman P. van Leeuwen
Laboratory for Physical
and Colloid Chemistry
Agricultural University
Wageningen, Netherlands

CRC Press
Taylor & Francis Group
Boca Raton London New York

CRC Press is an imprint of the
Taylor & Francis Group, an **informa** business

First published 1993 by Lewis Publishers
Taylor & Francis Group
6000 Broken Sound Parkway NW, Suite 300
Boca Raton, FL 33487-2742

Reissued 2018 by CRC Press

A Library of Congress record exists under LC control number: 91035911

Publisher's Note
The publisher has gone to great lengths to ensure the quality of this reprint but points out that some imperfections in the original copies may be apparent.

Disclaimer
The publisher has made every effort to trace copyright holders and welcomes correspondence from those they have been unable to contact.

ISBN 13: 978-1-138-57620-9 (hbk)
ISBN 13: 978-1-138-55411-5 (pbk)
ISBN 13: 978-1-351-27080-9 (ebk)

Visit the Taylor & Francis Web site at http://www.taylorandfrancis.com and the CRC Press Web site at http://www.crcpress.com

About the Editors

Jacques Buffle is Professor of analytical and environmental chemistry. He joined the staff of the Department of Inorganic, Analytical, and Applied Chemistry of the University of Geneva, Switzerland in 1969. He is also an Invited Professor at INRS-Eau, University of Quebec, Canada, since 1982 and has been Associate Professor of Electroanalytical Chemistry at the University of Lausanne Switzerland. He is presently the Chairman of the IUPAC Commission of Environmental Analytical Chemistry and a member of the Research Council of the Swiss National Foundation. He has authored 100 research papers, 2 books, and is a co-editor of 3 other monographs. He teaches under- and postgraduate courses in Analytical Chemistry and Environmental Chemistry and is actively involved in organizing postgraduate courses in limnology.

Dr. Buffle has received diplomas in biological chemistry, chemical engineering, and numerical calculation, and a Ph.D. degree in analytical chemistry. His research interests are the understanding of physico-chemical processes regulating the circulation of chemical components in environmental compartments, the relative influences of these processes and the development of biota, and the development of *in situ* sensors for the measurement of the corresponding key, physico-chemical parameters. He is particularly interested in contributing to the understanding of the control of macroscale effects, at the level of environmental compartments such as soils, lakes, etc., by microscale phenomena, at the molecular, colloidal, and microbial level, by considering multidisciplinary approaches.

Herman P. van Leeuwen is an electrochemist who received his education at the State University of Utrecht, The Netherlands. He obtained his degree in chemistry, with a main specialization in electrochemistry, in 1969. He then joined the electrochemistry group of Professor J. H. Sluyters where he prepared his thesis in the field of pulse methods in electrode kinetics. The Ph.D. degree was awarded cum laude in 1972. In the period from 1968 until 1973 he also was a part-time teacher of chemistry. In 1972 he joined the colloid chemistry and electrochemistry group of Professor J. Lyklema at the Wageningen Agricultural University, where he became a senior scientist, in 1978, and associate professor of electrochemistry, in 1986. He has been active in teaching analytical chemistry, physical chemistry, and electrochemistry.

His current research activities include electrodynamics of colloids (in relation to colloid stability), voltammetric speciation of heavy metals in environmental systems, and ion binding by synthetic and natural polyelectrolytes. He has published some 70 research papars, reviews, and book chapters in these fields. He has an intensive cooperation with colleagues from Czechoslovakia, Portugal, Spain, Switzerland, and the U.K. He is currently the secretary of the Electrochemistry Working Group of the Royal Dutch Chemical Society, associate member of the IUPAC Commission on Electroanalytical Chemistry, and titular member and secretary of the IUPAC Commission on Environmental Analytical Chemistry.

Preface

At the time of preparing Volume 1 of this series, it was already clear that the topic of environmental particles has a very broad scope and impact. Further volumes would be necessary to cover at least a reasonable number of the relevant aspects. At an early stage of the preparation of Volume 1, we therefore decided to initiate a second volume. The intention was to place some emphasis on spectroscopic and colloid chemical aspects. We have been fortunate enough to find the internationally acknowledged experts in these fields willing to contribute critical, authoritative reviews. And, as with Volume 1, we attained significant knowledge from the plenary discussion of the draft chapters between all the contributing authors (Davos, October, 1991).

Volume 2 continues the series in the sense that we try to contribute to the development of environmental analytical and physical chemistry. We deliberately extend the treatment from purely analytical aspects of sampling and characterization to the initiation and development of new concepts in the analysis of physicochemical processes and interactions in real environmental systems. Thus, we feel, environmental chemistry is best served.

The preparation of this volume was realized within the frame of our activities as an IUPAC commission. And we thank the responsible IUPAC officers for their most constructive reviews and their prompt help in bringing the various chapters of this volume to their final form. In particular, we would like to express our gratitude to Professors Hulanicki, Svehla, and Grasserbauer, who carefully read through all the draft chapters. We would like to thank the International Council of Scientific Unions (ICSU) for financial support of the scientific work of the commission. Thanks are also due to the European Environmental Research Organization (EERO) for supporting the plenary discussion meeting in Davos, Switzerland.

J. Buffle
H. P. van Leeuwen

Contributors

Ronald Beckett
Water Studies Centre and Department of Chemistry
Monash University
Melbourne, Australia

Jacques Buffle
Department of Inorganic, Analytical, and Applied Chemistry
University of Geneva
Geneva, Switzerland

Laurent Charlet
Laboratoire de Geophysique Interne et Tectonophysique
Université Joseph Fourier
Grenoble, France

William Davison
Institute of Environmental and Biological Science
Lancaster University
Lancaster, England

Richard R. De Vitre
Department of Inorganic, Analytical, and Applied Chemistry
University of Geneva
Geneva, Switzerland

Claude Deguelde
Paul Sherrer Institut
Wurenlingen/Villigen, Switzerland

W. L. Earl
Los Alamos National Laboratory
Los Alamos, New Mexico

Barry T. Hart
Water Studies Centre and Department of Chemistry
Monash University
Melbourne, Australia

George A. Jackson
Department of Oceanography
Texas A & M University
College Station, Texas

Clifford T. Johnston
Department of Soil Science
University of Florida
Gainesville, Florida

Steve Lochman
Department of Oceanography
Texas A & M University
College Station, Texas

Alain Manceau
Laboratoire de Minéralogie – Cristallographie
University of Paris
Paris, France

John F. McCarthy
Environmental Sciences Division
Oak Ridge National Laboratory
Oak Ridge, Tennessee

Meredith E. Newman
Department of Inorganic, Analytical, and Applied Chemistry
University of Geneva
Geneva, Switzerland

Charles R. O'Melia
Department of Geography and
Environmental Engineering
Johns Hopkins University
Baltimore, Maryland

Peter Schurtenberger
Polymer Institute
Swiss Federal Institute of
Technology
Zurich, Switzerland

Garrison Sposito
Department of Soil Science
University of California
Berkeley, California

Christine L. Tiller
Department of Geography and
Engineering
Johns Hopkins University
Baltimore, Maryland

R. Van Grieken
Micro- and Trace Analysis Centre
Department of Chemistry
University of Antwerp
Antwerp, Belgium

C. Xhoffer
Micro- and Trace Analysis Centre
Department of Chemistry
University of Antwerp
Antwerp, Belgium

Herman P. van Leeuwen
Laboratory for Physical and
Colloid Chemistry
Agricultural University
Wageningen, Netherlands

Contents

CHAPTER 1

SURFACE SPECTROSCOPY OF ENVIRONMENTAL PARTICLES BY FOURIER-TRANSFORM INFRARED AND NUCLEAR MAGNETIC RESONANCE SPECTROSCOPY

Clifford T. Johnston, Garrison Sposito, and W. L. Earl

TABLE OF CONTENTS

0-87371-895-X/93/$0.00+$.50

INTRODUCTION

Need for Mechanistic Information

Historically, most studies of the reactions between environmental particle surfaces and aqueous solutions have been limited to providing chemical composition data.[1,2] These kinds of data are essential to characterize the stoichiometry and, to some extent, the overall pathways of a surface reaction, but they are not usually sensitive to the details of molecular configuration needed in order to deduce the reaction mechanism. To do this, molecular spatial and temporal scales on particle surfaces must be resolved through spectroscopic methods adapted to heterogeneous systems containing water.

The molecular spectroscopy of adsorbed chemical species has two principal subdivisions: invasive methods, such as X-ray photoelectron or secondary ion mass spectrometry, that require sample desiccation and high vacuum, and noninvasive methods that require little or no alteration of a sample from its natural state. Invasive surface spectroscopies have an important role to play in the characterization of solid particle surfaces, but to use them for resolving surface speciation on particles normally suspended in aqueous solution simply begs the question. Noninvasive surface spectroscopies are those that can be applied in the presence of liquid water; most involve the input and detection of photons. The best known examples are nuclear magnetic resonance (NMR), electron spin resonance (ESR), Raman, Fourier-transform infrared (FTIR), UV-visible fluorescence, X-ray absorption, and Mössbauer spectroscopies, although Brown[3] has enumerated many others that are available to detect adsorbed molecules. These methods, two of which are the topic of this chapter, share the important feature that they can be used not only noninvasively but also in conjunction with *in situ* molecular probes. This means that, in principle, a homologous adsorptive probe can be introduced into a sample, without significantly affecting the indigenous surface characteristics, while permitting or enhancing the use of a noninvasive spectroscopic technique. The fundamental premise of this approach is that the particle surface will respond to the probe following the same molecular mechanisms that exist and are utilized in the absence of the probe. This broad experimental strategy and the prototypical results expected from it will be illustrated in the following sections with a variety of data for aqueous colloidal systems.

Vibrational and NMR Spectroscopy of Surfaces

FTIR Spectroscopy

During the past 15 years, FTIR spectrometers have replaced dispersive instrumentation for most applications. The optical throughput of an FTIR spectrometer is intrinsically greater ("the Jacquinot advantage") than for a dispersive monochromator, where most of the incident energy is discarded and only a narrow spectral element is transmitted through the instrument. A second advantage of FTIR spectroscopy is that radiation at all wavenumbers is measured simultaneously as opposed to sequentially with a grating monochromator. Quantitatively, this advantage (the multiplex or "Fellgett's advantage") is directly proportional to the square root of the number of spectral elements (i.e., the optical resolution) obtained. A thorough comparison of the performance and optical efficiency of FTIR to those of dispersive IR spectrometers is made in a number of recent monographs.[4-7] The overall efficiency advantage of FTIR over dispersive IR has been estimated to be between 10 and 200 times, depending on the spectral region.[5,6]

Application of IR methods to the study of adsorbed species has a long history and is the subject of numerous reviews.[4-6,8-12] A number of sample presentation methods designed for surface studies have been developed over the past 10 years. Most require the higher throughput of modern FTIR spectrometers. Of most consequence to IR studies of environmental particles are the sample presentation methods of diffuse reflectance (DR),[5,6,12-20] photoacoustic (PAS),[6,19,21-25] and attenuated-total-reflectance (ATR).[26,27] The most common sample presentation method for both IR and FTIR studies of mineral surfaces, however, continues to be transmission spectroscopy. Included in this category are methods using KBr pellets, self-supporting clay films, deposits on IR windows, and compressed wafers. Surface studies and descriptions of transmission cells using this technique have been reviewed extensively.[4-7,9,10,12,28]

NMR Spectroscopy

The use of NMR spectroscopy to study surfaces has a shorter history and fewer applications than vibrational spectroscopies, the primary reason being that the sensitivity of NMR is intrinsically much lower than IR.[29] There are several NMR techniques for measuring pore sizes and surface areas in porous solids that have been exploited over the years (see below). Early NMR studies of surfaces concentrated on ^{1}H nuclei because of sensitivity. These studies frequently relied on measurements of relaxation times, which also made them sensitive to the method of interpretation (i.e., the model used to describe relaxation). The advent of high-resolution, solid-state NMR techniques, such as magic angle sample spinning (MAS) and cross-polarization (CP), along with more sensitive, high magnetic field, user-friendly, pulsed NMR spectrometers, has brought increased applications to heterogeneous aqueous systems. In particular, ^{27}Al and ^{29}Si NMR in zeolites and other minerals have proven valuable for the structural elucidation of samples whose disorder has prevented diffraction techniques from being very useful.[30-41] However, NMR is essentially a bulk spectroscopic technique, which implies that there must be some property of a particle surface that allows discrimination between spectral peaks from surface species and those from species in bulk. Properties that might be exploited are the "chemical shift," NMR relaxation times, and magnetic couplings to nuclei that are characteristic of a surface.[38] The low intrinsic sensitivity and bulk-probe character of NMR spectroscopy also dictate that surface studies be performed on samples with relatively high specific surface areas.

Spectroscopic Characterization of Environmental Particles

Surface studies of natural colloids are still in their infancy, so most of the samples to be discussed in this chapter were prepared from purified adsorbents, such as specimen clay minerals, hydrous metal oxides, and zeolites. The details of NMR and IR spectroscopy applied to these model systems must be resolved before attempting work on natural samples with multiple adsorbents and adsorbates and extremely complex spectra.

There are three main reasons for emphasizing both IR and NMR spectroscopic studies of both natural and model particles. The first is that both techniques offer potential for nondestructive analyses. Aquatic surface chemistry occurs at hydrated interfaces, and there is a need for analytical methods that can probe surface species in the presence of water. Vibrational spectroscopy is recognized as a sensitive method, but interferences from water, trace impurities, and surface hydroxyls have limited its use on environmental particles. Recent developments in ATR-FTIR as well as Raman spectroscopy have alleviated these problems.[27,42-44] Like IR, NMR spectroscopy is a nondestructive method and has the additional advantage of utilizing radiofrequency radiation that does not require the samples to be transparent to infrared light. The primary disadvantage of NMR spectroscopy is that the signal is attenuated by conducting samples and is very strongly perturbed by paramagnetic species. For this reason, researchers have shied away from NMR methods for mineral particles that contain large amounts of Fe(III).

A second reason for using IR and NMR spectroscopies lies in the many recent improvements in sensitivity and spectral discrimination for these techniques. The introduction of continuous wave near-IR lasers and quantum mid-IR detectors, and the development of high-field magnets all have provided improvements in sensitivity, allowing the detection of adsorbates at submonolayer coverage. Novel techniques also have contributed by enhancing the spectroscopic signature of surface species and by eliminating interferences from species such as water molecules or those in the bulk adsorbent.[45] There are also improved techniques for spectral resolution, such as two-dimensional NMR spectroscopy.

Finally, the use of combined IR and NMR spectroscopic techniques is emerging as a significant advantage in studying heterogeneous systems. A particular spectroscopy has strengths and limitations that are different from those for another method because spectroscopic information always is related critically to the space, time, and energy scales probed. Thus, NMR is sensitive to motions with time scales in the range 10^{-3} to 10^{-6} s, whereas IR is sensitive to much faster motions on the smaller time scale of about 10^{-12} s. Since particle surfaces have been probed via macroscopic measurements traditionally, it is critical to correlate these spectroscopic scales with the scales of macroscopic information.[46] Thus, a multiple-technique approach provides a more complete picture of the particle system of interest.

Molecular Probes and Reporter Groups

Generally speaking, surface studies answer questions relating to the chemical identity of adsorbed species, their location on an adsorbent, including what functional groups are involved, and their molecular dynamics. One approach to characterize adsorbent surfaces is to use *in situ* molecular probes that can explore surface functionality or adsorption mechanisms. In this chapter, *in situ* molecular probes are defined as any solute or solvent species that is capable of interacting with a solid surface and is sensitive to changes in its local environment through diagnostic NMR or IR properties. Examples of these diagnostic properties are the vibrational modes of coordinated functional groups and the NMR chemical shift. When a probe interacts with a surface, its diagnostic properties are perturbed (e.g., changes in the intensity of a vibrational mode or in the chemical shift), and these changes are a powerful way to explore solid-solution and solid-vapor interfaces to gain insight into their chemical and physical properties.

Reporter groups provide another way of obtaining information about a particle surface. As defined in this chapter, reporter groups are part of the molecular structure of the adsorbent itself, but have diagnostic properties that are perturbed by reactions at an interface. Examples of reporter groups are structural hydroxyl groups and NMR-active structural nuclei, such as ^{29}Si and ^{27}Al. Their diagnostic properties include vibrational modes (e.g., O-H stretching and bending motions), chemical shifts, and motional behavior evidenced in NMR relaxation. For the purposes of this chapter, vibrational and NMR studies of natural particles will be distinguished on the basis of whether they use either molecular probes or reporter groups. Many of the studies reviewed, however, have used both.

The distinction between molecular probes and reporter groups can be illustrated by a proposed sorption mechanism for phosphate on goethite,[47-51] shown in Figure 1. Phosphate is a molecular probe of the goethite surface, in that its diagnostic vibrational modes are perturbed by inner-sphere complexation on the goethite surface. On the other hand, the surface hydroxyl groups on goethite function as reporter groups. If inner-sphere complexation reactions occur at the interface, the vibrational stretching bands of the surface hydroxyl groups should decrease in intensity because of replacement of Fe-OH by $Fe\text{-}PO_4$ via ligand exchange.

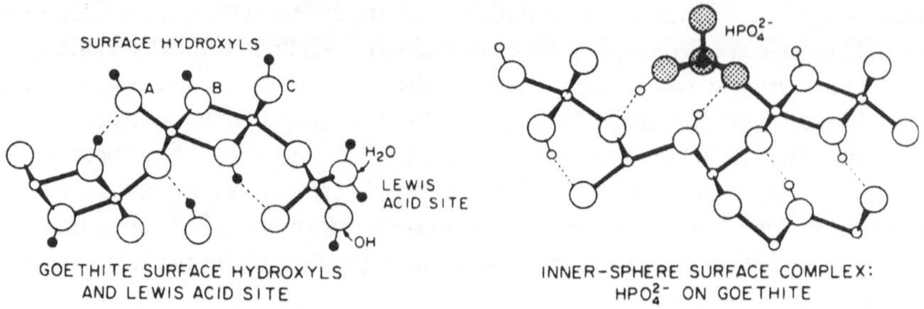

Figure 1. (a and b) Types of surface hydroxyl groups on goethite. Type A, B, and C groups are singly, triply, and doubly coordinated to Fe(III) ions (one Fe-O bond not represented for type B and C groups) and a Lewis acid site. (b) Phosphate adsorbed onto a Type A site. (From Sposito, G. *The Surface Chemistry of Soils.* New York: Oxford University Press, 1984. With permission.)

MOLECULAR PROBE STUDIES OF ENVIRONMENTAL SURFACES

Surface Acidity

The ability of mineral surfaces to donate protons to (Brønsted acidity) or to accept electrons from (Lewis acidity) interfacial species has direct application to the fate and transport of chemicals in soil and other natural environments, to heterogeneous catalysis, and to clay mineralogy. The surface acidity of colloids can be investigated with temperature-programmed desorption and titration methods and with IR, Raman, or NMR spectroscopy, utilizing surface probe molecules in order to characterize both the nature of the active sites (e.g., Lewis versus Brønsted sites) and their total number.

Two vibrational probe molecules used commonly to characterize Lewis and Brønsted acidity on mineral surfaces are pyridine and ammonia,[52-56] which exhibit vibrational modes that are influenced by the acidic nature of a surface. The ammonium cation, NH_4^+, for example, can be distinguished readily from NH_3 by the IR-active ν_4 deformation band of NH_4^+ at 1430 cm^{-1}. Mortland and Raman[55] used this band to measure the equilibrium constant for the conversion of ammonia to ammonium as a function of adsorbed water content in smectites containing different exchangeable metal cations, $\equiv M(H_2O)_x^{n+}$:

$$M(H_2O)_x^{n+} + NH_3 =\equiv M(H_2O)_{x-1}OH_{n-1} + NH_4^+ \quad (M = Na, Ca, Al) \quad (1)$$

They found that the proton-donating ability of the surface increased as the water content decreased and the hydration energy of the exchangeable cation increased, in the order: Al > Mg > Ca = Li > Na > K. Thus, Al^{3+}-smectite was a much better proton donor than hydrated Na^+-smectite. This increase in Brønsted acidity was attributed to increased polarization effects of the cation on coordinated water molecules[46,55] with increased ionic potential.

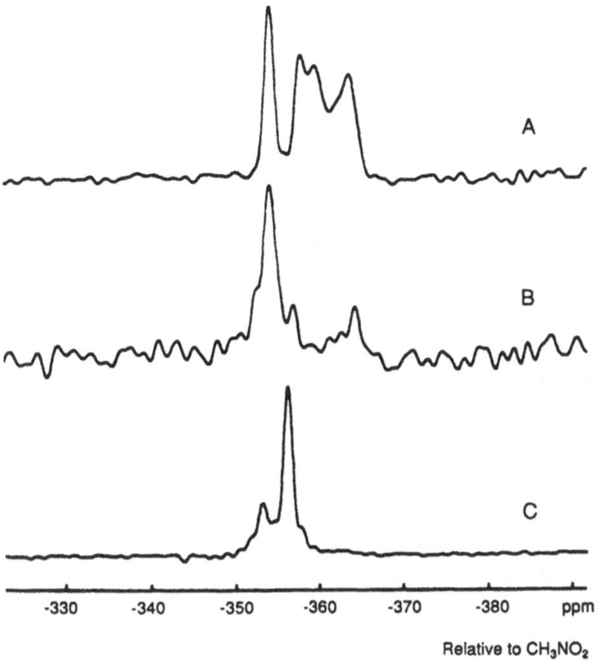

Figure 2. Spectra demonstrating the effect of evacuation and thermal treatment on the CP MAS NMR of $^{15}NH_4$ adsorbed on H-Y zeolites: (A) spectrum taken promptly after the NH_3 saturated zeolite was transferred to the rotor in an Ar atmosphere; (B) spectrum taken after the saturated sample was evacuated for 16 h at 25°C; (C) spectrum of a similar sample taken after the sample was evacuated for 1.5 h at 160°C. (From Earl, W.L. et al. *J. Phys. Chem.* 91:2091–2095 (1987). With permission.)

The use of NH_3 to characterize acidic sites on mineral surfaces can also benefit from NMR spectroscopy. In this case, the most sensitive methodology is to use ^{15}N-enriched ammonia and study the ^{15}N NMR resonance. Earl et al.[56] combined IR and CP MAS NMR spectroscopic methods to study the acidic sites on a H-Y zeolite. Although the presence of the ν_4 band confirmed the presence of NH_4^+, the ^{15}N CP MAS NMR spectra provided more definitive information about chemical differences between the adsorbed species. Figure 2 demonstrates the many different shifts obtained because of multiple sites on the zeolite. The upper trace immediately after adsorption of the NH_4^+ indicates the presence of at least five sites. After evacuation and heating, some of the resonances change in intensity indicating different stabilities for the NH_4^+ — surface complex at the different sites. A number of other studies of ammonia or amine adsorption on particle surfaces have focused primarily on zeolites. Michel et al.[57] have made excellent use of both solution and solid state ^{15}N NMR techniques to characterize acid sites on a variety of zeolites, and others have used the ^{13}C NMR of the tetramethylammonium ion to investigate site occupancy in zeolites.[58,59] This latter

Table 1. Position of IR- and Raman-Active Bands of Sorbed Pyridine

	ν_1 (R)	ν_{12} (R)	ν_{19b} (IR)	ν_{8a} (IR)
H-bonded pyridine	1002	1035	1440–1447	1580–1600
Pyridinium ion (Brønsted acidity)	1010	1029	1535–1550	1640
Coordinatively bound (Lewis acidity)	1025	1050	1440–1464	1600–1634

Sources: Blanco et al.,[52] Morterra et al.,[53] Egerton.[54]

technique offers some disadvantages over ^{15}N NMR spectroscopy alone because the major chemical differences occur at the nitrogen center, however, the intrinsic sensitivity of ^{13}C NMR is much greater than ^{15}N NMR. It is easier and cheaper to work with natural-abundance compounds instead of enriched materials. Scholle et al.[60] combined proton NMR and temperature-programmed desorption to study acidic sites in ZSM-5 and boralite.

The vibrational spectrum of sorbed pyridine exhibits several diagnostic bands that are sensitive to the acidic nature of a surface.[52-54,61-64] Raman-active bands occur at 1000 and 1035 cm^{-1} (ring breathing modes), and the corresponding IR-active bands are found in the 1440 to 1640 cm^{-1} region (Table 1). Adsorbents characterized using IR spectra of sorbed pyridine include goethite,[53] palygorskite,[52] layer silicates,[61,62] silica gel,[54] alumina,[63] and zeolites.[63,65] There are several ^{15}N NMR studies of pyridine adsorption by zeolites[66] and silica-alumina.[67] Derouane and Nagy[68] review early work using high-resolution ^{13}C NMR techniques to study reactions on zeolites and oxides, including both adsorption and catalytic reaction experiments. Water also may be used as a probe of surface acidity via its O-H stretching and bending modes. This technique has been used to study water sorption on zeolites exchanged with different metal cations.[69] The positions of the O-H stretching and bending bands of sorbed water were used to distinguish strong Lewis from strong Brønsted sites as a function of water content and the type of exchangeable cation.

Abiotic organic reactions on mineral surfaces are influenced strongly by surface Lewis and Brønsted acidity,[70] and the ability of a surface to promote a particular abiotic transformation can provide useful information about the nature of its acidic sites. Single electron transfer (SET) reactions of unsaturated organic compounds on 2:1 clay mineral surfaces represent an important class of abiotic transformations which can enhance the degradation of toxic organic chemicals in soils and sediments.[70-75] Mortland, Doner, and Pinnavaia[76-79] are among the first to show that simple arenes (e.g., benzene, toluene, and p-xylene) are sorbed onto transition metal (e.g., Cu^{2+}, Fe^{3+}, and Ru^{3+})-exchanged montmorillonites at low water content, forming strongly-colored complexes whose molecular properties are quite distinct from those of the unperturbed arene. Although this earlier work was restricted to simple arenes, it has since been extended to a much broader class

Table 2. Compounds Susceptible to Clay Mediated SET Reactions

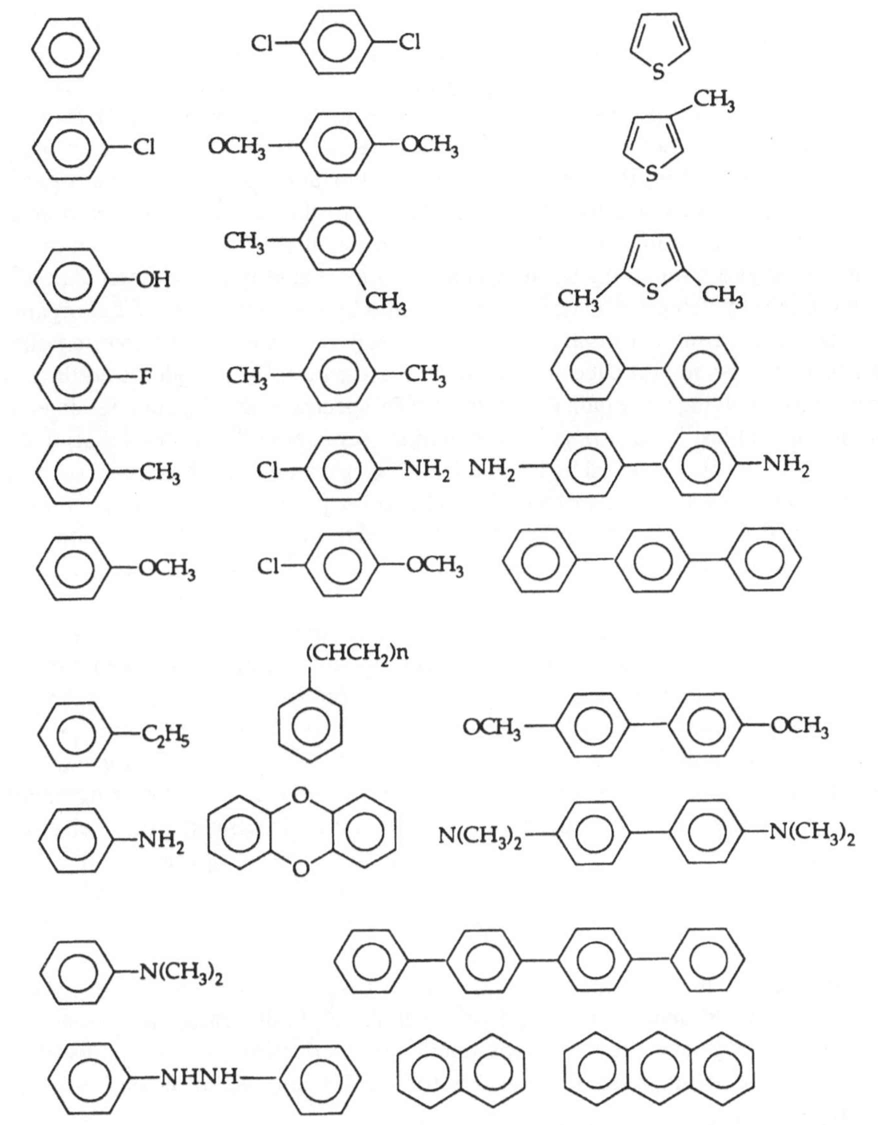

of unsaturated organic compounds, including dioxins,[80] chloroanisoles,[81] and chloroethenes.[82] Table 2 presents a partial listing of the compounds which are known to participate in SET reactions on transition metal-exchanged montmorillonites.

The SET reaction proceeds through a single electron transfer from the unsaturated organic compound to the transition metal cation.[78,79,83-85] In order for the

SET reaction to occur, the organic molecule must coordinate with the transition metal cation by assuming an inner-sphere ligand position. Under hydrated conditions, hydrophobic organic molecules have little, if any, affinity for the hydrophyllic, interlamellar environment on a clay mineral surface. Thus, minimal sorption of the hydrophobic solute occurs.[72,86] Dehydration of the transition-metal exchanged 2:1 clay mineral, however, enhances the sorption of the organic molecule by removal of interlamellar water molecules that otherwise would compete strongly for coordination with the metal cation. If the ionization potential of the organic molecule is < 10 eV and the reduction potential of the metal cation is favorable, an SET reaction can occur. Apparently, the negatively-charged clay mineral surface plays a critical role by increasing the Lewis acidity of the metal cation,[55] which, in turn, promotes the SET reaction, and by stabilizing the radical organic cation. Once a radical organic cation has been generated on the interlamellar surface of a clay mineral, it can either remain as a cation[28,87] or undergo additional reactions involving other radical cations, neutral species on the clay, or interlamellar water molecules.[88] For aromatic compounds, Soma et al.[89] proposed that, if the para positions are occupied by a suitable blocking group, then the radical organic cation is stabilized on the surface of the clay (e.g., p-dimethoxybenzene[87]); if the para positions are not occupied, however, then subsequent dimerization and polymerization reactions can occur (e.g., benzene,[85,90] anisole,[91] aniline,[92,93] and phenols[74]).

Clay-mediated SET reactions are characterized, in general, by the formation of a darkly colored complex;[78] loss of the ESR signal from the paramagnetic metal cation[83] (e.g, reduction of Cu^{2+} to Cu^{1+}); a concomitant appearance of a "new" ESR signal from the generation of an organic radical cation;[84] and strong perturbations in both the IR[91] and the Raman[85,89] spectra of the sorbed complex. The vibrational bands most sensitive to SET reactions are ring C-C stretching bands (ν_{19a}) and out-of-plane C-H deformation bands (ν_{11}). The transfer of an electron from a transition metal cation to an unsaturated organic species results in a decrease in the ν_{19} C-C stretching band and an increase in the ν_{11} band. If the radical organic cation undergoes subsequent dimerization or polymerization reactions, additional bands will appear for the sorbed species.

An example of a clay-mediated SET reaction is provided by the chemisorption of 1,4-dimethoxybenzene on Cu-montmorillonite.[28] In this case, the presence of the methoxy groups prevents subsequent polymerization of the chemisorbed radical cation. A proposed SET reaction for 1,4-dimethoxybenzene on Cu-montmorillonite is:

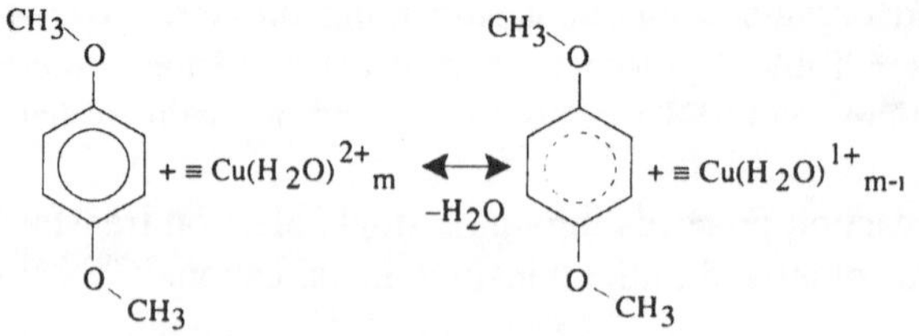

(2)

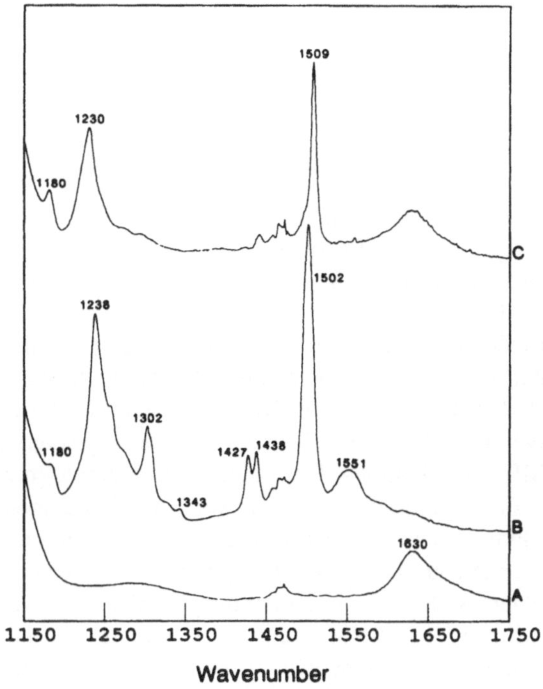

Figure 3. FTIR spectra of Cu-montmorillonite deposited onto a sheet of polyethylene (a) and of 1,4-dimethoxybenzene sorbed onto Cu-montmorillonite neutral and radical cation (b);. spectrum of the neutral species; (c) was obtained after exposure of the sample (b) to water vapor. (From Johnston, C.T., et al. *Langmuir* 7:289–296 (1991). With permission.)

The FTIR spectrum of the radical organic cation (Figure 3) is characterized by a "red shift" of the ν_{19} band from 1509 to 1502 cm^{-1} and a "blue shift" of the ν_{13} band from 1230 to 1238 cm^{-1}, as compared to the spectrum of the neutral species (Figure 3). The decrease in frequency of the ν_{19} band reflects a net decrease in the average C-C force constant of the C-C bonds within the ring. Conversely, the increase in frequency of the ν_{13} band indicates that the C-O bonds are strengthened upon oxidation. In addition to these perturbed bands, a number of new bands are observed in the spectrum of the radical organic cation, at 1302, 1308, 1342, 1427, 1438, and 1551 cm^{-1}, which disappear quickly upon rehydration of the clay-organic complex.[28]

Electron-transfer mechanisms have also been invoked to account for the oxidation of phenolic compounds by iron, manganese, and aluminum oxides.[14,15] McBride[14,15] has used catechol and related phenols as molecular probes to characterize active sorption sites on metal oxide surfaces. Several of the IR-active modes of catechol provide diagnostic information about its sorbed forms. In particular, sorption of catechol results in a decrease in wavenumber of the aromatic ring C-C stretching vibrations as compared to nonsorbed catechol. This decrease in wavenumber indicates that the bond order of the ring C-C bonds is decreased as a result of chelation to Al^{3+}, Fe^{3+}, or Mn^{2+}.

Cation Exchange Sites

NMR Probe Studies of Exchangeable Metal Cations

A straightforward method of studying cation adsorption is to use a spectroscopic technique that can give information about the cation itself, e.g., ultraviolet (UV), EXAFS, or NMR spectroscopy. Modern NMR spectrometers are equipped to detect virtually all of the magnetic nuclei in the periodic table. In order to reduce the influence of some of the intrinsic line broadening mechanisms that occur in solids and at interfaces, a spectrometer may also be equipped with high-power amplifiers and MAS probes to cover the NMR frequencies of the nuclei of interest. Most metals have at least one NMR-active isotope, but some nuclei are more amenable to NMR investigations than others. For example, nuclei with a spin quantum number of $^1/_2$ generally have narrower NMR lineshapes than those with higher spin quantum numbers because of the nuclear electric-quadrupole interactions that often broaden the spectra of higher-spin nuclei. (For a complete description of the advantages and disadvantages of quadrupolar versus dipolar nuclei, see Engelhardt and Michel.[29]) The NMR sensitivity of the sorbed nucleus of interest is determined by the natural abundance of the nucleus, its resonance frequency, and the number of nuclei in the NMR sample tube, i.e., surface coverage. In the final analysis, most nuclei can be observed by NMR either by isotopic enrichment above natural abundance, by using high surface coverages, or by using longer times for signal averaging.

A primary problem in studying metal cation exchange on particles is the fact that it is virtually impossible to obtain any useful molecular information by observing the nucleus of a paramagnetic metal directly. Thus, studies of cation exchange have focused on diamagnetic metals, such as Cd^{2+}, which has a spin of $^1/_2$ and an acceptable natural abundance (12 and 13%, respectively, for the two NMR-active isotopes, ^{113}Cd and ^{111}Cd). Bank et al.[94] investigated ^{113}Cd adsorption by several specimen clay minerals as influenced by chemical factors such as solution pH and the presence of structural Fe. They interpreted their NMR spectra as evidence for two types of exchange site, both involving O ligands but with one involving shorter Cd-O bonds than the other. Weiss et al.[95] have investigated ^{133}Cs adsorption by the trioctahedral smectite, hectorite, using variable-temperature MAS NMR. Their data permitted the assignment of Cs^+ to surface complexes and to the diffuse layer. The NMR peak assigned to surface complexes was more narrow and shielded, and it was not affected by changing Cs^+ concentration in solution as much as was the peak assigned to the diffuse layer. Laperche et al.[96,97] studied ^{23}Na, ^{111}Ca, and ^{133}Cs adsorption by trioctahedral vermiculite with simultaneous measurement of MAS NMR spectra, X-ray diffraction, and water sorption. Their results showed clearly that the hydration of an exchangeable cation plays a major role in determining its configuration in surface

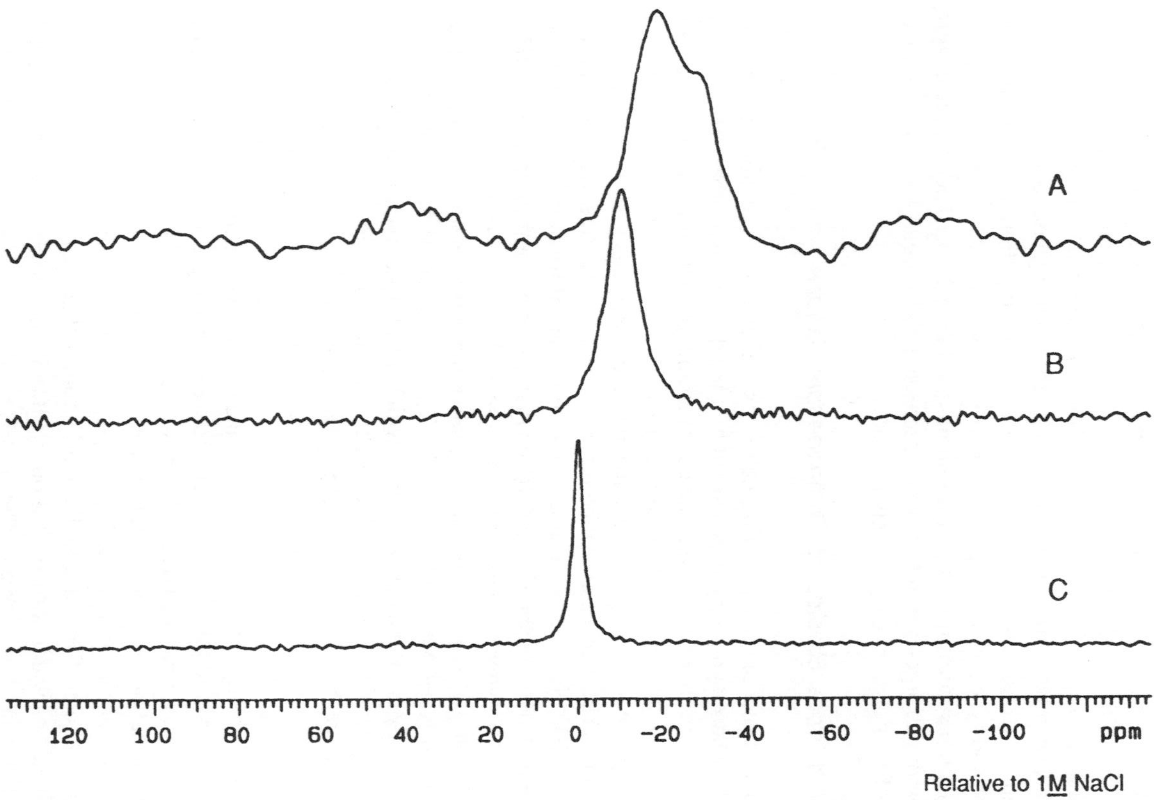

Figure 4. MAS NMR ^{23}Na spectra of Na-exchanged montmorillonite. The exchanged clay was dried at 110°C, then stored under differing "hydration atmospheres;" (A) was stored over P_2O_5; (B) was stored at 22% relative humidity, and (C) was kept over water, i.e., at 100% relative humidity. (Adapted from Johnston, C.T. et al. *Proceedings of the Metal Speciation Workshop*. Chelsea, MI: Lewis Publishers, Inc., in press. With permission.)

complexes. Chemical shifts for the three cations in varying hydration states of the clay mineral were a simple, universal function of the ionization potential of the cation, the number of exchange sites, and the number of O ligands (including those in H_2O) binding to the cation. Recently,^{23}Na NMR spectroscopy was used to investigate the effects of hydration on exchangeable Na^+ in montmorillonite.[98] Figure 4 shows the ^{23}Na NMR spectra obtained as a function of hydration of the clay. The bottom spectrum, obtained from a water-saturated clay, shows a narrow peak with a chemical shift very close to that observed in aqueous solutions of Na^+. The middle spectrum was obtained from a sample equilibrated at about 22% relative humidity, which corresponds to one monolayer of water in the interlamellar space. The top spectrum is that of a sample dried over P_2O_5. The chemical shift and linewidth changes are indicative of decreased Na^+ mobility and, in the dried sample, of two distinct cation exchange sites.

Vibrational Probe Studies of Exchangeable Cations

Another means of characterizing cation exchange sites is to study molecular adsorbates whose symmetry is perturbed by the adsorbate-absorbent interaction. For molecular probes possessing a center of symmetry, vibrational modes cannot be both Raman- and IR-active. Sorption processes can reduce the symmetry of a molecular probe, resulting in the appearance of previously forbidden IR and Raman bands. The molecular symmetry of adsorbed NH_4^+ (the symmetry group of isolated NH_4^+ is T_d), for example, can provide information about its local environment. Chourabi and Fripiat[99] found that the tetrahedral symmetry of NH_4^+ is reduced to C_{3v} when it resides on a smectite exchange site resulting from Si $\rightarrow$ Al substitution, whereas it remains as T_d when the cation is on an exchange site resulting from Al $\rightarrow$ Mg substitution. The C_{3v} species can be distinguished from the T_d species by an IR band at 3030 cm^{-1} which is not IR-active for the T_d species.

Water molecules sorbed on mineral surfaces are strongly influenced by the type of exchangeable metal cations and by the nature of the cation exchange sites. Spectroscopic studies of water sorbed on smectite surfaces have demonstrated that two distinct molecular environments are present: (1) water molecules coordinated directly to exchangeable metal cations and (2) physisorbed water molecules occupying interstitial pores, void spaces between exchangeable metal cations, or polar sites on external surfaces. The position and molar absorptivities of the H-O-H deformation (ν_2) and stretching (ν_1 and ν_3) modes of sorbed water molecules in the interlayer region of smectites, in particular, are perturbed by exchangeable cations. Russell and Farmer[100] found that the O-H stretching and bending bands of water sorbed on montmorillonite and saponite were perturbed upon lowering the water content. The H-O-H bending band (ν_2 mode, in the 1610 to 1640 cm^{-1} region) decreased in wavenumber and in relative intensity considerably more slowly than did the O-H stretching band upon dehydration. A similar result was reported by Poinsignon et al.,[101] who observed that the molar absorptivities of the O-H stretching and bending bands of sorbed water depended

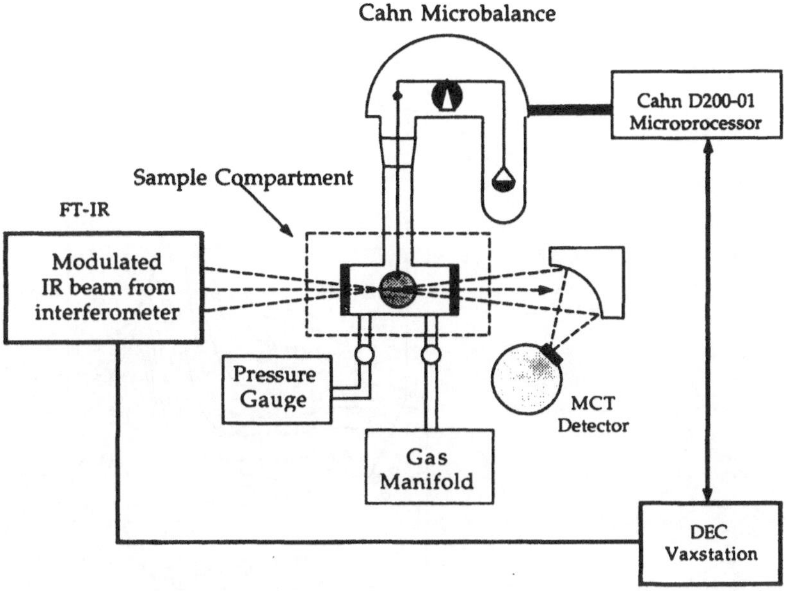

Figure 5. Schematic of the *in- situ* FT-IR/gravimetric cell used to collect FTIR spectra and record gravimetric data simultaneously. (From Johnston, C.T. et al. *Langmuir* 7:289–296 (1991). With permission.)

strongly on the hydration energy of the exchangeable metal cation, the number of water molecules coordinated to the metal cation, and the surface charge density of the clay mineral.

Johnston and co-workers[28] have developed a spectroscopic cell (Figure 5) to measure water desorption isotherms and collect FTIR spectra concurrently. Desorption isotherms and FTIR spectra have been obtained for self-supporting montmorillonite clay films exchanged with Na^+, K^+, Cu^{2+}, and Co^{2+}. The desorption isotherm for water from Na-exchanged montmorillonite is shown on the left side of Figure 6. For each data point shown in the desorption isotherm, a corresponding FTIR spectrum was obtained. The three-dimensional plot of the FTIR spectra in the 1400 to 1800 cm^{-1} region, shown on the right side of Figure 6, is plotted with the z-axis representing the relative vapor pressure of water. The intensity of the ν_2 band at 1640 cm^{-1} is directly proportional to the amount of water sorbed on the clay. Thus, simultaneous collection of the gravimetric and spectroscopic data provides a direct method to correlate macroscopic water sorption characteristics with IR vibrational data.

The position of the ν_2 band of water sorbed on montmorillonite exchanged with Na^+, K^+, Co^{2+}, and Cu^{2+} is plotted in Figure 7 as a function of the average number of water molecules per exchangeable metal cation. At high water content, the position of the ν_2 band is relatively stable. In the case of Na^+, the high water content limit of the position of the ν_2 band was 1640 cm^{-1}, which compares well to 1635 cm^{-1} for Cu^{2+}. Upon reducing the water content to less than 10 water

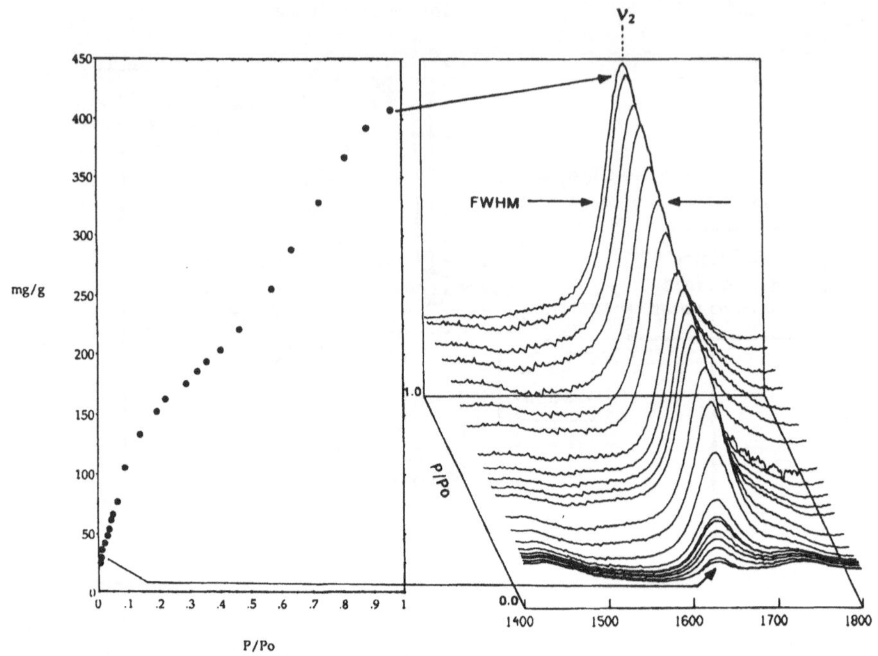

Figure 6. Comparison of the desorption isotherm of water from a self-supporting clay film of Na-SAz1 obtained at 24°C to the FTIR spectra (right side) of water sorbed on the Na-SAz1 clay film. Each spectrum shown on the right side corresponds to one data point on the desorption isotherm. (From Johnston, C.T. et al. *Clays Clay Miner.* in press (1993). With permission.)

molecules per Cu^{2+} ion, or to 6 water molecules per Na^{+} ion, the position of the ν_2 band shifts to lower wavenumber, in agreement with previous IR studies of water sorbed on smectite.[100,101] The shift in the ν_2 band as a function of water content indicates that water molecules coordinated directly to exchangeable metal cations are chemically different from those in bulk water.

Upon dehydration, the ratio of coordinated water to physisorbed water increases, the coordination number of water molecules around a metal cation decreases, and the influence of the clay mineral surface increases. Thus, it is difficult to separate the influence of metal cations on the behavior of interlayer water from the direct effect of the clay mineral surface. The position of the O-H (or O-D) stretching bands of water (or D_2O) can provide information about the strength of hydrogen bonding to a clay mineral surface. Farmer and Russell[62] showed that, as substitution of Al for Si in the clay mineral structure increased, the O-H stretching bands of sorbed water were shifted to lower wavenumbers. In the case of the trioctahedral smectite, hectorite, where no Al substitution occurs, the position of the predominant O-H stretching band of water occurs at 3480 to 3495 cm^{-1}. Water sorbed on saponite and vermiculite, by contrast, exhibits several O-H stretching bands below 3400 cm^{-1}. Thus, the vibrational properties of water

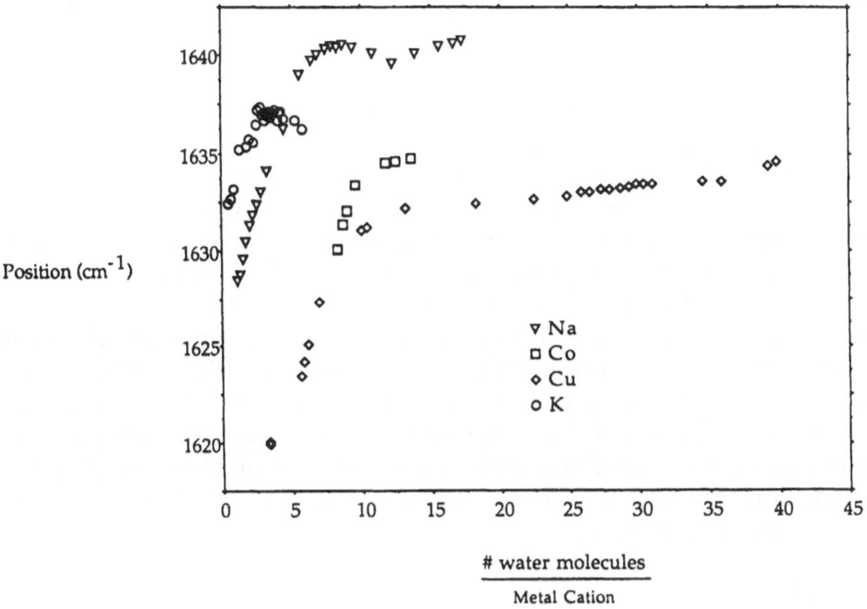

Figure 7. Position of the ν_2 band of water sorbed on SAz-1 montmorillonite exchanged with Na^+, K^+, Co^{+2} and $Cu+^2$, plotted as a function of water content expressed as the number of water molecules sorbed per exchangeable metal cation. (From Johnston, C.T. et al. *Clays Clay Miner.* in press (1993). With permission.)

provide diagnostic information about the local molecular environment around an exchangeable metal cation and about the extent of hydrogen bonding of water to the clay mineral surface.

Oxanion Adsorption Sites

Infrared spectroscopic methods have been used to characterize the sorption of phosphate and sulfate anions on hydroxylated metal oxides surfaces.[43,47-51,102-109] One approach, which utilizes the IR-active stretching vibrations of exposed, surface hydroxyl groups as reporter groups, will be discussed in a later section. A second approach examines the changes that occur in the vibrational modes of the oxyanion itself upon sorption. Similar to the reduction in symmetry of NH_4^+ from T_d to C_{3v} upon interaction with certain adsorption sites on smectite, the symmetry of PO_4^{3-} is reduced upon sorption to goethite (α-FeOOH). In general, sorption of PO_4^{3-} on metal oxides results in a splitting of the P-O stretching bands, indicative of lower anion symmetry. The two regions of interest are the P=O stretching region (1180 to 1200 cm^{-1}) and the P-O(Fe) region (990 to 1110 cm^{-1}). Determining the specific symmetry of the sorbed species is difficult, since assignment of the multiple P-O stretching bands is not straightforward. A bidentate bridging complex (C_{2v} symmetry) has been proposed for phosphate sorbed by goethite on the basis of IR spectroscopy and adsorption isotherms.[47-49,102,103] Similar spectroscopic results have been obtained for SO_4^{2-} and SeO_3^{2-} sorbed on goethite,[48,50,103,104] with the S-O and Se-O stretching bands, respectively, providing diagnostic information about the mechanism of sorption.

In addition to its use in anion sorption studies on goethite, IR spectroscopy has been used to examine the chemistry of anion interactions with hematite,[104] lepidocrocite,[104] akaganeite,[104] amorphous ferric hydroxide,[104] alumina gel,[105] iron hydroxide gel,[110] allophanic soils,[109] and pumice.[109] One significant restriction in these spectroscopic studies and the sorption mechanisms inferred from them is the fact that only dried samples were analyzed. Tejedor-Tejedor and Anderson[27,43,44] have avoided this restriction by using a cylindrical internal reflection (CIR) cell (a type of ATR-IR) to observe the interaction of phosphate with goethite in aqueous suspension using FTIR spectroscopy. A thin coating (<50 nm) of goethite was deposited on the internal reflection element of the CIR cell, and the interaction of the hydrated surface with phosphate was examined *in situ*. This approach has been extended recently to characterize the protonation of phosphate on the surface of goethite,[44] and to study the interaction of salicylate with goethite in aqueous suspension.

REPORTER GROUP STUDIES OF ENVIRONMENTAL PARTICLE SURFACES

^{29}Si and ^{27}Al NMR Studies

Lippmaa et al.,[111] in the first comprehensive ^{29}Si NMR study of silicates, demonstrated that the ^{29}Si chemical shift could be correlated with the extent of condensation of the Si-O moiety. They also demonstrated that the chemical shift of ^{29}Si in aluminosilicates is related to the number of next-nearest-neighbor Al in the structure and proposed the "Q notation" for Si condensation, which has since become standard nomenclature. This work engendered a flurry of measurements of ^{29}Si chemical shifts for a large number of silicates. In later work, Magi et al.[112,113] suggested that, within a given type of silicate structure, the chemical shift of ^{29}Si is correlated with the Si-O bond lengths and angles. Researchers studying zeolites have used these results to determine structural features and to demonstrate changes in the chemical shift as a function of adsorption or the population of molecules in zeolite "cages"; however, these are actually caused by changes in bond angles as the overall structure of a zeolite changes with adsorption. Maciel et al.[67,114] used ^{29}Si NMR and chemical modification of surfaces to investigate phenomena in several silica gels. They utilized the fact that the bulk Si in most silicates have very long longitudinal relaxation times, whereas Si at or near a surface are motionally more flexible and, therefore, have much shorter relaxation times. This permits spectral discrimination, dependent upon the time delays used in an NMR pulse sequence. They also exploited the fact that, since surface Si are usually in silanol groups and couple to protons or, alternatively, on chemically-modified surfaces, and couple to protons in the organic substituent, ^{29}Si CP techniques can be used to obtain enhanced signals from surface Si. These studies gave an excellent, qualitative picture of a surface, but unfortunately could not be used to quantify surface interactions. They have not yet been extended to environmental particles. Relaxation-time discrimination may not be possible for most environmental particles, at any rate, because paramagnetic impurities, which shorten the relaxation time, will likely attenuate the differences between Si nuclei on the surface and those in structure. On the other hand, the use of CP to enhance signals from surface Si should be applicable to silicates and aluminosilicates in environmental particles. Earl and co-workers[56] have investigated ^{29}Si NMR spectra for several 2:1 clay minerals exchanged with different metal cations and have been unable to unequivocally discriminate peaks. Even clay minerals exchanged with paramagnetic cations show very small spectral differences from those exchanged with Na^+. Presumably this is because the cations are separated from the Si nuclei by at least two to three bond lengths and additionally are highly mobile in the interlayer region.

Peaks from ^{27}Al NMR, which should be broadened only by second-order quadrupolar interactions and, therefore, in relatively symmetrical aluminosilicates, should be quite narrow, in practice are fairly broad and cannot be used for fine chemical distinctions as can ^{29}Si NMR peaks.[31,34,37,41] The main utility of ^{27}Al NMR spectroscopy has been in quantifying fourfold, fivefold, or sixfold coordination of Al in mineral structures. Figure 8 demonstrates an interesting aspect of MAS NMR in general, one which is accentuated for quadrupolar nuclei. The lower spectrum is the ^{27}Al NMR of Na^+-montmorillonite, and the upper two spectra are the same montmorillonite exchanged with (paramagnetic) Fe^{3+}. The spinning sidebands are the result of coherent spatial interactions and are invariably present in ^{27}Al NMR spectra. The increase in intensity and number of these sidebands seen in the upper two spectra is from long-range paramagnetic perturbations. The fact that these are long-range effects and that Fe^{3+} does not produce a significant paramagnetic shift makes these sidebands an interesting curiosity but not very useful in understanding either surface interactions or the structure of the solid.

Structural OH Groups

Structural hydroxyl groups exposed on surfaces are among the most abundant and reactive functional groups found on particles in soil and other subsurface environments.[115] The vibrational modes associated with the stretching and bending motions of these functional groups are very sensitive to the presence of sorbed species in the interfacial region. In studies of hydrogen-bonded solids (e.g., goethite, α-FeOOH), the change in wavenumber of the hydroxyl stretching band per unit change in the internuclear O···H···O bond distance is approximately 11 cm^{-1} pm^{-1} (1 pm = 0.001 nm) in the range of internuclear bond distances of 310 to 280 pm.[116] Surface hydroxyl groups can form inner-sphere complexes with metal species, hydrogen-bond to sorbed species or solvent molecules accumulated at

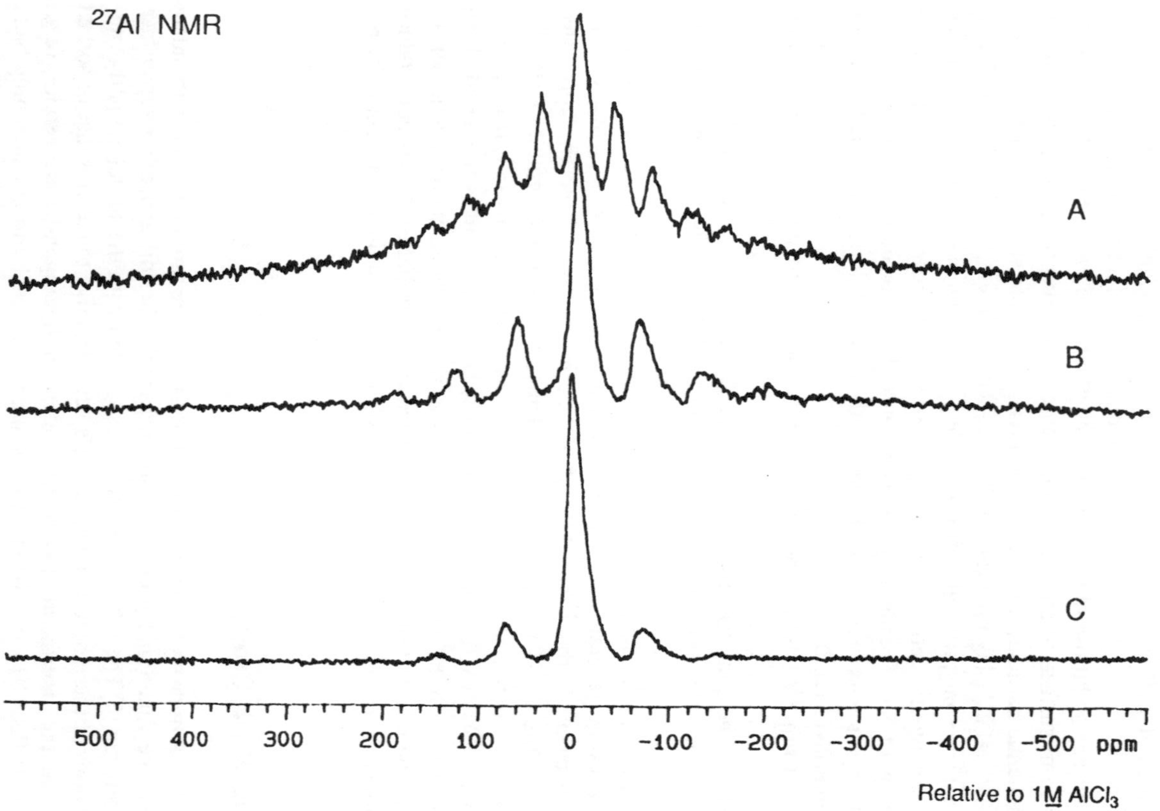

Figure 8. MAS NMR ^{27}Al spectra of SAz-1 montmorillonite exchanged with different cations and at different MAS spinning speeds; spectrum A was obtained of a sample exchanged with Fe^{3+} and rotated at 5500 Hz; spectrum B is the same sample as A, but the spinning rate was increased to 7500 Hz; spectrum C is a sample exchanged with Na^+ and rotated at 7500 Hz. (From Earl, W.L. Unpublished results.)

an interface, undergo replacement reactions with deuterons (^2H), tritium (^3H), and F, or be influenced by metal cations through electrostatic interactions. Any of these processes can produce a large perturbation in O-H vibrational modes. Examples of IR studies of surface hydroxyl groups that have been used to obtain information about sorption mechanisms, interlayer bonding, adsorbent disorder, dehydration and dehydroxylation process, as well as the behavior of exchangeable cations near surfaces, are listed in Table 3.

Selective Deuteration Studies

A universal problem encountered in studies of surface hydroxyl groups is that of distinguishing their vibrational model from those of bulk structural hydroxyl groups. The relative proportion of bulk to surface hydroxyl groups typically is quite large; thus, spectroscopic techniques must discriminate between two very disparate populations of OH. A surface-sensitive vibrational technique, such as IR reflection-absorption spectroscopy (IRRAS), surface enhanced Raman scattering (SERS), or electron energy loss spectroscopy (EELS), can be used to overcome this problem, but unfortunately, restrictions apply to each method at present which severely limit its application to the study of natural particles. A more useful approach is to label surface hydroxyl groups selectively with $^2H(D)$ by isotopic exchange between ^1H in the solid and D_2O (liquid or vapor). Although other isotopes can shift the vibrational frequency of the hydroxyl group, the advantage of using deuterium is that similar surface reactions are expected for OD and OH groups. If experimental conditions are arranged suitably, bulk hydroxyl groups cannot exchange with D and no change in the frequency of their stretching or bending bands will occur, allowing them to be distinguished spectroscopically from surface OH groups.

A useful illustration of selective deuteration techniques comes in the work of Rochester and Topham,[117] who studied the isotopic exchange of hydroxyl groups on goethite. At least five types of surface hydroxyl groups are expected to occur on the surface of goethite on the basis of its crystal structure and morphology. Bulk hydroxyl groups are characterized by an asymmetric, broad ν(O-H) band centered around 3200 cm^{-1}, as shown in Figure 9. Under hydrated conditions, the presence of surface hydroxyl groups is indicated by ν(O-H) bands at 3660 and 3486 cm^{-1} (Figure 9), whose assignment was confirmed by isotopic exchange with D_2O vapor. Upon exposure to the vapor, the 3660 and 3486 cm^{-1} bands were replaced rapidly by the corresponding ν(O-D) bands, at 2701 and 2584 cm^{-1} (Figure 9). Rochester and Topham[117] assigned these two latter bands to surface OH groups at site 'C' and at site 'D' or 'E', respectively. This assignment, however, does not agree with those made previously by others.[47,49,102,103] Despite this lack of agreement on the assignment of the OH stretching bands to specific mineral surface sites, selective deuteration provides a useful method to study surface hydroxyl groups. In addition to surface studies of goethite, selective deuteration methods have been used to study the surface structures of kaolinite[118-123] and gibbsite.[47]

Table 3. Reporter Group Studies of Environmental Particles

Application	Ref.
Structural/crystallographic	
Orientation of OH groups in layer silicates	150,151
Orientation and reactivity of OH groups in kaolinite	121,152
Sorption mechanisms	
Adsorption on gibbsite	153
Phosphate, Ca^{2+} and SiO_2 sorption on gibbsite	154
Exchange of surface hydroxyl with deuterium oxide	117
Anion adsorption by soils and soil materials	48
Adsorption on geothite	49,50,102–104
Selective deuteration of gibbsite	47
Selective deuteration of layer silicates	62,118,120,122
Interlayer bonding	
Nature of interlayer bonding in kaolin minerals	124,155
Disorder/crystallinity	
Chromium substitution in halloysite and dickite	156,157
Crystallinity of kaolinite, dickite, and nacrite	139,158,159
Dehydration/dehydroxylation	
Isothermal dehydroxylation of kaolinite	160
Proton delocalization in kaolinite	123
Hydration mechanism of smectites	100,161,162
Dehydroxylation of montmorillonite	148,149
Dehydroxylation of sepiolite	163
Behavior of exchangeable cations	
Nature of inorganic and organic cations on phyllosilicates	129,145,147
Cation migration into empty octahedral sites on clays	164
Identification of inorganic phases	
Identification of gibbsite and bayerite	153,165
Characterization of hydroxy-aluminum interlayer material	166
Identification of crystalline and amorphous Al-OH phases	167
Identification of kaolinite, dickite, and nacrite	153,168,169

Ligand Exchange

Formation of inner-sphere complexes via ligand exchange between Lewis acid sites ($Fe(III)\equiv H_2O$) on goethite and phosphate, sulfate, and selenite anions has been proposed to account for the high adsorptive retention characteristics of iron-rich soils and sediments for these anions (Figure 1). Replacement of surface hydroxyl groups through ligand-exchange reactions (e.g., Fe-O-H replaced by Fe-PO_3H) is evidenced, in part, by a loss in intensity in the ν(O-H) and δ(O-H) bands associated with the group. Infrared studies of this type have been used, for example, to determine the sorption mechanisms of oxyanions on gibbsite, goethite, and related metal oxides. Upon contact of these oxides with phosphate or sulfate anions in aqueous solution, the intensity of several of the surface hydroxyl ν(O-H) (or ν(O-D)) bands are attenuated sharply, indicating that an inner-sphere complex was formed between Fe or Al and the oxyanion.[47-51]

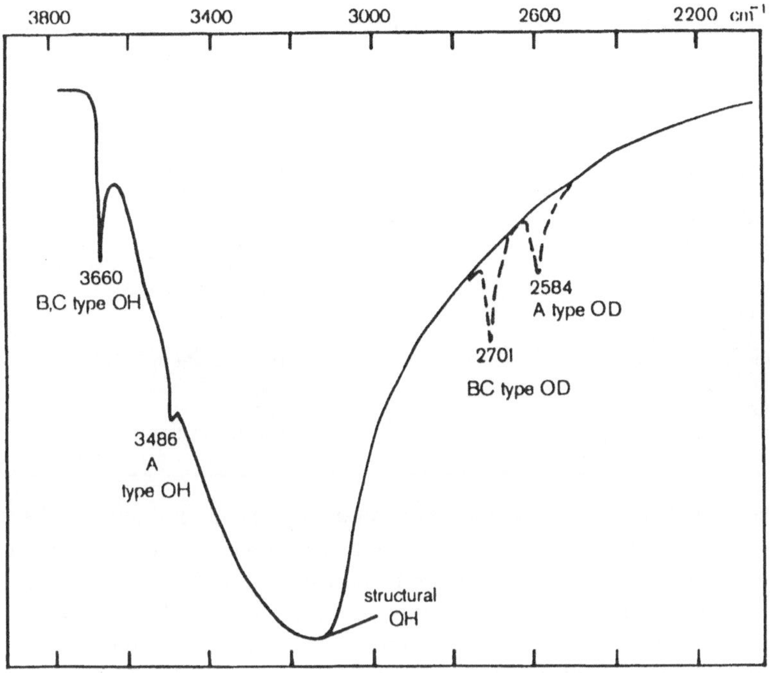

Figure 9. Infrared spectra of goethite showing the surface OH bands which exchange with D_2O, giving ν(OD) bands. (From Parfitt, R.L. *Adv. Agron.* 30:1–50 (1978). With permission.)

Kaolinite-Intercalation Chemistry

Although restricted to a select group of small, polar, organic molecules, vibrational studies of kaolinite-intercalation complexes have provided a unique insight into the chemical and physical properties of the kaolinite surface. Compounds capable of insertion between kaolinite layers include dimethyl-sulfoxide (DMSO), hydrazine, formamide, and N-methyl formamide.[118,124-144] A list of compounds that can intercalate into kaolinite is given in Table 4. Each compound is capable of disrupting the cohesive interlayer electrostatic bonds, permitting intercalation of the guest species. A direct method to determine the extent of interaction between the intercalated probe molecule and the surface of the mineral is to examine the change in wavenumber of the ν(O-H) bands of the "inner-surface" hydroxyl groups in kaolinite. The positions of the inner-surface hydroxyl stretching bands decreases upon intercalation by 150 to 250 cm^{-1}, depending on the strength of the hydrogen bonds formed with the adsorbate.

Inner Hydroxyl Spectroscopy

In addition to the use of the inner-surface hydroxyl stretching bands as reporter groups, the vibrations of the "inner" hydroxyl group can provide diagnostic information about an interface. In both 1:1 and 2:1 clay minerals, this hydroxyl group is located in the octahedral sheet and is not directly accessible to adsorbate

Table 4. List of Kaolinite Intercalation Compounds

Compound	Ref.
Acetamide	125
Dimethyl urea	124
Dimethyl acetamide	125
Dimethyl formamide	124,126,127
Dimethyl selenoxide	128,129
Dimethyl sulfoxide	124,128–136
Formamide	124,126,127,130,134,135,137,138
Hydrazine	118,134,138–140
Methyl urea	124,138
N-methylacetamide	125
N-methylformamide	126,127,135
Potassium acetate	138
Pyridine N-oxide	134,141
Water	136,142–144

species. Penetration of a molecular probe into the siloxane ditrigonal cavity of the tetrahedral sheet is required in order for interactions with this OH group to occur. Despite its recessed location, some probe molecules can, in fact, accomplish this task. Thermal treatment of montmorillonite exchanged with Li^+, for example, results in the migration of Li^+ into empty octahedral sites in this mineral. The frequency of the inner hydroxyl group, at 3630 cm^{-1}, is perturbed by the penetration of Li^+ into the clay structure, reflected by the appearance of two new bands at 3670 and 3700 cm^{-1}. Similar perturbations of the inner hydroxyl groups of layer silicates by other inorganic and organic cations have been reported by Fernandez et al.[145]

The position of the ν(O-H) band for the inner hydroxyl group of kaolinite, at 3620 cm^{-1}, is not generally influenced by the presence of intercalated organic molecules and, consequently, has not been considered a reporter group. Recently, however, a number of researchers[130,140,146] have observed that the frequency of this band is influenced by some intercalates when the kaolinite complex is partially collapsed. A good illustration of this is provided by the kaolinite-hydrazine intercalation complex.[103,104] Hydrazine intercalates readily into kaolinite, resulting in an expanded kaolinite-hydrazine complex with a basal spacing of 1.03 nm. The inner-surface hydroxyl stretching bands are "red-shifted" as a result of hydrogen-bond formation with the intercalated hydrazine species. The inner hydroxyl stretching band, at 3620 cm^{-1}, is not perturbed by this intercalation process. Upon desiccation of the kaolinite-hydrazine complex by vacuum, however, the inner-hydroxyl stretching band is perturbed and a new band appears at 3628 cm^{-1} due to penetration of the siloxane ditrigonal cavity resulting from the partial collapse of the kaolinite-hydrazine intercalate.

In addition to the use of structural OH stretching bands as spectral signatures of reporter groups, OH bending modes can also provide diagnostic information about interfacial reactions. Sposito et al.[147] observed that the intensity of structural OH bending modes in smectite (montmorillonite) increased as the water content of the clay mineral increased. They suggested that, at low water contents, the reduced intensity of the bending modes resulted from electrostatic interactions with exchangeable Na^+ residing near the base of the ditrigonal cavity. Assignment of the structural OH bending modes of montmorillonite has been established by correlating band positions with chemical composition. In the OH bending region, four bands are observed, at 920, 89, 845, and 800 cm^{-1}, which are assigned to Al-OH-Al, Fe(III)-OH-Al, Mg-OH-Al, and Fe(II)-OH-Fe(III), respectively. Malhotra and Ogloza[148,149] observed that the rates of decrease in intensity of these OH bending bands differed upon heating. On the basis of the IR data, they determined that there are four distinct regions of dehydroxylation in montmorillonite. The first involves removal of water from the interlayer cations, and some of the Al-OH-Fe groups and surface hydroxyls also are lost between $290 < T < 553$ K. From 553 to 773 K, both the Al-OH-Mg and the Al-OH-Al groups are involved. The rate of loss in intensity of the Al-OH-Al band is increased in the range 773 to 823 K. Above 823 K, the remaining OH groups are dissociated into H atoms and O centers.[149]

CONCLUSIONS

Vibrational and NMR spectroscopy provide powerful tools to examine the surface chemistry of naturally occurring inorganic particles. Combined application of the two types of spectroscopy always results in a more complete picture of the system of interest because the two methods are characterized by different fundamental time scales and each method has a particular set of advantages and disadvantages associated with it. In general, the sensitivity of vibrational spectroscopy is higher than that of NMR; however, each component of the system (i.e., solute, solvent, sorbent) absorbs infrared radiation, making it difficult to distinguish between the component or process of interest from the bulk sorbent or solvent (e.g., water). Although the sensitivity of NMR is somewhat less than that of vibrational methods, multinuclear NMR probe studies are only sensitive to the nuclei of interest, assuming the effect of paramagnetic impurities can be tolerated. Both methods are sensitive to changes which occur within about 1 nm of a particular vibrational or magnetic chromophore of interest; thus, interpretation of these perturbations can provide molecular-level insight into the structure and reactivity of natural colloids.

Technological advances such as the development of quantum detectors and high-field magnets have significantly improved the sensitivity of the spectroscopic methods and have served to minimize the obstacles which restrict the application of these methods to the study of aqueous colloidal particles. The two most formidable problems in surface spectroscopy of adsorbed species on natural colloids, however, continue to be the limited sensitivity of the spectroscopic methods and the interferences presented by water and paramagnetic impurities. In order for IR and NMR spectroscopy to be applied to the aqueous surface chemistry of natural colloids at environmentally relevant concentration levels, adsorbed species must be analyzed at low concentrations (i.e., $\ll$ monolayer coverage) on difficult-to-analyze substrates in the presence of water and paramagnetic impurities. Thus, the current trend in surface spectroscopy of natural colloids has been towards the development of nondestructive sampling techniques. Considerable progress has been made in this area, so *in situ* surface studies are being reported with increasing frequency in the literature. Of particular interest for aqueous geochemistry applications have been the development and application of methods which can observe colloids in aqueous suspension.

During the past 10 years, the number and diversity of molecular probes and reporter groups that have been used to characterize natural colloids have increased considerably. Molecular probes such as basic probe molecules (e.g., ammonia, pyridine) and water are providing direct information about the surface Lewis and Brønsted acidity of natural colloids. Surface reactivity can also be explored using molecular probes susceptible to chemisorption reactions. In this case, surface spectroscopy can provide information about the reaction intermediates, identity of the chemisorbed species, and functionality of the surface. In addition, exchangeable cations such as ^{3}Li, ^{23}Na, ^{39}K, ^{113}Cd, and ^{133}Cs can be used as probes of active sites on mineral surfaces. In a related way, anions can also be used as adsorptive molecular probes to study the sorption on oxyanions on mineral surfaces. Reporter groups provide another means of obtaining information about changes which occur near the surface. Two examples of reporter group studies reviewed here are solid-state NMR studies of structural ^{29}Si atoms and infrared studies of inorganic structure hydroxyl groups.

In order to bridge the gap between the environmentally relevant concentrations found in natural systems and the traditionally higher concentrations employed in spectroscopic studies, vibrational and NMR studies must be conducted at the low concentration levels which approach those found in natural systems. Furthermore, vibrational and NMR spectroscopic techniques should be combined with macroscopic sorption measurements so that the molecular-scale information can be correlated directly with thermodynamic properties. For example, one way of establishing a connection between the spectroscopic results and the macroscopic properties would be to conduct adsorption isotherms on the same samples being used in spectroscopic investigations. Many of the studies reviewed here correspond to specimen sorbents with a minimal number of solutes present. As information about the surface chemistry of colloidal particles is gained from these fundamental studies and as technological advances improve our ability to analyze difficult samples, experimentalists will be provided with increasingly powerful tools to examine the aqueous surface chemistry of naturally occurring colloids.

ACKNOWLEDGMENTS

We would like to thank Dave Bish and George Guthrie for their critical evaluation of the manuscript and helpful suggestions. In addition, one of the authors (CTJ) would like to express appreciation to Associated Western Universities and to the Research Council of the Katholieke Universiteit in Leuven for their financial support of a faculty development leave. In addition, we would like to thank Ms. C. Erickson for preparing the final figures.

REFERENCES

1. Sposito, G. "Distinguishing Adsorption from Surface Precipitation," in *Chemical Processes at Mineral Surfaces*, J.A. Davis and K.F. Hayes, Eds. (Washington, D.C.: American Chemical Society, 1986), pp. 217–228.
2. Johnston, C.T. and G. Sposito. "Disorder and Early Sorrow: Progress in the Chemical Speciation of Soil Surfaces," in *Future Developments in Soil Science Research*, L.L. Boersma, Ed. (Madison, WI: Soil Science Society of America, 1987), pp. 89–100.
3. Brown, G.E. "Spectroscopic Studies of Chemisorption Reaction Mechanisms at Oxide-Water Interfaces," in *Mineral-Water Interface Geochemistry*, M.F. Hochella and A.F. White, Eds. (Washington, D.C.: Mineralogical Society of America, 1990), pp. 309–363.
4. Hair, M.L. "Transmission Infrared Spectroscopy for High Surface Area Oxides," in *Vibrational Spectroscopies for Adsorbed Species*, A.T. Bell and M.L. Hair, Eds. (Washington, D.C.: ACS Symposium Series No. 137, 1980), pp. 1–11.
5. Bell, A.T. "Applications of Fourier Transform Infrared Spectroscopy to Studies of Adsorbed Species," in *Vibrational Spectroscopies for Adsorbed Species*, A.T. Bell and M.L. Hair, Eds. (Washington, D.C.: American Chemical Society Symposium Series 137, 1980), pp. 13–35.
6. Bell, A.T. "Infrared Spectroscopy of High-Area Catalytic Surfaces," in *Vibrational Spectroscopy of Molecules on Surfaces*, J.T. Yates and T.E. Madey, Eds. (New York: Plenum Publishing Corp., 1987), pp. 105–134.
7. Griffiths, P.R. and J.A. deHaseth. *Fourier Transform Infrared Spectrometry* (New York: John Wiley & Sons, 1986), pp. 1–656.
8. Buswell, A.M., K. Krebs, and W.H. Rodebush. "Infrared Studies. III. Absorption Bands of Hydrogels Between 2.5 and 3.5 Micrometers," *J. Am. Chem. Soc.* 59:2603–2605 (1937).
9. Little, L.H., A.V. Kiselev, and V.I. Lygin. *Infrared Spectra of Adsorbed Species* (London: Academic Press, Inc., 1966), pp. 1–428.
10. Hair, M.L. *Infrared Spectroscopy in Surface Chemistry*, (New York: Marcel Dekker, Inc., 1967).
11. White, J.L. "Proton Migration in Kaolinite," *9th Int. Congr. Soil Sci. Trans.* 701–707 (1968).
12. Delgass, W.N., G.L. Haller, R. Kellerman, and J.H. Lunsford. *Spectroscopy in Heterogeneous Catalysis* (New York: Academic Press, Inc., 1988), pp. 1–341.
13. Nguyen, T.T., L.J. Janik, and M. Raupach. "Diffuse Reflectance Infrared Fourier Transform (DRIFT) Spectroscopy in Soil Studies," *Aust. J. Soil Res.* 29:49–67 (1991).
14. McBride, M.B. and L.G. Wesselink. "Chemisorption of Catechol on Gibbsite, Boehmite, and Noncrystalline Alumina Surfaces," *Environ. Sci. Technol.* 22:703–708 (1988).

15. McBride, M.B. "Adsorption and Oxidation of Phenolic Compounds by Iron and Manganese Oxides," *Soil Sci. Soc. Am. J.* 51:1466–1472 (1987).
16. Hamadeh, I.M., D. King, and P.R. Griffiths. "Heatable–Evacuable Cell and Optical System for Diffuse Reflectance FT-IR Spectrometry of Adsorbed Species," *J. Catal.* 88:264–272 (1984).
17. Culler, S.R., M.T. McKenzie, L.J. Fina, H. Ishida, and J.L. Koenig. "Fourier Transform Diffuse Reflectance Infrared Study of Polymer Films and Coatings: A Method for Studying Polymer Surfaces," *Appl. Spectrosc.* 38:791–795 (1984).
18. Fuller, M.P. and P.R. Griffiths. "Diffuse Reflectance Measurements by Infrared Fourier Transform Spectrometry," *Anal. Chem.* 50:1906–1910 (1978).
19. Bowen, J.M., S.V. Compton, and M.S. Blance. "Comparison of Sample-Preparation Methods for the Fourier-Transform-Infrared Analysis of Organo-Clay Mineral Sorption Mechanism," *Anal. Chem.* 61:2047–2050 (1989).
20. Baes, A.U. and P.R. Bloom. "Diffuse Reflectance and Transmission Fourier Transform Infrared (Drift) Spectroscopy of Humic and Fulvic Acids," *Soil Sci. Soc. Am. J.* 53:695–700 (1989).
21. Vidrine, D.W. "Photoacoustic Fourier Transform Infrared Spectroscopy of Solids and Liquids," *Four. Trans. Inf. Spec.* 3:125–148 (1982).
22. Street, K.W., H.B. Mark, S. Vasireddy, R.A.L. Filio, C.W. Anderson, M.P. Fuller, and S.J. Simon. "Cadmium(II) Exchanged Zeolite as a Solid Sorbent for the Preconcentration and Determination of Atmospheric Hydrogen Sulfide II: Spectroscopic Techniques," *Appl. Spectrosc.* 39:68–72 (1985).
23. Yang, C.Q. and W.G. Fateley. "The Effect of Particle Size on Peak Intensities of FTIR Photoacoustic Spectra," *J. Mol. Struct.* 141:279–284 (1986).
24. Rockley, M.G., D.M. Davis, and H.H. Richardson. "Quantitative Analysis of a Binary Mixture by Fourier Transform Infrared Photoacoustic Spectroscopy," *Appl. Spectrosc.* 35:185–186 (1981).
25. Donini, J.C. and K.H. Michaelian. "Low–Frequency Photoacoustic Spectroscopy of Solids," *Appl. Spectrosc.* 42:289–292 (1988).
26. Sperline, R.P., S. Muralidharan, and H. Freiser. "In Situ Determination of Species Adsorbed at a Solid-Liquid Interface by Quantitative Infrared Attenuated Total Reflectance Spectrophotometry," *Langmuir* 3:198–202 (1987).
27. Yost, E.C., M.I. Tejedor-Tejedor, and M.A. Anderson. "In-Situ CIR-FTIR Characterization of Salicylate Complexes at the Goethite/Aqueous Solution Interface," *Environ. Sci. Technol.* 24:822–828 (1990).
28. Johnston, C.T., T. Tipton, D.A. Stone, C. Erickson, and S.L. Trabue. "Chemisorption of p-Dimethoxybenzene on Cu-montmorillonite," *Langmuir* 7:289–296 (1991).
29. Engelhardt, G. and D. Michel. *High-Resolution Solid-State NMR of Silicates and Zeolites, 2nd ed.* (Rochester, NY: John Wiley & Sons, 1987), pp. 1–256.
30. Watanabe, T., H. Shimizu, A. Masuda, and H. Saito. "Studies of ^{29}Si Spin-Lattice Relaxation Times and Paramagnetic Impurities in Clay Minerals by Magic Angle Spinning ^{29}Si NMR and EPR," *Chem. Lett.* 1293–1296 (1983).
31. Sanz, J. and J.M. Serratosa. "^{29}SI and ^{27}Al High Resolution MAS NMR Spectra of Phyllosilicates," *J. Am. Chem. Soc.* 106:4790–4793 (1984).
32. Lipsicas, M., R. H. Raythatha, T. J. Pinnavaia, I. D. Johnson, R.F. Giese, P.M. Costanzo, and J.L. Robert. "Silicon and Aluminium Site Distributions in 2:1 Layered Silicate CLays," *Nature* 309:604–607 (1984).
33. Thompson, J.G. "Two Possible Interpretations of ^{29}Si Nuclear Magnetic Resonance Spectra of Kaolin-Group Minerals," *Clays Clay Miner.* 32:233–234 (1984).

34. Thompson, J.G. "^{29}SI and ^{27}Al Nuclear Magnetic Resonance Spectrosocpy of 2:1 Clay Minerals," *Clay Miner.* 19:229–236 (1984).
35. Herrero, C.P., J. Sanz, and J.M. Serratosa. "Tetrahedral Cation Ordering in Layer Silicates by ^{29}Si NMR Spectroscopy," *Solid State Comm.* 53:151–154 (1985).
36. Kinsey, R.A., R.J. Kirkpatrick, J. Hower, K.A. Smith, and E. Oldfield. "High Resolution Aluminum-27 and Silicon-29 Nuclear Magnetic Resonance Spectroscopic Study of Layer Silicates, including Clay Minerals," *Am. Mineral.* 70:537–548 (1985).
37. Komarneni, S., C.A. Fyfe, G.J. Kennedy, and H. Strobl. "Characterization of Synthetic and Naturally Occurring Clays by ^{27}Al and ^{29}Si Magic-Angle-Spinning NMR Spectroscopy," *J. Am. Ceram. Soc.* 69:C45–C47 (1986).
38. Weiss, C.A., S.P. Altaner and R.J. Kirkpatrick. "High-Resolution Silicon 29 NMR Spectroscopy of 2:1 Layer Silicates: Correlation among Chemical Shift, Structural Distortions, and Chemical Variations," *Am. Mineral.* 72:935–942 (1987).
39. Altaner, S.P., C.A. Weiss, and R.J. Kirkpatrick. "Evidence from ^{29}Si NMR for the Structure of Mixed-Layer Illite/Smectite Clay Minerals," *Nature* 331:699–702 (1988).
40. Herrero, C.P., J. Sanz, and J.M. Serratosa. "Dispersion of Charge Deficits in the Tetrahedral Sheet of Phyllosilicates. Analysis from ^{29}Si NMR Spectra," *J. Phys. Chem.* 93:4311–4315 (1989).
41. Woessner, D.E. "Characterization of Clay Minerals by 27Al Nuclear Magnetic Resonance Spectroscopy," *Am. Mineral.* 74:203–215 (1989).
42. Johnston, C.T., G. Sposito, and R.R. Birge. "Raman Spectroscopic Study of Kaolinite in Aqueous Suspension," *Clays Clay Miner.* 33:483–489 (1985).
43. Tejedor-Tejedor, M.I. and M.A. Anderson. 'In Situ' Attenuated Total Reflection Fourier Transform Infrared Studies of the Goethite (a-FeOOH)-Aqueous Solution Interface," *Langmuir* 2:203–210 (1986).
44. Tejedor-Tejedor, M.I. and M.A. Anderson. "Protonation of Phosphate on the Surface of Goethite as Studied by CIR-FTIR and Electrophoretic Mobility," *Langmuir* 6:602–609 (1990).
45. Johnston, C.T. "Fourier Transform Infrared and Raman Spectroscopy," in *Instrumental Surface Analysis of Geologic Materials*, D.L. Perry, Ed. (New York: VCH, 1990), pp. 121–155.
46. Sposito, G. and R. Prost. "Structure of Water Adsorbed on Smectites," *Chem. Rev.* 82:553–573 (1982).
47. Russell, J.D., R.L. Parfitt, A.R. Fraser, and V.C. Farmer. "Surface Structure of Gibbsite, Goethite and Phosphated Goethite," *Nature* 248:220–221 (1974).
48. Parfitt, R.L. "Anion Adsorption by Soils and Soil Materials," *Adv. Agron.* 30:1–50 (1978).
49. Parfitt, R.L., J.D. Russell, and V.C. Farmer. "Confirmation of the Surface Structures of Geothite (Alpha-FeOOH) and Phosphated Goethite by Infrared Spectroscopy," *J. Chem. Soc. Faraday Trans. 1* 72:1082–1087 (1976).
50. Parfitt, R.L. and R.S.C. Smart. "Infrared Spectra from Binuclear Bridging Complexes of Sulphate Adsorbed on Goethite (Alpha-FeOOH)," *J. Chem. Soc. Faraday Trans. 1* 73:796–802 (1977).
51. Parfitt, R.L. "Competitive Adsorption of Phosphate and Sulphate on Goethite (Alpha-FeOOH): A Note," *N. Z. J. Sci.* 25:147–148 (1982).
52. Blanco, C., J. Herrero, S. Mendioroz, and J.A. Pajares. "Infrared Studies of Surface-Acidity and Reversible Folding in Palygorskite," *Clays Clay Miner.* 36:364–368 (1988).

53. Morterra, C., C. Mirra, and E. Borello. "IR Spectroscopic Study of Pyridine Adsorption onto Alpha-FeOOH (Goethite)," *Mat. Chem. Phys.* 10:139–154 (1984).
54. Egerton, T.A. "The Application of Raman Spectroscopy to Surface Chemical Studies," *Catal. Rev. Sci. Eng.* 11:71–116 (1975).
55. Mortland, M.M. and K.V. Raman. "Surface Acidities of Smectites in Relation to Hydration, Exchangeable-Cation and Structure," *Clays Clay Miner.* 16:393–398 (1968).
56. Earl, W.L., P.O. Fritz, A.A.V. Gibson, and J.H. Lunsford. "A Solid-State NMR Study of Acid Sites in Zeolite Y Using Ammonia and Trimethylamine as Probe Molecules," *J. Phys. Chem.* 91:2091–2095 (1987).
57. Michel, D., A. Germanus, and H. Pfeiffer. *J. Chem. Soc. Faraday Trans. 1* 78:237 (1982).
58. Jarman, R.H. and M.T. Melchior. *J. Chem. Soc. Chem. Commun.* 414–416 (1984).
59. Hayashi, S., K. Suzuki, S. Shin, K. Hayamizu, and O. Yamamoto. *Chem. Phys. Lett.* 113:368–371 (1985).
60. Scholle, K.F., A.P.M. Kentgens, W.S. Veeman, P. Frenken, and G.P.M. van der Velden. "Proton Magic Angle Spinning Nuclear Magnetic Resonance and Temperature Programmed Desorption Studies on the Acidity of the Framework Hydroxyl Groups in the Zeolite H-ZSM-5 and in H-Boralite," *J. Phys. Chem.* 88:5–8 (1984).
61. Farmer, V.C. and M.M. Mortland. "Infrared Study of the Coordination of Pyridine and Water to Exchangeable Cations in Montmorillonite and Saponite," *J. Chem. Soc.* 1966A:344–351 (1966).
62. Farmer, V.C. and J.D. Russell. "Interlayer Complexes in Layer Silicates: The Structure of Water in Lameller Ionic Solutions," *Trans. Faraday Soc.* 67:2737–2749 (1971).
63. Cooney, R.P., G. Curthoys, and N.T. Tam. "Laser Raman Spectroscopy and Its Application to the Study of Adsorbed Species," *Adv. Catal.* 24:293–342 (1975).
64. Breen, C. "Thermogravimetric and Infrared Study of the Desorption of Butylamine, Cyclohexylamine and Pyridine from Ni- and Co-Exchanged Montmorillonite," *Clay Miner.* 26:487–496 (1991).
65. Karge, H.G., V. Dondur, and J. Weitkamp. "Investigation of the Distribution of Acidity Strength in Zeolites by Temperature Programmed Desorption of Probe Molecules. II. Dealuminated Y-Type Zeolites," *J. Phys. Chem.* 95:283–288 (1991).
66. Ripmeester, J.A. "Surface Acid Site Characterization by Means of CP/MAS Nitrogen-15 NMR," *J. Am. Chem. Soc.* 105:2925–2927 (1983).
67. Maciel, G.E., J.F. Haw, I.S. Chuang, B.L. Hawkins, T.A. Early, D.R. McKay, and L. Petrakis."NMR Studies of Pyridine on Silica-Alumina," *J. Am. Chem. Soc.* 105:5529–5535 (1983).
68. Derouane, E.G. and J.B. Nagy. "Applications of High-Resolution Carbon-13-NMR and Magic-Angle Spinning NMR to Reactions on Zeolites and Oxides," in *Relationship Between Structure and Reactivity*, T.E. Whyte, R.A. Dalla Betta, E.G. Derouane, and R.T.K. Baker, Eds. (Washington, D.C.: American Chemical Society, 1984), Symposium Series No. 248, p. 101.
69. Jentys, A., G. Warecka, M. Derewinski, and J.A. Lercher. "Adsorption of Water on ZSM5 Zeolite," *J. Phys. Chem.* 93:4837–4843 (1989).
70. Voudrias, E.A. and M. Reinhard. "Abiotic Organic Reactions at Mineral Surfaces," in *Geochemical Processes at Mineral Surfaces. ACS Symposium Series Vol. 323*, J. Davis and K.F. Hayes, Eds. (Washington, D.C.: American Chemical Society, 1986), pp. 462–486.
71. Mortland, M.M. "Clay-Organic Complexes and Interactions," *Adv. Agron.* 22:75–117 (1970).

72. Chiou, C.T. and T.D. Shoup. "Soil Sorption of Organic-Vapors and Effects of Humidity on Sorptive Mechanism and Capacity," *Environ. Sci. Technol.* 19:1196–1200 (1985).
73. Zielke, R.C., T.J. Pinnavaia, and M.M. Mortland. "Adsorption and Reactions of Selected Organic-Molecules on Clay Mineral Surfaces," in *Reactions and Movement of Organic Chemicals in Soils. SSSA Special Publication Number 22*, B.L. Sawhney and K. Brown, Eds. (Madison, WI: Soil Sci. Soc. Am., 1989), pp. 81–98.
74. Sawhney, B.L. "Vapor-Phase Sorption and Polymerization of Phenols by Smectite in Air and Nitrogen," *Clays Clay Miner*. 33(2):123–127 (1985).
75. Yamagishi, A. "Optical Resolution and Asymmetric Syntheses by Use of Adsorption on Clay Minerals," *J. Coord. Chem.* 16:132–211 (1987).
76. Doner, H.E. and M.M. Mortland. "Benzene Complexes with Copper(II) Montmorillonite," *Science* 166:1406–1407 (1969).
77. Mortland, M.M. and T.J. Pinnavaia. "Formation of Copper(II) Arene Complexes on the Interlamellar Surfaces of Montmorillonite," *Nature* 229:75–77 (1971).
78. Pinnavaia, T.J. and M.M. Mortland. "Interlamellar Metal Complexes on Layer Silicates. I. Copper(II)-Arene Complexes on Montmorillonite ," *J. Phys. Chem.* 75 (26):3957–3962 (1971).
79. Pinnavaia, T.J., P.L. Hall, S.S. Cady, and M.M. Mortland. "Aromatic Radical Cation Formation on the Intracrystal Surfaces of Transition-Metal Layer Lattice Silicates," *J. Phys. Chem.* 78:994–999 (1974).
80. Boyd, S.A. and M.M. Mortland. "Dioxin Radical Formation and Polymerization on Cu(II)-Smectite," *Nature* 316:532–535 (1985).
81. Govindaraj, N., M.M. Mortland, and S.A. Boyd. "Single-Electron-Transfer Mechanism of Oxidative Dechlorination of 4-Chloroanisole on Copper(II)-Smectite," *Environ. Sci. Technol.* 21:1119–1123 (1987).
82. Mortland, M.M. and S.A. Boyd. "Polymerization and Dechlorination of Chloroethenes on Cu(II)-Smectite via Radical-Cation Intermediates," *Environ. Sci. Technol.* 23:223–227 (1989).
83. Rupert, J.P. "Electron-Spin-Resonance Spectra of Interlamellar Copper(II) - Arene Complexes on Montmorillonite," *J. Phys. Chem.* 77:784–790 (1973).
84. Eastman, M.P., D.E. Patterson, and K.H. Pannell. "Reaction of Benzene with Cu(II)- and Fe(III)- Exchanged Hectorites," *Clays Clay Miner*. 32:327–333 (1984).
85. Soma, Y., M. Soma, and I. Harada. "The Reaction of Aromatic Molecules in the Interlayer of Transition-Metal Ion-Exchanged Montmorillonite Studied by Resonance Raman Spectroscopy. I. Benzene and p-Phenylenes," *J. Phys. Chem.* 88:3034–3038 (1984).
86. Rhue, R.D., P.S.C. Rao, and R.E. Smith. "Vapor-Phase Adsorption of Alkylbenzenes and Water on Soils and Clay," *Chemosphere* 17:727–741 (1988).
87. Soma, Y. and M. Soma. "Raman Spectroscopic Evidence of Formation of p-Dimethoxybenzene Cation on Cu- and Ru-Montmorillonite," *Chem. Phys. Lett.* 94:475–478 (1983).
88. Mortland, M.M. and L.J. Halloran. "Polymerization of Aromatic Molecules on Smectite," *Soil Sci. Soc. Am. J.* 40:367–370 (1976).
89. Soma, Y., M. Soma, and I. Harada. "Reactions of Aromatic Molecules in the Interlayer of Transition Metal Ion-Exchanged Montmorillonite Studied by Resonance Raman Spectroscopy. II. Monosubstitued Benzenes and 4,4′ Disubstitued Biphenyls," *J. Phys. Chem.* 89:738–742 (1985).

90. Van de Poel, D., P. Cloos, J. Helsen, and E. Jannini. "Comportement Particuleir du Benzene Adsorbe sur la Montmorillonite Cuivrique," *Bull. Groupe. franc. Argiles* 25:114–126 (1973).
91. Fenn, D.B., M.M. Mortland, and T.J. Pinnavaia. "The Chemisorption of Anisole on Cu(II)-Hectorite," *Clays Clay Miner.* 21:315–322 (1973).
92. Moreale, A., P. Cloos, and C. Badot. "Differential Behavior of Fe(III)- and Cu(II)-Montmorillonite with Aniline: I. Suspensions with Constant Solid:Liquid Ratio," *Clay Miner.* 20:29–37 (1985).
93. Soma, Y. and M. Soma. "Adsorption of Benzidines and Anilines on Cu- and Fe-Montmorillonites Studied by Resonance Raman Spectroscopy," *Clay Miner.* 23:1–12 (1988).
94. Bank, S., J.F. Bank, and P.D. Ellis. "Solid-State ^{113}Cd Nuclear Magnetic Resonance Study of Exchanged Montmorillonites," *J. Phys. Chem.* 93:4847–4855 (1989).
95. Weiss, C.A., R.J. Kirkpatrick, and S.P. Altaner. "The Structural Environments of Cations Adsorbed onto Clays: Cesium-133 Variable Temperature MAS NMR Spectroscopy of Hectorite," *Geochimica et Cosmochimica Acta* 54:1655–1669 (1990).
96. Laperche, V., J.F. Lambert, R. Prost, and J.J. Fripiat. "High-Resolution Solid-State NMR of Exchangeable Cations in the Interlayer Surface of a Swelling Mica: ^{23}Na, ^{111}Cd, and ^{133}Cs Vermiculites," *J. Phys. Chem.* 94:8821–8831 (1990).
97. Laperche, V. "Etude de l'État et de la Localisation des Cations Compensateurs dans les Phyllosilicates par des Méthodes Spectrométriques," PhD Thesis, Universite de Paris (1991), pp. 1–104.
98. Luca, V., C.M. Cardile, and R.H. Meinhold. "High-Resolution Multinuclear NMR Study of Cation Migration in Montmorillonite," *Clay Miner.* 24:115–119 (1989).
99. Chourabi, B. and J.J. Fripiat. "Determination of Tetrahedral Substitutions and Interlayer Surface Heterogeneity from Vibrational Spectra of Ammonium in Smectites," *Clays Clay Miner.* 29:260–268 (1981).
100. Russell, J.D. and V.C. Farmer. "Infra-Red Spectroscopic Study of the Dehydration of Montmorillonite and Saponite," *Clay Miner. Bull.* 5:443–464 (1964).
101. Poinsignon, C., J.M. Cases, and J.J. Fripiat. "Electrical-Polarization of Water Molecules Adsorbed by Smectites. An Infrared Study," *J. Phys. Chem.* 82:1855–1860 (1978).
102. Parfitt, R.L. and R.J. Atkinson. "Phosphate Adsorption on Goethite (Alpha-FeOOH)," *Nature* 264:740–741 (1976).
103. Parfitt, R.L. and J.D. Russell. "Adsorption on Hydrous Oxides. IV Mechanisms of Adsorption of Various Ions on Goethite," *J. Soil Sci.* 28:297–305 (1988).
104. Parfitt, R.L. and R.S.C. Smart. "The Mechanism of Sulfate Adsorption on Iron Oxides," *Soil Sci. Soc. Am. J.* 42:48–50 (1978).
105. Nanzyo, M. and Y. Watanabe. "Mechanisms of Phosphate Sorption on Soil Components with High Phosphate Retention Capacity," *Soil Sci. Soc. Am. J.* 42:48–50 (1978).
106. Nanzyo, M. and Y. Watanabe. "Mechanisms of Phosphate Sorption on Soil Components with High Phosphate Retention Capacity," *JARQ* 18:87–91 (1984).
107. Nanzyo, M. "Influence of Relative Humidity on the Reaction Products of Phosphates and Noncrystalline Hydroxides of Aluminum and Iron," *Clays Clay Miner.* 35:228–231 (1987).
108. Nanzyo, M. "Formation of Noncrystalline Aluminum Phosphate through Phosphate Sorption on Allophanic Ando Soils," *Commun. Soil Sci. Plant Anal.* 18:735–742 (1987).

109. Nanzyo, M. "Phosphate Sorption on the Clay Fraction of Kanuma Pumice," *Clay Sci.* 7:89–96 (1988).
110. Nanzyo, M. "Infrared Spectra of Phosphate Sorbed on Iron Hydroxide Gel and the Sorption Products," *Soil Sci. Plant Nutr.* 32:51–58 (1986).
111. Lippmaa, E., M. Magi, A. Samosan, G. Engelhardt, and A.R. Grimmer. *J. Am. Chem. Soc.* 102:4889–4893 (1980).
112. Lippmaa, E., M. Magi, A. Samosan, M. Tarmak, and G. Engelhardt. *J. Am. Chem. Soc.* 103:4992–4996 (1981).
113. Magi, M., E. Lippmaa, A. Samosan, G. Engelhardt, and A.R. Grimmer. "Solid-State High-Resolution Silicon-29 Chemical-Shifts in Silicates," *J. Phys. Chem.* 88:1518–1522 (1984).
114. Maciel, G.E., J.F. Haw, I.S. Chang, B.L. Hawkins, T.A. Early, D.R. McKay, and L. Petrakis. "NMR Studies of Pyridine on Silica-Alumina," *J. Am. Chem. Soc.* 105:5529–5535 (1983).
115. Sposito, G. *The Surface Chemistry of Soils* (New York: Oxford University Press, 1984), pp. 1–234.
116. Lippincott, E.R. and R. Schroeder. "One-Dimensional Model of the Hydrogen Bond," *J. Chem. Phys.* 23(6):1099–1106 (1955).
117. Rochester, C.H. and S.A. Topham. "Infrared Study of Surface Hydroxyl Groups on Goethite," *J. Chem.* 591–602 (1978).
118. Ledoux, R.L. and J.L. White. "Infrared Study of Selective Deuteration of Kaolinite and Halloysite at Room Temperature," *Science* 145:47–49 (1964).
119. Anton, O. and P.G. Rouxhet. "Note on the Intercalation of Kaolinite, Dickite, and Halloysite by Dimethylsulfoxide," *Clays Clay Miner.* 25:259–263 (1977).
120. Wada, K. "A Study of Hydroxyl Groups in Kaolin Minerals Utilizing Selective-Deuteration and Infrared Spectroscopy," *Clay Miner.* 7:51–61 (1967).
121. Rouxhet, P.G., N. Samudacheata, H. Jacobs, and O. Anton. "Attribution of the OH Stretching Bands of Kaolinite," *Clay Miner.* 12:171–178 (1977).
122. Russell, J.D., V.C. Farmer, and B. Velde. "Replacement of OH by OD in Layer Silicates and Identification of the Vibrations of These Groups in Infra-Red Spectra," *Min. Mag.* 37(292):869–879 (1970).
123. White, J.L., A. Laycock, and M.I. Cruz. "Infrared Studies of Proton Delocalization in Kaolinite," *Bull. Groupe. Fr. Argiles* 22:157–165 (1970).
124. Cruz, M.I., H. Jacobs, and J.J. Fripiat. "The Nature of Interlayer Bonding in Kaolin Minerals," *Int. Clay Conf.*, Madrid 59–70 (1972).
125. Olejnik, S., A.M. Posner, and J.P. Quirk. "The Infrared Spectra of Interlamellar Kaolinite-Amide Complexes," *J. Colloid Interface Sci.* 37:536–547 (1971).
126. Cruz, M.I., A. Laycock, and J.L. White. "Perturbation of OH Groups in Intercalated Kaolinite Donor-Acceptor Complexes. I. Formamide-, Methylformamide, and Dimethylformamide Kaolinite Complexes," *Int. Clay Conf. Proc. Tokyo* 775–789 (1969).
127. Olejnik, S., A.M. Posner, and J.P. Quirk. "The IR Spectra of Interlamellar Kaolinite-Amide Complexes. I. The Complexes of Formamide, N-methylformamide and Dimethylformamide," *Clays Clay Miner.* 19:83–94 (1971).
128. Raupach, M., P.F. Barron, and J.G. Thompson. "Nuclear Magnetic Resonance, Infrared, and X-Ray Powder Diffraction Study of Dimethylsulfoxide and Dimethylselenoxide Intercalates with Kaolinite," *Clays Clay Miner.* 35:208–219 (1987).

129. Raupach, M. "Infrared Band Frequency Shifts and Dipole Interactions at Surfaces of Inorganic and Organic Clay Intercalates," *J. Colloid Interface Sci.* 121:476–485 (1988).
130. Costanzo, P.M. and R.F. Giese. "Ordered and Disordered Organic Intercalates of 8.4-A Synthetically Hydrated Kaolinite," *Clays Clay Miner.* 38:160–170 (1990).
131. Adams, J.M. and G. Waltl. "Thermal-Decomposition of a Kaolinite Dimethylsulfoxide Intercalate," *Clays Clay Miner.* 28:130–134 (1980).
132. Johnston, C.T., G. Sposito, D.F. Bocian, and R.R. Birge. "Vibrational Spectroscopic Study of the Interlamellar Kaolinite-Dimethylsulfoxide Complex," *J. Phys. Chem.* 88:5959–5964 (1984).
133. Olejnik, S., L.A.G. Aylmore, A.M. Posner, and J.P. Quirk. "Infrared Spectra of Kaolin Mineral Dimethylsulfoxide Complexes," *J. Phys. Chem.* 72:241–249 (1968).
134. Thompson, J.G. "Interpretation of Solid State ^{13}C and ^{29}Si Nuclear Magnetic Resonance Spectra of Kaolinite Intercalates," *Clays Clay Miner.* 33:173–180 (1985).
135. Lipsicas, M., R. Raythatha, R.F. Giese, and P.M. Costanzo. "Molecular Motions, Surface Interactions, and Stacking Disorder in Kaolinite Intercalates," *Clays Clay Miner.* 34:635–644 (1986).
136. Costanzo, P.M., R.F. Giese, and M. Lipsicas. "Static and Dynamic Structures of Water in Hydrated Kaolinites. I. The Static Structure," *Clays Clay Miner.* 32:419–428 (1984).
137. Adams, J.M., P.I. Reid, J.M. Thomas, and M.J. Walters. "On the Hydrogen Atom Positions in a Kaolinite: Formamide Intercalate," *Clays Clay Miner.* 24:267–269 (1976).
138. Ledoux, R.L. and J.L. White. "Infrared Studies of Hydrogen Bonding Interaction between Kaolinite Surfaces and Intercalated Potassium Acetate, Hydrazine, Formamide, and Urea," *J. Colloid Interface Sci.* 21:127–152 (1966).
139. Barrios, J., A. Plancon, M.I. Cruz, and C. Tchoubar. "Qualitative and Quantitative Study of Stacking Faults in a Hydrazine Treated Kaolinite. Relationship with the Infrared Spectra," *Clays Clay Miner.* 25:422–429 (1977).
140. Johnston, C.T. "Raman and FTIR Spectra of the Kaolinite-Hydrazine Intercalate," in *Spectroscopic Characterization of Minerals and their Surfaces*, L.M. Coyne, S.W.S. McKeever, and D.F. Blake, Eds. (Washington, D.C.: American Chemical Society, 1990), Symposium Series No. 415, pp. 432–454.
141. Olejnik, S., A.M. Posner, and J.P. Quirk. "Infrared Spectrum of the Kaolinite-Pyridine-N-Oxide Complex," *Spectrochim. Acta, Part A* 27A:2005–2009 (1971).
142. Lipsicas, M., C. Straley, P.M. Costanzo, and R.F. Giese. "Static and Dynamic Structure of Water in Hydrated Kaolinites," *J. Colloid Interface Sci.* 107:221–230 (1985).
143. Costanzo, P.M., R.F. Giese, M. Lipsicas, and C. Straley. "Synthesis of a Quasi-Stable Kaolinite and Heat-Capacity of Interlayer Water," *Nature* 296:549–551 (1982).
144. Costanzo, P.M. and R.F. Giese. "Dehydration of Synthetic Hydrated Kaolinites: a Model for the Dehydration of Halloysite (10 A)," *Clays Clay Miner.* 33:415–423 (1985).
145. Fernandez, M., J.M. Serratosa, and W.D. Johns. "Perturbation of the Stretching Vibration of OH Groups in Phyullosillicates by the Interlayer Cations," *Reunion Hispano-Belga de Minerales de la arcilla* 163–167 (1970).
146. Johnston, C.T. and D.A. Stone. "Influence of Hydrazine on the Vibrational Modes of Kaolinite," *Clays Clay Miner.* 38:121–128 (1990).

147. Sposito, G., R. Prost, and J.P. Gaultier. "Infrared Spectroscopic Study of Adsorbed Water on Reduced-Charge Na/Li Montmorillonites," *Clays Clay Miner.* 31:9–16 (1983).
148. Ogloza, A.A. and V.M. Malhotra. "Dehydroxylation Induced Structural Transformations in Montmorillonite: an Isothermal FTIR Study," *Phys. Chem. Mineral.* 16:378–385 (1989).
149. Malhotra, V.M. and A.A. Ogloza. "FTIR Spectra of Hydroxyls and Dehydroxylation Kinetics Mechanism in Montmorillonite," *Phys. Chem. Min.* 16:386–393 (1989).
150. MacCarthy, P., R.W. Klusman, and J.A. Rice. "Water Analysis," *Anal. Chem.* 59:308R–337R (1987).
151. Serratosa, J.M. and W.F. Bradley. "Determination of the Orientation of the OH Bond Axes in Layer Silicates by Infrared Absorption," *J. Chem.* 62:1164–1167 (1958).
152. Serratosa, J.M., A. Hidalgo, and J.M. Vinas. "Orientation of OH bonds in Kaolinite," *Nature* 195:486–487 (1962).
153. Alvarez, R., R.E. Cramer, and J.A. Siva. "Laser Raman Spectroscopy: A Technique for Studying Adsorption on Aluminum Sesquioxide, Gibbsite," *Soil Sci. Soc. Am. J.* 40:317–319 (1976).
154. Cunningham, K.M. and M.C. Goldberg. "A Reexamination of the Effects of Adsorbates on the Raman Spectrum of Gibbsite," *Soil Sci.* 136(2):102–110 (1983).
155. Wieckowski, T. and A. Wiewiora. "New Approach to the Problem of the Interlayer Bonding in Kaolinite," *Clays Clay Miner.* 24:219–223 (1976).
156. Maksimovic, Z. and J.L. White. "Infrared Study of Chromium-Bearing Halloysites," *Proc. 1972 Int. Clay Conf.* 61–73 (1972).
157. Maskimovic, Z., J.L. White, and M. Logar. "Chromium-Bearing Dickite and Chromium-Bearing Kaolinite from Teslic, Yugoslavia," *Clays Clay Miner.* 29:213–218 (1981).
158. Prost, R., A. Dameme, E. Huard, J. Driard, and J.P. Leydecker. "Infrared Study of Structural OH in Kaolinite, Dickite, Nacrite, and Poorly Crystalline Kaolinite at 5 to 600 K," *Clays Clay Miner.* 37:464–468 (1989).
159. Lobartini, J.C. and K.H. Tan. "Differences in Humic Acid Characteristics as Determined by Carbon-13 Nuclear Magnetic Resonance, Scanning Electron Microscopy, and Infrared Analysis," *Soil Sci. Soc. Am. J.* 52:125–130 (1988).
160. Miller, J.G. "An Infrared Spectroscopic Study of the Isothermal Dehydroxylation of Kaolinite at 470 Degrees," *J. Chem.* 65:800–804 (1961).
161. Prost, R. "Interactions between Adsorbed Water Molecules and the Structure of Clay Minerals: Hydration Mechanism of Smectites," *Proc. Int. Clay Conf.* 351–359 (1975).
162. Serratosa, J.M. "Dehydration Studies by Infrared Spectroscopy," *Am. Mineral.* 45:1101–1104 (1960).
163. Cannings, F.R. "An Infrared Study of Hydroxyl Groups on Sepiolite," *J. Phys. Chem.* 72:1072–1074 (1968).
164. Calvet, R. and R. Prost. "Cation Migration Into Empty Octahedral Sites and Surface Properties of Clays," *Clays Clay Miner.* 19:175–186 (1971).
165. Huneke, J.T., R.E. Cramer, R. Alvarez, and S.A. El-Swaify. "The Identification of Gibbsite and Bayerite by Laser Raman Spectroscopy," *Soil Sci. Soc. Am. J.* 44:131–134 (1980).
166. Weismiller, R.A., J.L. Ahlrichs, and J.L. White. "Infrared Studies of Hydroxy-Aluminum Interlayer Material," *Soil Sci. Soc. Am. Proc.* 31:459–463 (1967).

167. White, J.L., S.L. Nail, and S.L. Hem. "Infrared Technique for Distinguishing between Amorphous and Crystalline Aluminum Hydroxide Phases," *7th Conf. Clay Mineral. Petrol.* 51–59 (1976).
168. Parker, T.W. "A Classification of Kaolinites by Infrared Spectroscopy," *Clay Miner.* 8:135–141 (1969).
169. Wiewiora, A., T. Wieckowski, and A. Sokolowska. "The Raman Spectra of Kaolinite Sub-group Minerals and of Pyrophyllite," *Archiwum Mineralogiczne* 35:5–14 (1979).
170. Johnston, C.T., G. Sposito, and C. Erickson. "Vibrational Probe Studies of Water Interactions with Montmorillonite," *Clays Clay Miner.* in press (1993).
171. Johnston, C. T., W. L. Earl, and C. Erickson. "Vibrational and NMR probe studies of SAz-1 Montmorillonite," in *Proceedings of the Metal Speciation Workshop* (Chelsea, MI, Lewis Publishers, Inc., in press).

CHAPTER 2

CHARACTERIZATION OF BIOLOGICAL AND ENVIRONMENTAL PARTICLES USING STATIC AND DYNAMIC LIGHT SCATTERING

Peter Schurtenberger and Meredith E. Newman

TABLE OF CONTENTS

0-87371-895-X/93/$0.00+$.50

INTRODUCTION

Enormous theoretical and instrumentational progress has been made in both static light scattering (SLS) and dynamic light scattering, also called quasielastic light scattering (QLS), within the last 20 years. Applications of SLS and QLS have primarily focused on colloid and polymer physics.[1,2] Although numerous biological applications exist, they have been restricted primarily to determinations of mean molar masses and average particle sizes. Amazingly few applications of SLS and QLS to environmental samples currently exist, although there is an increasing interest in the characteristics of submicron particles in environmental systems. Both biological and environmental applications may benefit greatly from theoretical and instrumental improvements, and information beyond molar masses and size distributions may be obtained.

In this chapter we shall summarize the necessary theoretical background of SLS and QLS and discuss recent experimental trends and open questions. While the basic principles of dynamic and static light scattering from colloidal suspensions are first presented with a minimum of mathematical formalism, we realize that the second part of the theoretical section of this chapter is a bit "mathematical." However, we believe that a detailed knowledge of the underlying physical and mathematical principles is necessary for an understanding of many phenomena observed with light scattering. We have attempted to provide a qualitative explanation and interpretation of the important equations in the theoretical section and to point out their most important implications to biological and environmental applications.

For example, it is clear that the relatively detailed treatment of the influence of interparticle interactions on the results from dynamic and static light scattering experiments may not be very relevant for scientists interested primarily in aquatic environmental samples, due to the usually very low particle concentrations in these systems, but the following "readers guide" should make it possible to obtain the requested information selectively:

Section 2.1 qualitatively describes the basic principles of dynamic and static light scattering from colloidal suspensions and the information which can be obtained by means of scattering experiments. The most important relations used in a quantitative analysis and interpretation of scattering experiments are then introduced.

In Section 2.2, the effect of intraparticle interference (i.e., size and shape effects) on the angular dependence of the scattering intensity are discussed for monodisperse particles, and the influence of interparticle interactions on both the static and the dynamic light scattering experiments are considered. This is particularly important for investigations of relatively well-defined and monodisperse biological macromolecules such as proteins or viruses, as is demonstrated in Section 4.1 with a discussion of recent results from an experimental study of lens proteins.

Section 2.3 then considers the influence of polydispersity on the static and the dynamic light scattering experiments for both noninteracting and interacting colloidal particles in suspension. While the discussion of polydispersity in noninteracting systems provides the basis for an investigation of aquatic environmental samples, as is shown later in the review of currently existing experimental studies in Section 5.1, the influence of polydispersity in interacting systems is considered as a general basis for a characterization of most biological samples and soil particles.

A number of biological macromolecules and environmental particles (such as colloidal clays, for example) are known to have nonspherical shape and a behavior which is reminiscent of polymer molecules. While a complete theoretical treatment of the scattering of light by anisotropic particles is well beyond the scope of this chapter, in Section 2.4 we give a brief summary of some of the most important theoretical results and their implications, primarily in studies of polymerlike biological macromolecules. This section serves as a basis for our discussion of selected experimental studies with polydisperse biopolymers and biological micelles in Sections 4.1 and 4.2.

Section 2.5 shows how a combination of static and dynamic light scattering and concepts from fractal geometry can be used to study aggregation processes found, for example, in biological samples (such as antigen-antibody systems or protein solutions) or environmental samples. The now existing literature on aggregating colloidal suspensions is then reviewed in Section 5.2.

In Section 3 we discuss instrumentation (3.1) and measurement procedures (3.2) for static and dynamic light scattering experiments. Particular attention is given to the analysis and interpretation of dynamic light scattering data from polydisperse colloidal suspensions in Section 3.2.2.

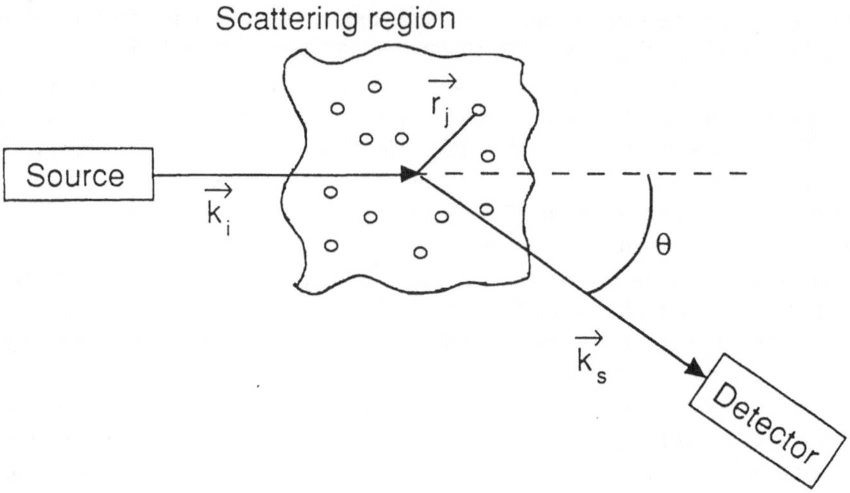

Figure 1. Schematic representation of a light scattering experiment.

THEORY OF LIGHT SCATTERING FROM PARTICLE SUSPENSIONS

Static and Dynamic Light Scattering — Basic Concepts

In this section, the theory of light scattering from colloidal suspensions is summarized. First, we qualitatively describe the basic principles of dynamic and static light scattering from colloidal suspensions and the information which can be obtained by means of scattering experiments. While this is done with a minimum of mathematical formalism, we then try to list and discuss the most important relations used in a quantitative analysis and interpretation of scattering experiments. However, we restrict ourselves to those features which are essential to the following discussion of recent experimental results, and readers interested in additional details should consult the numerous excellent textbooks and reviews on the theory of light scattering currently available.[1-4]

A typical scattering geometry is shown schematically in Figure 1. A vertically polarized and monochromatic incident laser light beam of angular frequency ω_i (or vacuum wavelength λ_i) and wavevector $\vec{k}_i$ (where $|\vec{k}_i| = 2\pi n/\lambda_i$ and n is the index of refraction of the solution) illuminates the scattering volume V_s containing N particles. The scattered light with wavevector $\vec{k}_s$ (where $|\vec{k}_s| = 2\pi n/\lambda_s$) is then observed at a scattering angle θ and at a (large) distance R from the scattering volume. First we shall consider the simple case of a single particle, before we see what happens if there are many particles in the scattering volume. We can divide the particle into many small subregions, which are all polarized by the alternating incident electric field $\vec{E}_i$ (which mathematically can be described as a plane wave) and will consequently act as radiation sources. If the largest dimension of the particle is small compared to the wavelength of the light, all subregions see the same incident electric field. The amplitude $\vec{E}_s$ of the scattered field can then

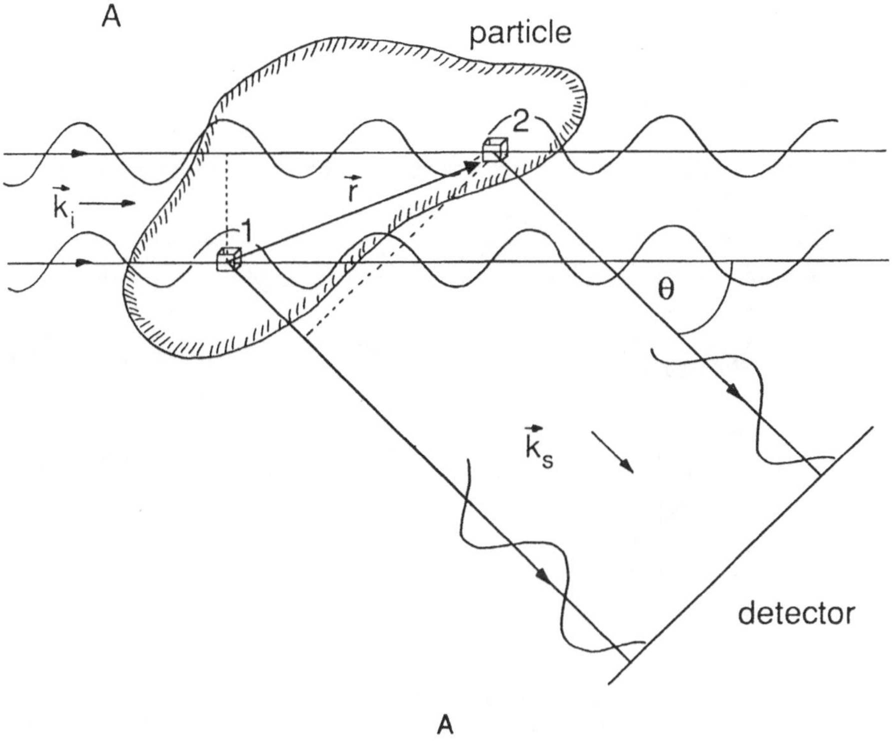

Figure 2. Schematic representation of scattering from a particle that is large compared to the wavelength λ_i. The phase of the radiation (and thus of the induced dipoles) at the two volume elements shown is different. The path difference for the light scattered by volume elements 1 and 2, respectively, can lead to destructive interference at the detector as shown in B. (A) scattering angle θ = 45°; (B) scattering angle θ = 90°.

be written as a simple sum of the contributions from the individual scatterers, and is thus proportional to the polarizability α of the particle or the particle volume V (or molar mass *M*).[5,6] The experimentally measured scattering intensity I_s depends on the square of the field amplitude and is thus given by $I_s = |\vec{E}_s|^2$, i.e., is proportional to α^2 (and thus M^2).

However, if the largest dimension of the particle is comparable or larger than the wavelength of the light, the situation becomes more complicated and the scattering experiment will yield additional information. Scattering from a single large particle is schematically illustrated in Figure 2. The dipoles induced by the incident electric field within a given large particle are oscillating out of phase, and we must therefore take into account interference effects in the calculation of $\vec{E}_s$. This can be done by calculating the path length differences for the light scattered by all possible pairs of scattering centers within the particle, and comparing it with the wavelength λ. The optical path difference between the two volume elements shown in Figure 2A is given by $\Delta s = \vec{r}\cdot\vec{k}_i - \vec{r}\cdot\vec{k}_s = \vec{Q}\cdot\vec{r}$, where $\vec{Q} = \vec{k}_i - \vec{k}_s$ is the

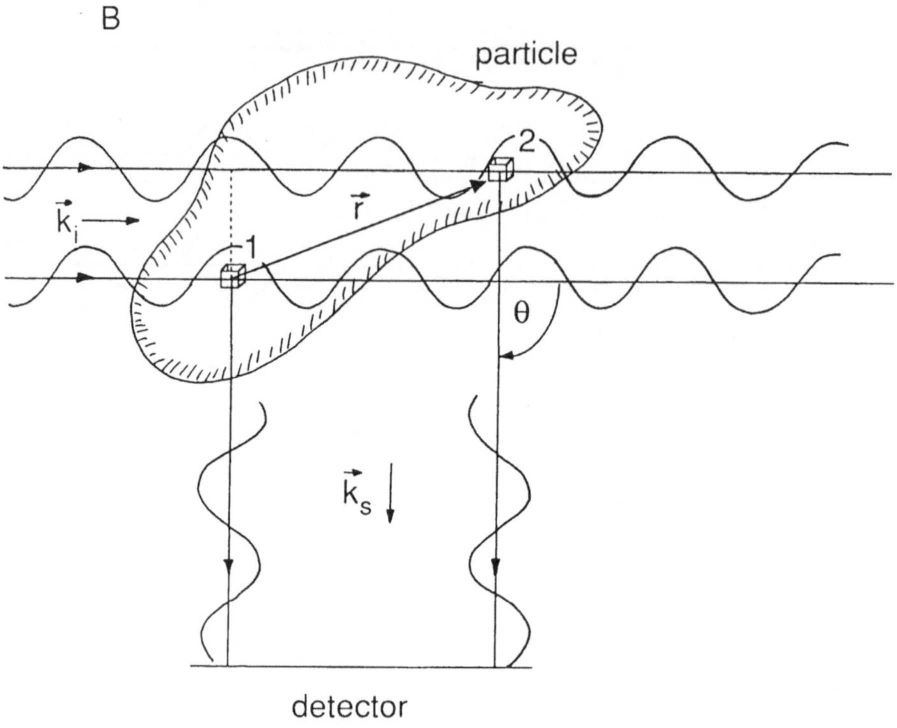

Figure 2(B).

scattering vector. For elastic or quasielastic scattering, where $|\vec{k}_i| \cong |\vec{k}_s|$, the magnitude of $\vec{Q}$ is given by $Q = (4\pi n/\lambda_i)\cdot\sin(\theta/2)$. For $Q = 0$, there is no path length difference and thus no phase difference for the light scattered by the different scattering centers. With increasing scattering angle, the path length differences increase, and the effects from destructive interference will become more important. This is illustrated in Figure 2A and B, which provide a qualitative explanation of several important features of scattering experiments:

1. We can expect to find a strong angular dependence of the scattering intensity I_s for particles with dimensions comparable to or larger than λ_i. The angular dependence of I_s is usually described by a so-called particle form factor $P(\theta)$. $P(\theta)$ is defined as the ratio of the scattered intensity for a large particle at angle θ, $I_s(\theta)$, and the scattered intensity without interference, $I_s(\theta = 0°)$, which may be less than 1 when θ is large, but will increase with decreasing θ and become unity for $\theta \rightarrow 0°$.
2. At angles greater than zero, $P(\theta)$ will always be smaller for an extended particle than for a compact particle of the same molar mass and polarizability. We thus see that important information on the size and shape of particles can be obtained from the angular dependence of the scattering intensity.

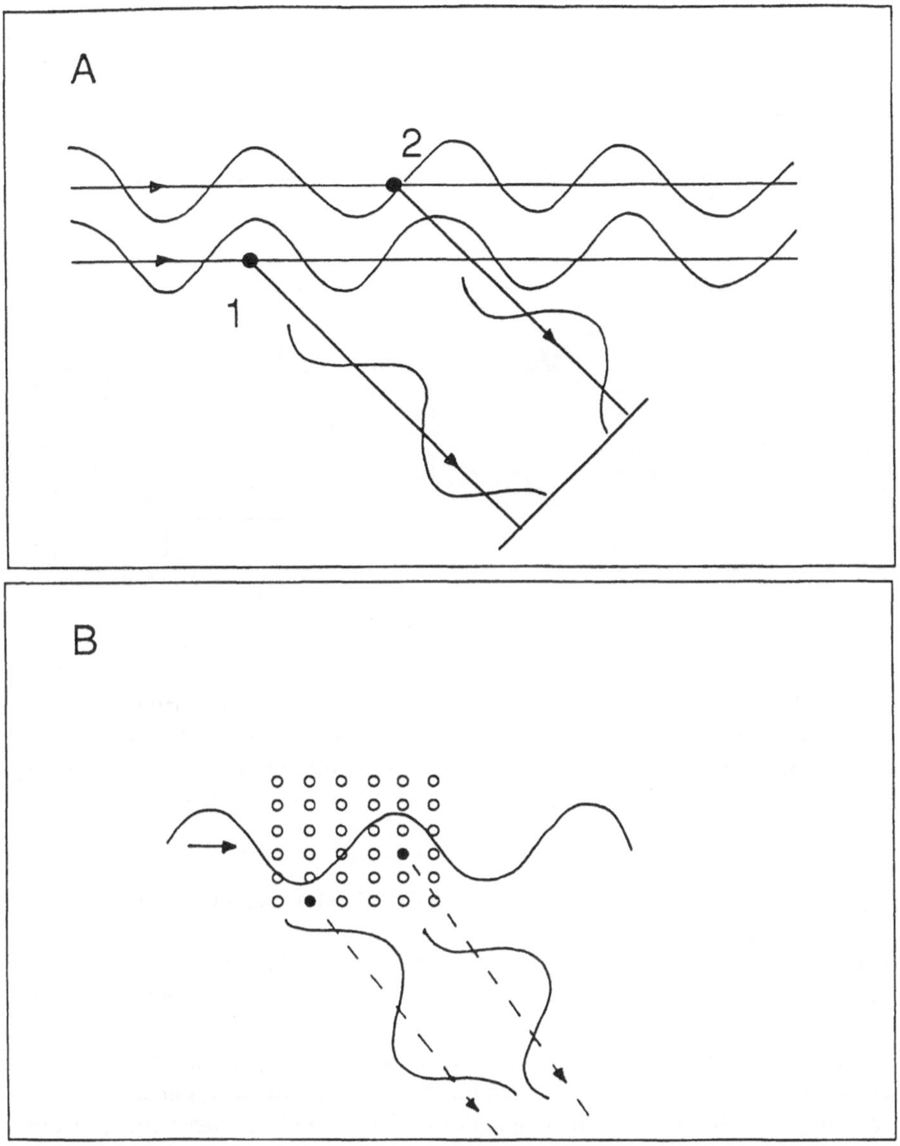

Figure 3. (A) Schematic representation of scattering from two particles fixed in space (see text for details); (B) Schematic representation of scattering from a perfect crystal (see text for details); and (C) Schematic representation of scattering from a particle suspension and the resulting fluctuations of the scattered intensity (see text for details).

3. The influence of the destructive intraparticle interference effects on the scattering intensity depends on the quantity Q·r. Interference effects are not important for Q·r << 1. Therefore 1/Q is often called the 'spatial resolution' of a scattering experiment, since we can only obtain additional information

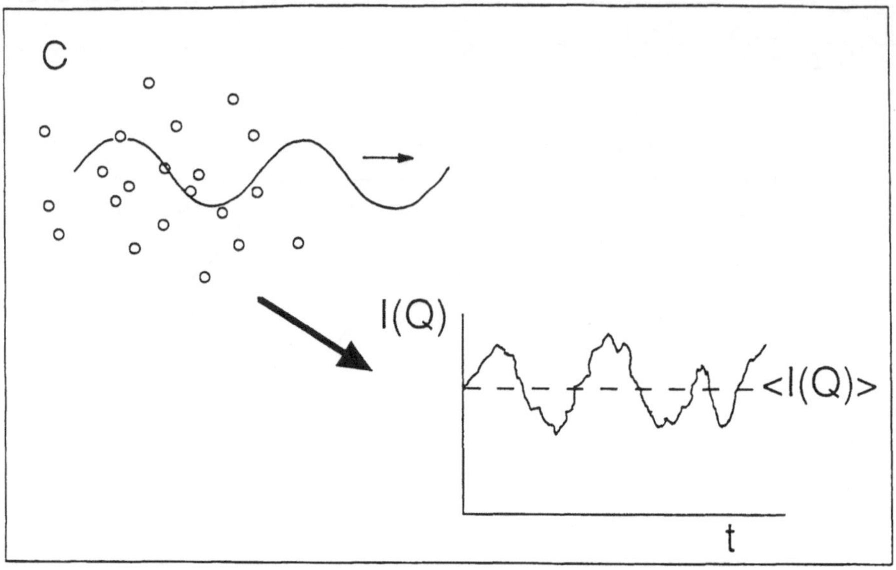

Figure 3(C).

on the particle size and structure from P(θ) for particle dimensions r comparable or larger than 1/Q. We shall in fact use P(Q) instead of P(θ) in the coming sections of the text, since this will permit us to discuss **intraparticle** interference or shape effects in a more general way, independent of wave length λ.

To see what happens when scattering by more than one particle occurs, we now consider the simplest case of two particles as illustrated in Figure 3A. The particles are assumed to be identical, fixed in space, and illuminated by the same incident plane wave. We may ask ourselves what the amplitude and intensity of the scattered light observed at a distant point R will be. In general, there will be a phase difference between the two scattered waves from particles 1 and 2. If the particles are now able to move at random with respect to one another, all values for the phase difference will be equally probable, and the measured and time averaged scattering intensity $\langle I_s \rangle$ is simply the sum of the two individual scattering intensities.[5] This result may easily be generalized to a situation with a large number of particles: if all particles are able to move at random with respect to one another, the total average scattering intensity will then be the sum of all the individual scattering intensities. Together with the above mentioned results for the scattering intensity from one particle we thus obtain $\langle I_s \rangle \sim N \cdot M^2 \cdot P(\theta)$ for an ideal suspension of monodisperse particles.

However, this may be quite different for strongly interacting particles, which are no longer able to diffuse at random with respect to one another. We can qualitatively understand this by considering the simple (and extreme) case of scattering from a perfect crystal, as illustrated in Figure 3B. If the lattice spacing is small compared to the wavelength of the light, the scattering intensity will be zero except in the forward direction. This is easy to understand, since for any point i in the crystal we can always find a scattering center at some point j for which the phases of the scattered waves differ by some odd multiple of 180°. We can thus think of the crystal as being made up of scattering elements that cancel one another for $\theta > 0°$. In scattering from real liquids or colloidal suspensions, the positions of the particles change due to Brownian motion, leading to concentration fluctuations, nonuniform distribution of scattering centers, and therefore incomplete destructive interference between scattered light of different phases. It is, however, obvious that strong repulsive interparticle interactions and high particle concentrations will induce increasing 'order' in the solution, and significantly influence (i.e., decrease) the average total scattering intensity measured in a static light scattering experiment due to **interparticle** interference effects.

The concentration fluctuations are also the basis for dynamic light scattering experiments. The continuous fluctuations in the number of molecules per volume element due to Brownian motion result in fluctuations in the light scattering intensity with time, which are superimposed on a background of scattering (i.e., on $\langle I_s \rangle$ measured in the time-averaged static light scattering experiment), which arises from the fact that at any instant the particle concentration is likely to be different in the different volume elements (see Figure 3C). The kinetics of these intensity fluctuations will depend on the size (diffusion coefficient) of the particles: small and rapidly diffusing particles will produce rapid fluctuations, and large and slowly moving particles will cause slow fluctuations in the scattering intensity. These fluctuations are usually and conveniently analyzed by recording the so-called intensity autocorrelation function $G_2(\tau) = \langle I_s(t) \cdot I_s(t+\tau) \rangle$, and a subsequent analysis of G_2 permits a fast and noninvasive determination of the diffusion coefficient of the particles in solution. It is quite obvious that interparticle interaction effects will also be important in a dynamic light scattering experiment, since the diffusion of colloidal particles in suspension can be strongly modified by direct (such as electrostatic or steric) or hydrodynamic interactions.

We can now make a next step and formulate simple mathematical relations which summarize the qualitative discussion of the basic principles of light scattering from **monodisperse** particles. These definitions and equations represent the basis for the analysis of 'real' experimental data and for a sound discussion of important aspects such as the influence of polydisperse size distributions or anisotropic particle shapes. The starting point is again a typical light scattering experiment as outlined in Figure 1, for which it is straightforward to obtain a simple quantitative expression for the resulting amplitude of the scattered field. If the particles are optically homogeneous and spherical and multiple scattering effects can be ignored, the scattering amplitudes of the individual particles are time independent and the amplitude $\vec{E}_s$ of the scattered field can be written as a sum of the contributions from the individual scatterers, taking properly into account the phase differences between waves scattered by different particles:

$$E_s(Q,t) = \sum_{j=1}^{N} A_j(Q) e^{i\vec{Q}\vec{r}_j(t)} \tag{1}$$

where

Aj(Q) is the scattering amplitude of particle j
$\vec{r}_j(t)$ is the position of the center of particle j at time t

The average scattering intensity is then given by

$$I_s(Q,t) = < \left| E_s(Q,t) \right|^2 > = < \sum_i \sum_j A_i(Q) A_j(Q)_e^{i\vec{Q}\left(k_i(t) - \vec{r}_j(t)\right)} > \tag{2}$$

where < > denotes an ensemble average. The scattering amplitudes of the individual particles have been written as $A_j(Q)$ in order to indicate that a Q-dependence (and thus a dependence on the scattering angle) will be seen for larger particles (particle radius $r \geq \lambda/30$). $A_j(Q)$ is obtained through a summation of the contributions from small scattering elements dV over the particle volume V_j, and by properly taking into account intraparticle interference effects and all possible particle orientations. A_j thus corresponds to the Fourier transform of the scattering density (which in the case of light scattering is due to the polarizability of matter, i.e., corresponds to the dielectric susceptibility of the particles).

Static Light Scattering

In a static light scattering experiment (SLS), the intensity of the scattered light is measured at different scattering angles. For a quantitative analysis of the SLS experiment, a reduced intensity, the so-called excess Rayleigh ratio, $\Delta\mathbf{R}(\theta)$, (i.e., the differential cross section of the scattering process) is generally used. $\Delta\mathbf{R}(\theta)$ is defined as[5]

$$\Delta\mathbf{R}(\theta) = \frac{\Delta I_s(\theta)}{I_o} \frac{R^2}{V_s} \tag{3}$$

where

$DI_s(q)$ is the difference between the intensity of the light scattered by the suspension and by the pure solvent at a distance R from the scattering volume, respectively, and I_o is the intensity of the incident beam

For identical particles, $\Delta\mathbf{R}(\theta)$ can then be written as[5]

$$\Delta\mathbf{R}(\theta) = K \cdot C \cdot M \cdot P(Q) \cdot S(Q) \tag{4}$$

where

K = $4\pi^2 n^2 (dn/dc)^2/(N_A \cdot \lambda_o^4)$
dn/dc = the refractive index increment
C = particle concentration (mass/volume)
M = molar mass of the particles in solution
P(Q) = particle form factor which describes the influence of **intraparticle** interference effects on the angular dependence of the scattered intensity (see Figure 2), and thus provides information on the macromolecular shape and structure
S(Q) = the time-averaged structure factor which describes **interparticle** interference effects (see Figure 3), i.e., it contains important information on the interparticle interaction potential[7]

Equation 4 represents the key to an analysis of static light scattering experiments, and it will always be important to remember that static light scattering data contains contributions from both single particle properties (M, P(Q)) as well as interparticle interaction effects (S(Q)). Summarizing Equation 4, static light scattering permits in principle a noninvasive and *in situ* study of the particle molar mass, the particle shape (from P(Q)), and the interparticle interaction potential (from S(Q)). In Sections 2.2 and 2.3, we shall see how the influence of interparticle interactions can be modeled theoretically, and how polydispersity can be taken into account.

Dynamic Light Scattering

In a dynamic or quasielastic light scattering experiment (a frequently used abbreviation is also PCS, which is derived from 'photon correlation spectroscopy'), the temporal fluctuations of the scattered intensity, which are due to the Brownian motion of the particles, are analyzed. The basis for this technique is in fact already summarized in Equation 1: while the scattering amplitude A_j can only change in time if particle j changes its structure or is anisotropic and changes its orientation with time (see Section 2.4), the phase factors $\exp\left[\vec{Q} \cdot \vec{r}_j(t)\right]$ will change in time with the motion of the scatterer's center of mass and produce temporal fluctuations of the scattering intensity. The analysis of these fluctuations is usually achieved through an estimate of the normalized temporal autocorrelation function of the scattered field amplitude[4]

$$g_1(Q,\tau) = \frac{< E_s(Q,t) E_s^*(Q,t+\tau) >}{< \left| E_s(Q,t) \right|^2 >} \tag{5}$$

or the normalized temporal autocorrelation function of the scattered intensity

$$g_2(Q,\tau) = \frac{< I_s(Q,t) I_s(Q,t+\tau) >}{< I_s(Q,t) >^2} \tag{6}$$

Under the assumption of a Gaussian distribution[8] of the intensity profile, $g_1(\tau)$ and $g_2(\tau)$ are related through the Siegert relation

$$g_2(Q,\tau) = 1 + \left| g_1(Q,\tau) \right|^2 \tag{7}$$

and the experimentally measured intensity autocorrelation function $C(Q,\tau)$ in a 'homodyne' (see Section 3.2 for details) QLS experiment can be written as

$$C(Q,\tau) = B\left[1 + a\left| g_1(Q,\tau) \right|^2\right] \tag{8}$$

where the 'signal to base line ratio' $a \leq 1$ is a (geometrical) factor which is primarily determined by the number of coherence areas seen by the photomultiplier, i.e., by the size of the scattering volume.[4,9]

For monodisperse solutions of N optically homogeneous and spherical particles, $E_s(Q,\tau)$ is given by Equation 1, and $g_1(Q,\tau)$ can then be written as

$$g_1(Q,\tau) = \frac{F(Q,\tau)}{S(Q)} \tag{9}$$

where the dynamic structure factor $F(Q,\tau)$ is given by

$$F(Q,\tau) = \left(N < A^2 >\right)^{-1} < \sum_j \sum_{j'} A_j A_{j'} e^{i\vec{Q}\left(\vec{r}_j(0) - \vec{r}_{j'}(\tau)\right)} > \tag{10}$$

In the hydrodynamic limit, where $2\pi/Q$, the characteristic 'probing length' of the scattering experiment, is much larger than the most probable interparticle distance, $F(Q,\tau)$ describes (long range) spatial fluctuations of the particle concentrations. These concentation fluctuations are expected to decay through a diffusive mechanism, and $F(Q,\tau)$ then decays exponentially with time as [10,11]

$$F(Q,\tau) = S(O) e^{-D_c Q^2 \tau} \tag{11}$$

where D_c is the collective diffusion coefficient of the particles in solution. The experimentally determined autocorrelation function $C(Q,\tau)$ for such a system thus decays like a single exponential with

$$C(Q,\tau) = B \cdot \left[1 + a \cdot e^{-2D_c Q^2 \tau}\right] \tag{12}$$

Dynamic light scattering thus permits a fast and noninvasive *in situ* measurement of the collective diffusion coefficient of colloidal particles and biological macromolecules in solution. In Sections 2.2 and 2.3 we shall see how interparticle interaction effects and polydispersity can be taken into account when interpreting light scattering results.

Suspensions of Monodisperse Spherical Particles

In this section we shall look more carefully at the effect of intraparticle interference on the angular dependence of the scattering intensity and consider the influence of interparticle interactions on both the static and the dynamic light scattering experiments.

Static Light Scattering

In the limit of low scattering angles, i.e., for $Qr \ll 1$, where r is the particle radius, and low particle concentrations, we obtain simple approximations for P(Q) and S(O). P(Q) can then be represented, independently of the specific form of the particles, by[3,5]

$$P(Q) = 1 - \frac{Q^2 \cdot R_g^2}{3} + \ldots \tag{13}$$

where R_g is the radius of gyration of a particle. The radius of gyration of a particle is essentially the root mean-square radius of a particle. The particle can be considered as an assembly of mass elements of mass m_i, each located a distance r_i from the center of mass. The radius of gyration for a given particle structure is then defined as the square root of the weight average of r_i^2 for all the mass elements, i.e.,[6]

$$R_g = \left[\frac{\sum_i m_i r_i^2}{\sum_i m_i}\right]^{1/2} \tag{14}$$

The evaluation of the angular dependence of the scattering intensity $\Delta\mathbf{R}(\theta)$ will only be useful for a determination of R_g if $R_g > \lambda/40$ (that is, $R_g > 10$ nm for the commonly used modern argon ion lasers).

The structure factor S(Q) can be written as[7]

$$S(Q) = 1 + 4\pi\rho \int_0^\infty r^2 [g(r) - 1] \frac{\sin Qr}{Qr} dr \tag{15}$$

where

ρ is the number density of particles
g(r) is the radial distribution function
(the total correlation function, h(r) = g(r) – 1, is the relative mean number density deviation at a distance r from the center of a given particle)

In the limit of Q → 0, S(O) is related to the osmotic compressibility $(\partial\Pi/\partial C)^{-1}$, i.e., to the derivative of the osmotic pressure Π with respect to the concentration, by[7]

$$S(O) = \frac{N_A}{M} k_B T \left(\frac{\partial \Pi}{\partial C} \right)^{-1} \tag{16}$$

Simple analytical expressions and approximations exist for the thermodynamic quantity $(\partial\Pi/\partial C)^{-1}$ for a number of model systems, such as ideal macromolecular solutions or hard sphere suspensions. S(0) is often expressed via a virial expansion of the osmotic pressure Π in Equation 16 as[5,7]

$$S(O) = 1 - 2A_2 MC + \ldots \tag{17}$$

where A_2 is the second virial coefficient, which is related to the pair potential V(r) through

$$A_2 = \frac{2\pi N_A}{M^2} \int_0^\infty \left[1 - \exp\left(\frac{-V(r)}{k_B T} \right) \right] r^2 dr \tag{18}$$

For a suspension of hard spheres with volume V and molar mass *M*, it is well established that $A_2 = 4VN_A/M^2$, and Equation 17 simplifies to

$$S(0) = \frac{1}{1 + 8\Phi} \tag{19}$$

where Φ is the volume fraction of the particles.[12] Similarly, A_2 can be estimated for repulsive electrostatic interactions or additional attractive interactions by using the

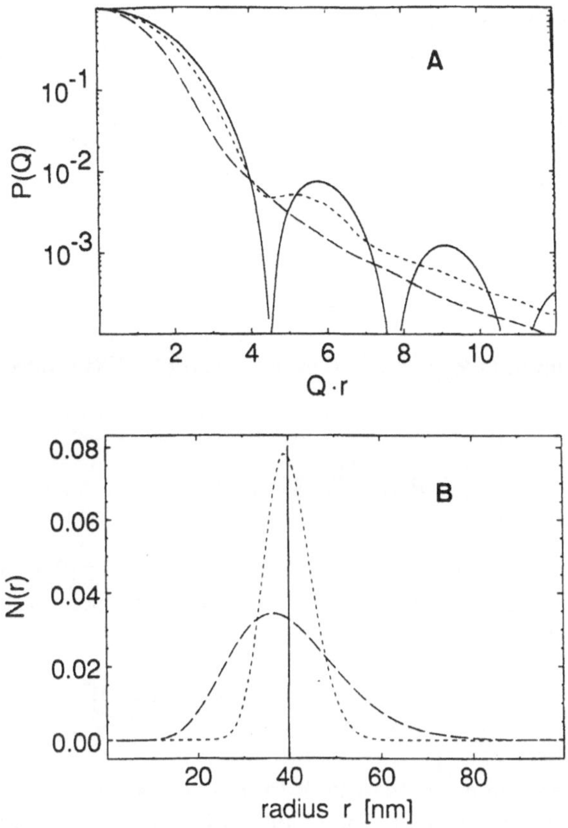

Figure 4. Scattering form factor P(Q) for homogeneous spherical particles as a function of the product Q·r, where r is the radius of the particle. (A) Monodisperse and polydisperse spheres calculated using a Schultz distribution, (—): monodisperse; (- -): z = 60; and (– –): z = 10 (B) Schultz number distribution, <r> = 40 nm, (—): monodisperse; (- -): z = 60; and (– –): z = 10.

corresponding form of the interaction potential in Equation 18. A classical DLVO potential has been used, for example, to calculate the influence of interparticle interactions in the virial expansion approximation.[13] More recently, the potential of mean force V(r) was determined from 'direct force measurements' between macroscopic surfaces directly using the Derjaguin approximation, and A_2 was then calculated from Equation 18 and compared with the results from light scattering measurements.[14]

Using Equations 13 and 17, Equation 4 now simplifies to

$$\frac{K \cdot C}{\Delta \mathbf{R}(\theta)} = \frac{1}{M} \cdot \left(1 + \frac{Q^2 \cdot R_g^2}{3}\right) + 2A_2C \tag{20}$$

and we see that a measurement of the concentration and Q dependence of $\Delta\mathbf{R}(\theta)$ provides information on the particle size and shape and the interaction potential.

However, the expansion of P(Q) used in Equation 13 is valid for $Q \cdot R_g \ll 1$ only. If additional information on the particle shape is available, better approximations of P(Q) are known for various geometrical forms. For example, for spheres we can use ('Guinier-approximation')

$$P(Q) = e^{\frac{-Q^2 R_g^2}{3}} \tag{21}$$

which is valid up to $Q \cdot R_g \approx 1$.[15,16] For larger particles, P(Q) must be calculated from Equation (22)

$$P(Q) = \left[1/V \cdot \int e^{i\vec{Q}\cdot\vec{r}'} dV\right]^2 \tag{22}$$

where V is the volume of the particle, and the evaluation involves averaging over all possible orientations of the particle. This leads for spheres with radius r to[3]

$$P(Q) = \frac{9}{(Qr)^6}\left[\sin(Qr) - Qr\cos(Qr)\right]^2 \tag{23}$$

Figure 4A presents an example of P(Q) for monodisperse homogeneous spheres as a function of the product Qr. The scattering intensity exhibits distinct minima and maxima, and their position permits a very accurate estimate of the particle dimensions, provided the particles are sufficiently large ($r \geq 150$ nm for $\lambda \approx 500$ nm). Furthermore, we shall see later that the depth of the intensity minima strongly depends on the polydispersity of the particles.

In the above treatment, we have used the so-called Rayleigh-Gans approximation for a description of the angular dependence of the scattering intensity, i.e., the assumption that each volume element of a particle is illuminated by the same incident plane wave. However, the Rayleigh-Gans approximation is valid only for particles with radius $Q \cdot r \ll 1/|m - 1|$, where m is the ratio of the refractive indices of particle and solvent, respectively. For very large particles (and/or very large values of m), the difference between the index of refraction of the particles and that of the solvent causes a distortion of the electric field of the incident radiation. An application of 'Mie scattering theory' is then necessary. (Example solutions for spherical particles or oriented cylinders can be found in van de Hulst.[17])

For concentrated suspensions, the virial expansion of Π used in Equation 17 requires an increasing number of higher order terms, which are generally unknown. On the other hand, no exact theoretical solution is available for Π or g(r), even for the case of a simple hard sphere potential. However, an application of modern concepts from liquid state theory has yielded good approximations, such as the Percus-Yevick Equation or the Carnahan-Starling Equation.[7,12] In particular, the semiempirical extension of the Percus-Yevick theory derived by Carnahan and Starling[18] provides us with a very accurate analytical expression for the osmotic pressure of a hard sphere suspension, which together with Equation 16 leads to

$$S(0) = \frac{(1-\Phi)^4}{1+4\Phi+4\Phi^2-4\Phi^3+\Phi^4} \tag{24}$$

A simple hard sphere potential may not be appropriate for most experimental systems, but a more realistic approach can be used in which an additional attractive or repulsive potential V′(r) is treated as a perturbation and added. This leads to[19]

$$S(0) = \frac{(1-\Phi)^4}{1+4\Phi+4\Phi^2-4\Phi^3+\Phi^4} + A'\cdot\Phi \tag{25}$$

where $A' \sim \int V'(r)\, r^2\, dr$ represents the additional contribution from attractive or repulsive interactions. While Equation 25 provides us with a simple analytical expression for S(0), it is not appropriate for highly charged colloidal particles or at very low ionic strength, where the (screened) Coulomb potential completely dominates. In this case, one has to go back to Equation 15, which requires a theoretical description of g(r). Based on the Ornstein-Zernike relationship, several approximations, such as the mean spherical approximation (MSA) or the renormalized mean spherical approximation (RMSA), have been used in order to interpret light and neutron scattering data from charged colloidal solutions.[20-22]

So far we have only looked at S(0), assuming that we can determine $\Delta\mathbf{R}(0)$ from a simple extrapolation of the scattering intensity. However, at high particle concentrations and strong interparticle interactions, a nonmonotonic Q-dependence of the intensity with a characteristic intensity maximum can be found, which makes an extrapolation to Q = 0 much more difficult. We can then try to extract S(Q) over the Q-range accessible to light scattering and interpret it using Equation 15 together with a suitable approximation of g(r).[21]

Dynamic Light Scattering

We have seen from Equation 12 that, for a suspension of monodisperse spherical particles, a measurement of the intensity autocorrelation function permits a deduction of the collective diffusion coefficient D_c. In the limit of $Q \to 0$, the concentration dependence of the collective diffusion coefficient can be described by a 'generalized Stokes-Einstein relation' of the form[11]

$$D_c = \frac{k_B T}{S(0) f_c} \tag{26}$$

where f_c is the friction coefficient, i.e., the frictional resistance which particles undergo while moving in the same direction. We see from Equation 26 that D_c depends both on hydrodynamic (f_c) and static (S(0)) interparticle interactions. In the limit of low concentrations, we can use a simple expansion of f_c and S(0) in the volume fraction Φ in order to account for interparticle interactions,[11]

$$f_c = \frac{f_o}{1 - k_f \Phi + \ldots} \tag{27}$$

$$S(0) = \frac{1}{1 + k_I \Phi + \ldots} \tag{28}$$

where

$f_o = 6\pi\eta_o R_h$ is the infinite dilution value of the friction factor
η_o is the viscosity of the solvent
R_h is the so-called hydrodynamic radius of the particles (which for ideal spheres corresponds to their geometrical radius r)
k_I and k_f represent the coefficients for the first order correction term to the infinite dilution values of f_o and S(0), respectively

A combination of Equations 26–28 then leads to[11]

$$D_c = D_o\left[1 + k_D \phi + \ldots\right] \tag{29}$$

where

$k_D = k_I - k_f$
D_o = the diffusion coefficient at infinite dilution

given by

$$D_o = \frac{k_B T}{f_o} = \frac{k_B T}{6\pi\eta_o R_h} \tag{30}$$

However, even for hard spheres there are some discrepancies between the theoretical values of k_f in the literature.[10,11,23,24] A quite complete treatment of the complicated problem of many body hydrodynamic interactions between hard spheres by Batchelor has led to a value of $k_f = 6.55$, which together with Equations 19 and 28 leads to $k_D = 1.45$.[23] This value is in good agreement with recent experimental results.[25] In addition to the simple hard sphere case, expressions for k_I and k_f that take into account an attractive van der Waals potential and a repulsive screened Coulomb potential have been proposed, but their validity is restricted to dilute solutions and low ionic strength.[11,13] In general, the description of the hydrodynamic properties is very complicated for concentrated systems, and the theory is not as advanced as in the case of the static (thermodynamic) properties, although some interesting contributions from Brownian dynamics simulations have appeared recently.[26,27]

Suspensions of Polydisperse Spherical Particles

In this section we shall consider the influence of polydispersity on the static and the dynamic light scattering experiments for both noninteracting and interacting colloidal particles in suspension.

Static Light Scattering

In the case of a suspension of polydisperse spherical particles with a normalized size distribution N(r), Equation 4 can be written as[1]

$$\Delta R(\theta) = KC < M >< P(Q) >< S(Q) > \tag{31}$$

where $<M>$, <P(Q)> and <S(Q)> represent average effective molar mass, particle form factor, and structure factor taking into account the size distribution, respectively. $<M>$ and <P(Q)> are given by

$$< M >= \frac{\int_0^\infty N(r) \cdot M(r)^2 \, dr}{\int_0^\infty N(r) \cdot M(r) dr} \tag{32}$$

and

$$< P(Q) >= \frac{\int_0^\infty N(r) \cdot M(r)^2 \cdot P(Qr) dr}{\int_0^\infty N(r) \cdot M(r)^2 \, dr} \tag{33}$$

where $<M>$ corresponds to the so-called weight average molar mass (one frequently finds the symbol M_w for the weight average molar mass). Equation 32 shows why light scattering is so extremely sensitive to the presence of only trace amounts of very large particles, high molar mass impurities, or dust.

Examples of <P(Q)> for polydisperse spheres with increasing polydispersity index σ, where

$$\sigma^2 = \frac{<r^2>}{<r>^2} - 1 \tag{34}$$

are given in Figure 4A for a Schultz distribution, i.e., a generalized exponential size distribution which is commonly used in the colloid literature,

$$N(r) = \frac{r^z}{z!}\left(\frac{z+1}{<r>}\right)^{(z+1)} \cdot e^{-\frac{r}{<r>}(z+1)} \tag{35}$$

where σ^2 is related to the polydispersity parameter z by $\sigma^2 = 1/(z + 1)$.[28-30] The corresponding distribution functions are shown in Figure 4B. We see from Figure 4 that the higher order minima and maxima are smeared out with increasing polydispersity, and that we lose the ability to accurately determine the particle size from P(Q) for large particles at high values of σ.

While it is relatively straightforward to obtain average values for the scattering amplitude and the particle form factor for suspensions of polydisperse spheres, an extremely difficult problem is represented by a calculation of the effective structure factor <S(Q)>, given by [22,30,31]

$$<S(Q)> = \frac{\sum_{ik=1}^{N} A_i(Q)A_k(Q)\left(\rho_i\rho_k\right)^{1/2}\left[\delta_{ik} + S_{ik}(Q)\right]}{\sum_{i=1}^{N}\rho_i A_i(Q)^2} \tag{36}$$

where

ρ_i is the number density of particles i with scattering amplitude $A_i(Q)$
δ_{ik} is the Kronecker symbol

$$S_{ik}(Q) = \left(\rho_i\rho_k\right)^{1/2}\int_0^\infty 4\pi r^2\left[g_{ik}(r) - 1\right]\frac{\sin Qr}{Qr}\,dr \tag{37}$$

In particular we have to keep in mind that <S(Q)> cannot be defined independently of the intraparticle interference factors (i.e., the ratio $A_j(Q)/A_j(0)$) as in the case of monodisperse particles. It is often assumed that the interaction potentials between polydisperse particles are identical, although their scattering properties may vary greatly. However, pair potentials between different particles will show variations in both magnitude and range, and it would thus be desirable to have at least an approximate theoretical model which could account for the influence of polydispersity on <S(Q)>. Closed expressions exist, for example, for the scattering of polydisperse hard spheres in the Percus-Yevick approximation. They are, however, mathematically very complex and beyond the scope of this chapter, and the interested reader should look up the original references.[32,33]

A comparison of the Percus-Yevick approximation with Monte-Carlo computer simulations for polydisperse hard spheres has yielded good agreement up to volume fractions as high as $\Phi = 0.3$.[31] These simulations have also shown that a frequently used assumption,[34] in which positional polydispersity and polydispersity in the form factors are uncorrelated (i.e., that the particle size and orientation are uncorrelated with the particles' positions), and which leads to a considerably simplified expression for <S(Q)>, provides a very poor description of the Monte Carlo results even at low values of σ. This has also been demonstrated by Pusey et al., who compared the Percus-Yevick approximation for <S(Q)> with this so-called decoupling approximation for three different types of size distributions and various concentrations, ranging from $\Phi = 0.1$ to $\Phi = 0.4$.[35]

Additional work has been done by Klein et al., who used the MSA solution of the multicomponent version of the Ornstein-Zernike equation in order to calculate <S(Q)> for charged polydisperse solutions.[22] The calculations were performed as a function of the ionic strength, particle concentration, and polydispersity, and the results were compared with those from the decoupling approximation. The authors also came to the conclusion that, despite its popularity among experimentalists, the decoupling approximation can lead to rather unreliable results.

In the limit of low concentrations and low Q, we obtain from Equation 31 an approximate expression for $\Delta\mathbf{R}(\theta)$, which is analogous to Equation 20,

$$\frac{K \cdot C}{\Delta\mathbf{R}(\theta)} = \frac{1}{<M>} \cdot \left(1 + \frac{Q^2 \cdot <R_g^2>}{3}\right) + 2A_2C \tag{38}$$

but where M has been replaced by its weight average $<M>$ and $R_g{}^2$ by its z-average $<R_g^2>$, which is given by

$$<R_g^2> = \frac{\int_0^\infty N(r) \cdot M^2(r) \cdot R_g^2(r)dr}{\int_0^\infty N(r) \cdot M^2(r)dr} \tag{39}$$

Equation 38 is routinely used to analyze SLS data from suspensions of biological macromolecules, polymers, or colloids. It clearly shows that for polydisperse solutions, static light scattering provides a measurement of average quantities

($\langle M \rangle$ and $\langle R_g^2 \rangle$) only. Taking into account the proper weighting with M_i and M_i^2, respectively, is particularly important when we try to compare light scattering results with those from other types of experiments such as electron microscopy or size exclusion chromatography.

Dynamic Light Scattering, Noninteracting Systems

For polydisperse noninteracting systems, the dynamic structure factor F(Q,τ) (Equation 10) is given by a sum of exponentials, one for each species of particles. Replacing the sum with an integral over the particle size distribution (i.e., using a continuous distribution rather than a discrete form), $g_1(Q,\tau)$ can then be written as[1,4]

$$g_1(Q,\tau) = \frac{\int_0^\infty N(r)A^2(Q,r)e^{-D_o(r)Q^2\tau}dr}{\int_0^\infty N(r)A^2(Q,r)dr} \tag{40}$$

where A(Q,r) and $D_o(r)$ represent the scattering amplitude and diffusion coefficient of particles with radius r, respectively. For spherical particles we can use $A^2(Q,r) \sim M^2P(Qr) \sim r^6P(Qr)$, which leads to

$$g_1(Q,\tau) = \frac{\int_0^\infty N(r)r^6P(Qr)e^{-D_o(r)Q^2\tau}dr}{\int_0^\infty N(r)r^6P(Qr)dr} \tag{41}$$

Equation 41 is frequently written as

$$g_1(Q,\tau) = \int_0^\infty G(\Gamma)e^{-\Gamma\tau}d\Gamma \tag{42}$$

where the weight function

$$G(\Gamma) = \frac{N(r)r^6P(Qr)}{\int_0^\infty N(r)r^6P(Qr)dr} \tag{43}$$

gives the relative scattering contribution of a component with decay rate $\Gamma = D_oQ^2$.

For narrow size distributions, the exponentials in Equation 42 can be expanded around an average decay rate $\langle\Gamma\rangle$, which leads to[36,37]

$$g_1(Q,\tau) = e^{-<\Gamma>\tau}\left[1+\frac{\sigma}{2}<\Gamma>^2\tau^2+\ldots\right] \quad (44)$$

We can thus obtain an average decay rate $<\Gamma>$ and a coefficient of variation σ from a second order polynomial fit in τ to $\ln[g_1(\tau)]$, a so-called cumulant analysis. From the average decay rate $<\Gamma>$, obtained from the initial slope of $\ln[g_1(\tau)]$, we can then calculate an intensity weighted diffusion coefficient $<D_o(Q)>$[28,29]

$$<D_o(Q)> = \frac{<\Gamma>}{Q^2} = \frac{\int_0^\infty N(r)r^6 P(Qr)D_o(r)dr}{\int_0^\infty N(r)r^6 P(Qr)dr} \quad (45)$$

The polydispersity index σ can be deduced from the second cumulant, and is given by

$$\sigma = \frac{\int_0^\infty N(r)r^6 P(Qr)D_o^2(r)dr}{<D_o(Q)>^2 \int_0^\infty N(r)r^6 P(Qr)dr} - 1 = \frac{<D_o^2(Q)>}{<D_o(Q)>^2} - 1 \quad (46)$$

A mean hydrodynamic radius $<R_h>$ is often calculated from $<D_o(Q)>$, whereby the Stokes-Einstein relation (Equation 30) is used and the values D_o and R_h are replaced by the corresponding average values $<D_o(Q)>$ and $<R_h>$.

In the limit of low Q, a simple second order cumulant analysis of a QLS experiment provides us thus with a measurement of the z-average collective diffusion coefficient, and an estimate of the coefficient of variation of the intensity-weighted size distribution. If we use the Stokes-Einstein relation for ideal spheres, we see that[29]

$$\lim_{Q\to 0} <D_o(Q)> = \frac{k_B T}{6\pi\eta_o\left(<r^6>/<r^5>\right)} \quad (47)$$

$$\lim_{Q\to 0} \sigma = \frac{<r^4>\cdot<r^6>}{<r^5>^2} - 1 \quad (48)$$

i.e., the experimentally determined average diffusion coefficient is heavily weighted towards the large sizes/high molar mass components (see also Section 3.2.2 for details).

Dynamic Light Scattering, Interacting Systems

For polydisperse and interacting systems, the dynamic structure factor F(Q,τ) in Equation 10 can in general not be reduced further, since the scattering amplitudes A_j and $A_{j'}$ are correlated with the particle positions $\vec{r}_j$ and $\vec{r}_{j'}$, and F(Q,τ) is thus a complicated function of Q and τ. However, simple expressions have been presented by Pusey et al. for several limiting conditions.[11,35] Although they are only valid under well-defined and restricted assumptions, they provide us with a theoretical framework in order to interpret some interesting experimental results presented in this chapter. The basic assumption for this treatment is that particle correlation and dynamics are not affected by the size distribution, i.e., that the average in Equation 10 can be separated into independent averages over particle position and scattering amplitudes. This would correspond to a case of scattering-power polydispersity, in which the particles are identical in terms of size and interactions, and differ in scattering power only. Such an approach should also be valid for so-called 'paucidisperse' systems, systems for which the standard deviation of the particle size is much less than the mean. This leads to[11,35]

$$F(Q,\tau)=\frac{<A>^2}{<A^2>}\cdot S^1(O)\cdot e^{-D_cQ^2\tau}+\frac{<A^2>-<A>^2}{<A^2>}\cdot e^{-D_sQ^2\tau} \tag{49}$$

where

$S^1(O)$ is the 'ideal' static structure factor (i.e., of a system with particles identical in size and scattering amplitude)

D_s is the self diffusion coefficient (which describes the random motion of individual particles)

and $<A>$ and $<A^2>$ are defined as

$$<A>=\frac{1}{N}\sum_{i=1}^{N}A_i \tag{50}$$

and

$$<A^2>=\frac{1}{N}\sum_{i=1}^{N}A_i^2 \tag{51}$$

respectively. Equation 49 shows that at relatively high volume fractions, the measured correlation function in a QLS experiment will be composed of two independent modes with well-separated decay constants, even for fairly narrow size distributions. While the concentration dependence of the collective diffusion coefficient can be described approximately by Equation 29, we can use a similar expression for D_s,

$$D_s = D_o\left[1 + k_s\Phi + \ldots\right] \quad (52)$$

where $k_s = -2.10$ for hard spheres.[11,23] Therefore we can expect a much more hindered self-diffusion process in concentrated suspensions when compared to the collective diffusion, and thus $D_s \ll D_c$. In particular for strongly interacting particles where $S(Q) \ll 1$, the amplitude of the second term, the 'slow mode,' will become comparable to the first term, the 'fast mode,' even for narrow size distributions and a correspondingly small value of $(\langle A^2\rangle - \langle A\rangle^2)/\langle A^2\rangle$. Two characteristic decay times have in fact been found for a number of model studies in concentrated colloidal suspensions (see References 11 and 23). However, as much as one needs to be aware of the implications of Equation 49 for an analysis of bimodal correlation functions in terms of the particle size distribution, one also must be very cautious when interpreting a nonexponential correlation function as due to contributions from collective and self diffusion. Clearly a careful study of the concentration and Q-dependence is required before one can make a safe conclusion.

Suspensions of Anisotropic Particles

Almost all the results summarized above were deduced under the explicit or implicit assumption that the particles are spherical. However, a number of biological macromolecules and environmental particles (such as colloidal clays) are known to be nonspherical and to have a behavior that is reminiscent of polymer molecules. While a complete theoretical treatment of the scattering of light by anisotropic particles is well beyond the scope of this article, we give a brief summary of some of the most important results and their implications, primarily in studies of polymerlike biological macromolecules.

Dilute Solutions of Anisotropic Particles

It is possible to obtain valuable information on the particle shape from a combination of static and dynamic light scattering measurements. For particles with a longest dimension much smaller than the wavelength of the incident light, the scattering intensity is quite insensitive to the particle shape and depends primarily on the particle volume (or molar mass). This is quite in contrast to the hydrodynamic radius R_h, which strongly depends on both the volume and the shape of the particle. Analytical expressions for R_h as a function of the geometrical dimensions of a particle have been developed for a number of different geometrical models such as prolate or oblate ellipsoids, rigid cylinders, semiflexible ('wormlike'), or flexible chains.[4,38-45] Some of them are summarized in Table 1.

Additional information can be obtained, if the particles are large enough to show a measurable angular dependence of the scattering intensity. It is then

Table 1. Formulas for Hydrodynamic Radius R_h, Radius of Gyration R_g, and Form Factor P(Q) for Various Geometrical Shapes

Sphere:[49] $R_h = r \quad R_g = \sqrt{\frac{3}{5}}\, r \quad P(Q) = \frac{9}{(Qr)^6}\left[\sin Qr - Qr\cos(Qr)\right]^2$

Ellipsoid of Revolution:[5,39,39,49]

$$R_h = a \cdot G(b/a)^{-1} \qquad R_g = \left[\frac{a^2}{5} + \frac{2b^2}{5}\right]^{1/2} \qquad P(Q) = \frac{9\pi}{2}\int_0^{\pi/2} \frac{J^2(V)}{V^3}\cos\beta d\beta$$

where a, b, b = semiaxis of ellipsoid (b < a ; prolate, b > a : oblate)

$$G = \left[1 - \left(\frac{b}{a}\right)^2\right]^{-1/2} in\left[\frac{1 + \left[1 - \left(\frac{b}{a}\right)^2\right]^{1/2}}{\frac{b}{a}}\right]; \frac{b}{a} < 1$$

$$G = \left[\left(\frac{b}{a}\right)^2 - 1\right]^{-1/2} \arctan\left[\left(\frac{b}{a}\right)^2 - 1\right]^{1/2}; \frac{b}{a} > 1$$

J = Bessel function of order 0
$V^2 = Q^2(a^2\cos^2\beta + b^2\sin^2\beta)$

Cylinder:[5,49,160]

$$R_h = \frac{3}{2}r\left(\left[1 + \left(\frac{L}{2r}\right)^2\right]^{1/2} + \frac{2r}{L} In\left(\frac{L}{2r} + \left[1 + \left(\frac{L}{2r}\right)^2\right]^{1/2}\right) - \left(\frac{L}{2r}\right)\right)^{-1}$$

$$R_g = \left[\frac{L^2}{12} + \frac{r^2}{2}\right]^{1/2}$$

$$P(Q) = \int_0^{\pi/2} \frac{\pi}{LQ\cos\beta} f(r, L, Q, \beta)\sin\beta d\beta$$

where L = cylinder length
r = cylinder radius

$$f(r,L,Q,\beta)=\left[J\left(\frac{LQ\cos\beta}{2}\right)\cdot\frac{2J_1(rQ\sin\beta)}{rQ\sin\beta}\right]^2$$

L → 0: infinitely thin disc;[49] $P(Q)=\frac{2}{(Qr)^2}\left[1-\frac{J_1(2Qr)}{Qr}\right]$;

where J_1 = Bessel function of order 1

r → 0: infinitely thin rod;[41,49]

$$R_h=\frac{L}{2s-0.19-\frac{8.24}{s}+\frac{12}{s^2}};P(Q)=\frac{2}{QL}\left[\int_0^{QL}\frac{\sin z}{z}dz\right]-\left(\frac{\sin(QL/2)}{QL/2}\right)^2;$$

where $s=in\left(\frac{L}{r}\right)$

Wormlike chain:[343,44] $R_h=f\left(L,d,l_p\right)$ $R_g^2=\frac{l_pL}{3}-l_p^2+\frac{2l_p^3}{L}-\frac{2l_p^4}{L^2}\left[1-\exp\left(\frac{-L}{l_p}\right)\right]$

$$P(Q)=\frac{2}{Q^2Rg^2}\left(Q^2Rg^2-1+e^{-Q^2Rg^2}\right)$$

where $f(L,d,l_p)$ represents Equations 49 through 52 in Reference 52
L = chain contour length
d = chain diameter
l_p = persistence length

Sources: Burchard,[49] van Holde,[5] Perrin,[38] Perrin,[39] Mazer et al,[160] van de Sande and Persoons,[41] Yamakawa and Fujii,[43] and Yamakawa and Fujii.[44]

possible to determine from the Q-dependence of the scattered intensity $\Delta\mathbf{R}(\theta)$, in the limit of small Q, the radius of gyration R_g independent of the shape of the particle through Equation 20. This ability to obtain an experimental value for R_g provides a powerful tool in the determination of macromolecular structure, since macromolecules of different configurational types (e.g., solid spheres, rods, disks, flexible chains, random coils) have quite different values of the radius of gyration for the same value of the molecular weight.[46,47] Again, a number of exact or approximate expressions for R_g as a function of the geometrical particle dimensions have been developed for different geometrical models, and some are listed in Table 1. In particular the ratio R_g/R_h is a quantity which has distinctly different values for different particle shapes (see Table 1). A measurement of R_g/R_h thus provides us with a relatively unambiguous test for the particle shape. However, it must be emphasized that polydispersity may strongly alter R_g/R_h and should be considered carefully.[48,49]

We cannot only calculate R_g for various geometrical models, but there also exist a number of analytical expressions for the single-particle form factor P(Q) for simple particle shapes.[3] This is particularly helpful for large particles. However, it is once again very important to mention that polydispersity strongly affects the shape of P(Q) for various particle geometries. Even moderate polydispersity tends to smear out the otherwise well-defined intensity minima and maxima at high Q values. Moreover, for polydisperse solutions it is often not possible to unambiguously distinguish between different geometrical models. For example, the scattering from polydisperse spheres is often indistinguishable from the scattering from prolate ellipsoids.[16]

If the particles are anisotropic, the scattering amplitudes A_j in Equation 10 for the dynamic structure factor will be time dependent due to the rotational motion of the particles. For dilute noninteracting particles, where we can assume that orientation and position are uncorrelated, we can then write $g_1(Q,\tau)$ as

$$g_1(Q,\tau) = \left(N < A^2 >\right)^{-1} < A(0)A(\tau) > e^{-D_c Q^2 \tau} \tag{53}$$

The amplitude correlation function $<A(0)A(\tau)>$ for cylindrical particles with length L undergoing rotational diffusion is according to Berne and Pecora[4]

$$< A(0)A(\tau) >= 4\pi N < A^2 > \sum_{j=0}^{\infty} B_j(QL) e^{-j(j+1)D_R \tau} \tag{54}$$

where D_R is the rotational diffusion coefficient, and the coefficients $B_j(QL)$ depend on the particle size and the scattering vector according to

$$B_j(QL) = \left[\frac{1}{QL} \int_{-QL/2}^{+QL/2} j_j(Qx)d(Qx) \right]^2 \tag{55}$$

with $j_j(QL)$ being spherical Bessel functions. Combining Equations 53 and 54 then leads to the following expansion for $g_1(Q,\tau)$:

$$g_1(Q,\tau) = B_0(QL)e^{-D_c Q^2 \tau} + B_2(QL)e^{-\left(D_c Q^2 + 6D_R\right)\tau} + B_4 e^{-\left(D_c Q^2 + 20D_R\right)\tau} + \ldots \tag{56}$$

For small particles or larger particles at sufficiently low values of Q, $B_0(QL)$ >>

$B_2(QL)$, $B_4(QL)$, ..., and Equation 56 reduces to a single exponential decay (Equation 12). For larger particles and higher values of Q, however, we can expect to find an additional fast decay time due to the contribution from rotational motion of the particles. This is illustrated for the case where $QL \leq 10$, where only $B_0(QL)$ and $B_2(QL)$ need to be considered. D_c can then be determined from $\lim_{Q\to 0} C(Q,\tau)$ using Equation 12, and $B_2(QL)/B_0(QL)$ and D_R can be obtained from fits to the approximate form[50]

$$\frac{C(Q,\tau)-B}{B} = a \cdot e^{-2D_c Q^2 \tau}\left[1 + \frac{B_2(QL)}{B_0(QL)} \cdot e^{-6D_r\tau}\right]^2 \tag{57}$$

Additional contributions to the measured autocorrelation function cannot only be due to rotational diffusion of the particles, but may also occur for large polymerlike particles due to internal chain dynamics.[49,51,52] Depending on the product of $Q{\cdot}R_g$, we will be able to look at different properties of the biopolymer. For $Q{\cdot}R_g \ll 1$, the QLS experiment is sensitive to concentration fluctuations whose spatial Fourier wavelength Q^{-1} is large compared to the size of a single particle. The dominating relaxation process is thus translational motion of the center of mass of the particles.[52] On the other hand, for $Q{\cdot}R_g \gg 1$, the translational diffusion has negligible influence, and the dynamic structure factor is completely determined by the internal chain dynamics of the biopolymer.[52-54] For polymer molecules, one finds an intermediate regime $Q{\cdot}R_g \approx 1$ for which the initial decay of the measured intensity autocorrelation function from a cumulant analysis can be described by

$$<\Gamma>/Q^2 = <D>_z \cdot\left(1 + C \cdot Q^2 \cdot <R_g^2>_z + \ldots\right) \tag{58}$$

where

$<\Gamma>$ is the first cumulant

C is a dimensionless quantity which strongly depends on the structure, flexibility, and polydispersity of the macromolecule[49]

Values of C have been calculated for a number of structural models and various degrees of polydispersity, and can be found in References 48 and 55 to 57. C-values are generally between 0.1 and 0.3. Equation 58 clearly shows that, for high molecular mass biopolymers, internal chain dynamics will contribute significantly to the first cumulant at higher scattering angles. This points out again how important measurements at different values of θ are, and how easy it is to pick up an artefact in 'simply' determining an average <D>-value, in what one considers to be a 'routine experiment.'

Interacting Anisotropic Particles

The results summarized above are valid for dilute, noninteracting particles only. For higher concentrations, the situation becomes much more complicated, and in particular the problem of hydrodynamic interactions between anisotropic molecules, which would be vital for an understanding of the QLS results, remains still unsolved. However, there are a number of approximate phenomenological relations which can be used at least at lower concentrations, where virial expansions of $\partial\Pi/\partial\Phi$ and D_c can be applied, i.e., where Equations 20 and 29 are valid.

For anisotropic particles with volume V and molar mass *M* which interact through excluded volume interactions only, we can express A_2 by

$$A_2 = \frac{4VN_A}{M^2} \cdot f \tag{59}$$

where the parameter f corrects for anisotropy and polydispersity. For example, analytical expressions for f have been derived for monodisperse prolate or oblate ellipsoids, for spherocylinders by Onsager, or for flexible chains.[58-61] For rodlike particles, the influence of polydispersity on A_2 can be estimated based on Onsager's result for the orientationally averaged excluded volume of two prolate spherocylinders. One can show that, for polydisperse rodlike systems, f simply corresponds to the value that one would obtain for monodisperse particles having the weight-average axial ratio and is insensitive to the precise form of the particle size distribution.[62] However, while there exist a number of simple and fairly accurate expressions for excluded volume interactions between anisotropic particles, the situation is much more complicated for charged macromolecules. Rather than going into the details of these recent theoretical developments, we refer the interested reader to the literature.[63-65]

The concentration dependence of the collective diffusion coefficient of anisotropic particles can be expressed to a first approximation using Equation 29.[48,49,57] Most theoretical progress has been made for rod-like and flexible chain systems, where a number of theoretical models have been proposed.[66-72] However, the agreement between the theoretical predictions and the existing experimental results is not as good as for the much simpler case of spherical particles, and substantial theoretical work is clearly required.

Light Scattering from Aggregating Systems and Fractal Objects

In recent years, there has been a growing interest in the study of aggregation phenomena, due to the discovery of the fractal nature of the aggregates. A very clear and detailed summary of the theory of aggregation and aggregate structure is given in Reference 73. The structure of the aggregates is controlled in particular by their association kinetics as defined by the kinetic kernels (or rate constants) $K_{i,j}$ and $K_{k,j}$ of the Smoluchowski equation for irreversible aggregation

Table 2. Fractal Dimension d_F from Simple Aggregation Models

	Fractal Dimension d_F		
		Cluster-cluster	
Path	**Monomer-cluster**	**Polydisperse**	**Hierarchical**
Diffusion limited (DLA)	2.50	1.80	1.78
Ballistic	3.0	1.95	1.89
Reaction limited (RLA)	3.0	2.09	1.99

Sources: Brinker and Scherer[74] and Meakin.[75]

$$\frac{dC_k}{dt} = \frac{1}{2}\sum_{i+j=k} K_{i,j}C_iC_j - C_k\sum_{j=1}^{\infty}K_{k;j}C_j \tag{60}$$

where

the first term represents the generation of clusters of size k from smaller clusters i and jthe second term represents the disappearance of clusters of size k to form larger clusters k+j

Large repulsive interaction between the clusters lead for example to slow aggregation kinetics (reaction controlled) and to compact structures, whereas systems with weakly repulsive interaction potentials lead to fast kinetics (diffusion controlled) and to 'open' structures.[74] Extensive large-scale computer simulations of various aggregation models such as diffusion (DLA) or reaction (RLA) limited aggregation or diffusion or reaction limited cluster-cluster aggregation (DLCCA and RLCCA, respectively) have been used in the study of the structure and the kinetic and equilibrium size distribution of the forming clusters.[73-75] The fact that these simple aggregation models and real aggregation processes lead to structures that can be described using the concepts of fractal geometry has stimulated this research area considerably. Fractal objects can be described by a fractal dimension d_F, which relates the mass M (or number of primary particles N) of the (mass) fractal to its radius r by

$$M \sim r^{d_F} \tag{61}$$

Due to the self-similarity of the fractal objects, d_F has also been called the 'similarity exponent,' since it tells how the mass changes after a change of length scale: if all lengths are multiplied by a factor λ, masses must be multiplied by λ^{dF}.

The fractal dimension d_F strongly depends on the aggregation model used, and a number of simulation results are summarized in Table 2. Attempts to gain information on the probable mechanism of aggregate formation by comparing the results from various computer simulations of colloidal aggregation with the particle growth observed in 'real' systems have been made in material sciences.[74,76] The interest in equilibrium and kinetic cluster size distribution and the structure of the formed aggregates is not only restricted to theoretical physics or material sciences, but has had quite an impact on the study of biological systems (antibody-antigen complexes) or environmental particles.

The fractal dimensionality d_F can in fact be determined by measuring a variety of mass-length scaling relationships with scattering techniques. The intensity $I_M(Q)$ scattered by one randomly oriented fractal object of mass M, which is formed by N identical spherical primary particles of radius a, averaged over all orientations of the aggregate, is given by a simple Fourier transform on all particles of the aggregate[73,77]

$$I_M(Q) = I_o(Qa) < \sum_i \sum_j e^{i\vec{Q}(\vec{r}_i - \vec{r}_j)} > \tag{62}$$

where

$I_o(Qa)$ is the scattering intensity from one primary particle
< > denotes averaging over all orientations

If we assume spherical symmetry, which is correct for a collection of randomly oriented aggregates, we can describe the ratio $I_M(Q)/I_o(Qa)$ using the correlation function h(r) between the centers of the particles in the aggregate[73]

$$I_M(Q) / I_o(Qa) \sim N^2 \int_0^\infty r^2 h(r) \frac{\sin Qr}{Qr} dr = N^2 S'(Q) \tag{63}$$

where $S'(Q)$, the intraparticle structure factor, describes the spatial arrangement of the particles. It is important to remember that the characteristic length scale probed in a scattering experiment is given by $1/Q$, or that the information content that can be obtained depends on the product $Q \cdot L$, where L is a characteristic length scale of the particle. For $Qr \approx 1$, where r is the radius of the aggregate, we can estimate the overall dimension of the aggregate from a measurement of its radius of gyration R_g from

$$I_M(Q) / I_o(Qa) = N^2 e^{-\frac{Q^2 R_g^2}{3}} \tag{64}$$

which can be obtained from a simple expansion in Q up to second order terms of Equation 63.[73]

The fractal properties of the aggregate can be studied in the regime Qr >> 1 and Qa << 1. For fractal objects, the correlation function h(r) has a form of [73]

$$h(r) \sim f(r/\xi)\cdot r^{(d_F-3)} \tag{65}$$

where

f(r/ξ) is a cut-off function which describes the behavior of h(r) at large r
the cut-off value ξ is the characteristic distance above which the mass distribution does not follow the fractal law anymore

Together with Equation 63 this leads to

$$I_M(Q)I_o(Qa) \sim Q^{-d_F} \tag{66}$$

i.e., we observe a typical power law dependence of $I_s(Q)$.

Finally, details of the structure of the primary particles can be obtained for Qa ≥ 1 (which, however, is usually difficult to reach with light scattering). According to Equation 66, we can thus determine the fractal dimension directly from the Q-dependence of the scattering intensity, provided a << 1/Q << r. However, the range of validity for the power law in Equation 66 can be very limited because of a saturation at low Q due to the Guinier regime, and the influence from the Porod regime ($I_s(Q) \sim Q^{-4}$) at high Q.[73] A power law dependence according to Equation 66, a linear behavior in a log-log plot of $I_s(Q)$ versus Q, should thus be observed over at least one decade before one can have some confidence in the exponent determined in this way.

Another consequence of the fractal nature of the aggregates is that any measure of the radius r(*M*) of a single cluster is related to the total mass of the cluster by Equation 61, which should thus also be valid for R_g and R_h measured in static and dynamic light scattering:

$$<M> \sim <R_h>^{d_h} \tag{67a}$$

$$<M> \sim <R_g>^{d_g} \tag{67b}$$

It is important to remember that real aggregating systems are polydisperse as indicated by the brackets < > in Relations 67, and that the measured quantities R_h and R_g correspond to different weighted averages, as does the molar mass determined with SLS. However, the cluster size distributions resulting from the different aggregation models seem to show so-called scale invariance, and Relations 67 thus hold, even for polydisperse systems, with $d_h = d_g = d_F$.[78,79] When trying to

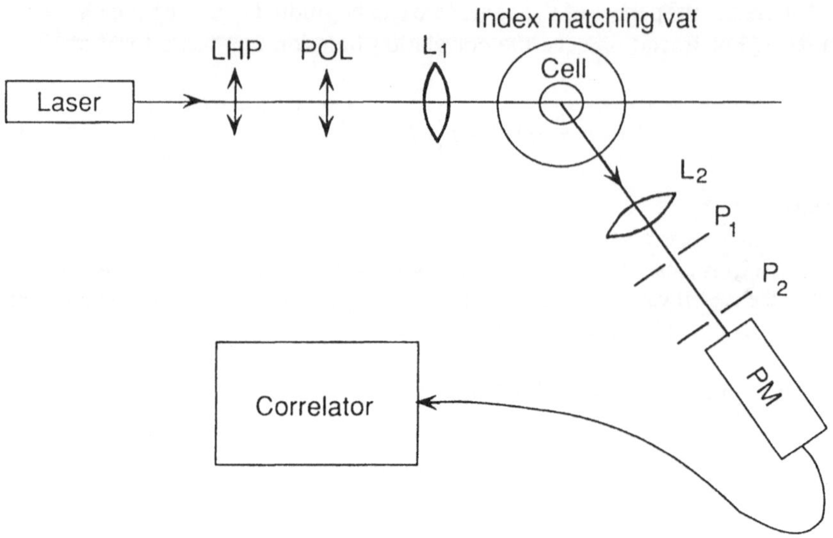

Figure 5. Schematic view of a typical configuration of a light scattering spectrometer (see text for details).

determine the fractal dimension of aggregates with dynamic light scattering, one should also be aware that for Qr > 1, we also probe internal motion and rotation of the aggregates.[77]

EXPERIMENTAL METHODS

Instrumentation

A thorough discussion of the instrumental details involved in dynamic and static light scattering experiments is given elsewhere.[9] A number of commercial instruments which are perfectly suitable for particle size distribution analysis are currently available (Malvern, Amtec-Brookhaven, ALV, Nicomp, Coulter Electronics, Otsuka Electronics). Here we only review some of the more important points concerning instrumentation. A typical instrument configuration is shown in Figure 5.

Dynamic light scattering requires a light source of high coherence, and thus the use of a laser is necessary for these measurements. Generally, two types of lasers are used for light scattering measurements: helium-neon (HeNe) and argon-ion (Ar) gas lasers. While HeNe lasers are inexpensive but very stable air cooled light sources at a wavelength λ_o = 632.8 nm, their use in QLS experiments is limited due to their relatively low power (generally 2 to 15 mW). Dynamic light scattering experiments with environmental particle suspensions or dilute suspensions of small biological macromolecules such as enzymes or proteins normally require a laser power of a few hundred mW, which necessitates the use of a (water cooled)

Ar laser at $\lambda_0 = 488$ nm or 514.5 nm. If one intends to perform very accurate static light scattering experiments, one needs lasers with a high beam pointing stability, and we thus suggest the use of a temperature-controlled cooling water system for Ar lasers.

Several inexpensive instruments are offered with only 90° scattering or a very limited range of scattering angles. However, while this may be adequate for simple quality control measurements, it is clearly not sufficient for a serious characterization of samples with unknown particle size distribution and structure. We have seen in Section 2 that very important information can be obtained from the angular dependence of $< I_s >$ and $C(Q,\tau)$. In particular, the large and polydisperse environmental particles with radii of a few hundred nm would require measurements at θ-values as low as 15°. Measurements of the Q-dependence are conveniently done with a computer-controlled stepping motor driven detection system. Instruments with fiber optics detection at fixed scattering angles represent a very interesting and sophisticated development, since such instruments allow for a much better detection scheme and, due to the lack of any moveable parts, offer a much higher optical stability and increased size resolution in determining R_g.[47,81,82]

As shown in Figure 5, QLS experiments require the use of a focusing lens (L_1) in front of the sample cell, which reduces the laser beam diameter of nominal 1 to 2 mm to about 100 μm in the scattering volume. This is done because the time-dependent part of the signal in a QLS experiment (signal to base line ratio a in Equation 8) is proportional to the coherence area (i.e., the area over which the phase of the scattered light is correlated), and because the coherence area can be increased by a reduction of the scattering volume.[4,9] The optimum detection limit is obtained with one coherence area, and the number of coherence areas and the accuracy of Q can be adjusted with the size of the pinholes P_2 and P_1, respectively.[9,83,84] In contrast to QLS, larger scattering volumes are preferred for SLS measurements. It is thus generally better to have photomultiplier (PM) optics, which have different selectable pinhole diameters, in order to optimize the signal detection for the respective task. This is also quite helpful for QLS experiments with diluted samples and low scattering intensity, where it may be better to decrease the signal-to-base-line ratio a in favor of a higher number of counts per sample time interval.

Stray light represents a serious problem for both SLS and QLS, and the scattering cells are therefore normally immersed in a vat which contains an index-matching fluid. The use of a good index matching fluid such as toluene helps to reduce the stray light, which otherwise occurs at the entrance and exit window of the vat and is due to surface imperfections of the scattering cells.

Since SLS experiments are done with a larger scattering volume and by detecting light from several coherence areas, one needs to work at much lower incident beam intensity. Rather than decreasing the laser power, which normally results in a shift of the laser beam point, one could adjust the beam intensity very reproducibly over a wide range with the combination of a so-called $\lambda/2$-plate

(LHP), which is able to rotate the polarization of the laser, and a high quality polarizer (POL), which selects vertically polarized light only.[81,82]

For SLS studies of larger particles, several instruments are available for low angle light scattering, where so-called Fourier or inverse Fourier optics are combined with a diode array detector and on-line data analysis using Mie theory, thus providing a relatively fast and convenient estimate of the particle size distribution.

Measurement Procedures

Static Light Scattering

The key features of a measurement protocol for static measurements are

1. A large number of short individual measurements of the scattered intensity, $I_s(\theta)$, at different angles which are then averaged for each angle (yielding $<I_s(\theta)>$). The averaging can be done with a dust discrimination procedure as described in Reference 81, for example;
2. The contributions $<I_b(\theta)>$ from background stray light and solvent are measured using the same cell containing the filtered solvent only and subtracted from $<I_s(\theta)>$, thus yielding $\Delta<I_s(\theta)>$;
3. To obtain absolute values of the scattered intensity, the average values $\Delta<I_s(\theta)>$ are finally normalized with respect to the intensity $<I_{ref}(\theta)>$ from pure, isotropically scattering reference solvents such as toluene, benzene, or water. The excess Rayleigh ratio of the sample $\Delta\mathbf{R}(\theta)$ is then calculated from[85]

$$\Delta\mathbf{R}(\theta) = \frac{\Delta < I_s(\theta) >}{< I_{ref}(\theta) >} \cdot \mathbf{R}_{ref}(\theta) \cdot \left(\frac{n}{n_{ref}} \right)^2 \tag{68}$$

where

$\mathbf{R}_{ref}(\theta)$ is the Rayleigh ratio of the reference solvent
n and n_{ref} are the index of refraction of the solution and the reference solvent, respectively

Tabulated values of $\mathbf{R}_{ref}(\theta)$ for a vertically polarized incident beam at λ_o = 488 nm and T = 25°C are $\mathbf{R}_{ref}(\theta) = 39.6 \cdot 10^{-4}\ m^{-1}$ for toluene, $\mathbf{R}_{ref}(\theta) = 35.4 \cdot 10^{-4}\ m^{-1}$ for benzene, and $\mathbf{R}_{ref}(\theta) = 2.49 \cdot 10^{-4}\ m^{-1}$ for water, respectively.[82,86] An apparent weight-average molecular weight $<M>$ and a z-average mean-square radius of gyration $<R_g^2>$ of the particles in solution can now be obtained from a least-squares fit of Equation 38 to the data for $\Delta\mathbf{R}(\theta)$. A value for dn/dc is required in order to calculate $<M>$, which can easily be measured with a good refractometer.

An improved measurement protocol can be used with the fiber optics instruments which permit a fast sequential readout of the scattering intensity $I_s(\theta)$ and the intensity I_o of the transmitted beam.[47,81] Step 1 in the above outlined procedure can then be replaced by a rapid and many times repeated sequential sampling of the transmitted beam intensity I_o and the scattered light intensity at each angle $I_s(\theta)$, normalization of $I_s(\theta)$ with I_o, and averaging of $I_s(\theta)/I_o$ using a dust-discrimination procedure, thus yielding $\langle I_s(\theta)/I_o \rangle$ instead of $\langle I_s(\theta) \rangle$. This protocol reduces inaccuracies due to drifts in laser power and photomultiplier response and automatically corrects $I_s(\theta)$ for sample turbidity when using cylindrical scattering cells. The background correction and the normalization with the reference solvent is then made with $\langle I_b(\theta)/I_o \rangle$ and $\langle I_{ref}(\theta)/I_o \rangle$.

Dynamic Light Scattering

Two different detection schemes can be used in a dynamic light scattering experiment:[9]

In a heterodyne experiment, the scattered field is mixed coherently with a fraction of the incident field. In the limit of $I_{lo} >> I_s(\theta)$, where I_{lo} is the intensity of the reference beam, the normalized heterodyne autocorrelation function $g_{2,het}(Q,\tau)$ can then be written as

$$g_{2,het}(Q,\tau) = 1 + 2a \cdot a_{het} \cdot \frac{I_s(Q) I_{lo}}{\left(I_s(Q) + I_{lo}\right)^2} \cdot g_1(Q,\tau) \tag{69}$$

where a_{het} is the heterodyne efficiency. Ideally, the ratio $I_{lo}/I_s(Q)$ is chosen between 10 and 30, but practically this is quite difficult to control for different samples and angular measurements.

In a homodyne experiment, which is much more conveniently done, the intensity autocorrelation function (Equation 8) associated with the scattered light detected by the PM is measured. As we can see from a comparison of Equations 69 and 8 (or 7), in the homodyne experiment we expect to find an additional factor of two in the exponent describing the time dependence of the measured intensity autocorrelation function, i.e., an exponential decay of $-2\Gamma\tau$ in a homodyne versus $-\Gamma\tau$ in a heterodyne measurement. This explains why a very low level of stray light is not only important for SLS experiments, but also for QLS measurements. Stray light would act as a local oscillator and lead to an additional term in Equation 12 and thus to an apparent decrease in D_c and a significant increase of the polydispersity obtained from a second order cumulant fit.

Dynamic light scattering experiments are frequently performed at a scattering angle of 90° only. A single measurement and a cumulant analysis of the experimentally obtained intensity autocorrelation function can then be completed in a few minutes, which makes QLS a particularly interesting technique for routine size analysis. However, we believe that precise measurements of the particle size distribution require multiple measurements using different values of the so-called correlator sample time and the scattering angle θ. This is particularly important if no *a priori* knowledge of the size distribution is available, since the presence of large particles may otherwise not be correctly detected at a single value of θ due the fact that the particle form factor P(Q) can be very low for certain combinations of particle radius r and angle θ (see Figure 4).

Analysis of nonexponential correlation functions. The dominant application of dynamic light scattering is particle sizing. The measured intensity autocorrelation function obtained for a suspension of polydisperse noninteracting particles can be written as[1]

$$C(Q,\tau) = a \cdot B\left[\int_0^{\infty} G(\Gamma)e^{-\Gamma\tau}d\Gamma\right]^2 + B + \varepsilon(t) \tag{70}$$

where $\varepsilon(t)$ represents a random noise contribution. We have seen from Equations 44 through 48 that information on the z-average diffusion coefficient <D> and the polydispersity index σ can be obtained from a simple moments expansion of $C(Q,\tau)$. Such an approach has been and still is very popular among experimentalists, since it only requires an easy-to-perform polynomial fit to $\ln[C(Q,\tau)-B]^{1/2}$. However, several remarks should be made with regard to such a cumulant analysis. If the suspension contains large particles, for which P(Q) can deviate significantly from 1 over the Q-range accessible to light scattering, an extrapolation to Q = 0 needs to be made in order to determine the 'true' z-averages of <D> or $\langle R_h^{-1}\rangle^{-1}$ and the correct value of σ. Furthermore, these values cannot be compared directly with those obtained from other techniques such as electron microscopy or size exclusion chromatography without taking into account the intensity weighting and the inverse relation between D and r (Equations 30, 45, and 46). Frequently, the polydispersity index has been used in order to quantitatively characterize the width of the particle size distribution. However, σ is very insensitive to a small degree of polydispersity. It can only be used quantitatively for $0.03 \leq \sigma \leq 0.4$ and may otherwise at lower values of σ lead to an overestimation and at higher values of σ to an underestimation of the 'true' polydispersity.[37,87]

Several theoretical approaches have been used in the past in order to obtain more than just an average <D> and an estimate of the polydispersity σ from a cumulant analysis, and to extract the full intensity distribution $G(\Gamma)$ from the measured autocorrelation $C(Q,\tau)$.[88-91] The basic problem in the data analysis results from the ill-conditioned nature of the inversion of Equation 70, the fact that even a small amount of noise on the experimental correlation function, together with the limited accessible of the τ-range, may greatly distort $G(\Gamma)$. There thus exists an entire set of solutions to Equation 70 that lie within the experimental noise level ε. While some of them can be discarded immediately because they have no 'physical significance' (such as those with negative amplitudes), we still end up with the substantial problem of having to choose the 'correct' or most probable solution among all the feasible ones.

One possible way of reducing the effect of noise is to permit positive amplitudes only through a non-negatively constrained least squares algorithm and to employ a smoothing procedure which restricts curvature in the function $G(\Gamma)$. Such an approach has been implemented by Provencher in his widely used Laplace inversion program CONTIN.[89] CONTIN is a relatively large package of 66 subroutines, which runs in a noninteractive way (with a few exceptions, such as the initial setting of the integration limits and the selection of a particle form factor) and requires considerable computational effort. (Preferably it would be run on a fast workstation or mainframe computer, although some implementations on a transputer board for PC's are available.) It runs through two cycles, where the first performs an unweighted analysis and selects a trial set of parameters, and the second does a weighted data analysis and proposes an 'optimal' set of parameters based on sophisticated statistical tests. While CONTIN is a very stable and reliable fit program, it nevertheless has its shortcomings. The smoothing algorithm clearly affects its ability to resolve bi- and multimodal size distributions or to reproduce narrow size distributions correctly.

A particularly powerful method to deal with the ill-conditioned inverse Laplace transformation and to extract the maximum information content from $C(Q,\tau)$ has been proposed by Ostrowsky et al.[88] This so-called exponential sampling method is based on an expansion of $G(\Gamma)$ using the recently discovered eigenfunctions and eigenvalues of the Laplace transformation.[92] This expansion is truncated according to a resolution criterion which depends on the experimental noise level. (A band-limited Laplace transform is constructed so that any eigenvalues below the noise level ε are not considered in the eigenvalue expansion.) It is thus possible to reconstruct a $G(\Gamma)$ which contains the most information that can be recovered safely from the experimental data. The main problem of this approach is a correct estimate of the noise level of the data, the 'resolution limit' or the number of components in the eigenfunction expansion which can be recovered from the experimental data. If the resolution is increased too much, this may result in the appearance of a fine structure in the reconstructed distribution function $G(\Gamma)$ that has no physical significance.

The tendency of the exponential sampling algorithm to produce spurious peaks and oscillations in the reconstructed $G(\Gamma)$ can be avoided by including a non-negatively constrained least squares analysis procedure,[93] which greatly enhances the numerical stability and prevents arbitrary oscillations in $G(\Gamma)$.[87,94] Such a constraint has been incorporated in several modern commercial implementations of the exponential sampling algorithm, although it is frequently and unfortunately quite difficult to find out what these analysis software packages do with the data.

A vital parameter of Laplace inversion routines, such as the exponential sampling method described above, is the resolution criterion, which depends on the experimental noise level. It is thus important to optimize the detection optics in order to achieve a high amplitude to base line ratio (a in Equation 70), and to perform the measurements at sufficiently high count rates and collect the intensity autocorrelation function for a sufficiently long time to obtain high resolution. We generally analyze a number of different correlation functions from the same sample individually and average the resulting intensity distributions G(Γ), rather than analyzing a single (averaged) high resolution correlation function which has been collected for a long time in order to achieve a high signal to noise ratio. Such an analysis of multiple data sets from the same sample reduces or eliminates the tendency of the inversion programs to produce spurious peaks or to report multiple peaks when the sample actually has a broad unimodal distribution.[87,94,95]

However, it must always be kept in mind that a Laplace inversion of an experimentally determined correlation function requires extreme precautions. One should always try to verify that a change in the resolution or integration limit does not significantly alter the resulting G(Γ). A variation of the scattering angle θ is always suggested, since the suspension may contain large particles for which their form factor P(Q) vanishes at a given value of θ (see below for further details). Measurements at different θ values are not only important because larger particles may be masked due to a minimum in P(Q), but they permit a test of whether an observed multimodal distribution G(Γ) is due to a multimodal size distribution (Equation 41), the appearance of a self-diffusive mode (Equation 49), or caused by some other effects such as internal chain dynamics (important for biopolymers) or contributions from particle anisotropy (Equation 57). A diffusive process should, in general, result in a Q^2-dependence of the decay constant Γ, therefore we can easily distinguish between multimodal size distributions (or contributions from D_s and D_c) and internal modes, which for polymers exhibit a Q^3-dependence.[52] In addition, for multimodal intensity distributions we can test whether the Q-dependence of the relative scattering intensities G(Γ) is consistent with the influence of P(Q) for a given size class and (assumed) particle shape. We can furthermore easily distinguish between a 'truly' bimodal size distribution and contributions from self and collective diffusion by measuring the concentration dependence of the peaks in G(Γ). Based on Equations 29 and 52, we would expect a very different concentration dependence for the two peaks in G(Γ) corresponding to D_c and D_s, respectively, quite in contrast to the concentration dependence of the peak locations expected for a bimodal size distribution.

The intensity distribution G(Γ) obtained with an inverse Laplace transform algorithm for multimodal size distributions can be used to estimate the mass (C(r)) or number (N(r)) distribution using Equation 40, which for homogeneous spherical particles leads directly to Equation 43. It is quite evident from Equation 43 that one cannot directly use a plot of G(Γ) versus R_h as an estimate of the particle size distribution, but that one must take into account the weighting of N(r) by the sixth power of the radius. A trace amount of large particles may thus completely dominate the intensity distribution G(Γ). This is illustrated in Figure 6, where number and intensity weighted distributions for a monomodal and bimodal size distribution are plotted. Figure 6A shows that the intensity weighted distribution G(Γ) is significantly shifted to larger values of r with respect to the number distribution N(r) even for a relatively monodisperse suspension of particles. A much more complicated situation can be observed for a bimodal size distribution (Figure 6B through D), in particular when one peak (or even both peaks) is in a

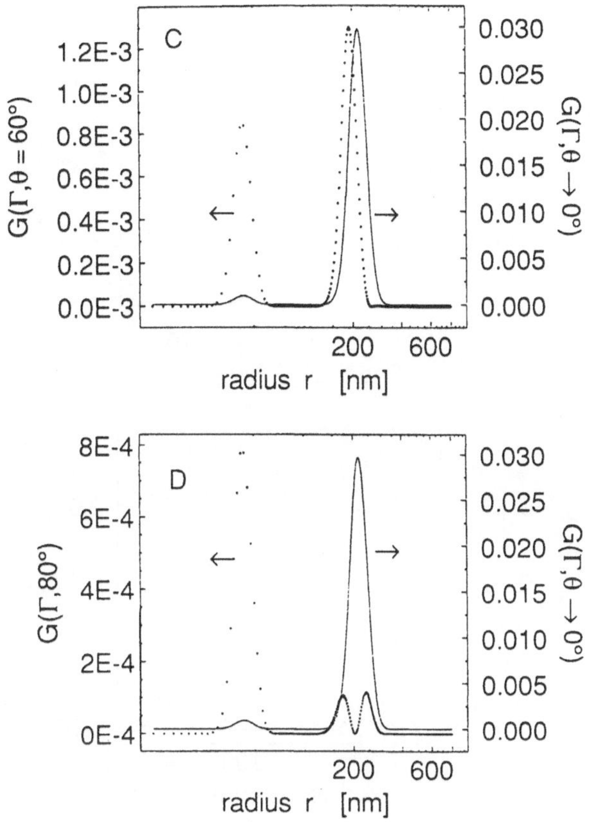

Figure 6. Number and intensity weighted monomodal and bimodal size distribution. The number distributions N(R) were calculated using a Schultz distribution. Intensity weighted distributions G(Γ) were calculated from Equation 43 using the number distributions shown in (A) and (B) and for scattering angles θ of 60°, 80° and $\theta \to 0$, respectively, using the form factor P(Q) for homogeneous spheres (Equation 23) and λ_o = 488 nm: (A) $< r >$ = 40 nm and z = 60; (. . .): number distribution N(r); (——): intensity weighted distribution $G(\Gamma, \theta \to 0°)$; (B) $< r_1 >$ = 40 nm, $< r_2 >$ = 200 nm, z = 60; (. . .): number distribution N(r); (- - -): mass distribution C(r); (——): intensity weighted distribution $G(\Gamma, \theta \to 0°)$; (C) Same number distribution as in (B); (——): intensity weighted distribution $G(\Gamma, \theta \to 0°)$; (. . .): intensity weighted distribution G(Γ,60°); and (D) same number distribution as in (B); (——): intensity weighted distribution $G(\Gamma, \theta \to 0°)$;(. . .): intensity weighted distribution G(Γ,80°).

size range where P(Q) exhibits distinct minima and maxima over the Q-range accessible to light scattering. Figure 6B shows N(r), C(r), and $G(\Gamma, \theta \to 0°)$ for a bimodal size distribution with $< r_1 >$ = 40 nm, $< r_2 >$ = 200 nm, $z_1 = z_2 = 60$, and a number ratio of $N(< r_1 >)/N(< r_2 >) = 500$ using a Schultz distribution (Equation 35). We clearly see that the intensity weighted distribution $G(\Gamma, \theta \to 0°)$ is completely dominated by the trace amount of larger particles, and that the small peak around 40 nm is well below the resolution limit of a standard QLS-

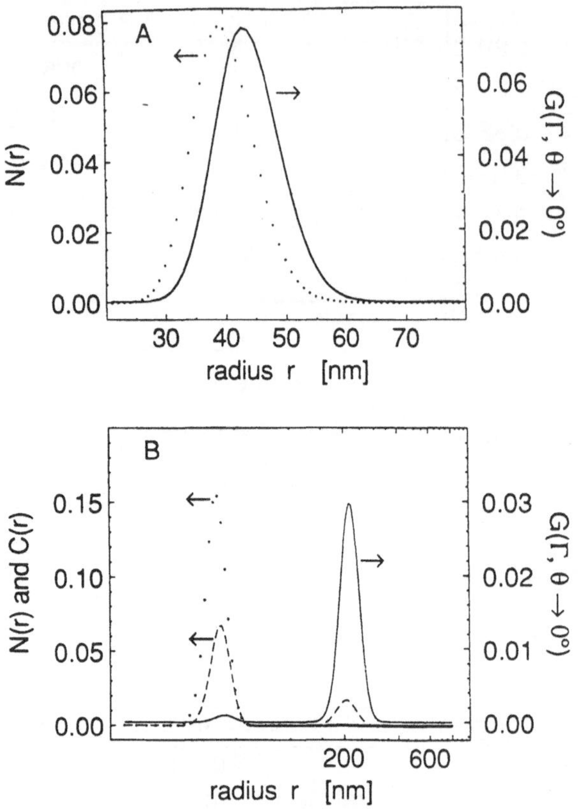

Figure 6(C, D).

experiment. It is also quite obvious that a very different picture would emerge from measurements with two different techniques. For example, a size analysis of such a suspension with electron microscopy would yield a unimodal and narrow size distribution with $<r> = 40$ nm, and the trace amount of large particles present would either not be seen at all or interpreted as artifacts by the operator.

Figure 6C and D demonstrate the importance of multiangle measurements. At a scattering angle of $\theta = 60°$, the scattering contribution from the larger particles has decreased due to the effect of P(Q), which has only a value of approximately 0.05 for r = 200 nm. At this value of θ, QLS would thus reveal a bimodal size distribution, whereby the second peak is shifted considerably to lower values of r with respect to $G(\Gamma, \theta \rightarrow 0°)$. At a scattering angle of $\theta = 80°$, the first minimum in P(Q) (see Figure 4A) has now been reached for particles with r = 203 nm; they will not contribute at all to $G(\Gamma)$. This results in a drastic suppression of the scattering contribution from the larger particles, and the appearance of three peaks in $G(\Gamma)$. This figure demonstrates how apparently bi- or multimodal size distributions can be observed with QLS for suspensions of large particles with unimodal but broad size distributions, due to the oscillations in P(Q). When analyzing particles with radii $r > 100$ nm, in a size range where, for example, the majority of aquatic environmental particles can be found (see Section 5.1), it is thus very important to perform measurements at different angles.

However, the relative scattering amplitudes must be interpreted with caution. While the peak positions Γ_i can be determined with reasonable accuracy for a multimodal size distribution, the corresponding relative scattering intensities $G(\Gamma_i)$ are associated with a much larger error. This is particularly severe for intensity ratios of more than a factor of two, where the errors in $G(\Gamma)$ can be as high as 50%.[87] Attempts to convert intensity distributions $G(\Gamma)$ to number distributions N(r) using Equation 43 may thus result in very large errors.

When using inverse Laplace transform routines, one should also be aware of the limited resolution of these programs. For bimodal size distributions, in general these algorithms cannot, even under optimal measurement conditions, such as high signal-to-baseline and signal-to-noise ratios, resolve peaks which are separated by less than a factor of 2 for the ratio of the two mean sizes. In addition to the lower resolution limit, there also exists an upper limit for the separation distance between two peaks. However, this upper resolution limit can be extended by the use of a so-called multiple tau option, which enables a measurement of the intensity autocorrelation function $C(Q,\tau)$ over a very wide range of time τ. The ability of an inverse Laplace transform program will be drastically limited by the presence of even trace amounts of dust or very large particles outside the resolution window of the method.[87]

Sample Preparation for Biological Systems

Light scattering is particularly sensitive to the presence of dust or a few large oligomeric particles (due to the M^2-dependence of the scattering intensity; see Equation 32). For an accurate estimate of the particle size distribution or the weight average molar mass it is thus crucial to work with dust-free samples. This can be achieved, for example, by either subjecting the samples to centrifugation directly in the scattering cells or by using a closed loop filtration system, where the sample is filtered continuously into the scattering cell until no dust particles can be detected at low scattering angles. However, one should be aware of the fact that filtration may not be possible with a number of biologically relevant macromolecules due to the interaction of the macromolecules with the filter membrane, and that the concentration after filtration should always be measured again. Solutions of a variety of high molecular mass systems, such as liposomes, viruses or antigen-antibody complexes, scatter light very strongly due to the high scattering cross section (large value of dn/dc) and the relatively large size of the particles. In order to avoid multiple scattering effects, the size analysis should thus generally be done at low concentrations. The presence of multiple scattering can easily be detected as an extended diffuse bright region around the laser beam profile in the sample, accompanied by a decrease in the signal-to-base-line ratio a in QLS experiments. However, it is important to realize that, with a number of self assembling systems such as the bile salt-lecithin solutions reviewed in Section 4.2, a dilution may not be possible without a subsequent change of the size distribution.

Practical Concerns and Sample Preparation for Environmental Samples

In aquatic environmental samples, the typically low colloidal concentrations ($\leq$ 1 to $\approx$ 500 mg/l total suspended solids)[96-124] and the extremely broad distribution of sizes present (1 nm to > 100 μm)[96-124] push the limits of QLS and most other techniques. Therefore, in considering the application of QLS for environmental samples, the colloidal concentration of the sample, the detection limit of the instrument, and a fractionation method for reducing the polydispersity of the sample must be considered.

The Detection Limit. The detection limit of QLS, the lowest particle concentration at which measurements can be made, is a function of particle size (see Equation 45), scattering angle (see Figure 6), detection optics of the instrument (in particular collection efficiency), and measurement duration.[9,125] The detection limit is controlled primarily by the signal-to-noise ratio. Methods for optimizing the signal-to-noise ratio have been discussed in detail in Reference 9, and a thorough discussion of the influence of statistical noise on the analysis of photon correlation data can be found in Reference 126. A practical limit is clearly reached when the light scattered by the solvent has approximately the same intensity as that scattered by the solute particles.

We tried to estimate the detection limit for a number of polystyrene beads differing in size using a Malvern Zeta Sizer III instrument equipped with a 5 mW HeNe (632.8 nm wavelength) laser. This (somewhat arbitrary) detection limit was determined at an angle of 90° and a maximum measurement duration of 12.5 min.[127] The results from this study are shown in Figure 7. A minimum of 0.25 mg/l was found for 280 nm polystyrene beads. For smaller particles, the detection limit increased dramatically, which is primarily due to the fact that the scattering intensity is proportional to the sixth power of the particle radius for particles much smaller than λ. This weak scattering of small particles may make their detection at typically low environmental concentrations difficult. The apparent increase in the detection limit for larger particles is due to the angular dependence of the scattering intensity, i.e., due to the influence of the form factor $P(\theta)$. The detection limit would thus be different at different angles.

The addition of a 1 W Ar laser (488 nm wavelength) to the Zeta Sizer III system lowered the detection limit to 0.04 mg/l for 280 nm polystyrene beads and 0.02 mg/l for 98 nm polystyrene beads (Figure 7). Again, these determinations were made at a scattering angle of 90°, and a maximum measurement duration of 12.5 min. The detection limit will, of course, vary from system to system. In particular, the use of an instrument which is primarily designed for dynamic light scattering experiments (and not for electrophoretic measurements) will lower the detection

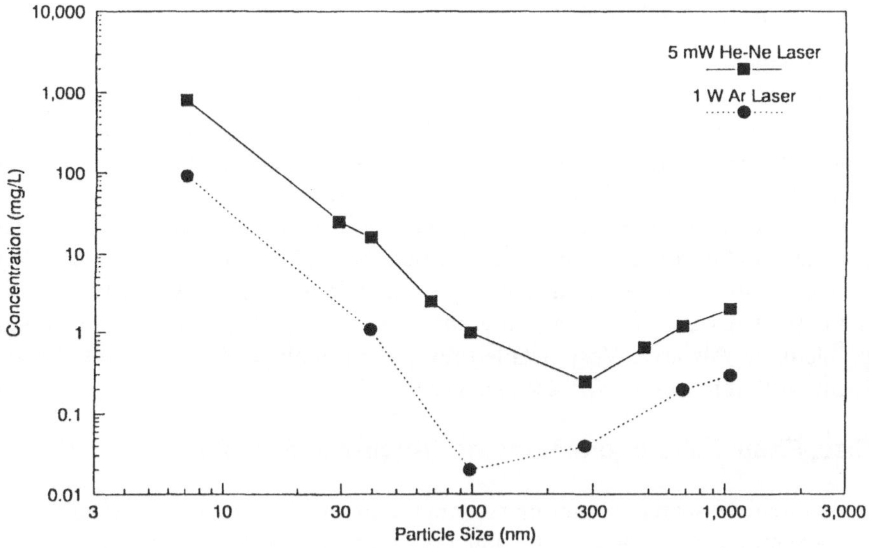

Figure 7. Detection limit versus particle size as determined with a Zetasizer III and latex particles.

limit by several orders of magnitude compared to the values shown in Figure 7. The detection limits reported by most manufacturers of QLS instruments appear to be based on idealized minimum values, rather than on those achievable for larger or smaller particles, and thus will probably not be achieved in natural systems.

Polydispersity. The effect of polydispersity on QLS analysis has been discussed in Section 2.3. From Equations 47 and 48 it can be seen that the estimated average size is heavily weighted towards large particles. The presence of a few large particles can dominate the scattered light signal, obscuring the presence of small particles (see also Figure 6B).[87,125,127-129] It is therefore necessary to remove all large particles (larger than a few microns) and probably to further fractionate the sample in order to minimize the polydispersity.

Various researchers have used sedimentation, centrifugation, filtration, or a combination of these techniques.[127,130-140] A preliminary sedimentation step of sufficient duration to remove particles larger than a few microns (depending on the particle densities) may be helpful in reducing the inadvertent removal of small particles with the large particles during subsequent fractionation steps.[127,130,139] This is particularly important for samples with low initial suspended solids concentrations. However, we have found sedimentation alone not to be efficient for samples containing significant quantities of planktonic organisms or microbial debris.[127] Subsequent fractionation by either centrifugation or filtration is generally necessary to reduce polydispersity adequately . For a review of the advantages and disadvantages of centrifugation and filtration, see References 130, 131, 139, 141, and 142.

BIOLOGICAL APPLICATIONS

Since the first application of QLS to biologically relevant systems was reported in 1967,[143] a rapidly increasing number of publications and review articles have reported numerous studies on topics such as molar mass, size and shape, conformational changes and denaturation, interactions, aggregation and polymerization of biological macromolecules such as proteins, viruses, nucleic acids, polysaccharides, vesicles, or micelles. In the present article we do not intend to provide a complete and historical summary of the field but rather to restrict ourselves to a critical review of some recent studies. We shall try to summarize briefly some of the current research trends and present a selection of still open questions and problems in this area. Very detailed reviews on biological applications can be found in References 1, 50, 144, and 145.

Size, Shape, and Interactions of Proteins in Solution

In protein research, scattering techniques are used primarily to determine the average molecular weight of individual proteins using SLS or their tertiary structure (conformation) in a crystal lattice by means of X-ray and neutron scattering. However, light scattering permits, in principle, a fast and noninvasive study of the size and shape of biopolymers in solution, especially when combined with an application of concepts from colloid and polymer physics. Furthermore, using modern liquid-state theories, interesting information about intermolecular interactions can be obtained, as outlined in Section 2. We shall look at the possibilities and limits of such an approach, using as an example a number of recent SLS and QLS studies on lens proteins,[146-151] and we shall furthermore try to point out where one might encounter difficulties with these biological systems.

The cytoplasma in eye lenses contains a highly concentrated solution of lens-specific proteins, the crystallins.[152,153] Their concentration increases from approximately 250 to 400 mg/ml across the lens, thus establishing the refractive index gradient required for the proper focusing of incident light. Despite the very high protein concentration, the lens cytoplasm is normally transparent because of short range ordering of the crystallins. Mammalian lenses contain primarily α-, β-, and γ-crystallins. Their relative proportions vary with age and location within the lens.

α-Crystallin, the major mammalian lens protein, is a large multisubunit protein. The protein can be obtained easily in large quantities, and an enormous range of chemical, hydrodynamic, and immunological studies on α-crystallin can be found in the literature (see Reference 47, and references therein). Its function appears to be only structural, as a part of the 'space filling material' of the eye lens. Clearly not only the size and structure of α-crystallin but also protein-protein and protein-solvent interactions and the resulting macroscopic properties of the eye lens medium (such as solution structure, S(Q)) are relevant for an understanding of questions which are related to the transparency of the eye lens and the loss of transparency during cataract formation. This was the rationale behind a number of investigations where the static and dynamic properties of α-crystallin solutions were studied by means of SLS and QLS as a function of protein concentration and ionic strength.[146-151]

Andries et al. published a detailed light scattering and fluorescence study of cortical bovine α-crystallin ('cortical' meaning prepared from the lens cortex, from the outer periphery of the lens).[146] Prepared under these conditions, the protein is known to be quite monodisperse and spherical. In a first paper these authors looked at the concentration dependence of the osmotic compressibility $(\partial\Pi/\partial C)^{-1}$ (Equation 16) and the collective and tracer diffusion coefficients D_c and D_s at moderate concentrations, volume fractions $0.005 \le \Phi \le 0.025$ (which correspond to 6.7 mg/ml $\le C \le$ 34 mg/ml). Most measurements were done at relatively low ionic strength ($I = 0.08$), and a linear concentration dependence was found for all three quantities as described by Equations 28, 29, and 52. The coefficients k_I, k_D, and k_S indicated the presence of strong contributions from electrostatic repulsions, which was also seen in the strong ionic strength dependence of k_S. Attempts were made to interpret the results in terms of simple two-body interaction potentials between spherical particles. Unfortunately, much of the discussion of the data is invalid due to the erroneous assumption that the "friction coefficient" f_s ($= k_BT/D_s$) measured in the self diffusion experiment is the same as f_c in collective diffusion (Equation 26). It has already been pointed out by Hess[10] that $f_s \ne f_c$. Therefore, the relation $D_c/D_s = (k_BT)^{-1}\cdot(\partial\Pi/\partial C)$ as used by Andries et al. is not valid. (This can also be seen from the coefficients k_I, k_D, and k_S derived for the simple hard sphere case; see Sections 2.2 and 2.3.) In a second paper, where Andries and Clauwaert extended their measurements to high concentrations, $\Phi \le 0.158$ (corresponding to $C \le$ 214 mg/ml), a second and slowly decaying component in the intensity correlation function was observed.[147] This was interpreted with the reversible formation of large clusters, giving rise to a strong increase of $I_s(\theta)$ and an increasing contribution from a slowly decaying mode in $C(Q,\tau)$ at low angles. However, especially in view of the results of Licino, Delaye, and coworkers presented below, this may not be the only explanation possible.

These authors published a series of four papers in which they analyzed SLS and QLS data from quite monodisperse cortical α-crystallin solutions up to very high protein concentrations of $C = 280$ mg/ml.[148-151] In a first article,[148] Licino et al. analyzed the intensity autocorrelation functions obtained for these solutions with the maximum entropy method.[91] Whereas at low concentrations ($C \approx 10$ mg/ml) the distribution of exponential decay times $G(\Gamma)$ thus obtained could be described by a single peak, $C(Q,\tau)$ became clearly nonexponential, and more peaks were resolved by the maximum entropy method for high concentrations. For $C \ge 80$ mg/ml, two components Γ_1 and Γ_2 could be separated, and for $100 \le C \le 280$ mg/ml, a third component was clearly resolved. An analysis of the angular dependence revealed a Q^2-dependence of Γ_1, Γ_2, and Γ_3, which is typical for diffusive processes. Values of the corresponding diffusion coefficients $D_1 \ge D_2 > D_3$ were then calculated, and their concentration dependence analyzed. D_1, which increased with increasing concentration, was interpreted as the collective diffusion coefficient D_c of α-crystallin. D_2 decreased with increasing C and extrapolated to D_1 in the limit of low concentrations. D_2 was interpreted as a self diffusion coefficient, as was suggested for example by Pusey for strongly interacting spheres with small polydispersity (Equation 49). The authors could in fact unambiguously prove this interpretation with a direct measurement of D_s in these solutions, using the technique of fluorescence recovery after fringe pattern photobleaching (FRAP). The third component, D_3, was related to the diffusion of a small number of reversible clusters or residual permanent aggregates by Licino et al. (see below for additional comments on this issue). Their data thus clearly shows how multicomponent distribution functions $G(\Gamma)$ can be obtained even for nearly monodisperse proteins at high concentrations, and that one needs to be very careful when interpreting the presence of a slow decay in $C(Q,\tau)$ measured with these solutions as due to the presence of high molecular weight components. Based on an effective hard-sphere model, they were subsequently able to develop a semiempirical model for calculating the relative amplitudes of the two modes $G(\Gamma_1)$ and $G(\Gamma_2)$.[151]

In their following articles, the static and dynamic data at low and high protein concentrations were analyzed using some of the theoretical models described in Section 2. The static data were in quantitative agreement with the rescaled mean spherical approximation (RMSA) model using a screened Coulomb potential for both low ($I = 0.017$) and high ($I = 0.15$) ionic strength up to concentrations of $C = 50$ mg/ml.[149] Furthermore, the parameters such as the hard sphere diameter, the charge, and the effective specific volume obtained by this analysis were in very good agreement with previous X-ray scattering experiments.[154] The concentration dependence of the collective diffusion coefficient was analyzed with the RMSA model for the direct interactions and with a description of two-body hydrodynamic interactions derived originally by Felderhof.[11] Very good agreement between theory and experimental data was obtained up to $C = 20$ mg/ml (i.e., $\Phi = 0.05$), whereas the theory clearly underestimated the effective D_c for higher concentrations and also was not capable to explain the concentration dependence of D_s, which can be measured for $C \geq 80$ mg/ml.[149,150] However, the collective diffusion coefficient quite unexpectedly exhibited a linear Φ-dependence (Equation 29) up to the highest concentrations studied, indicating that the theory for many-body hydrodynamic interactions needs further improvements.[150]

A different and very interesting study on interactions in protein solutions has been presented for bovine serum albumin (BSA) by Neal et al.[155] In this case, experiments were performed at low volume fractions ($\Phi \approx 0.005$) but very low ionic strength and thus very strong electrostatic repulsive interactions. The authors first analyzed their static light scattering data for various values of the protein charge and ionic strength, using the hypernetted chain approximation (HNC) [7] in order to calculate g(r) and subsequently S(Q) from Equation 15. They were able to show, from a comparison between Monte Carlo simulations and the HNC calculations, that the HNC results for g(r) are very accurate for 'soft potentials' such as a screened Coulomb potential under these experimental conditions. They obtained a very good agreement between measured and calculated S(Q) for all values of ionic strength and protein charge, without any adjustable parameters. However, when they tried to analyze the ionic strength and charge dependence of D_c measured with QLS, and using their previously determined accurate values of S(Q) to account for the direct interactions, they observed a clear failure of the currently existing dynamic theories, such as those by Ackerson or Felderhof[11,24,156] for the interaction dependence of D_c for sample conditions where $S(Q) \leq 0.5$. Their data was subsequently reanalyzed by Belloni and Drifford, who concluded that both small ion dynamics (finite relaxation time of the ion cloud around the protein) and hydrodynamic interactions can have a considerable effect on D_c for BSA solutions, and that a quite good agreement between the data of Neal et al. and their theoretical model can be obtained when such a 'dynamic attenuation' through noninstantaneous ion diffusion is taken into account.[157]

The systematic studies of Licino, Delaye, and coworkers and of Neal et al. show how a wealth of additional information on protein interactions and solution structure can be obtained from a combination of light scattering experiments and an analysis using theories from modern colloid physics. In particular, their studies show that, rather than avoiding interparticle interaction effects by always working at very low concentrations and high ionic strength, light scattering experiments can often be used to obtain high order information on the properties of and interactions between biological macromolecules from the concentration, pH, or ionic strength dependence of $I_s(Q)$ and $D_c(Q)$, provided that the limits of the theories used in such an approach are always kept in mind.

However, working with biological samples often requires much more caution than the classical model systems, such as latex spheres, commonly used in colloid physics. This is already shown in reference 150, where the authors mention that the collective diffusion coefficient determined by the maximum entropy analysis depends on the age of the protein sample at low ionic strength, indicating some release of bound water and ions or a loss of protein charge due to oxidation. Furthermore, if the lens proteins used by these authors are prepared not only from the lens cortex (the lens periphery) but also from the lens nucleus, a very different picture of the protein emerges. Under these preparative conditions, α-crystallin appears to be very polydisperse. In a recent study, gel filtration was used in combination with QLS, SLS, and electron microscopy (EM) in order to obtain additional information on the size and structure of the protein obtained from nuclear lens extract.[47] When subjected to gel filtration, the individual fractions of the protein preparation showed a strong and monotonic decrease of $< R_h >$ and $< M >$ with increasing elution volume, indicating a broad size distribution. SLS and QLS measurements were then used in order to obtain $< R_h >$, $< R_g^2 >^{1/2}$, and $< M >$ for individual fractions at very low protein concentrations and high salt content, where intermolecular interactions can be neglected. The experimental results, in particular the ratio $< R_g^2 >^{1/2} / < R_h >$ and the dependence of $< M >$ on $< R_h >$, were clearly not compatible with the presence of small monomeric spherical proteins, as in the studies of Licino et al. or Andries et al.,[146-151] but were quantitatively consistent with a polymerization of monomeric units into linear chains, which may have a certain degree of flexibility. Using theoretical expressions for $< R_g^2 >$ and R_h originally derived for semiflexible polymers in solution,[43,44] and by taking into account the appropriate weighting for polydisperse solutions (Equations 39 and 40), it was possible to calculate $<R_h>$ and $< R_g^2>^{1/2}$ from the size distribution observed with EM and to compare the results from such a calculation with the experimentally determined values from QLS and SLS. A similar approach, which permits a self-consistent analysis of the data from static and dynamic light scattering and from electron microscopy experiments, was also used successfully in an investigation of the size and structure of antigen-antibody complexes.[46]

The study of α-crystallin from nuclear lens extract demonstrates that, by slightly varying the preparation procedure, a very different quaternary structure can be observed. This shows how important a thorough characterization of biological macromolecules is when attempts are made to study intermolecular interactions by SLS and QLS. The presence of a few large oligomeric particles among many small monomeric protein molecules would drastically influence the scattering intensity in a strongly interacting system (for example, at high protein concentrations or at low salt content and high protein charge). The small number of oligomeric particles would correspond to a 'dilute solution,' their positions would thus be uncorrelated, and their contributions to the scattering intensity in a first approximation would be unaffected by the strong correlations between the highly interacting monomeric species.[155] This could lead to a serious error in the determination of S(0) and to a measurably slow contribution to the intensity autocorrelation function, as has been observed by Licino et al. in their maximum entropy analysis of the correlation functions from highly concentrated α-crystallin solutions.[148]

The investigations presented in References 46 and 47 also demonstrate how detailed information on the size and shape of biological macromolecules can be obtained fron a combination of SLS and QLS, provided that the particle size is large enough to enable a measurement of the radius of gyration of the particle. The application of models from polymer physics to SLS and QLS data has been particularly beneficial for studies of large anisotropic macromolecules such as polysaccharides.[158]

Size and Shape of Self-Assembling Macromolecules: Micelles and Vesicles

In recent years, extensive investigations of the physical chemistry of model and native biliary lipid systems have been performed. Bile salts, lecithin, and cholesterol, the major lipid components of bile, show a variety of aggregation phenomena in aqueous solutions. While the biologically important long-chain lecithins form liquid crystalline bilayer structures in water, they can be solubilized in the presence of bile salt molecules by forming thermodynamically stable mixed micelles or vesicles. The size and structure of the particles formed in bile salt-lecithin mixtures have been studied extensively by static and dynamic light scattering.[159-166] A strong dependence of the micellar size on the total lipid concentration, C_{tot}, and the mixing ratio between bile salt and lecithin, C_{BS}/C_L, was reported in these studies. In their pioneering article, Mazer et al. proposed the so-called "mixed disc" model for the structure of the bile salt-lecithin mixed micelles present in these solutions.[160] In this model, the mixed micelles were assumed to be disc-like, consisting of a lecithin bilayer with bile salt molecules incorporated within the disc in a fixed stoichiometry, and in addition forming a "ribbon" around the perimeter, thus preventing a contact between the water and the lecithin hydrocarbon chains. By taking into account the equilibrium partitioning of the bile salt molecules between the mixed micelles and the monomers in the "intermicellar solution," these authors were able to derive a simple geometrical model which provides a quantitative relation between the disc radius r_d and the concentrations C_{BS} and C_L.

In their article, Mazer et al. presented two sets of light scattering data which they thought provided convincing experimental evidence for their model.[160] In a first step, the experimental foundation of the mixed disc model was based on a deduction of the micellar shape using combined intensity measurements and QLS experiments at 90° scattering angle, and for different values of C_{BS}/C_L, C_{tot} and temperature T. A plot of the experimentally determined ratios $I_s(\theta = 90°)/C_{tot}$ versus R_h was made, and the experimental data points were compared with theoretical curves for the dependence of the scattered intensity from monodisperse spheres, discs and rods, and polydisperse discs on R_h. In this comparison, intermicellar interactions were neglected, and $I_s(90°)/C_{tot}$ was assumed to be proportional to $M_w \cdot P(\theta = 90°)$, where $P(\theta = 90°)$ is the particle scattering form factor at 90° scattering angle. Since no absolute intensity measurements were made, for the convenience of comparison all curves were normalized so as to have identical values of $I_s(90°)/C_{tot}$ for $R_h = 5$ nm. As shown in Figure 8A, closest agreement between experimental data and calculated curve was found for polydisperse discs, although some small but characteristic deviations are visible. In a next step, Mazer et al. could then use their geometrical model in order to quantitatively reproduce the strong dependence of the hydrodynamic radius upon variations of C_{BS}/C_L and C_{tot}.

The article of Mazer et al.[160] subsequently gained "reference status," and the model became widely accepted and was used in numerous studies of mixed micellar biliary lipid systems to interpret the experimental results obtained with a variety of different techniques. Using the mixed disc model, it was possible to quantitatively reproduce the dramatic increase of R_h measured with QLS upon a dilution of mixed micellar solutions, and correctly predict the existence of a phase boundary at which a micelle to vesicle transition can be observed.[165] However, a recent small angle neutron scattering (SANS) study then presented data which seemed in clear disagreement with the existence of disc-like mixed micelles, but proposed a rod-like morphology.[167,168]

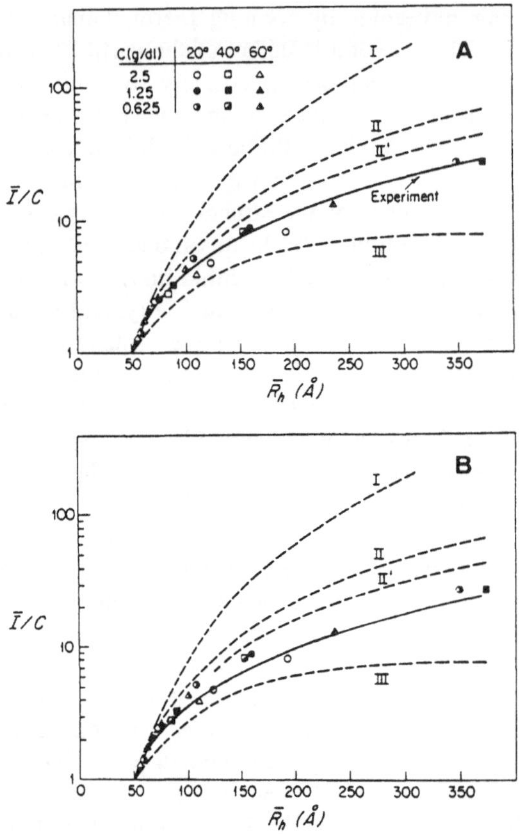

Figure 8. Experimental and theoretical dependence of the reduced intensity $I_s(90°)/C_{tot}$ on the mean hydrodynamic radius R_h of the bile salt-lecithin mixed micelles. (A) Data are derived from total lipid concentrations of 6.25 to 25 mg/ml at 20, 40, and 60°C (see insert). Theoretical curves represent different micellar shapes: I, monodisperse spheres; II, monodisperse discs (thickness 5 nm); II′, polydisperse discs; III, monodisperse rods (diameter 5 nm). The solid curve is drawn as a guide to the eye only. See Reference 160 for details. (From Mazer, N.A. et al. *Biochemistry* 19:601–615 (1980). With permission.) (B) Same as in (A), but the solid line is now a theoretical curve for monodisperse wormlike chains, diameter 5 nm, persistence length l_p = 15 nm.[52,53,91] The form factor P(90°) was calculated using Equation 21.

In response to this controversial situation, Chamberlin then reanalyzed the data presented by Mazer et al. and conducted additional static and dynamic light scattering experiments, in which he determined $\langle R_h \rangle$, $\langle M \rangle$, and $\langle R_g^2 \rangle$ using a sophisticated fiber optics instrument.[82] The results presented by Chamberlin could not be interpreted self-consistently with a disc model, but were in quantitative agreement with the model of a semiflexible (worm-like) chain originally derived for polymers.[43,44] A ratio of $\langle R_g^2 \rangle^{1/2} / \langle R_h \rangle \geq 1.5$ was obtained for all samples in the corresponding region of the phase diagram, which should be compared with the theoretical values of 0.77, 0.95, 1.6, and 2.3 for monodisperse spheres, discs, worm-like chains, and stiff rods, respectively, which can be calculated using the corresponding equations in Table 1. A reanalysis of the original data presented by Mazer et al.[160] does indeed show a quite good agreement between the experimental points and the theoretical curve of $M \cdot P(\theta = 90°)$ versus R_h for monodisperse wormlike chains with a persistence length $l_p = 15$ nm (Figure 8B). Based on his data, Chamberlin thus concluded that bile salt-lecithin mixed micelles do not grow into large disc-like structures upon dilution, as proposed by Mazer et al., but that they form flexible cylindrical aggregates.[82]

This recent controversy clearly demonstrates that light scattering, like X-ray or neutron scattering, is an indirect technique for determining size, shape, and structure of macromolecules, and only allows for a self-consistent data analysis and interpretation based on explicit structural models and assumptions for the intra- and intermolecular interaction effects. Thus, one can only decide whether a specific model agrees or disagrees with the data, but there may always exist other models which fit the data as well. It is therefore important to obtain as much information as possible from light scattering; in particular one should always try to determine absolute intensities and cover a wide range of scattering angles for both dynamic and static light scattering.

These light scattering studies of biliary lipid systems also demonstrate some of the most important problems connected with a size and shape determination for micellar solutions. Due to the dynamic nature of the micelles and the equilibrium between the surfactant monomer concentration and the micellar aggregates, one generally cannot use dilution in order to avoid interaction effects without changing the properties of the micelles. The data analysis in Figure 8 assumes that interaction effects are negligible, which may not be true for the highest concentrations studied (25 mg/ml). Furthermore, due to the self-assembling process responsible for micelle formation, these structures are inherently polydisperse, and polydispersity is always increasing with increasing micellar growth.[169] Therefore, one should try to include polydispersity in the theoretical model calculations used for the data interpretation, following for example the approach of Schmidt for semiflexible chains.[48] Since the data by Mazer et al. and Chamberlin were obtained near the phase boundary for a micelle to vesicle transition,[165] it could even be possible that two different particle morphologies could coexist in some of these solutions. It would thus be worthwhile to analyze the QLS data with modern inverse Laplace transform programs instead of a cumulant analysis only, and to obtain additional information on the size distribution from careful QLS measurements at different angles. Unfortunately, the instrument used by Chamberlin was optimized for SLS rather than QLS, and no attempts were made to obtain high signal-to-baseline and signal-to-noise ratios required for such an analysis.[82] While the recent articles by Hjelm et al.[167,168] and Chamberlin[82] indicate that a widely accepted model for the structure of biliary lipid mixed micelles may have to be revised, it is clear that in view of the physiological importance of this system, additional and complementary experiments should be done.

In contrast to micellar solutions, vesicles or liposomes are generally easier to characterize with light scattering. Their size and structure are usually independent of lipid concentration (as long as no detergents such as bile salts are present), and they have a negligible surface charge density. Due to their relatively large size and correspondingly high scattering cross section, ideal solution conditions can thus be obtained quite easily. QLS and SLS have been used regularly in the past to characterize the size distribution of vesicles in aqueous suspensions.[94,170-177] However, it is important to remember that the size information obtained with QLS is either in the form of intensity weighted average quantities such as $<R_h>$ or σ from a cumulant analysis (Equations 45 and 46) or in the form of intensity weighted size distributions $G(\Gamma)$ from an inverse Laplace transform program (Equation 43). Several recent articles have shown how mass or number distributions can be obtained for vesicle suspensions, and how these can be related to the size information obtained with other techniques such as gel filtration or electron microscopy.[94,177]

ENVIRONMENTAL APPLICATIONS

Although applications of static and dynamic light scattering for the analysis of atmospheric or soil particles do exist, we will restrict our discussion to the characterization of aquatic colloids.

Due to the large surface areas of submicron particles relative to their mass, they may sorb significant quantities of pollutants. The fate of these substances, sorbed to the colloids, is then linked directly to the fate of the colloids. Whether or not the colloids will remain suspended, coagulate with other colloids and/or sediment to the lake or river bed, or be removed by the aquifer matrix is a function of the particle size distribution, as well as the particle composition and morphology, the surrounding water chemistry, and the flow patterns of the water body. The size distribution, shape, and interactions of submicron particles in natural waters is, therefore, of interest to many researchers. However, as of the writing of this chapter, few researchers have applied QLS and/or SLS in the determination of particle size distributions in natural waters. This discrepancy may be due to

1. The high cost of QLS/SLS systems
2. The fact that the existence of QLS/SLS systems is not well known among aquatic chemists
3. The fact that the usefulness of QLS/SLS in natural systems has not yet been established
4. The intimidating mathematical theory involved in QLS/SLS analysis

Interest in aquatic colloids may be divided into two aspects. The first of these is the steady-state size distribution of colloids present in various water bodies, and the second is the formation, coagulation, and sedimentation processes which may control the steady-state size distribution of particles observed in various water bodies. QLS and SLS can be useful for investigating both of these aspects of aquatic colloids. Although SLS and QLS have in the past been used primarily for size estimation, a combination of SLS and QLS may also yield information concerning particle shape and particle interactions. The applicability of QLS and SLS for estimating the size distribution, shape, interactions, coagulation rates, and fractal dimensions of aquatic particles will be discussed in the following two sections.

Applicability of Static and Dynamic Light Scattering for the Analysis of Particle Size Distributions in Natural Waters

There has been a recent proliferation of articles concerning the determination of submicron particle size distributions in aquatic systems. For example, References 96 through 102 discuss particle size distributions in rivers, References 103 through 109 discuss particle size distributions in lakes, References 110 through 115 mention particle size distributions in oceans, seas, and estuaries, and References 116 through 124 discuss particle size distributions in groundwaters. The majority of these researchers have used either filtration and gravimetric determination or microscopy. Both of these techniques involve significant sample handling and processing. The determination of particle size distributions in natural water is indeed difficult with most currently available techniques, due to the low concentration of particles (≤ 1 to ≈ 500 mg/l), the wide distribution of sizes (1 nm to > 100 μm), and the large variation in morphologies present in natural water samples.[96-124] Although QLS, as with all other techniques, has drawbacks which should be considered, it has the advantage of being one of the few techniques which yields an estimate of the size distribution of suspended particles with a minimum of sample processing. QLS is therefore a promising technique for determining particle size distributions in natural waters. In this section, applications of QLS to the analysis of particle size distributions in natural waters will be reviewed and the usefulness of QLS for this purpose assessed. A summary of these applications is given in Table 3. Problems encountered during the investigations reviewed here which are common to most environmental applications of QLS will be pointed out.

One of the first applications of QLS to particle size determinations in natural systems was reported by Gallegos and Menzerl.[130] They carried out a series of experiments designed to assess the applicability of QLS for a variety of environmental problems, including determination of particle size distributions in soil suspensions, creek water, and water from a turbid impoundment, as well as coagulation rates and critical coagulation concentrations. These last two applications will be discussed in Section 5.2.

Gallegos and Menzerl[130] used an instrument equipped with a 5 mW HeNe laser. This instrument allowed analysis of measured autocorrelation curves using either the method of cumulants or a proprietary form of the exponential sampling technique of Ostrowski et al.[88] The authors did not adequately indicate which of these two techniques was employed for the reported data analysis. All measure-

Table 3. Summary of Applications of QLS for the Determination of Particle Size Distributions in Environmental Samples

Ref. No.	Sample Type	Instrumentation and Conditions of Analysis	Sample Pretreatment	Comments
130	Soil suspension, creek water, turbid impoundment water	5mW HeNe laser, fixed angle, 90° analysis, cumulants or exponential sampling	Sieving, sedimentation, centrifugation, varying storage time	Coagulation observed during storage Sample fractionation necessary
131	Sewage waste, contaminated groundwater	5 W laser used at 300 mW, fixed angle, 90° analysis, CONTIN	Filtration	Multiangle analysis necessary for multimodal distributions Error greatly magnified converting from intensity to number weighted distributions More quantitative comparisons of microscopy and QLS possible
133,134	Uncontaminated groundwater	5 mW HeNe laser, fixed angle, 90° analysis CONTIN, in field	None	Sample fractionation necessary Multiangle analysis necessary Field analysis possible comparison of different techniques desirable
136,137, 138	River water	5 mW HeNe laser, goniometer, 90° analysis,	Sedimentation, filtration	

		No method of data analysis given		Conversion to similar weighting (intensity, volume, number) necessary to compare techniques
140	Oligotrophic lake water	5 mW HeNe laser, goniometer, 90° analysis, no method of of data analysis given	None	Sample fractionation necessary More quantitative comparisons of QLS and microscopy possible Multiangle analysis necessary

Sources: Gallegos and Menzerl,[130] Gschwend and Reynolds,[131] Ryan, Jr.,[133] Ryan, Jr. and Gschwend,[134] Rees and Ranville,[136,137] Rees,[138] and Ranville et al.[140]

ments were taken at a fixed angle of 90°. All surface water samples were screened through 10 μm nylon mesh prior to analysis.

Soil suspensions were separated into size fractions by sedimentation followed by centrifugation. The range of particle size distributions of soil suspensions estimated by QLS agreed well with the range of sizes expected based on centrifugation speeds and times. For each fraction, more than 70% of the distribution estimated by QLS fell within the range expected. All distribution means fell well within the expected range. As pointed out by Gallegos and Menzerl,[130] the observed differences may have resulted from an inability to calculate the exact reduction in effective density of small clay particles due to sorbed water molecules, and therefore an inability to calculate the exact size range resulting from separation by centrifugation.

As discussed previously, the range over which the size distribution of a sample may be determined by QLS is strongly affected by the presence of larger particles (see Equations 47 and 48, and Figure 6). When large particles are present, their greater scattering intensity may mask that of smaller particles. This problem was observed by Gallegos and Menzerl[130] in soil suspensions prior to centrifugation. Smaller sizes, visible in the centrifuged fractions, were not observed in the initial suspension prior to centrifugation. As demonstrated by this example, reducing the polydispersity of the sample by a fractionation procedure, such as centrifugation, is often necessary to alleviate this problem. Analysis at varying scattering angles is also necessary to verify the resulting distribution (see Figure 6).

Particle size distributions measured in samples from Chuckwa Creek, Oklahoma, were quite reproducible when analyzed within 1 h after collection. The arithmetic distribution means determined for five successively collected samples ranged from 0.93 to 1.04 μm, with a coefficient of variation of 4.2%. Similar reproducibilities were obtained for the standard deviations and ranges of the five size distributions. However, particle size distributions appeared to change rapidly with storage time. The mean size estimated for a Chuckwa Creek sample was observed to increase 56% within 2.5 h and 117% within 1 d after collection. The probability of altering the initial size distribution during sample storage is a problem common to most aquatic samples and must always be considered.

Gschwend and Reynolds[131] used a QLS instrument fitted with a 5 W laser which was operated at 300 mW output to determine the particle size distribution of colloids in sewage waste and in groundwater contaminated by this waste from the Cape Cod aquifer in Massachusetts. The samples were filtered using a glass fiber filter (nominal pore size, 2.7 μm) prior to analysis. All analyses were made at a fixed angle of 90°. The autocorrelation curves obtained were analyzed using the Size Distribution Processor (a proprietary version of the CONTIN program developed by Provencher).[89,132]

Three distinct particle size classes were observed in the sewage, 6 ± 3 nm, 560 ± 320 nm, and 2800 ± 1100 nm. Although it was not possible with the instrument used in this study, verification of multimodal distributions by measurements at different scattering angles is always necessary for aquatic samples of unknown size distributions, as was emphasized in Sections 2.3 and 3.2.2. This is because aquatic samples may contain particles for which the form factor, P(Q), vanishes at certain angles (see Figures 4 and 7).

The 6 ± 3 nm sewage particles were estimated to be present in by far the greatest number, practically 100% of the particles. This result corresponded well with the percent of macromolecular organic material expected in sewage sludge. However, as pointed out in Section 3.2.2 and illustrated in Figure 6, number weighted distributions should be considered with extreme caution due to the magnification of measurement error which occurs when converting scattering intensity to particle number. Scanning electron microscopy (SEM) observations of the sewage waste showed primarily fine debris (<< 100 nm) along with several larger individual particles, thus qualitatively confirming the QLS results. Although environmental chemists and engineers generally have not taken advantage of the opportunity, it is possible, as was discussed previously in Biological Applications, to verify QLS results by microscopy in a much more quantitative manner. The z-average mean (see Equation 47) can be calculated for micrographs and compared to the value obtained by QLS (see Section 4 and References 46, 47, and 94).

Two distinct particle size classes were observed in each of two groundwaters collected from two different wells downgradient from the sewage input, 104 ± 26 nm and 648 ± 210 nm, and 102 ± 16 nm and 628 ± 190 nm, respectively. Again, multiangle analysis would have been necessary to verify the bimodal size distributions observed. The 104 nm colloids were thought to be a ferrous phosphate precipitate based on energy dispersive X-ray (EDX) analysis. Again the smaller, 104 nm, particles were estimated to account for practically 100% of the total particle number. However, as expected, although the larger particles were present in smaller numbers, they were estimated to account for almost half of the particle mass and approximately 40% of the total light scattering intensity. The concentration of the 100 nm particles in the two groundwaters was estimated to be 2 and 4.5 mg/l. The concentration of the larger colloids in the two groundwaters was estimated to be 0.9 and 2 mg/l.

These particle concentrations were obtained by comparing the scattering intensity of the samples to those of known concentrations of 90, 500, and 2000 nm polystyrene beads. The index of refraction of the polystyrene beads was 1.6, which corresponded reasonably well with that of ferrous phosphate (1.58 to 1.63) and clays (1.53 to 1.64).[178] However, concentrations estimated by comparison of scattering intensities from different types of particles should always be considered skeptically because particles with different refractive indexes have quite different scattering properties (see Section 2.1).

SEM analysis of the samples showed a virtually monodisperse population of approximately 100 nm spherical colloids mixed with a few microorganisms and mineral flakes of 500 to 1000 nm. Although the 100 nm colloids appeared to be aggregated on the filters, they were apparently observed as individual suspended particles by QLS. Again, a more quantitative use of SLS, QLS, and microscopy could have been used to check whether coagulation occurred in solution or during preparation of the samples for microscopy.

Ryan[133] and Ryan and Gschwend[134] used the same QLS instrument mentioned earlier, except with the 4 mW HeNe laser which is standard for the instrument and the addition of a high gain photomultiplier. Again, all analysis were taken at a fixed angle of 90°.

Ryan[133] attempted to determine the particle size distribution for two pristine groundwaters from sandy aquifers in the Delaware coastal plain. Analysis was performed in the field immediately after the samples were collected. Although the analysis duration and sample time were varied extensively, the heterogeneity of the sample, including the presence of large, slow moving particles, outside the capabilities of QLS, prevented the determination of accurate, reproducible results by QLS. SEM micrographs of filtered samples confirmed the heterogeneity of particle size and shape, as well as the presence of many particles larger than 3 μm. Again, removal of particles larger than a few microns, possibly further sample fractionation, and analysis at varying scattering angles are essential to obtain accurate results of environmental samples using QLS and probably would have made QLS analysis possible in this case.

The concentration of colloids estimated to be present in these groundwaters ranged from ≤ 1 to 60 mg/l. Particle concentration was estimated by comparing the scattering intensity of the sample to that of known concentrations of kaolinte (the mineral thought to be the primary colloidal component based on SEM/EDX analysis). Use of kaolinte in estimating the total colloidal concentration is an improvement over that of polystyrene beads, provided the scattering intensity of the sample is dominated by kaolinte particles.

Autocorrelation curves must be measured with minimum noise in order to achieve accurate size estimates because of the ill-conditioned nature of the mathematical transformations necessary to extract size data from measured autocorrelation curves (see Section 3.2.2, Equation 70). Therefore, in order to obtain accurate results by QLS, care must be taken to keep the sample and environment as free of "dust" as possible. Also it is necessary to maintain the stability of the laser current, and therefore of the power source, as well as the physical alignment of the instrument, including turbulence in the sample cell (see Section 3.1).[125,135] It is possible that QLS systems equipped with air cooled HeNe lasers may be kept adequately stable under field conditions. However, solutions of standard particles should be analyzed in the field to verify that an increased noise level is not observed (i.e., the system is stable). It is probable that the detection limit of a QLS system with a HeNe laser is not adequate for analysis of fractions of a sample with a total particle concentration as low as that estimated by Ryan (see Section 3.2.4).[133]

Rees and Ranville[136,137] and Rees[138] have compared the particle size distributions determined by QLS, photosedimentation analysis (a technique which utilizes the change in optical density of a suspension of colloidal material in a centrifugal field), and SEM for samples taken from the Illinois, Missouri, Arkansas, White, Ohio, and Mississippi Rivers. Samples were collected and processed as described by Leenheer et al.[139] In summary, all samples were allowed to sediment 6 to 12 h, which was estimated to be sufficiently long to remove particles larger than 3 μm with densities greater than 1.7 g/cm^3 from suspension. The colloidal fraction was then siphoned off and concentrated by tangential flow filtration.

QLS was performed using an instrument equipped with a motorized goniometer, and a 5 mW HeNe laser. However, all measurements were made at a scattering angle of 90°. Autocorrelation functions were analyzed by a commercial "Non-negatively Constrained Least Squares" analysis package. Although Rees and Ranville[136,137] and Rees[138] state that the software determined values for the mean and for the 15th and 85th percentile of the distribution, they completely lack a discussion of the method of data analysis employed; therefore, a critical evaluation of their results is not possible.

For 15 samples from the rivers mentioned earlier, QLS observed relatively monomodal distributions with means ranging from 260 to 480 nm. The 15th and 85th percentiles associated with these means ranged from 290 to 580 nm. The mean and standard deviations of these 15 mean values were 353 and 68 nm, respectively. The mean and standard deviations of the 15th and 85th percentile values were 390 and 73 nm, respectively.

A Student's t-test showed the means and 15th and 85th percentiles determined by QLS and photosedimentation not to vary significantly. Although comparison of results obtained by several analytical techniques is always a good idea, in this case the relatively large variation between the particle size distributions of the different rivers may have masked smaller differences between the two techniques. Also, QLS results in an intensity weighted distribution (a distribution heavily weighted toward larger sizes; see Equations 43 and 47) which cannot be directly compared to distributions weighted by volume (mass) or number without correcting for the different weighting methods.

SEM micrographs showed the river samples to be composed of a wide range of particle sizes, from below the resolution of the microscope to large aggregates. Most particles appeared to be aggregated to some degree. Again, a more quantitative use of QLS, SLS, and microscopy, as discussed in Section 4, could have been used to verify the QLS result and determine whether coagulation occurred in the sample suspension or during preparation for microscopy. Both the particle sizes and the concentrations estimated by Ranville and Rees[136,137] for these six rivers fall within the wide range of values obtained by different techniques in other rivers.[96-102]

Ranville et al.[140] used the instrument described previously to determine the particle size distribution in Pueblo Reservoir, an oligotrophic reservoir near Pueblo, Colorado. Analysis was performed on a raw water sample prior to settling. QLS yielded a mean of 2.6 μm and a variance of 1.2 μm, indicating a broad size distribution. The distribution of mass in the silt and colloid fractions, as determined by the separation and concentration technique of Leenheer et al.,[139] described previously, also indicated a mass distribution centered around 2 to 5 μm. The total suspended solids concentration for the raw water sample was 4 mg/l.

In this case, as mentioned previously, it is probable that smaller particles present in the sample were masked due to the presence of large particles. Particles larger than a few microns should always be removed prior to QLS analysis. Also, analysis at varying scattering angles is necessary to assure accurate results for samples of unknown size distribution. SEM analysis of the raw water sample indicated that the particulates were aggregated to a large extent. Aggregates up to 70 μm, composed of smaller particles, were observed. Again, a quantitative treatment of SEM, SLS, and QLS results could have been used to determine if coagulation occurred in the suspended sample or during sample preparation for microscopy (see Section 4). Other investigators, using either ultrafiltration or microscopy, have observed lake particles ranging in size from 1 nm to 100 μm.[103-109] Several investigators have reported peaks in particle concentration corresponding to approximately 100 nm and approximately 5 μm particles.[103-109]

In summary, a review of current research indicates that QLS may yield useful results for particle size distributions in natural aquatic environments if care is taken concerning sample storage, preparation, and analysis. More specifically, it appears that samples should be analyzed as quickly as possible after collection. Particles larger than 2 to 3 μm should be removed from the sample prior to analysis by sedimentation, filtration, and/or centrifugation. Further fractionation by either filtration or centrifugation may be required to reduce polydispersity for many samples (see Equations 43, 47, and 48). Although it does not appear to be the current practice among environmental chemists, samples should be analyzed at varying scattering angles to check for masking of smaller colloids by the greater scattering intensity of larger colloids (see Figure 6) and for the existence of particles whose form factor, P(Q), vanishes at certain angles (see Figures 4 and 7). Scanning the scattering intensity at varying angles may be quite useful in selecting the appropriate angles for size analysis. For environmental applications, QLS instruments with multiangle capabilities are thus essential.

As with all analytical techniques, QLS results should be carefully evaluated (do the results make sense physically and chemically?). It is generally best to compare the results obtained by QLS with results obtained by other techniques, such as microscopy. The combination of QLS and microscopy appears to be underused for environmental samples. By calculating the z-average mean from electron micrographs (see Equation 47) and comparing the calculated values to those obtained by QLS (see Section 4 and References 46, 47, and 94), the QLS results may be verified, and whether or not the sample coagulated during preparation for microscopy can be determined. Finally, QLS analysis results in size distributions heavily weighted toward larger particles (see Equation 43 and Figure 6) and cannot be directly compared to results obtained by other techniques which are weighted in a different manner without first correcting for the different weighting methods.

Applicability of Dynamic and Static Light Scattering in Determining Coagulation Rates and Fractal Dimensions

Two interrelated properties of aggregation are quite useful in characterizing the process. The first is the numerical value of the fractal dimension of the aggregates, and the second is the evolution of the mean aggregate radius with time. Knowledge of these properties may allow diffusion-limited coagulation to be distinguished from reaction-limited coagulation. As discussed in Section 2.5, both of these properties may be estimated by SLS and/or QLS.[73,77,179,180]

The Fractal Dimension

As discussed in Section 2.5, the fractal dimension is a number which quantitatively expresses the degree of openness within the structure of an aggregate. The fractal dimension is usually estimated by plotting the log of the scattering intensity, $I_s(Q)$ against the log of the scattering wavevector, Q. The theory of light scattering by fractal aggregates has been discussed previously in Section 2.5.

Numerous researchers have used SLS to determine fractal dimensions for aggregates of gold, polystyrene beads, and silica sols under a variety of conditions.[77,79,179-187] A detailed review of all such articles is not appropriate in the context of this chapter. However, two more environmentally applicable studies of fractal dimensions are summarized below.

Amal et al.[182] used an SLS/QLS system equipped with a 15 mW HeNe laser to estimate the fractal dimension of hematite aggregates. The mean diameter of the primary, spherical, hematite particles was < 100 nm. In order to assure that the aggregates were within the range $a \ll 1/Q \ll r$ (see Equation 66), they were allowed to aggregate to a mean size of approximately 1 μm before SLS was performed. At this stage, the aggregation was slow and significant sedimentation was not observed. This last point was evidenced by stable photon count rates throughout the SLS experiment. Amal et al.[182] obtained log-log plots of $I_s(Q)$ versus Q which were linear over more than one order of magnitude in Q. The fractal dimensions obtained were incorporated in a modified Smoluchowski model to predict aggregation kinetics. The resulting predictions agreed well with kinetic results obtained by QLS.

Hackley and Anderson[183] estimated fractal dimensions of geothite aggregates using a QLS/SLS system equipped with a 10 mW HeNe laser. The geothite primary particles were estimated to have dimensions of 60 nm × 18 nm × 11 nm. Again, in order to confirm that the aggregates were within the range $a \ll 1/Q \ll r$ (see Equation 66), SLS was performed on aggregates only after the scattering intensity of the geothite suspension was observed to be independent of time. Hackley and Anderson[183] also observed a linear region in Q of approximately one order of magnitude for log-log plots of $I_s(Q)$ versus Q. Although the fractal dimensions obtained were slightly lower than those previously obtained for spherical primary particles, they are reasonable when the face-to-face orientations possible between the plate-like geothite particles are considered.

Aggregation Rates

Dynamic light scattering is well suited for the determination of aggregation rates because it allows frequent analysis of the evolution of particle size over time within the reaction container (sample cell) without disturbing the aggregation process. Numerous such applications exist in the literature.[179-183,188-193] A detailed review of all such articles is beyond the scope of this chapter. For a detailed quantitative treatment of the advantages and limitations of QLS for investigating aggregation kinetics, see References 77, 179, 180, 194, and Section 2.5 of this chapter. A more general treatment of the experimental considerations will be presented here.

For the investigation of aggregation kinetics, the mean particle size is of interest, rather than a detailed size distribution. Therefore, a different set of considerations from that discussed for the analysis of detailed, steady-state, size distributions in aquatic samples is required to achieve accurate aggregation rate results.

During the initial stages of a kinetic study, both the scattering intensity and the mean particle size may change rapidly with time. It is subsequently necessary to adjust the measurement duration to minimize changes within a measurement.[182] Duplicate analysis at varying angles or sample times may not be possible during the initial stages of aggregation. Therefore, it is necessary to consider carefully the optimum conditions for analysis prior to beginning a kinetic experiment. Fortunately, prior knowledge of the initial size and shape of the primary particles is usually available in coagulation kinetic experiments. As aggregation continues, the scattering intensity and sample time must be altered in accordance with the increasing mean particle size.[181,182,188] At later stages in the aggregation process, rotational as well as translational diffusion will contribute to the decay of the autocorrelation function (see Section 2.4 and Reference 77). Also, at this stage the greater polydispersity in the sample may influence the estimated mean particle size.[77,179,180,194] The limit beyond which QLS can no longer follow aggregation is reached when a significant fraction of the aggregates are removed by sedimentation.[181]

Due to the contributions of both rotational and translational diffusion during the later stages of a kinetic experiment, the first moment of the autocorrelation function yields an effective hydrodynamic radius (R_{eff}). The 'true' z-average hydrodynamic radius (R_h), whose time dependence determines aggregation kinetics, is defined by $R_h = R_{eff}(Q = 0)$. However, due to the Q-dependence of R_{eff}, calibration curves of the sort $R_h = R_h(R_{eff}(Q))$ can be obtained.[77] Thus, by measuring the change in R_{eff} over time and using such calibration curves, the true R_h-based kinetics can be extracted.[77]

As discussed in Section 2.5, polydispersity of the sample may affect the estimated mean particle size. Cametti et al.[194] suggest that, for diffusion limited aggregation, the polydispersity is usually small and not significant. However, for reaction-limited aggregation, high polydispersities may shift the estimated mean size to lower values. Klein et al.[77] indicate that polydispersity may be important for both types of aggregation processes. This problem might be reduced by analyzing at lower angles, at which the scattering intensity would be dominated by the large aggregates.

As previously pointed out, it is necessary to consider the optimum conditions for analysis prior to beginning an aggregation experiment. Obviously the size and shape of the primary particles should be considered, and most previous studies appear to have based their conditions for analysis on these two factors.[182,183,188-193] However, the significance of rotational diffusion and polydispersity in the latter stages of the kinetic study should also be considered.[77]

Despite the considerations and difficulties discussed, many researchers report successful results from aggregation kinetic studies which agree well with current aggregation theory.[181,182,188-193] It appears that SLS and QLS are well suited to the investigation of aggregation processes.

CONCLUSIONS

A combination of static and dynamic light scattering allows information on the particle size distribution, shape (including the fractal dimension), and interaction effects to be obtained with a minimum of sample handling. This makes SLS and QLS useful techniques for biological and environmental applications.

However, in order to obtain useful information with SLS and QLS, analysis must be made carefully and correctly. The most commonly used routine analysis at a fixed scattering angle of 90° is not sufficient, and it is dangerous to accept all values produced by commercial computer programs as accurate, especially without an a priori knowledge of the sample. Multiangle measurements and careful use of different data analysis techniques are usually required to obtain accurate results.

Large particles must be eliminated from the sample and polydispersity reduced by a fractionation technique such as sedimentation, filtration, or centrifugation prior to sample analysis. This is especially true for environmental samples, which generally contain quite broad particle distributions. SLS and QLS have a limited ability to resolve broad size distributions.

In general, a "self-consistent" data analysis utilizing several techniques should be carried out. For example QLS and SLS measurements may be quantitatively verified by microscopy as demonstrated in the section on biological applications.

Useful information on particle shape (including the fractal dimension) and interaction effects can be obtained by a combination of SLS and QLS. This ability of QLS and SLS has not yet been utilized much and may open new areas of investigation in biological and environmental applications. It is our belief that an interdisciplinary approach, in which modern theoretical concepts are borrowed from scientific disciplines such as colloid and polymer physics or fractal geometry and included in the 'traditional' investigation of biological and environmental particles with scattering experiments, can be very helpful for the biophysicist and environmental scientist.

LIST OF SYMBOLS

a	"signal-to-baseline ratio" determined by the size of the scattering volume
a_{het}	heterodyne efficiency
A	scattering amplitude
$<A>$	number averaged scattering amplitude
A(Q,r)	scattering amplitude of particles with radius r
$A_j(Q)$, A_j	scattering amplitude of particle j
A_2	second viral coefficient
A′	additional contributions to the viral coefficient of attractive or repulsive interactions in systems other than hard spheres
B	base line in experimentally measured intensity autocorrelation function
B_i, B_o, B_2, B_4	amplitude coefficients of the correlation function for cylindrical particles
C	particle concentration (mass/volume)
$C(Q,\tau)$	experimentally measured intensity autocorrelation function
d_F	fractal dimension of a mass fractal
d_h, d_g	analogous to d_F but relating the particle mass to $<R_h>$ and $<R_g>$, respectively
dn/dc	refractive index increment
D_c	collective diffusion coefficient of particles in solution
D_o	diffusion coefficient at infinite dilution
$D_o(r)$	diffusion coefficient of particles with radius r
$<D_o>$	z-average diffusion coefficient
$<D_o(Q)>$	intensity weighted diffusion coefficient measured at scattering vector Q
D_R	rotational diffusion coefficient
D_s	self diffusion coefficient which describes the random motion of individual particles
$E_s(Q,t)$	amplitude of the scattering field
f	parameter which corrects A_2 for anisotropy and polydispersity
$f(M/M_c)$	cut-off function for the number distribution
$f(r,\xi)$	cut-off function which describes the behavior of h(r) for fractal objects at large r
f_c	friction coefficient
f_o	infinite dilution value of the friction coefficient f_c
$F(Q,\tau)$	dynamic structure factor

$g(r)$	radial distribution function
$g_1(Q,\tau)$	normalized temporal autocorrelation function of the scattered field amplitude
$g_2(Q,\tau)$	normalized temporal autocorrelation function of the scattered intensity
$g_{2,het}(Q,\tau)$	normalized heterodyne autocorrelation function
$G(\Gamma)$	weight function which gives the relative scattering contribution of a component with decay rate Γ
$h(r)$	total correlation function, i.e., the relative mean number density deviation at a distance r from the center of a given particle
I	ionic strength
I_s	scattered intensity
$I_s(Q,t)$	scattered intensity as a function of scattering vector Q and time t
$I_s(\theta)$	scattered intensity at angle θ
$\langle I_s(\theta)\rangle$, $\langle I_s(Q,t)\rangle$	time averaged scattered intensity $I_s(\theta)$ or $I_s(Q,t)$
$\langle I_b(\theta)\rangle$	time averaged background scattered intensity from the solvent and stray light
$\Delta I_s(\theta)$	excess scattering intensity, i.e., difference between the intensity scattered by the suspension and by the pure solvent
$\Delta\langle I_s(\theta)\rangle$	time averaged excess scattering intensity: $\langle I_s(\theta)\rangle - \langle I_b(\theta)\rangle$
$I_m(Q)$	scattered intensity of one randomly oriented fractal object
I_o	intensity of the incident laser beam
$I_o(Qa)$	scattered intensity from one primary particle with radius a in a fractal aggregate
$\langle I_{ref}(\theta)\rangle$	time averaged scattered intensity from pure, isotropically scattering reference solvents
$j_j(Q,L)$	spherical Bessel functions
k_D, k_f, k_I, k_s	first order correction constants for the interaction dependence of the infinite dilution values of D_c, f_o, S(0), and D_s, respectively
k_i	wavevector of the incident laser
k_s	wavevector of the scattered light
K	constant ($= 4\pi^2n^2(dn/dc)^2/(N_A\ \lambda_o^4)$) used in static light scattering experiments
$K_{i,j}$, $K_{k,j}$	kinetic kernels (or rate constants) of the Smoluchowski equation for irreversible aggregation
L	characteristic length scale of aggregates; or length of cylindrical particles
m	ratio of the refractive indices of particles and solvent, respectively
M	molar mass of the particles; or mass of the primary particles within a fractal aggregate
$\langle M\rangle$	weight average molar mass
n	index of refraction of the solution
n_{ref}	index of refraction of the reference solvent

N	number of particles in the scattering volume, or number of primary particles within a fractal aggregate
$N(r)$	normalized (number weighted) size distribution of polydisperse spherical particles
N_A	Avogadro's number
$P(Q)$	particle form factor which describes the influence of intraparticle interference effects on the angular dependence of the scattered intensity
$\langle P(Q) \rangle$	intensity weighted average particle form factor
Q	scattering vector
r	radius of a particle, or radius of a fractal aggregate
$\vec{r}'$	position of volume element
$\vec{r_j}(t)$	position of the center of particle j at time t
R	distance of the detector from the scattering volume
R_{eff}	effective hydrodynamic radius measured in QLS experiment with fractal aggregates
R_h	hydrodynamic radius of a particle which, for ideal spheres, corresponds to the geometric radius, r
R_g	radius of gyration of a particle
$\langle R_g^2 \rangle$	z-average mean square radius of gyration
$\mathbf{R}_{ref}(\theta)$	Rayleigh ratio of the reference solvent
$\Delta\mathbf{R}(\theta)$	excess Rayleigh ratio
$S(Q)$	time average structure factor which describes interparticle interference effects
$\langle S(Q) \rangle$	average structure factor which takes into account the particle size distribution
$S^I(Q)$	"ideal" static structure factor for a system with particles identical in size and scattering amplitude
$S'(Q)$	intraparticle structure factor which describes the spatial arrangement of the particles within a fractal aggregate
t	time
T	absolute temperature
$V(r)$	interaction pair potential (potential between a pair of particles)
$V'(r)$	additional attractive or repulsive potential for systems other than hard spheres
V_j	volume of particle j
V_s	scattering volume
z	polydispersity parameter in a Schultz distribution

LIST OF GREEK SYMBOLS

$\varepsilon(t)$	random noise contribution to the measured autocorrelation function
Γ	decay rate of the autocorrelation function

$<\Gamma>$	average decay rate
δ_{ik}	Kroneker symbol
η_o	viscosity of the solvent
θ	scattering angle
λ_i	wavelength of the incident laser light
ξ	the cut-off value, which is the characteristic distance above which the mass distribution does not continue to follow the fractal law
Π	osmotic pressure
$(d\Pi/dC)^{-1}$	osmotic compressibility
ρ	number density of particles
σ	polydispersity coefficient
τ	delay time in autocorrelation function
τ'	polydispersity exponent in a power-law size distribution
Φ	volume fraction of particles
ω_i	angular frequency of the laser

REFERENCES

1. Schmitz, K.S. *An Introduction to Dynamic Light Scattering by Macromolecules* (San Diego: Academic Press, Inc., 1990).
2. Pecora, R., Ed. *Dynamic Light Scattering* (New York: Plenum Publishing Corp., 1985).
3. Kerker, M. *The Scattering of Light and Other Electromagnetic Radiation* (New York: Academic Press, Inc., 1969).
4. Berne, B.J. and R. Pecora. *Dynamic Light Scattering* (New York: John Wiley & Sons, Inc., 1976).
5. van Holde, K.E. *Physical Biochemistry*, 2nd ed. (Englewood Cliffs, N.J.: Prentice-Hall, Inc., 1985).
6. Yamakawa, H. *Modern Theory of Polymer Solutions* (New York: Harper & Row Publishers, Inc., 1971).
7. Goodstein, D.L. *States of Matter* (Mineola, N.Y.: Dover Publications, Inc., 1985).
8. Pusey, P.N. "Statistical Properties of Scattered Radiation," in *Photon Correlation Spectroscopy and Velocimetry*, H.Z. Cummins and E.R. Pike, Eds. (New York: Plenum Publishing Corp., 1977).
9. Ford, N.C., Jr. "Light Scattering Apparatus," in *Dynamic Light Scattering*, R. Pecora, Ed. (New York: Plenum Publishing Corp., 1985), pp. 7–58.
10. Hess, W. "Diffusion Coefficients in Colloidal and Polymeric Solutions," in *Light Scattering in Liquids and Macromolecular Solutions*, V. Degiorgio, M. Corti, and M. Giglio, Eds. (New York: Plenum Publishing Corp., 1980), pp. 31–50.
11. Pusey, P.N. and R.J.A. Tough. "Particle Interactions," in *Dynamic Light Scattering*, R. Pecora, Ed. (New York: Plenum Publishing Corp., 1985), pp. 85–179.
12. Hansen, J.P. and I.R. McDonald. *Theory of Simple Liquids* (London: Academic Press, Inc., 1976).
13. Corti, M. and V. Degiorgio. "Quasi-Elastic Light Scattering Study of Intermicellar Interactions in Aqueous Sodium Dodecyl Sulfate Solutions," *J. Phys. Chem.* 85:711–717 (1981).

14. Gee, M.L., P. Tong, J.N. Israelachvili, and T.A Witten. "Comparison of Light Scattering of Colloidal Dispersions with Direct Force Measurements between Analogous Macroscopic Surfaces," *J. Chem. Phys.* 93:6057–6064 (1990).
15. Guinier, A. and G. Fournet. *Small Angle Scattering of X-rays* (New York: Wiley-Interscience, 1955).
16. Cabane, B. "Small Angle Scattering Methods," in *Surfactant Science Series, Vol. 22*, R. Zana, Ed. (New York: Marcel Dekker, Inc., 1987), pp. 57–145.
17. van de Hulst, H.C. *Light Scattering by Small Particles* (New York: John Wiley & Sons, Inc., 1957).
18. Carnahan, N.F. and K.E. Starling. "Equation of State for Nonattracting Rigid Spheres," *J. Chem. Phys.* 51:635–636 (1969).
19. Vrij, A., E.A. Nieuwenhuis, H.M. Fijnaut, and W.G.M. Agterof. "Application of Modern Concepts in Liquid State Theory to Concentrated Particle Dispersions," *Faraday Discuss. Chem. Soc.* 65:101–113 (1978).
20. Hayter, J.B. and J. Penfold. "An Analytic Structure Factor for Macroion Solutions," *Mol. Phys.* 42:109–118 (1981).
21. Hansen, J.P. and J.B. Hayter. "A Rescaled MSA Structure Factor for Dilute Charged Colloidal Dispersions," *Mol. Phys.* 46:651–656 (1982).
22. Klein, R., W. Hess, and G. Nägele. "Static and Dynamic Properties of Suspensions of Charged Spherical Particles," in *Physics of Complex and Supermolecular Fluids*, S.A. Safran and N.A. Clark, Eds. (New York: John Wiley & Sons, Inc., 1987).
23. Pusey, P. "Colloidal Suspensions," in *Liquids, Freezing and Class Transition*, J.P. Hansen, D. Levesque, and J. Zinn-Justin, Eds. (Amsterdam: North Holland, 1991), pp. 763–942.
24. Felderhof, B.U. "Hydrodynamic Interaction between Two Spheres," *Physica A* 89:373–384 (1978).
25. Kops-Werkhoven, M.M. and H.M. Fijnaut. "Light Scattering and Sedimentation Experiments on Silica Dispersions at Finite Concentrations," *J. Chem. Phys.* 74:1618–1625 (1981).
26. Van Megen, W. and I. Snook. "Brownian-Dynamics Simulation of Concentrated Charge-Stabilized Dispersions," *J. Chem. Soc. Faraday Trans.* 2:383–394 (1984).
27. Krause, R., G. Nägele, D. Karrer, J. Schneider, R. Klein, and R. Weber. "Structure and Self-Diffusion in Dilute Suspensions of Polystyrene Spheres: Experiment vs. Computer Simulation and Theory," *Physica A*, 153:400–419 (1988).
28. Aragon, S.R. and R. Pecora. "Theory of Dynamic Light Scattering from Polydisperse Systems," *J. Chem. Phys.* 64:2395–2404 (1976).
29. Pusey, P.N. and W. van Megen. "Detection of Small Polydispersity by Photon Correlation Spectroscopy," *J. Chem. Phys.* 80:3513–3520 (1984).
30. van Beurten, P. and A. Vrij. "Polydispersity Effects in the Small-Angle Scattering of Concentrated Solutions of Colloidal Spheres," *J. Chem. Phys.* 74:2744–2748 (1981).
31. Frenkel, D., R.J. Vos, C.G. de Kruif, and A. Vrij. "Structure Factor of Polydisperse Systems of Hard Spheres: A Comparison of Monte Carlo Simulations and Percus-Yevick Theory," *J. Chem. Phys.* 84:4625–4630 (1986).
32. Vrij, A. "Light Scattering of a Concentrated Multicomponent System of Hard Spheres in the Percus-Yevick Approximation," *J. Chem. Phys.* 69:1742–1747 (1978).
33. Vrij, A. "Mixtures of Hard Spheres in the Percus-Yevick Approximation. Light Scattering at Finite Angles," *J. Chem. Phys.* 71:3267–3270 (1979).

34. Kotlarchyk, M. and S.-H. Chen. "Analysis of Small Angle Neutron Scattering Spectra from Polydisperse Interacting Colloids," *J. Chem. Phys.* 79:2461–2469 (1983).
35. Pusey, P.N., H.M. Fijnaut, and A. Vrij."Mode Amplitudes in Dynamic Light Scattering by Concentrated Liquid Suspensions of Polydisperse Hard Spheres," *J. Chem. Phys.* 77:4270–4281 (1982).
36. Koppel, D.E. "Analysis of Macromolecular Polydispersity in Intensity Correlation Spectroscopy: The Methods of Cumulants," *J. Chem. Phys.* 57:4814–4820 (1972).
37. Candau, J.S. "Quasi-Elastic Light Scattering from Dilute Suspensions of Polydisperse Spherical Particles," in *Scientific Methods for the Study of Polymer Colloids and their Applications*, F. Candau and R.H. Ottewill, Eds. (Dordrecht: Kluwer Academic Publishers, 1990), pp. 329–347.
38. Perrin, F. "Mouvement brownien d'un ellipsoïde. I. Dispersion diélectrique pour des molécules ellipsoïdales," *J. de Phys. et Rad.* 5:497–511 (1934).
39. Perrin, F. "Mouvement brownien d'un ellipsoïde. II. Rotation libre et dépolarisation des fluorescences. Translation et diffusion de molécules ellipsoïdales," *J. de Phys. et Rad.* 7:1–11 (1936).
40. Flamberg, A. and R. Pecora. "Dynamic Light Scattering Study of Micelles in a High Ionic Strength Solution," *J. Phys. Chem.* 88:3026–3033 (1984).
41. van de Sande, W. and A. Persoons."The Size and Shape of Macromolecular Structure: Determination of the Radius, the Length, and the Persistence Length of Rodlike Micelles of Dodecyldimethylammonium Chloride and Bromide," *J. Phys. Chem.* 89:404–406 (1985).
42. Young, C.Y., P.J. Missel, N.A. Mazer, G.B. Benedek, and M.C. Carey. "Deduction of Micellar Shape from Angular Disymmetry Measurements of Light Scattered from Aqueous Sodium Dodecyl Sulfate Solutions at High NaCl Concentrations," *J. Phys. Chem.* 82:1375–1378 (1978).
43. Yamakawa H. and M. Fujii. "Translational Friction Coefficient of Wormlike Chains," *Macromolecules* 6:407–415 (1973) .
44. Yamakawa H. and M. Fujii. "Light Scattering from Wormlike Chains. Determination of the Shift Factor," *Macromolecules* 7:649–654 (1974).
45. de la Torre, J.G. and V. Bloomfield. "Hydrodynamic Properties of Complex, Rigid, Biological Macromolecules: Theory and Applications," *Q. Rev. Biophys.* 14:81–139 (1981).
46. Murphy, R.M., H. Slayter, P. Schurtenberger, R.A. Chamberlin, C.K. Colton, and M.L. Yarmush. "Size and Structure of Antigen-Antibody Complexes, Electron Microscopy and Light Scattering Studies," *Biophys. J.* 54:45–56 (1988).
47. Schurtenberger, P. and R.C. Augusteyn. "Structural Properties of Polydisperse Biopolymer Solutions – A Light Scattering Study of Bovine Alpha-Crystallin," *Biopolymers* 31:1229–1240 (1991).
48. Schmidt, M. "Combined Integrated and Dynamic Light Scattering by Poly(γ-benzyl glutamate) in a Helicogenic Solvent," *Macromolecules* 17:553–560 (1984).
49. Burchard, W. "Static and Dynamic Light Scattering from Branched Polymers and Biopolymers," *Adv. Polym. Sci.* 48:1–124 (1983).
50. Cummins, H.Z. "Applications of Light Beating Spectroscopy to Biology," in *Photon Correlation and Light Beating Spectroscopy*, H.Z. Cummins and E.R. Pike, Eds. (New York: Plenum Publishing Corp., 1975), pp. 285–330.

51. Schaefer, D.W., J.F. Joanny, and P. Pincus. "Dynamics of Flexible Polymers in Solution," *Macromolecules* 13:1280–1289 (1980).
52. Schaefer, D.W. and C.C. Han. "Quasielastic Light Scattering from Dilute and Semidilute Polymer Solutions," in *Dynamic Light Scattering*, R. Pecora, Ed. (New York: Plenum Publishing Corp., 1985), pp. 181–243.
53. Adam, M. and M. Delsanti. "Dynamical Properties of Polymer Solutions in Good Solvent by Rayleigh Scattering Experiments," *Macromolecules* 10:1229–1237 (1977).
54. Wiltzius, P. and D. S. Cannell. "Wave-Vector Dependence of the Initial Decay Rate of Fluctuations in Polymer Solutions," *Phys. Rev. Lett.* 56:61–64 (1986).
55. Stockmayer, W.H. and M. Schmidt. "Effects of Polydispersity, Branching and Chain Stiffness on Quasielastic Light Scattering," *Pure Appl. Chem.* 54:407–414 (1982).
56. Schmidt, M. and W.H. Stockmayer, "Quasi-Elastic Light Scattering by Semiflexible Chains," *Macromolecules* 17:509–514 (1984).
57. Burchard, W. and W. Richtering. "Dynamic Light Scattering from Polymer Solutions," *Prog. Colloid Polym. Sci.* 80:151–163 (1989).
58. Isihara, A. and T. Hayashida. "Theory of High Polymer Solutions. I. Second Virial Coefficient for Rigid Ovaloids Model," *J. Phys. Soc. Jpn.* 6:40–45 (1951).
59. Isihara, A. and T. Hayashida. "Theory of High Polymer Solutions. II. Special Forms of Second Osmotic Coefficient," *J. Phys. Soc. Jpn.* 6:46–50 (1951).
60. Onsager, L. "The Effect of Shape on the Interaction of Colloidal Particles," *Ann. N.Y. Acad. Sci.* 51:627–659 (1949).
61. Tanford, C. *Physical Chemistry of Macromolecules* (New York: John Wiley & Sons, Inc., 1961).
62. Schurtenberger, P., N. Mazer, and W. Känzig. "Static and Dynamic Light Scattering Studies of Micellar Growth and Interactions in Bile Salt Solutions," *J. Phys. Chem.* 87:308–315 (1983).
63. Schulz, S.F. "Untersuchung zur Struktur und Dynamik wässriger Lösungen stäbchenförmiger geladener fd-Virusteilchen mittels Statischer und Dynamischer Lichtstreuung," PhD Thesis, Universität Konstanz, Konstanz, Germany (1989).
64. Benmouna, M., G. Weill, H. Benoit, and Z. Akcasu. "Scattering from Charged Macromolecules," *J. Physique* 42:1079–1085 (1982).
65. Genz, U. and R. Klein. "On the Scattering from Interacting Polymer Systems," *J. Physique* 50:439–447 (1989).
66. Tanaka, G. and W. Stockmayer. "Excluded Volume Effect on Quasi-elastic Light Scattering by Flexible Macromolecules," *Proc. Natl. Acad. Sci. U.S.A.* 79:6401–6403 (1982).
67. Doi, M. and S.F. Edwards. "Dynamics of Rod-like Macromolecules in Concentrated Solutions," *J. Chem. Soc. Faraday Trans. 2: Chem. Phys.* 74:560–570 (1978).
68. Maeda, T. and S. Fujime. "Spectrum of Light Quasielastically Scattered from Solutions of Very Long Rods at Dilute and Semidilute Regimes," *Macromolecules* 17:1157–1167 (1984).
69. Maeda, T. and S. Fujime. "Spectrum of Light Quasi-Elastically Scattered from Solutions of Semiflexible Filaments in the Dilute and Semidilute Regimes," *Macromolecules* 17:2381–2391 (1984).
70. Doi, M., T. Shimada, and K. Okano. "Concentration Fluctuations of Stiff Polymers. II. Dynamical Structure Factor of Rod-like Polymers in the Isotropic Phase," *J. Chem. Phys.* 88:4070–4075 (1988).
71. Maeda, T. "Matrix Representation of the Dynamical Structure Factor of a Solution of Rodlike Polymers in the Isotropic Phase," *Macromolecules* 22:1881–1890 (1989).

72. Maeda, T. and M. Doi. "Dynamical Structure Factor of a Solution of Charged Rodlike Polymers in the Isotropic Phase," Proceedings of LALS (1990).
73. Jullien, R. and R. Botet. *Aggregation and Fractal Aggregates* (Singapore: World Scientific, 1987).
74. Brinker, C.J. and G.W. Scherer. *Sol-Gel Science* (San Diego: Academic Press, Inc., 1990).
75. Meakin, P. "Models for Colloidal Aggregation," *Annu. Rev. Phys. Chem.* 39:237–267 (1988).
76. Schaefer, D.W. "Polymers, Fractals, and Ceramic Materials," *Science* 243:1023–1027 (1989).
77. Klein, R., D.A. Weitz, M.Y. Lin, H.M. Lindsay, R.C. Ball, and P. Meakin. "Theory of Scattering from Colloidal Aggregates," *Prog. Colloid Polym. Sci.* 81:161–168 (1990).
78. Rarity, J.G., R.N. Seabrook, and R.J.G. Carr. "Light-Scattering Studies of Aggregation," *Proc. R. Soc. London A* 423:89–102 (1989).
79. Wiltzius, P. "Hydrodynamic Behavior of Fractal Aggregates," *Phys. Rev. Lett.* 58:710–713 (1987).
80. Martin, J. "Scattering Exponents for Polydisperse Surface and Mass Fractals," *J. Appl. Crystallogr.* 19:25–27 (1986).
81. Haller, H.R., C. Destor, and D.S. Cannell. "Photometer for Quasielastic and Classical Light Scattering," *Rev. Sci. Instrum.* 54:973–983 (1983).
82. Chamberlin, R.A. "Light Scattering Studies on Lecithin Micellar Solutions," PhD Thesis, Massachusetts Institute of Technology, Cambridge, MA (1991).
83. Haller, H.R. "The Application of Dynamic Light Scattering and Fluorescence Depolarization to the Determination of Macromolecular Binding Characteristics," PhD Thesis (6604), ETH, Zürich, Switzerland (1980).
84. Haller, H.R. "A Photon Detection System for Dynamic Light Scattering," *J. Phys. E: Sci. Instrum.* 14:1137–1138 (1981).
85. Coumou, D.J. "Apparatus for the Measurement of Light Scattered in Liquids. Measurement of the Rayleigh Factor of Benzene and of Some Other Liquids," *J. Colloid Sci.* 15:408–417 (1960).
86. Bender, T.M., R.J. Lewis, and R. Pecora. "Absolute Rayleigh Ratios of Four Solvents at 488 nm," *Macromolecules* 19:244–245 (1986).
87. Schurtenberger, P., S. Schurtenberger, S. Egelhaaf, and S. Christ. "A Light Scattering Study of the Size Distribution of Polydisperse Macromolecular Solutions," (manuscript in preparation).
88. Ostrowsky, N., D. Sornette, P. Parker, and R. Pike. "Exponential Sampling Method for Light Scattering Polydispersity Analysis," *Opt. Acta* 28:1059–1071 (1981).
89. Provencher, S.W. "A Constrained Regularization Method for Inverting Data Represented by Linear Algebraic or Integral Equations," *Comput. Phys. Commun.* 27: 213–227 (1982).
90. Bertero, M., P. Brianzi, E.R. Pike, G. De Villiers, K.H. Lan, and N. Ostrowsky, "Light Scattering Polydispersity Analysis of Molecular Diffusion by Laplace Transform Inversion in Weighted Spaces," *J. Chem. Phys.* 82:1551–1554 (1985).
91. Livesey, A. K., P. Licino, and M. Delaye. "Maximum Entropy Analysis of Quasielastic Light Scattering from Colloidal Dispersions," *J. Chem. Phys.* 84:5102–5107 (1986).
92. McWhirter, J.G. and E.R. Pike. "On the Numerical Inversion of the Laplace Transform and Similar Fredholm Integral Equations of the First Kind," *J. Phys. A* 11:1729–1745 (1978).

93. Lawson, C.L. and R.J. Hanson, in *Solving Least Squares Problems* (Englewood Cliffs, NJ: Prentice-Hall, Inc., 1974).
94. Schurtenberger, P. and H. Hauser. "Size Characterization of Liposomes," in *Liposome Technology,* 2nd ed., G. Gregoriadis, Ed. (Boca Raton, FL: CRC Press, Inc., 1993), pp. 253–270.
95. Morrison, I.D., E.F. Grabowski, and C.A. Herb. "Improved Techniques for Particle Size Determination by Quasi-Elastic Light Scattering," *Langmuir* 1:496–501 (1985).
96. Ford, T.E. and M. A. Lock. "A Temporal Study of Colloidal and Dissolved Organic Carbon in Rivers: Apparent Molecular Weight Spectra and Their Relationship to Bacterial Activity," *Oikos* 45:71–78 (1985).
97. Hiraide, M., T. Ueda, and A. Mizuike. "Humic and Other Negatively Charged Colloids of Iron and Copper in River Water," *Anal. Chim. Acta* 227:421–424 (1989).
98. Hoffmann, M.R., E.C. Yost, S.J. Eisenreich, and W.J. Maier. "Characterization of Soluble and Colloidal-Phase Metal Complexes in River Water by Ultrafiltration. A Mass-Balance Approach," *Environ. Sci. Technol.* 15(6):655–661 (1981).
99. Lock, M.A. and T.E. Ford. "Colloidal and Dissolved Organic Carbon Dynamics in Undisturbed Boreal Forest Catchments: A Seasonal Study of Apparent Molecular Weight Spectra," *Freshwater Biol.* 16:187–195 (1986).
100. Nomizu, T., K. Goto, and A. Mizuike. "Electron Microscopy of Nanometer Particles in Freshwater," *Anal. Chem.* 60:2653–2959 (1988).
101. Van de Meent, D., A. Los, J.W. Leeuw, P.A. Schenck, and J. Haverkamp. "Size Fractionation and Analytical Pyrolysis of Suspended Particles from the River Rhine Delta," *Adv. Org. Geochem.* 336–349 (1983).
102. Waber, U.E., C. Lienert, and H. R. Von Gunten. "Colloid-Related Infiltration of Trace Metals from a River to Shallow Groundwater," *J. Contam. Hydrol.* 6:251–265 (1990).
103. Barton, L.L. and G.V. Johnson. "Dynamics of Iron Movement Involving Colloids, Bacteria, and Siderophores in an Aqueous Alkaline Environment," *J. Plant Nutr.* 11(6–11):969–979 (1989).
104. Buffle, J., R.R. DeVitre, D. Perret, and G.G. Leppard. "Physico-Chemical Characteristics of a Colloidal Iron Phosphate Species Formed at the Oxic-Anoxic Interface of a Eutrophic Lake," *Geochim. Cosmochim. Acta.* 399–408 (1989).
105. Laxen, P.P.H. and I.M. Chandler. "Size Distribution of Iron and Manganese Species in Freshwaters," *Geochim. Cosmochim. Acta* 47:731–741 (1983).
106. Nomizu, T., T. Nozue, and A. Mizuike. "Electron Microscopy of Submicron Particles in Natural Waters — Morphology and Elemental Analysis of Particles in Fresh Water," *Mikro Chim. Acta.* 2:99–106 (1987).
107. Orlandini, K.A., W.R. Penrose, B.R. Harvey, M.B. Lovett,and M.W. Findlay. "Colloidal Behavior of Actinides in an Oligotrophic Lake," *Environ. Sci. Technol.* 24:706–712 (1990).
108. Salbu, B. and H.E. Bjornstaad. "Analytical Techniques for Determining Radionuclides Associated with Colloids in Waters," *J. Radioanal. Nucl. Chem.* 138(2):337–346 (1990).
109. Salbu, B., H.E. Bjornstaad, E. Lydersen, and A.C. Pappas. "Determination of Radionuclides Associated with Colloids in Natural Waters," *J. Radioanal. Nucl. Chem.* 115(1):113–123 (1987).
110. Macko, S.A. and A.C. Sigleo. "Chesapeake Bay Dissolved Colloidal Particulate Matter: Stable Isotope and Amino Acid Compositions," *Estuaries* 8:A103 (1985).

111. McCave, I.N. "Size Spectra and Aggregation of Suspended Particles in the Deep Ocean," *Deep Sea Res.* 31(4):329–352 (1984).
112. Moran, B.B. and R.M.Moore. "The Distribution of Colloidal Al and Organic Carbon in Coastal and Open Ocean Waters off Nova Scotia," *Geochim. Cosmochim. Acta* 53:2519–2526 (1989).
113. Pashkova, Y.A., N.I. Gulko, S.V. Lyutsarev, and S.I. Tsekhonya. "Colloidal and Dissolved Forms of Gold and Organic Carbon in the Water of the Bering Sea and the Northern Pacific Ocean," *Oceanology* 28(3):305–313 (1988).
114. Sigleo, A.C. and J.C. Means. "Organic and Inorganic Components of Estuarine Colloids: Implications of Sorption and Transport of Pollutants," *Rev. Environ. Cont. Toxic.* 112:123–147 (1990).
115. Wells, M.L. and E.D. Goldberg. "Marine Sub-Micron Particles," *Environ. Sci. Technol.* submitted (1990).
116. Buddemeier, R.W. and J.R. Hunt. "Transport of Colloidal Contaminants in Groundwater: Radionuclide Migration at the Nevada Test Site," *Appl. Geochem.* 3:535–548 (1988).
117. Degueldre, C., B. Baeyens, W. Goerlich, and P. Stadelmann. "Colloids in Water from a Subsurface Fracture in Granitic Rock, Grimsel Test Site, Switzerland," *Geochim. Cosmochim. Acta.* 53:603–610 (1989).
118. Kim, J.I., G. Buckau, H. Rommel, and B. Sohnius. "The Migration Behavior of Transuranium Elements in Gorleben Aquifer Systems: Colloid Generation and Retention Process," *Mater. Res. Soc. Symp. Proc.* 127:849–854 (1989).
119. Lieser, K.H., A. Ament, R. Hill, R.N. Singh, U. Stingl, and B. Thybusch. "Colloids in Groundwater and Their Influence on Migration of Trace Elements and Radionuclides," *Radiochim. Acta* 49:83–100 (1990).
120. Lieser, K.H., H. Gleitsmann, S. Peschke, and T.H. Steinkopf. "Colloid Formation and Sorption of Radionuclides in Natural Systems," *Radiochim. Acta* 40:39–47 (1986).
121. Maitia, T.C., M.R. Smith, and J.C. Laud. "Colloid Formation Study of U, Th, Ra, Pb, Po, Sr, Rb, and Cs in Briny (High Ionic Strength) Groundwaters: Analog Study for Waste Disposal," *Nucl. Tech.* 84:82–87 (1989).
122. Penrose, W.R., W.L. Polzer, E.H. Essington, D.M. Nelson, and K.A. Oelandini. "Mobility of Plutonium and Americium through a Shallow Aquifer in a Seimiarid Region," *Environ. Sci. Technol.* 24:228–234 (1990).
123. Short, S.A., R.T. Lowson, and J. Ellis. "$^{234}U/^{238}U$ and $^{230}Th/^{234}U$ Activity Ratios in the Colloidal Phases of Aquifers in Lateritic Weathered Zones," *Geochim. Cosmochim. Acta* 52:2555–2563 (1988).
124. Vilks, P., J.J. Cramer, T.A. Shewchuk, and J.P.A. Larocque. "Colloid and Particulate Matter Studies in the Cigar Lake Natural-Analog Program," *Radiochim. Acta* 44/45:305–310 (1988).
125. Van der Meeren, P.P.A., J. Vanderdeelen, and L. Baert. "Optimization of Quasi-Elastic Light Scattering Measurements," in *Particle Analysis 1988,* P.J. Lloyd, Ed. (New York: John Wiley & Sons, Inc., 1988), pp. 101–107.
126. Schätzel, K. "Noise on Photon Correlation Data. I. Autocorrelation Functions," *Quantum Opt.* 2:287–305 (1990).
127. Newman, M.E. and J. Buffle. Unpublished results (1990).
128. Bott, S.E. "Enhanced Resolution Particle Size Distributions by Multi Angle Photon Correlation Spectroscopy" in *Particle Size Analysis 1988,* P.J. Lloyd, Ed. (New York: John Wiley & Sons, Inc., 1988), pp. 77–89.

129. Cummins, P.G. and E.J. Staples. "Particle Size Distributions Determined by a 'Multiangle' Analysis of Photon Correlation Spectroscopy Data," *Langmuir* 3:1109–1113 (1987).
130. Gallegos, C.L. and R.G. Menzerl. "Submicron Size Distributions of Inorganic Suspended Solids in Turbid Waters by Photon Correlation Spectroscopy," *Water Resour. Res.* 23(4):596–602 (1987).
131. Gschwend, P.M. and M.D. Reynolds. "Monodisperse Ferrous Phosphate Colloids in an Anoxic Groundwater Plume," *J. Contam. Hydrol.* 1:309–327 (1987).
132. Provencher, S.W. "Inverse Problems in Polymer Characterization: Direct Analysis of Polydispersity with Photon Correlation Spectroscopy," *Makromol. Chem.* 180:201–209 (1979).
133. Ryan, J.N., Jr. "Groundwater Colloids in Two Atlantic Coastal Plain Aquifers: Colloid Formation and Stability," MS Thesis, Massachusetts Institute of Technology, Cambridge, MA (1988).
134. Ryan, J.N., Jr. and P.M. Gschwend. "Colloid Mobilization in Two Atlantic Coastal Plain Aquifers: Field Studies," *Water Resour. Res.* 26(2):307–322 (1990).
135. Ford, N.C. "Theory and Practice of Correlation Spectroscopy," in *Measurement of Suspended Particles by Quasi-Elastic Light Scattering*, B.E. Danke, Ed. (New York: John Wiley & Sons, Inc., 1983), pp. 31–77.
136. Rees, T.F. and J.F. Ranville, "Collection and Analysis of Colloidal Particles Transported in the Mississippi River , USA," *J. Contam. Hydrol.* 6:241–250 (1990).
137. Rees, T.F. and J.F. Ranville, "Characterization of Colloids in the Mississippi River and Its Major Tributaries," *J. Contam. Hydrol.* (1991).
138. Rees, T.F. "Comparison of Photon Correlation Spectroscopy with Photosedimentation Analysis for the Determination of Aqueous Colloid Size Distributions," *Water Resour. Res.* 26:2777–2781 (1990).
139. Leenheer, J.A., R.H. Meade, H.E. Taylor, and W.E. Pereira."Sampling, Fractionation, and Dewatering of Suspended Sediment from the Mississippi River for Geochemical and Trace-Contaminant Analysis," USGS Water-Resources Investigations Report 88-4220 (1988), pp. 501–511.
140. Ranville, J.F., R.A. Harnish, and D.M. McKnight. "Particulate and Colloidal Organic Material in Pueblo Reservoir, Colorado: Influence of Autochthonous Source on Chemical Composition," USGS Water Resources Investigations Report (1990).
141. Buffle, J., D. Perret, and M. Newman. "The Use of Filtration and Ultrafiltration for Size Fractionation of Aquatic Particles, Colloids and Macromolecules," in *Characterization of Environmental Particles*, J. Buffle and H.P. van Leeuwen, Eds., IUPAC Environmental Analytical Chemistry Series, Vol. I (Chelsea, MI: Lewis Publishers, Inc., in press).
142. Salbu, B., H.E. Bjornstad, N.S. Lindstrom, E. Lydersen, E.M. Brevik, J.P. Rambaek, and P.E. Paus. "Size Fractionation Techniques in the Determination of Elements Associated with Particulate or Colloidal Material in Natural Fresh Waters," *Talanta* 32(9): 907–913 (1985).
143. Dubin, S.B., J.H. Lunacek, and G.B. Benedek. "Observation of the Spectrum of Light Scattered by Solutions of Biological Macromolecules," *Proc. Natl. Acad. Sci. U.S.A.* 57:1164–1171 (1967).
144. Bloomfield, V.A. "Biological Applications," in *Dynamic Light Scattering*, R. Pecora, Ed. (New York: Plenum Publishing Corp., 1985), pp. 363–417.

145. Harding, S.E., Ed. "Laser Light Scattering in Biochemistry," *Biochem. Soc. Trans.* 19:477–516 (1991).
146. Andries, C., W. Guedens, J. Clauwaert, and H. Geerts. "Photon and Fluorescence Correlation Spectroscopy and Light Scattering of Eye-Lens Proteins at Moderate Concentrations," *Biophys. J.* 43:345–354 (1983).
147. Andries, C. and J. Clauwaert. "Photon Correlation Spectroscopy and Light Scattering of Eye Lens Proteins at High Concentrations," *Biophys. J.* 47:591–605 (1985).
148. Licino, P., M. Delaye, A.K. Livesey, and L. Léger. "Colloidal Dispersions of α-Crystallin Proteins. II. Dynamics: a Maximum Entropy Analysis of Photon Correlation Spectroscopy Data," *J. Phys. Fr.* 48:1217–1223 (1987).
149. Licino, P. and M. Delaye. "Direct and Hydrodynamic Interactions Between α-Crystallin Proteins in Dilute Colloidal Dispersions: A Light Scattering Study," *J. Colloid Interface Sci.* 123:105–116 (1988).
150. Licino, P. and M. Delaye. "Mutual and Self-Diffusion in Concentrated α-Crystallin Protein Dispersions. A Dynamic Light Scattering Study," *J. Phys. Fr.* 49:975–981 (1988).
151. Licino, P. and M. Delaye. "Polydispersity of Colloidal Particles: Intensities for Static and Dynamic Light Scattering," *J. Colloid Interface Sci.* 132:1–12 (1989).
152. Bloemendal, H., Ed. *Molecular and Cellular Biology of the Eye Lens* (New York: Interscience, 1981) pp. 1–47.
153. Bloemendal, H. "Lens Proteins," *CRC Crit. Rev. Biochem.* 12:1–38 (1982).
154. Tardieu, A., D. Laporte, and M. Delaye. "Colloidal Dispersions of α-Crystallin Proteins. I. Small Angle X-Ray Analysis of the Dispersion Structure," *J. Phys. Fr.* 48:1207–1215 (1987).
155. Neal, D.G., D. Purich, and D.S. Cannell. "Osmotic Susceptibility and Diffusion Coefficient of Charged Bovine Serum Albumin," *J. Chem. Phys.* 80:3469–3477 (1984).
156. Ackerson, B.J. "Correlations for Interacting Brownian Particles," *J. Chem. Phys.* 64:242–246 (1976).
157. Belloni, L. and M. Drifford. "On the Effect of Small Ions in the Dynamics of Polyelectrolyte Solutions," *J. Phys. Lett.* 46:L-1183–L-1189 (1985).
158. Coviello, T., K. Kajiwara, W. Burchard, M. Dentini, and V. Crescenzi. "Solution Properties of Xanthan. 1. Dynamic and Static Light Scattering from Native and Modified Xanthans in Dilute Solutions," *Macromolecules* 19:2826–2831 (1986).
159. Mazer, N.A. "Quasielastic Light Scattering Studies of Micelle Formation, Solubilization and Precipitation in Aqueous Biliary Lipid Systems," PhD Thesis, Massachusetts Institute of Technology, Cambridge, MA (1978).
160. Mazer, N.A., G.B. Benedek, and M.C. Carey. "Quasielastic Light Scattering Studies of Aqueous Biliary Lipid Systems. Mixed Micelle Formation in Bile Salt-Lecithin Solutions," *Biochemistry* 19:601–615 (1980).
161. Mazer, N.A. and M.C. Carey. "Quasielastic Light Scattering Studies of Aqueous Biliary Lipid Systems. Cholesterol Solubilization and Precipitation in Model Bile Solutions," *Biochemistry* 22:426–442 (1983).
162. Schurtenberger, P., N.A. Mazer, W. Känzig, and R. Preisig. "Quasielastic Light Scattering Studies of the Micelle to Vesicle Transition in Aqueous Solutions of Bile Salt and Lecithin," in *Surfactants in Solution, Vol. 2,* K.L. Mittal and B. Lindman, Eds. (New York: Plenum Press, 1984), pp. 841–855.

163. Mazer, N.A., P. Schurtenberger, M.C. Carey, R. Preisig, K. Weigand, and W. Känzig. "Quasi-Elastic Light Scattering Studies of Native Hepatic Bile from the Dog: Comparison with Aggregative Behavior of Model Biliary Lipid Systems," *Biochemistry* 23:1994–2005 (1984).
164. Mazer, N.A. and P. Schurtenberger. "Quasi-Elastic Light Scattering Studies of Aqueous Biliary Lipid Systems and Native Bile," in *Proceedings of the International School of Physics "Enrico Fermi": Physics of Amphiphiles, Micelles, Vesicles and Microemulsions*, V. Degiorgio and M. Corti, Eds. (Amsterdam: North Holland, 1985), pp. 587–606.
165. Schurtenberger, P., N.A. Mazer, and W. Känzig. "The Micelle to Vesicle Transition in Aqueous Solutions of Bile Salt and Lecithin," *J. Phys. Chem.* 89:1042–1049 (1985).
166. Stark, R.E., G.J. Gosselin, J.M. Donovan, M.C. Carey, and M.F. Roberts. "Influence of Dilution on the Physical State of Model Bile Systems: NMR and QLS Investigations," *Biochemistry* 24:5599–5605 (1985).
167. Hjelm, R.P., P. Thiyagaragan, D.S. Sivia, P. Lindner, H. Alkan, and D. Schwahn. "Small-Angle Neutron Scattering from Aqueous Mixed Colloids of Lecithin and Bile Salt," *Prog. Colloid Polym. Sci.* 81:225–231 (1990).
168. Hjelm, R.P., P. Thiyagaragan, and H. Alkan. "A Small-Angle Neutron Scattering Study of the Effects of Dilution on Particle Morphology in Mixtures of Glycocholate and Lecithin," *J. Appl. Crystallogr.* 21:858–863 (1988).
169. Israelachvili, J.N., D.J. Mitchell, and B.W. Ninham. "Theory of Self-Assembly of Hydrocarbon Amphiphiles into Micelles and Bilayers," *J. Chem. Soc. Faraday Trans. II* 72:1525–1568 (1976).
170. Selser, J.C., Y. Yeh, and R.J. Baskin. "A Light Scattering Characterization of Membrane Vesicles," *Biophys. J.* 16:337–356 (1976).
171. Chen, F.C., A. Chrzseszczyk, and B. Chu. "Quasielastic Laser Light Scattering of Monolayer Vesicles," *J. Chem. Phys.* 66:2237–2238 (1977).
172. Ostrowsky, N. and Ch. Hesse-Bezot. "Dynamic Light Scattering Study of the Conformational Change and Fusion Phenomenon of Phospholipid Vesicles," *Chem. Phys. Lett.* 52:141–144 (1977).
173. Goll, J., F.D. Carlson, Y. Barenholz, B.J. Litman, and T.E. Thompson. "Photon Correlation Spectroscopy Study of the Size Distribution of Phospholipid Vesicles," *Biophys. J.* 38:7–13 (1982).
174. Schurtenberger, P. and H. Hauser. "Characterization of the Size Distribution of Unilamellar Vesicles by Gel Filtration, Quasi-Elastic Light Scattering and Electron Microscopy," *Biochim. Biophys. Acta* 778:470–480 (1984).
175. Sun, S.-T. , A. Milon, T. Tanaka, G. Ourisson, and Y. Nakatani. "Osmotic Swelling of Unilamellar Vesicles by the Stopped-flow Light Scattering Method. Elastic Properties of Vesicles," *Biochim. Biophys. Acta* 860:525–530 (1986).
176. Sornette, D., C. Hesse-Bezot, and N. Ostrowsky. "Fusion Kinetics of Dimyristoyl Phosphatidylcholine Vesicles around the Phase Transition of the Aliphatic Chains," *Biochimie* 63:955–959 (1981).
177. Hallett, F. R., J. Watton, and P. Krygsman. "Vesicle Sizing. Number Distribution by Dynamic Light Scattering," *Biophys. J.* 59:357–362 (1991).
178. *Handbook of Chemistry and Physics*, 67th Ed. (Boca Raton, FL: CRC Press, Inc., 1986).

179. Lindsay, H.M., M.Y. Lin, D.A. Weitz, P.Sheng, and Z.Chen. "Properties of Fractal Colloid Aggregates," *Faraday Discuss. Chem. Soc.* 83:153–165 (1987).
180. Skood, A.K. "Light Scattering from Colloids," *Hyperfine Interac.* 37:365–384 (1987).
181. Ramsay, J.D. and M. Scanlon. "Structure and Interactions in Aqueous Colloidal Dispersions," *Colloid Surf.* 18:207–221 (1987).
182. Amal, R., J.A. Raper, and T.D. Waite. "Fractal Structure of Hematite Aggregates," *J. Colloid Interface Sci.* 140(1):158–168 (1990).
183. Hackley, V.A. and M.A. Anderson. "Effects of Short-Range Forces on the Long-Range Structure of Hydrous Iron-Oxide Aggregates," *Langmuir* 5:191–198 (1989).
184. Martin, L.E. and D.W. Schaefer. "Dynamics of Fractal Colloidal Aggregates," *Phys. Rev. Lett.* 53:2457–2460 (1984).
185. Schaefer, D.W., J.E. Martin, P. Wiltzius, and D.S. Cannell. "Fractal Geometry of Colloidal Aggregates," *Phys. Rev. Lett.* 52:2371–2374 (1983).
186. Weitz, D.A., J.S. Huang, M.Y. Lin, and J. Sung. "Dynamics of Diffusion-Limited Kinetic Aggregation," *Phys. Rev. Lett.* 53:1657–1660 (1984).
187. Weitz, D.A., J.S. Huang, M.Y. Lin, and J. Sung. "Limits of the Fractal Dimension for Irreversible Kinetic Aggregation of Gold Colloids," *Phys. Rev. Lett.* 54:1416–1419 (1985).
188. Amal, R., J.R. Coury, J.A. Raper, W.P. Walsh, and T.D. Waite. "Structure and Kinetics of Aggregating Colloidal Hematite," *Colloids Surf.* 46:1–19 (1990).
189. Higashitani, K., K. Masahiro, and S. Hatade. "Effect of Particle Size on Coagulation Rate of Ultrafine Colloidal Particles," *J. Colloid Interface Sci.* 142(1):204–213 (1991).
190. Lin, M.Y., H.M. Lindsay, D.A. Weitz, R.C. Ball, R. Klein, and P. Meakin. "Universality of Colloid Aggregation," *Nature* 339(1):360–362 (1989).
191. Oliver, B.J. and M. Sorensen. "Evolution of the Cluster Size Distribution During Slow Aggregation," *J. Colloid Interface Sci.* 134(1):139–145 (1990).
192. Pefferkorn, E. and S. Stoll. "Aggregation/Fragmentation Processes in Unstable Latex Suspensions," *J. Colloid Interface Sci.* 138(1):261–272 (1990).
193. Strehlow, P. "Thermodynamic Stability of Monodispersed Particles in Solution," *J. Non-Cryst. Solids* 107:55–60 (1988).
194. Cametti, C., P. Codastefano, and P. Tartaglia. "Aggregation Kinetics in Model Colloidal Systems: A Light Scattering Study," *J. Colloid Interface Sci.* 131(2):409–422 (1989).

CHAPTER 3

STRUCTURE, FORMATION, AND REACTIVITY OF HYDROUS OXIDE PARTICLES: INSIGHTS FROM X-RAY ABSORPTION SPECTROSCOPY

Laurent Charlet and Alain Manceau

TABLE OF CONTENTS

0-87371-895-X/93/$0.00+$.50

INTRODUCTION

Recent advances in the development of noninvasive, *in situ* spectroscopic techniques have been applied successfully to the study of natural colloids in aqueous suspensions.[1,2] Among emerging spectroscopic methods that recently have been developed, X-ray absorption spectroscopy (XAS) is one of the few methods capable of providing, at room temperature and possibly in aqueous media, direct structural information on the short range order of "amorphous" natural colloids and of dilute metal species sorbed on them.

The purpose of this review is threefold. First, we will present a brief description of the physics of X-ray absorption spectroscopy and of the data analysis. This will be followed by a discussion of the formation of environmental, typically poorly crystallized, particles using the example of Fe and Mn (oxyhydr)oxide colloids. Well-crystallized mineral particles shall not be considered here, as their study by XAS has been reviewed at length.[3,4] Finally, we will present a selective review of some applications of X-ray absorption spectroscopy to the study of sorption phenomena, which include the formation of surface complexes, heterogeneous nucleation, coprecipitation, and solid state diffusion.

XAS: BASIC PRINCIPLES, POSSIBILITIES, AND LIMITATIONS

X-rays with high intensity are required for X-ray absorption spectroscopy (XAS) and are produced by electrons/positrons moving in a storage ring at relativistic energies (1 to 6 GeV) in paths curved by a magnetic field. These charged particles radiate a collimated white beam of photons tangentially to their curved path.

XAS consists of recording the absorption by a given sample of X-rays as a function of the wavelength. The white beam is monochromatized by a diffraction on single crystals (Figure 1). The spectral scan is performed in the vicinity of an X-ray absorption edge (K, L, or M) of the chosen target element (Figure 2). XAS is therefore a bulk element specific, spectroscopic method. Unlike most spectroscopic methods, however, most elements are spectroscopically active, and their spectral features do not overlap since K edges are separated by several hundred eV. XAS can therefore be used to study compositionally complex materials by

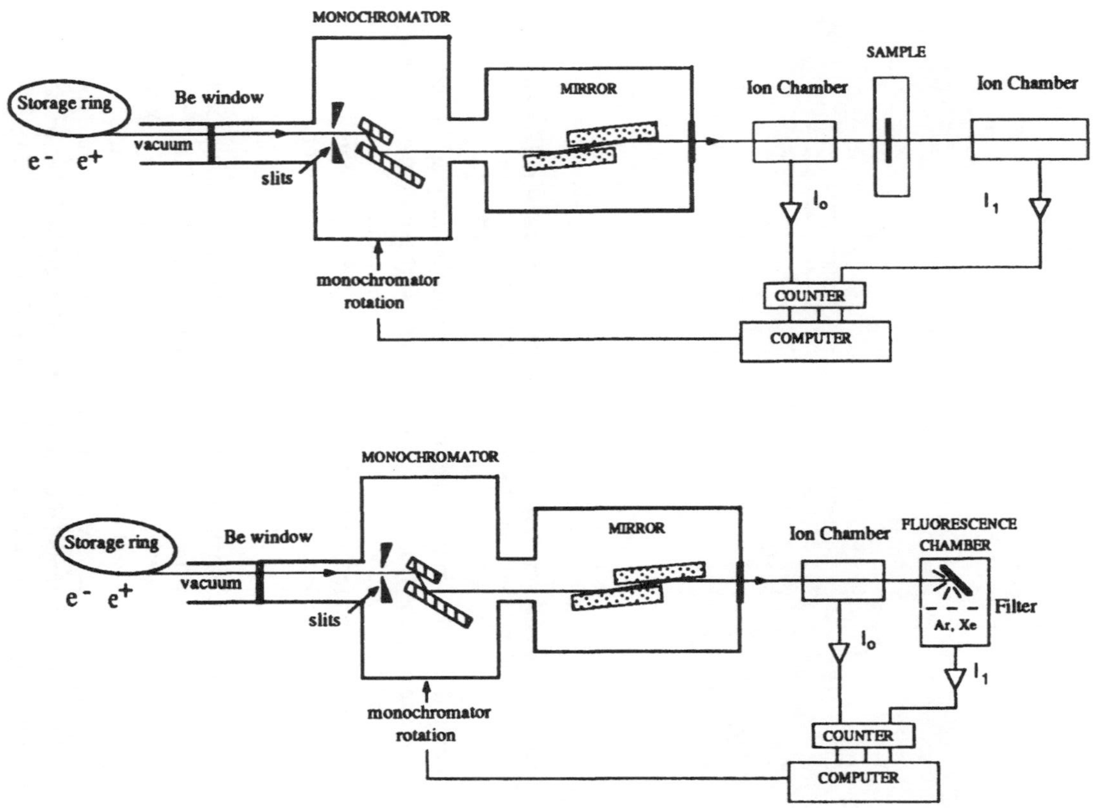

Figure 1. Line out of a XAS spectrometer. Detection in transmission and fluorescence mode. Sizes of the storage ring and spectrometer are not scaled. The incident white beam emitted by electrons or positrons is monochromatized by a double diffraction on single crystals (Si, Ge). High order harmonics can be rejected by reflection on mirrors at grazing angle. X-ray detectors are used to measure intensities of the incident monochromatic beam (I_o) and of the transmitted beam or fluorescence X-ray emission (I). Fluorescence detection is used for dilute samples, samples with high matrix absorption, or samples that are very small.

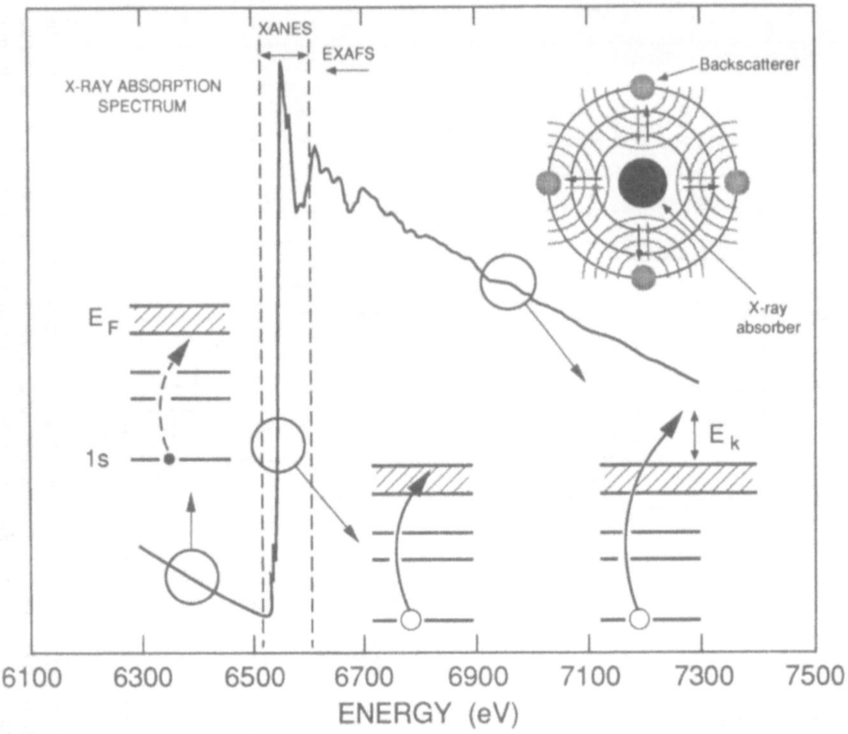

Figure 2. X-ray absorption spectroscopy, physical principle. 1s = Inner shell at the K-edge; E_F: Fermi level; E_k: kinetic energy of the photoelectron.

successively tuning to the absorption edge of each spectroscopically active atom present within the sample. The K-edge energy of elements whose atomic number (Z) is higher than about 25 (i.e., Mn) is high enough for the incident beam not to be absorbed too much by water. Thus, XAS spectra of these elements can be recorded *in situ* on particles suspended in water. The main limitation of XAS is its lack of spatial resolution, which limits uses to cases where the X-ray absorber-containing phase is unique.

By convention, XAS spectra are divided in two regions: (1) the X-ray absorption near-edge structure (XANES), in the range from few eV below to about 70 eV above the absorption edge, and (2) the extended X-ray absorption fine structure (EXAFS) for higher energies up to about 800 eV beyond this edge. The discussion below covers only the highlights of the XAS spectroscopy, and in particular the type of information in which an environmental chemist might be interested. The reader is referred to recent monographs for a complete technical discussion of XAS.[5,6]

XANES Spectroscopy

In XANES regime, an electron is ejected from a core level (e.g., From 1s at the K edge) of the X-ray absorber toward bond, or delocalized, empty states (Figure

2). Electronic transitions obey dipole selection rules. For instance, at the K edge final states wave functions responsible for the absorption edge discontinuity must have a p symmetry. For first row transition elements, electronic transitions to empty 3d bond states are possible if these have a certain p character through hybridization. These 1s → 3d transitions take place at the bottom of the steeply rising absorption edge giving rise to "pre-edge" features (Figures 2 and 3). For a centrosymmetric site (e.g., $Mn_{Oct.}(II)$ in $MnCO_3$), the hybridization is at a minimum, and the pre-edge peak is weak. The reverse situation is observed for a noncentrosymmetric site (e.g., $Mn_{Tet.}(II)$ in $MnCr_2O_4$) where the intensity of the pre-edge peak is maximum. A five to seven enhancement of the pre-edge intensity is thus observed in going from $Me_{Oct.}$ to $Me_{Tet.}$ species, which permits the site occupancy of a given ion to be derived with a good accuracy.

The second information which can be derived from XANES spectroscopy is the oxidation state of the X-ray absorber. A positive energy shift of the main absorption edge and of the pre-edge is observed in going from Me^{n+} to $Me^{(n+1)+}$. This positive energy shift is understood conceptually to be due to an increase in the attractive potential of the nucleus on the core level. However, this spectral shift is not only determined by core level effects, but final state wave functions are to be taken into account. Final states of 3d character (pre-edge) are more tightly bound and, hence, less sensitive to changes in ionicity and solid-state effects than the delocalized higher energy final states (main absorption edge features). These differences explain why the oxidation state is more reliably determined from the energy position of the pre-edge peak than from that of the main absorption edge spectrum.

The sensitivity of pre-edge spectroscopy to changes in the valence and coordination number of metal ions is illustrated in Figure 3 with Mn oxides. The chemical shift between Mn^{2+} and Mn^{4+} reaches 1.5 eV and is high enough to distinguish Mn ions with different oxidation states and site occupations as, for instance, the two oxidation states of Mn (II and III) in Mn_3O_4 or the two Mn (tetrahedral and octahedral) sites in γMn_2O_3. This capability is especially interesting as this distinction is not presently possible with other spectroscopic techniques such as X-ray photoelectron spectroscopy (XPS), energy electron loss spectroscopy (EELS), or UV-visible reflectance spectroscopy. In the field of environmental particles, XANES spectroscopy has already been used for assessing the site occupation of Fe and Mn in hydrous oxides[7,8] and for following the kinetics of the surface assisted oxidation of Cr(III) to Cr(VI)[9] and Ce(III) to Ce(IV).[10]

EXAFS Spectroscopy

Basic Principles

In the EXAFS regime, photons with an energy (E) equal to or larger than the threshold energy (E_o) are absorbed by the target atom, which in turn ejects an ionized photoelectron. The kinetic energy of this electron ($E_k = E - E_o$) allows it to reach up to three to four first shells of neighboring atoms and to be backscattered

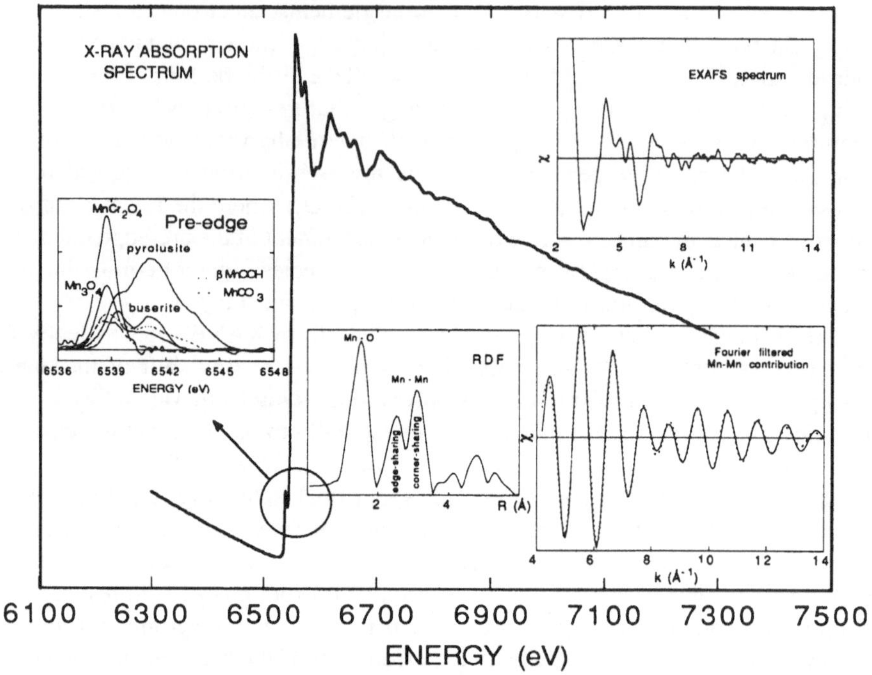

Figure 3. X-ray absorption spectroscopy, data reduction and type of information: the example is that of pyrolusite (MnO_2), whose structure consists of single chains of edge-sharing octahedra cross-linked by corners (see also Figure 5a). RDF is the radial distribution function obtained by Fourier transforming of the EXAFS (insert) spectrum. Distances reported on RDF differ from true values because these functions are usually not corrected for phase shifts.

by them. Interferences between outgoing and incoming electronic waves influence the probability of absorption of the incident X-ray photon, and likewise the probability of emission of a decay fluorescence photon. These interferences thus modulate the absorption coefficient, resulting in the existence of oscillations in the X-ray absorption spectrum (Figures 2 and 3). These oscillations, called EXAFS, provide information on the local structure around the X-ray absorbing atom. From this simple description of the physical process, it is evident that this method only requires the presence of one atom in the vicinity of the absorber to "work." In addition, given the short mean free path of the photoelectron (few Å for a kinetic energy of a few hundred eV), this method yields structural information on the first atomic shells surrounding the X-ray absorber. These characteristics make it a unique tool for elucidating the local structure of poorly crystallized particles and the stereochemistry of surface complexes.

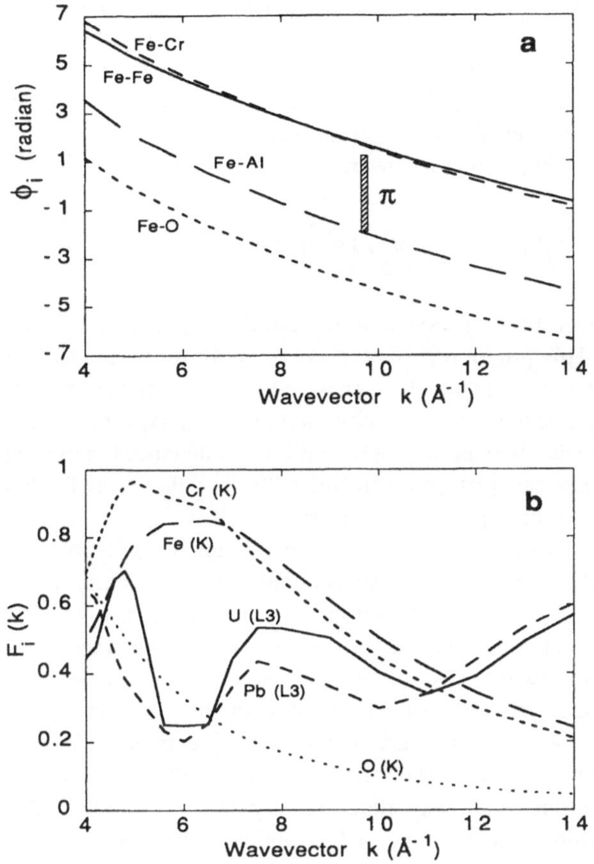

Figure 4. Theoretical phase shift (a) and amplitude (b) functions calculated by McKale et al.[112]

In the single-scattering plane wave approximation, the contribution to the EXAFS spectrum of each atomic shell j is formulated as:

$$\chi_j(k) = A_j(k)\sin\left(2kR_j + \phi_{ij}(k)\right) \quad (1)$$

and the whole EXAFS spectrum is obtained by adding each individual contribution

$$\chi(k) = \sum_j \chi_j(k) \quad (2)$$

where k is the modulus of the wavevector of the photoelectron. The phase of $\chi_j(k)$ depends on the absorber (i) to backscatterer (j) distance (R_j), and on the phase shift function $\phi_{ij}(k)$ which is characteristic of a given absorber-backscatterer pair. $\phi_{ij}(k)$ for Fe-O, Fe-Al, Fe-Cr, and Fe-Fe pairs is plotted in Figure 4a. These functions vary monotonically with the atomic number (Z). For instance, Fe-Cr and Fe-Fe pairs have nearly identical phase shift functions, while the phase contrast reaches its maximum between Fe-Fe and Fe-Al since $\phi_{Fe-Fe} - \phi_{Fe-Al} \simeq \pi$. The amplitude of $\chi_j(k)$ includes the following terms:

$$A_j(k) = F_j(k)\left[N_j / kR_j^2\right]\exp\left[-2\sigma_j^2 k^2 - 2R_j / l(k)\right] \tag{3}$$

$F_j(k)$ is the backscattering wave amplitude and is characteristic of the atom present in the shell j. This function is plotted in Figure 4b for O, Fe, Cr, Pb, and U. N_j is the number of atoms in the j shell. σ_j is a Debye-Waller factor accounting for the wave damping due to the thermal vibration of atoms (σ_T) and to the static disorder (σ_S). The more the incoherency of the individual distances in the j shell, the larger the σ_S value. l is the electron mean free path. Its value must be determined from the analysis of a reference compound for which N_j is known.

The validity of the theoretical F_j and $\phi_{ij}(k)$ functions has been established for many compounds. However, more accurate structural parameters are obtained directly using experimental functions. $\phi_{ij}(k)$ can be easily calculated from a reference material containing the i–j pair at a known distance R_j. Finding such a reference is not always feasible when a heteroatom i is sorbed onto a substrate containing j (e.g., Pb sorbed on Fe oxide). In this case, theoretical phase shift functions of i (Pb) and j (Fe) are used, and the precision on R_j is not better than a few hundredths of one Å (0.02 to 0.05Å). This difficulty does not exist with F_j since the amplitude of the wavefunction only depends on the backscatterer j. But obtaining F_j requires knowledge of l, σ_j and N_j, which is currently not possible. Usually the entire expression

$$F_j(k)\exp\left[-2\sigma_j^2 k^2 - 2R_j / l(k)\right] \tag{4}$$

is deduced from the reference. Reference materials must be chemically and structurally as close as possible to the sample under study in order to assure that the transferability of l be physically valid. This requirement is generally fulfilled for the study of natural particles, since their structures are similar to those of well-crystallized minerals. This is the case, for example, with hydrous ferric and manganese oxides, whose local structures have been shown to be related to those of well crystallized oxides.[11-13]

Analysis of EXAFS Spectra

The first operation consists of extracting the EXAFS spectrum from the raw X-ray absorption spectrum (Figure 3, central curve) and to pass from the energy E (eV) to the momentum k ($Å^{-1}$) space by using the equation:

$$k^2 = \frac{8\pi^2 m E_k}{h^2} \tag{5}$$

in which m is the mass of the electron and h is the Planck's constant. The EXAFS spectrum $\chi(k)$ thus obtained is the sum of individual atomic pair contributions ($\chi_j(k)$). To a first approximation, one may consider that each atomic shell gives a single damped sine wave. The basic principle of the data reduction consists of filtering these different sine waves. For this purpose a double Fourier transform is performed. First, a Fourier transform from the momentum k ($Å^{-1}$) to real R (Å) space yields a radial distribution function (RDF), i.e., a one dimensional representation of the local structure around the X-ray absorber. Each RDF peak corresponds to one, or possibly several, atomic shell(s). For instance, in the case of pyrolusite (MnO_2), three RDF peaks are observed and correspond to the nearest six O atoms at 1.88 Å, the nearest 2 Mn across edges at 2.87 Å, and the next-nearest 8 Mn at 3.43 Å, respectively (Figure 3). Usually the RDF is not corrected for phase shift functions, and thus each peak is displaced from its crystallographic position. Additional data reduction is needed for extracting structural information. This extraction is usually accomplished by Fourier back-transforming in momentum space (k) one or several peak(s). This yields the contribution to the EXAFS spectrum of the atomic shell(s) attached to the selected RDF peak(s) (Figure 3). This experimental Fourier filtered partial contribution to the EXAFS spectrum is then least-square fitted by a theoretical curve.

Identification of Backscatterers

In heterogeneous systems such as a mineral surface covered with adsorbed atoms, three methods have been used to identify the backscatterer present in a given shell. The first method is based on the contrast of amplitude and phase shift functions between two possible atomic neighbors, the second applies when the two possible neighbors give distinct local structures, and the third applies when the two neighbors have different ionic radii.

Contrast of amplitude and phase shift functions. Let us postulate a system where an atom i has been adsorbed on a metal oxide adsorbent containing j. Atom i can then be surrounded by atoms of j, i, or a combination of both. In the two first hypotheses, amplitude and phase shift functions to be considered in the data analysis are ϕ_{ij} (k) plus F_j, or $\phi_{ii}(k)$ plus F_i. These two structural possibilities could then be discriminated if $\phi_{ij}(k)$ and $\phi_{ii}(k)$, and if F_j and F_i are notably different. This requirement is satisfied whenever i and j have a Z difference of at least 10 (Figure 4). But while the method works well when only one type of atom is present in a given neighbor atomic shell, as in the case of isolated adsorbed atoms, care has to be taken when two types of neighbors coexist, as in the case of a multinuclear surface complex. Equation 1 shows that each new contribution introduces three new parameters in the $\chi(k)$ equation, namely R, N, and σ. Therefore, one is faced with three new degrees of freedom when fitting the Fourier filtered EXAFS contribution. Before deducing the presence of i and j in this shell (i.e., formation of a multinuclear *surface* complex), one should demonstrate that two subshells of atom i could not improve the fit just as well. But given the increase in the number of adjustable parameters which results from the introduction of a second and sometimes a third atomic shell, the distinction between i–i + i–j and i–i + i–i models based on amplitude and phase shift contrasts remains difficult and has generally little experimental support when the Z contrast is less than 10. Based on this discussion, data pertaining to the existence of multinuclear metal-ion complexes bonded to the sorbent appear not to be always well sustained.[14,15]

Polyhedral approach. If the prerequisite of the first method is not fulfilled, as in the adsorption of Cr (Z = 24) on Fe (Z = 26) or Mn (Z = 25) hydrous oxides[9,12] or in the structure analysis of multielement minerals,[13] the "polyhedral approach"[16] can be used to determine the nature of cations in the two to three nearest shells. This approach is based on the recognition that in many closely packed structures bond angles depend primarily on the way octahedra are linked to each other. Metal-metal distances are thus determined by the type of linkage and the ionic radius of the metal atom. From the knowledge of the Me-Me distances, it is possible to determine the types of linkages and, consequently, the way a metal sorbs on the adsorbent. Fe(III) and Mn(IV) (oxyhydr)oxides will serve as examples to illustrate this point. In these minerals, Me octahedra can be linked in several ways to build 2D and 3D frameworks. As depicted in Figures 5a and 5b, octahedra can be linked by

- a Single Corner (SC), as in a monodentate mononuclear complex. A given octahedron shares in this case one oxygen with another octahedron.
- a Double Corner (DC), as in a bidentate binuclear complex. A given octahedron shares two oxygens with two different octahedra linked themselves by edges.
- an Edge (E), as in a bidentate mononuclear complex. A given octahedron shares in this case two nearest oxygens with another octahedron.
- a Face (F), as in a tridentate mononuclear complex. An octahedron shares then three nearest neighbors with another octahedron.

These different linkages correspond to decreasing metal-metal distances. For Fe(III) oxyhydroxides they are respectively:[16] 3.87 Å, 3.46 to 3.7 Å, 3.01 to 3.34 Å, and 2.89 Å. The influence of the ionic radius on R_j is easily perceived by comparing Me-Me distances across edges in MnO_2 and FeOOH structures: these are equal to 2.80 to 2.95 Å and 3.01 to 3.34 Å, respectively. In addition, and with the exception of ramsdellite and hollandite, each of these structures can be distinguished according to the number (N_j) and type of octahedral linkages. For instance, goethite has a 4E + 4DC local structure, and hematite has a 1F + 3E + 9DC local structure[17] (see Figure 5b). It results from this analysis that the structure of Fe and Mn oxides can be differentiated on the basis of their Me-Me distances (R_j) and number of Me neighbors (N_j). Accordingly, each of these structures has a distinct RDF (Figure 5).

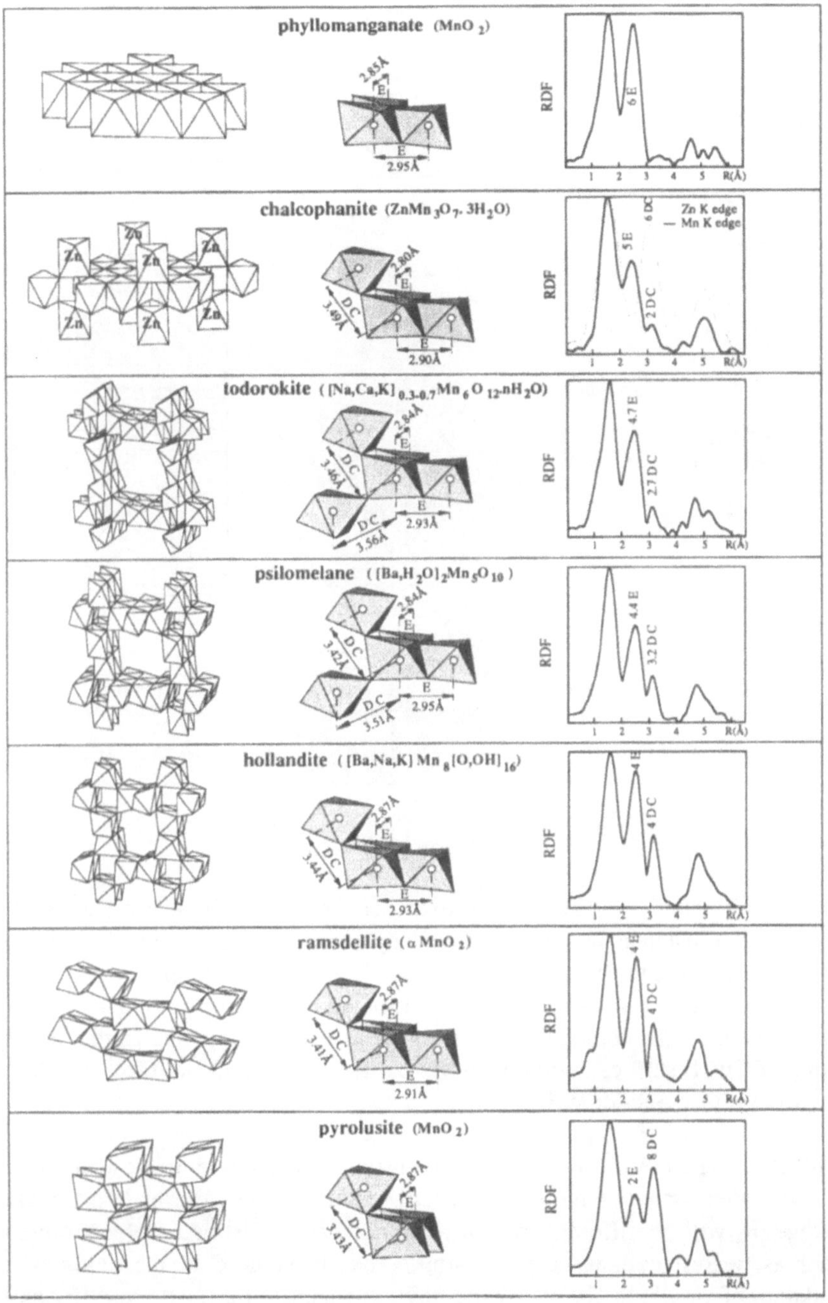

Figure 5. Polyhedral approach of the structure of (a) Mn(IV) and (b) Fe(III) (oxyhydr)oxides. The different structures can be differentiated at the local scale by EXAFS owing to their distinct octahedral linkages and/or metal-metal distances. (Modified from Manceau and Charlet[9] and Charlet and Manceau.[12])

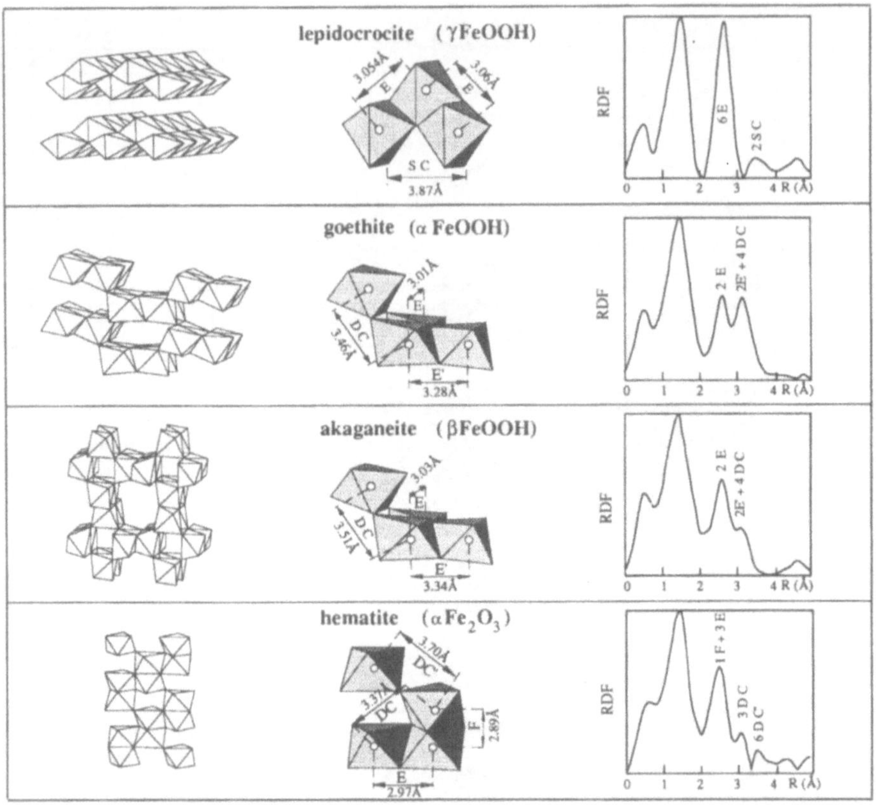

Figure 5(b).

This polyhedral approach can be extended from the bulk to surface structure of solids. For instance, in the case of chalcophanite where Zn atoms lie above and below Mn(IV) vacancies of a MnO_2 layer (Figure 6), the presence of six nearest cations at 3.5Å (Figure 5a) is logically attributed to DC linkages between Zn and Mn octahedra. This approach has also been successfully applied to the determination of Cu(II) surface complex on Mn oxides[18] and Cr(III) surface complex on Fe oxides[12] (see Section 5.3.2).

Contrast in radii. A third method for identifying the nature of nearest cations can be used when i and j have different ionic radii. In that case, and for identical linkage, R_j will be different from R_i, making i and j distinguishable from each other as nearest cations. This possibility has been used in the study of the oxidation of Cr(III) to Cr(VI) by the phyllomanganate birnessite to differentiate Cr(III)-Cr(III) from Cr(III)-Mn(IV) pairs (see Section 5.8.1).[9] In Cr(III) (oxyhydr)oxides, the Cr(III)-Cr(III) distance across an edge is equal to 2.98 to

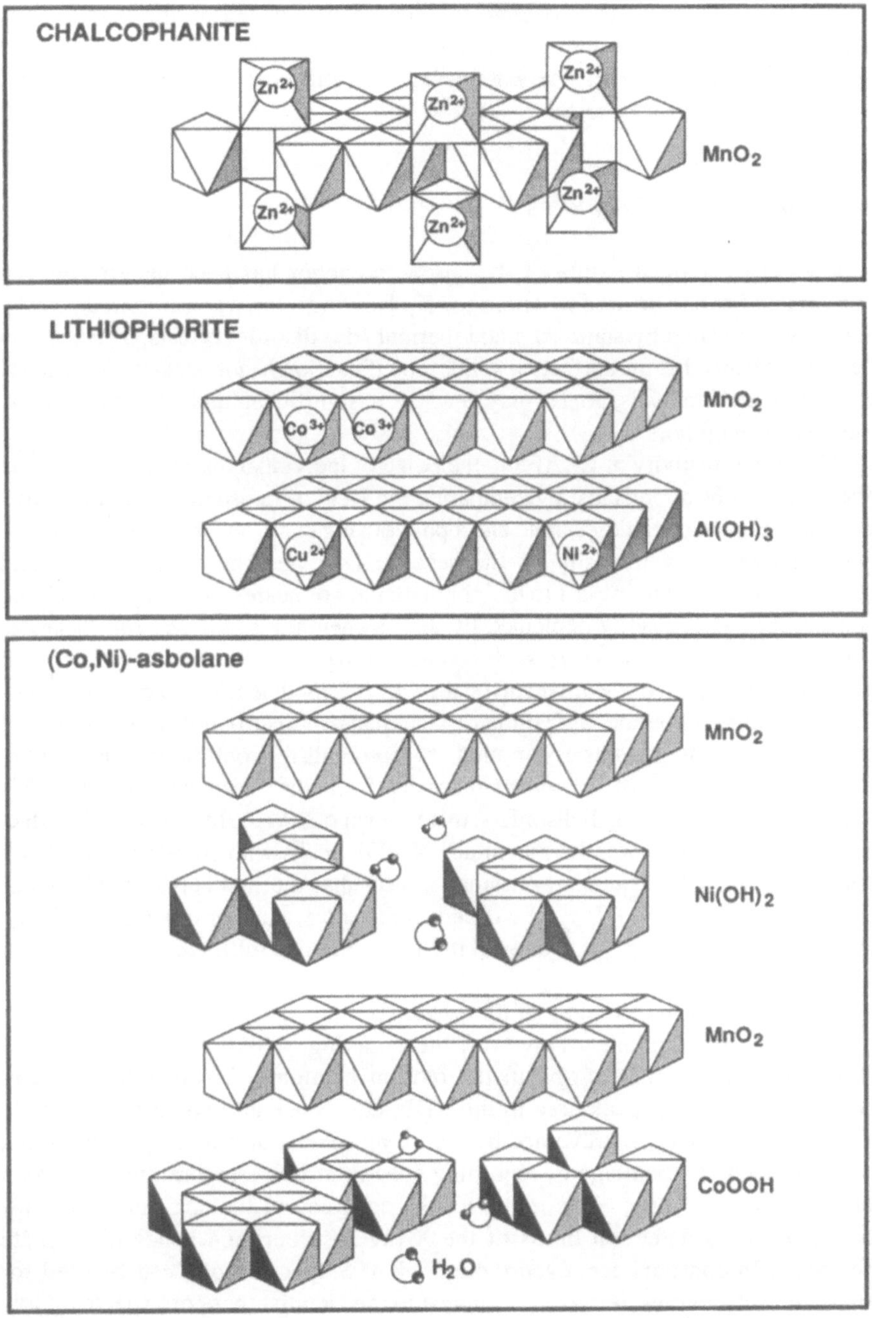

Figure 6. Structure of chalcophanite, lithiophorite, and (Co,Ni) asbolane.[53,58,59]

3.00 Å, which is to be compared to the one known for tetravalent manganates: $\bar{d}$(Mn-Mn) ≃ 2.90 Å (Figure 5a). After sorption, a Cr(III)-Me distance equal to 2.90 Å was found, which was interpreted by a diffusion of Cr^{3+} ions towards Mn(IV) octahedral vacancies.

Number of Atomic Neighbors (N_j)

Once the chemical nature of the atomic neighbor has been determined, the amplitude of χ_j can be used to determine N_j. However, the wave amplitude can be severely attenuated by static (σ_S) and thermal (σ_T) disorder effects. These disorders are enhanced at the surface of solids, and if caution is not taken in the spectral recording and analysis, these may lead to a significant underestimate of the number of neighbors.

The high sensitivity of EXAFS to the range of individual distances in an atomic shell (σ_S) can be appreciated by comparing the RDF of ramsdellite and hollandite, on one hand, and of chalcophanite and a phyllomanganate, on the other (Figure 5a). The relatively low intensity of the second RDF peak of chalcophanite and phyllomanganate with respect to their high number of nearest Mn neighbors (5 and 6, respectively) is readily explained by the incoherency in the Mn-Mn distances across edges (ΔR = 0.1 Å). In such disordered systems, the precision on N_j will depend on the extent of disorder and on the way it is modeled. The expression $\exp(-2\sigma_j^2k^2)$ assumes a continuous symmetric Gaussian pair distribution of the individual distances. According to this assumption, σ_S represents the root-mean-square deviation or displacement of individual distances from the mean distance R_j.[5] This expression is valid for small disorders, that is when $\sigma \leq 0.1$ Å. In that case, the effect of the disorder is to lower the magnitude of RDF peaks with conservation of their peak area. In such a situation one may expect that the number of neighbors as determined by EXAFS (N_{EXAFS}) will reflect the actual number of neighbors in the shell under study ($N_{structural}$). But even if this condition is fulfilled, the precision on N_j rarely exceeds ≃ 20%.

In the general expression of the disorder, the vibration of atoms (σ_T) is assumed to be harmonic.[5] However, the breaking of bonds at solid surfaces results in a severe anisotropy in the motion of atoms.[19] This anharmonicity causes a decrease in peak area in the RDF, and hence an underestimate of N_j. This effect can only be prevented by recording spectra at low temperature (e.g., 77K). A second advantage of recording spectra at low temperature is the gain in signal-to-noise ratio. This improves the accuracy of the structural analysis, but also lowers detection limits of the X-ray absorber, and hence the sorbate coverage. In consequence, even though EXAFS spectroscopy can be used for studying wet samples, it is also of interest to the chemist to record spectra at low temperature. De facto, best results are achieved when recording spectra on wet and cooled samples.

Summary of the Method

EXAFS spectroscopy is a structural method from which distance (R_j), number (N_j), and in favorable cases nature of atoms j located in the vicinity (2 to 4 Å) of the X-ray absorber can be obtained. The precisions on R_j and N_j are of the order of ± 0.02 Å and ± 20%, respectively. Main specificities of the method in regards to environmental particles can be summarized as follows:

- XAS is an element-specific method which can be used for studying compositionally complex particles. Even though any element of the periodic classification can be theoretically studied, in practice only those elements whose edge (K,L) ranges between 1.5 keV and 30 keV can be studied routinely. For K edge, this energy range includes elements from Z = 13 (Al) to Z = 26 (Sb); XAS spectra of heavier elements up to actinides can be recorded at the L (EXAFS, XANES) or M (XANES) edges.
- Structural information extends beyond the first coordination shell. This characteristic explains the growing success of EXAFS for studying the structure of surface complexes. However, the amplitude of the signal is very sensitive to disorder effects, and the experimentalist should be aware that complexes with a strong distribution of Me-Me distances (i.e., distribution in interpolyhedral angles) may not be detected. For geometric reasons, the fewer the number of surface linkages, the greater the degree of freedom of surface complexes. So face (F), edge (E), and double corner (DC) sharings are predominantly detected over single corner sharings (SC). In addition, because of the $1/R_j^2$ dependence of the $\chi_j(K)$ amplitude (Equation 3), and of the intrinsically low value of the photoelectron mean free path (l), the signal associated with single corner linkages ($R_j \simeq 4Å$) is always weak and would require good spectra (i.e., spectra recorded at 77K) to be detected confidently. But this conflicts with one of the main advantages of the methods for the study of environmental systems, i.e., the possibility to study aqueous suspensions. Indeed, the results obtained with dried samples cannot be extrapolated automatically to aqueous solution conditions since drying may lead to a rearrangement of surface complexes, and specifically, to the transformation of monodentate (SC) to bidentate (DC) coordination.
- Solutions as well as poorly- or well-crystallized materials can be studied, since this method does not require any long range order to work. This sensitivity to local order makes it a valuable tool for probing the structure of hydrous particles that are amorphous through X-ray diffraction.[13,17,20]
- In certain conditions, it can be an *in situ* noninvasive method. Material can be studied in its natural aqueous environment (e.g., as wet pastes or in suspension) if the energy of the edge is high enough for the incident beam

not to be absorbed too much by water molecules. High absorption by water decreases the beam intensity (I), and thus, the signal-to-noise ratio. To provide the reader with some orders of magnitude water thickness of 0.45 mm and 18 mm reduce the beam intensity by a factor of two at the Fe (7 keV), and Cd (27 keV) K-edges, respectively. For energies lower than about 4 keV ($Z \leq 20$ at the K edge and $Z \leq 51$ at the L_{III} edge), the absorption by air becomes significant, and experiments are conducted in vacuum.

- Detection limits depend on many factors and cannot be given in an absolute manner. Some parameters which determine the detection limit are (1) the intensity of the X-ray beam (the intensity can vary by several orders of magnitude from a beam line to another); (2) the chemical composition of the matrix, which determines the background noise; (3) the nature of the element (fluorescence yield is a monotonically increasing function of Z); and (4) efficiency, energy resolution, and counting rate of the detector. To give some values, with the present technology detection thresholds of high Z elements can be as low as 0.03 to 0.05% per weight (metal adsorbed per metal in the adsorbent). Less favorable cases correspond to systems where a Z element is diluted in a matrix containing a Z-1 element, because at the Z edge the fluorescence of the Z-1 element drastically increases the background noise.
- Collecting a spectrum takes from a few tenths of seconds to several hours depending on the experimental arrangement and the concentration of the element. Classically, for concentrations higher than 2 to 3%, collecting a spectrum lasts 15 to 20 min. New experimental spectrometers with faster data collection have emerged recently. But the gain in speed has been realized at the expense of the detection threshold. In the dispersive mode,[21] the incident beam is focused on a point smaller than 1 mm^2 which allows study of small samples (few µg compared to few mg in the classical mode). In addition, this method speeds data collection greatly allowing study of transient phenomena down to 10^{-1} s. The quick-EXAFS (Q-EXAFS) experiment makes use of a classical step-scanning monochromator, but the electronics of data collection are fast; the time required for measuring a spectrum is between 1 and 60 s.[22]
- One of the major limitations of XAS is its lack of spatial resolution in comparison to electron diffraction, for example. This limits its use to materials where the X-ray absorber phase is unique, but not necessarily to pure samples. If X-ray absorber sites are several, such as in a sorption experiment where solely 50% of the sorbate would have been sorbed, the XAS spectrum would be a weighted average of all the local structures of the sorbate.

In conclusion, relevance of XAS spectroscopy to the study of environmental particles is twofold. On the one hand, particles formed in low temperature environments are often poorly crystallized. Their local structure is not accessible to X-ray diffraction but can be obtained for colloids suspended in water at room temperature from XAS data interpreted by the polyhedral approach. On the other hand, environmental particles often contain a variety of elements as a result of coprecipitation, adsorption, substitution, or diffusion phenomena. Through the atomic selectivity, information can be obtained on the way foreign elements are trapped by host particles.

FORMATION AND STRUCTURE OF PARTICLES

Information on the shape, size, and density of particles is usually derived from scattering methods such as light and small angle X-ray scattering.[23,24] But the determination of their structure is frequently hindered by their long range disorder. Indeed, these particles often appear as "amorphous" through XRD since the size of their coherent scattering domains[25] is not high enough (< to a few tens of thousands of $Å^3$) to give rise to well-defined *hkl* reflections by X-ray diffraction lines. Because of its intrinsically high sensitivity to short range order, XAS is the method of choice to probe the short range (a few Å) organization of these particles and to follow structural changes during their evolution in low temperature environments. This scale of investigation is important, as most geochemical processes, such as nucleation, particle growth, dissolution, and trace element trapping, proceed by elementary chemical reactions.

Structure of Fe-Based Colloids

Little is known about the structure of soluble metastable species formed in oxic environments during the hydrolysis of Fe^{3+} ions and in reducing conditions during the formation of Fe(II) sulfide minerals.

Owing to the abundance of iron in the surficial environment and to the high surface reactivity of its metastable species, the elucidation of the local structure of iron (III) hydroxide polymers deserves particular attention. The Fe hydroxy case study provides a good example of the possibilities offered by XAS spectroscopy, in particular when it is combined with other techniques. XAS studies have shown that Fe(III) remains sixfold coordinated up to completion of the hydrolysis process.[11,20,26] As hydrolysis starts, and as the hydrolysis ratio ($[OH^-]_T/[Fe]_T$) remains below 1, no structural order is detected beyond the first coordination sphere of the Fe atoms. At larger hydrolysis ratios, two Fe-Fe contributions at 3.05 Å and 3.44 Å are detected. These indicate the presence of small polymers in which Fe octahedra are sharing edges and double corners. These low molecular weight Fe polymers are well-ordered at the local scale and possess a local structure similar to that in α-FeOOH/β-FeOOH (Figures 5b and 7a). This arrangement is preserved up to completion of the hydrolysis and during the ripening and crystallization of this gel.

Longer-range changes in the binding of the Fe(III) low molecular weight polymers has been investigated by small angle X-ray scattering (SAXS).[23,24] A

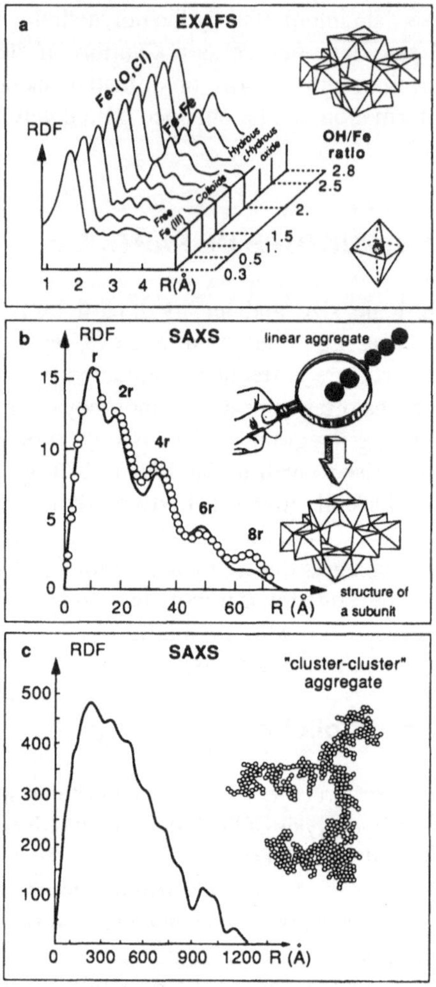

Figure 7. Homogeneous nucleation of hydrous ferric oxide from a ferric solution. (a) XAS radial distribution function (uncorrected for phase shift) showing changes in the short range order around iron during the hydrolysis, and structural model of the local structure of Fe polymers. (b) and (c) Small angle X-ray scattering-derived distance distribution function of Fe colloids at different hydrolysis ratio. (b) NaOH/Fe = 1.0, (c) NaOH/Fe = 2.6.[11,23,24]

Fourier transform of the SAXS spectra yields a radial distribution function which now represents the probability of finding a particle in an aggregate at a distance R from another particle. RDF of low-weight polymers shows several peaks, revealing the presence of subunits within aggregates (Figure 7b). Analysis of data in both the inverse ($Å^{-1}$) and direct (Å) spaces leads to the conclusion that subunits possess a spherical shape whose diameter is equal to 16 Å. In low weight polymers, these subunits are arranged linearly, the stability of aggregates being due to long-range magnetic dipolar interactions. With increasing hydrolysis, and as the fractal dimension reaches 1.7 to 1.8, the aggregation leads to "cluster-cluster" type polymers (Figure 7c) owing to the decrease of surface charge. When the fractal dimension reaches the "percolation threshold," flocculation occurs, leading to a hydrous ferric oxide (HFO) phase. The two complementary structural approaches, i.e., EXAFS and SAXS, have confirmed the existence of a structural continuity between aqueous Fe^{3+} species and ferric hydrous oxide gels all along the hydrolysis process.[27] Recently the same approach has been applied to follow the formation of Fe(II) sulfide minerals. Amorphous 'FeS' is a metastable precursor to the formation of mackinawite: $(Fe,Ni)_9S_8$.[28] An XAS study has shown that both solid phases are characterized by the same local order, i.e., by Fe tetrahedra sharing edges in a sheet structure.[29] But, in contrast to the pathway leading to goethite, a structural gap exists between the low temperature precursors, 'FeS' and mackinawite, and the product of their burial diagenesis, pyrite. In this mineral, Fe(II) is no more in tetrahedral coordination but is instead found in an octahedral environment.[29]

Structure of Fe and Mn Hydrous Oxides

Hydrous gels of iron and manganese are ubiquitous near the Earth's surface. They are found in temperate and tropical soils, in stream and groundwaters, and in lake and marine sediments. For a long time, hydrous ferric and manganese oxides were thought to be "amorphous" in their highest disordered state since their XRD patterns display only two broad bands at about 2.5 Å and 1.4 to 1.5 Å. Owing to the limited number of X-ray reflections, XRD has very low sensitivity to their actual structure, and several contradicting models have been proposed.[30-33] The similarity of XRD curves for all these compounds results from the small size of their coherent scattering domains. These domains are composed of mixed cubic and hexagonal anionic packings where each pair of anionic layers contains, on average, the same number of cations.[34] The main disagreement among proposed models concerns the distribution of Fe or Mn atoms among available tetrahedral and octahedral sites.

XANES spectroscopy at both the Mn and Fe K-edges has shown that cations only fill octahedral sites.[7,8] But more importantly, EXAFS studies have shown that these featureless XRD patterns may correspond to at least five distinct local structures owing to the way in which the gel has been formed. Freshly precipitated ferric gels obtained after the complete hydrolysis and oxidation at neutral pH of ferrous sulphate or chloride solutions possess a lepidocrocite-like local structure.[26] Instead, when "2-line" gels are synthesized from a nitrate or chlorite ferric salt, these possess in suspension a goethite-like (αFeOOH) or akaganeite-like (βFeOOH) local structure, respectively, and they rapidly convert upon aging at neutral pH or heating at 92°C into a feroxyhite-like (δFeOOH) "2-line" gel, which further transforms into hematite.[11,17,20] This last local structure has been encountered in all of the natural "2-line" Fe-gels (the so-called protoferrihydrite) analyzed up to now. 2-line hydrous manganese oxides, called vernadite (δMnO_2), are also well ordered at the local scale.[13] But in contrast to hydrous Fe oxides, their local structure does not seem to be related to that of a well-crystallized MnO_2 polymorph. It most probably consists of a 3D-framework of randomly distributed edge- and corner-sharing MnO_2 octahedra.[13] This fundamental difference between the structure of Fe and Mn gels might explain why Mn^{4+} gels may further transform into large varieties of MnO_2 polymorphs (Figure 5a).

These XAS results clearly point to the limit of the XRD techniques for analyzing the structure of hydrous oxides. Structural interpretations of these featureless diffractograms are hardly feasible and not always unique. Access to local structures provide severe constrains for interpreting XRD features, and new structural models that satisfy both XRD and XAS results have emerged recently.[13,34]

ADSORPTION AND THE FORMATION OF ISOLATED SURFACE COMPLEXES

Sorption on a mineral surface may result from various microscopic mechanisms (Figure 8):[35] (1) when an ion is attracted to the surface via long range coulombic forces but retains its water of hydration, it either stands within the diffuse ion swarm or forms an outer-sphere complex with the surface reactive group; (2) when the ion loses some of its hydration water molecules and becomes directly bound to the surface by short-range forces (chemical bond), it forms an inner-sphere surface complex with the surface reactive groups; (3) when this inner-sphere complex involves sorbed polymers, surface nucleation and eventually precipitation has occurred; finally, (4) when the sorbed metal ion is found within the sorbent matrix, lattice diffusion and/or coprecipitation may have occurred. These different mechanisms are often indistinguishable by macroscopic studies,[36,37] but can be discriminated by EXAFS, as is exemplified hereafter.

Outer-Sphere Complexes

Ions in outer-sphere complexes differ from those in the diffuse swarm only by the time the ion remains immobilized on the particle surface. The former species remains immobilized over time scales that are long when compared with the ca. 10 ps required for a diffusive step by a solvated ion in aqueous solution.[35] Since the EXAFS phenomenon takes place in a time scale (about 10^{-16} s) much shorter than the time scale for diffusion of the sorbate ion on a length equal to its diameter, EXAFS works like an instantaneous snapshot of atomic configurations. However, EXAFS experiments require a much longer time than diffusion, which means that all atomic configurations are averaged, and the two types of species cannot be distinguished. Furthermore, since the method is a bulk method, it cannot lead to a description of the double layer structure, such as that recently obtained by X-ray standing waves.[38] However, EXAFS allows an unambiguous distinction to be made between the above two types of species and the surface species which involve the formation of a chemical bond with the surface.

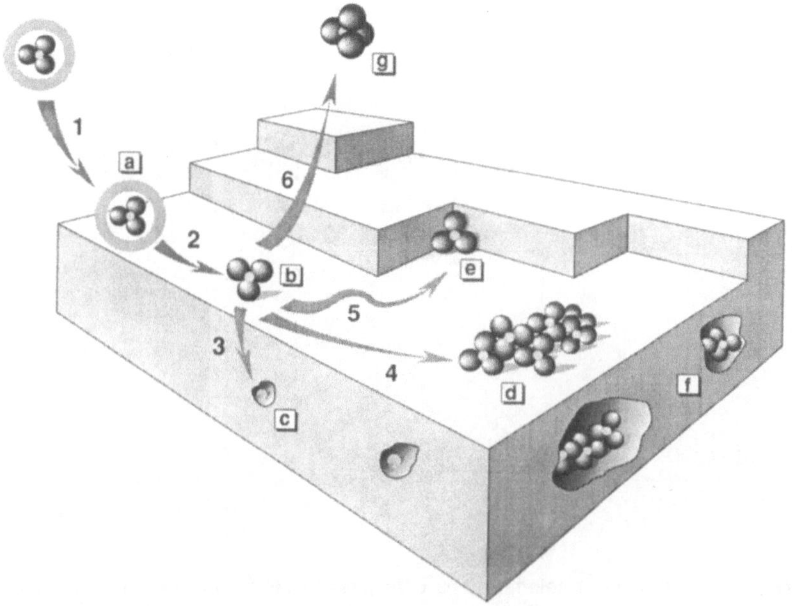

Figure 8. Various mechanisms of sorption of an ion at the mineral/water interface: (1) adsorption of an ion via formation of an outer-sphere complex (a); (2) loss of hydration water and formation of an inner-sphere complex (b); (3) lattice diffusion and isomorphic substitution within the mineral lattice (c); (4) and (5) rapid lateral diffusion and formation either of a surface polymer (d), or adsorption on a ledge (which maximizes the number of bonds to the adatom) (e). Upon particle growth, surface polymers end up embedded in the lattice structure (f); finally, the adsorbed ion can diffuse back in solution, either as a result of dynamic equilibrium or as a product of surface redox reactions (g).

Anions such as selenate (SeO_4^{2-}) do not seem to have an organized inner hydration sphere. The RDF of selenate ions sorbed on goethite contains only one peak, that of the oxyanion oxygen atoms (Figure 9).[39] This anion is retained at the surface, either as an outer-sphere surface complex or within the diffuse ion swarm.

Adsorption in the diffuse ion swarm or via formation of the outer-sphere surface complex occurs almost exclusively by electrostatic interactions. According to mass action laws, these "nonspecific" adsorptions will generally be ionic strength dependent. The higher the ionic strength becomes, the higher the concentration of background ions and the lower the amount of selenate ions being sorbed.[40] On the contrary, adsorption which results from the formation of inner-sphere surface complexes is to a large extent ionic strength independent.

Sorption of uranyl ions on montmorillonite[41] is the only documented case of nonspecific adsorption. According to the hard and soft acid and base principle, "nonspecific" adsorption concerns usually hard acid. Monovalent and divalent hard acid cations are formed with light elements, which are the most difficult to study by EXAFS. Readers interested in the study of such adsorption by other more appropriate spectroscopic methods may wish to refer to Sposito and Prost.[42]

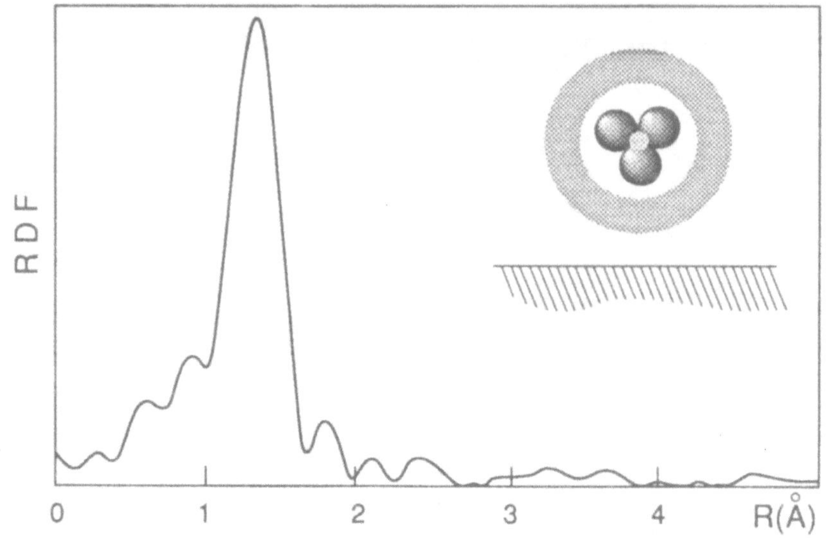

Figure 9. Adsorption of selenate ions onto goethite. RDF and model of outer-sphere complex and/or adsorption within the diffuse layer. (Modified from Hayes et al.[39])

Isolated Inner-Sphere Surface Complexes

Sorption sites of anions and cations. The presence in the RDF of additional peaks beyond the first or, eventually, the two first oxygen one(s) is indicative of an organized local structure which extends beyond the hydration sphere. Evidence for the formation of isolated inner-sphere surface complexes in which the sorbed ions are bound to the surface without polymerizing to each other has been obtained in the following adsorption case studies: SeO_3^{2-} on goethite,[39] SeO_3^{2-},[36] AsO_4^{2-},[43] UO_2^{2+},[36,44] and Pb^{2+} [36] on hydrous ferric oxide, and Cu^{2+} on MnO_2.[18]

XAS experiments conducted on selenite and arsenate complexes indicate the formation of binuclear bidentate complexes on hydrous ferric oxides, at least at high surface coverage (DC linkage). On ferrihydrite and at low surface coverage, arsenate partly forms mononuclear monodentate complexes (SC linkage). XAS[36,39,43] and infrared spectroscopy[45,46,51] have now clearly established that, on both goethite and ferrihydrite, anions mainly attach to two singly coordinated OH groups (A-type OH groups) through a double corner (DC) linkage (Figure 10). Surface OH groups linked to two (C-type) or three (B-type) edge-sharing Fe octahedra are not active with respect to anions.

Hydrous ferric oxides and hydrous manganese oxides are thought to be important scavengers of U and Pb in many natural systems.[47,48] Laboratory experiments have shown that the enrichment factor[49] of uranyl is about 500 times greater on Fe gels than on well-crystallized goethite. The difference of surface area between ferrihydrite (350 ± 50 m^2/g) and goethite (25 ± 15 m^2/g) is not high enough to explain alone this very large variation of affinity. Furthermore this result could be

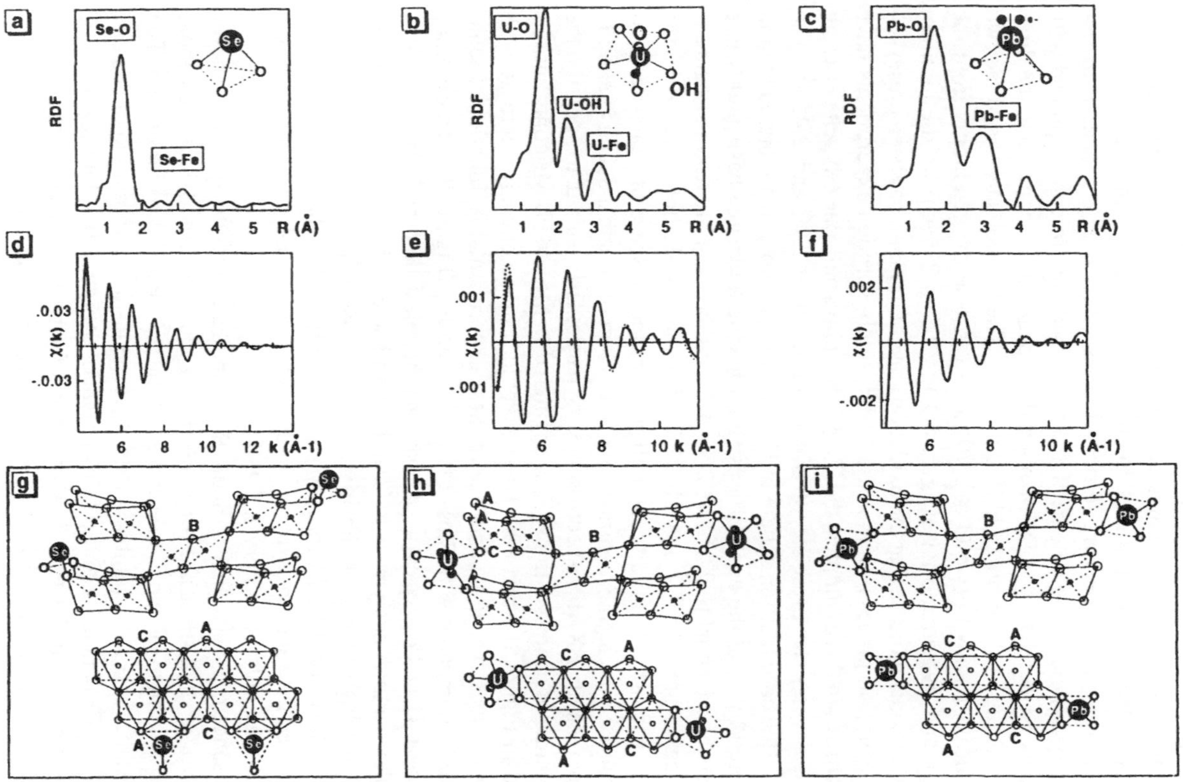

Figure 10. From left to right: adsorption of SeO_3^{2-}, UO_2^{2+}, and Pb^{2+} on hydrous ferric oxide. From top to bottom: Se-, U-, and Pb-RDF; Fourier filtered Se-Fe, U-Fe, and Pb-Fe contributions to EXAFS; modelization of the sorption mechanism. A, C, and B are respectively 1-, 2-, and 3-coordinated OH groups. (Modified from Manceau et al.[36])

considered at variance with the αFeOOH/βFeOOH-like local structure of hydrous ferric oxides. We will see later on how the knowledge of the sorption mechanism of UO_2^{2+} and Pb^{2+} ions sheds direct light on this apparent contradiction and accounts for the tremendous cation affinity of hydrous ferric oxides.

In the case of uranium, two kinds of experiments have been conducted:[44] (1) sorption onto freshly precipitated hydrous ferric oxides and (2) sorption/coprecipitation in uranyl-bearing ferric solutions. Only minor structural differences have been detected in whether UO_2^{2+} is sorbed onto or coprecipitated with hydrous ferric oxides. The U-RDF displays three peaks corresponding to the three nearest coordination shells about U (Figure 10b). The Fourier analysis of the two first RDF peaks results in two O at 1.79 to 1.81 Å (the oxygen atoms from the uranyl ion), and five equatorial oxygens (O_e) at 2.3 to 2.4 Å from the uranium atom. These results indicate that the oxidation state of uranium has been preserved during the sorption process. In the case of Pb, solely sorption experiments have been performed. In these samples, the Pb-RDF displays only one oxygen peak, but its analysis reveals the presence of two O subshells at 2.22 and 2.42 Å. The geometry of the coordination sphere of Pb is extremely close to the one known in βPbO where Pb^{2+} is at the apex of a pyramid having a lozenge basis and where Pb-O distances are equal to 2.21 and 2.49 Å.[50] But the most interesting feature is the presence in both the U- and Pb-RDF of a third peak, not detected in the hydrous Fe oxide-free solutions, and indicative of the presence of nearest cations (Me). Because of the high Z contrast between Fe, on the one hand, and U and Pb, on the other hand, EXAFS spectroscopy is particularly suited to identify the chemical nature of nearest cations. Fourier filtered U-Me and Pb-Me contributions are plotted in Figure 10e and 10f. Their envelopes monotonically decrease with increasing k values. The k dependence of these experimental A_j functions (Equation 3) is to be compared with those of F_{Fe}, F_U, and F_{Pb} (Figure 4b): the presence of multinuclear U or Pb species would have resulted in a deep node beat at 6 $Å^{-1}$. This result unambiguously rejects U and Pb and selects Fe as a nearest cation. Owing to this high chemical sensitivity, surface precipitates and multinuclear surface complexes of the adsorbate ion are excluded. This shows that the complexes involved in U and Pb sorption are distinct from the polynuclear complexes found during hydrolysis and crystallization.

The presence of a wave beat at about 9 $Å^{-1}$ indicates that the Fe shell is never unique. Two Fe shell spectral fittings result in about 0.5 Fe at 3.3_1 Å plus about 1 Fe at 3.4_8 Å around U, and 0.3 Fe at 3.2_9 Å and 0.4 Fe at 3.4_5 Å around Pb. The shortest distance at 3.3 Å unambiguously corresponds to the formation of an edge bond between Fe octahedra and (U,Pb) polyhedra (Figure 10). Interpretation of the second distance at 3.4 to 3.5 Å in terms of type of surface linkage is not straightforward. This distance has been attributed in the past to the existence of a second, structurally distinct, edge surface site.[36] But recent works performed on the sorption of Cd on goethite and hydrous ferric oxides at varying surface coverages have led to revise this interpretation and to favor the formation of double corner linkages (Spadini et al., in preparation).

Effect of adsorption on crystallinity. Cr(III) and Fe(III) ions have similar radii, and their octahedra may be bound in the same way. The presence of Cr(III) ions does not affect the precipitation of hydrous ferric oxide. This "coprecipitation" process leads to a $(Cr,Fe)(OH)_3$ solid solution, which, after a few days, evolves into a well-crystallized chromiferous goethite.[12] The dissolution rate of this phase is smaller than for pure goethite.[52]

Conversely, adsorbed ions which do not isomorphically substitute for Fe(III) may affect the transformation both of the 16 Å polymers into hydrous ferric oxide gel and of this gel into well-crystallized goethite. This poisoning by sorbed species is thought to be different for cations and anions. Anions, which bond to the apices of Fe octahedra (DC linkage), should block the 3-D condensation of octahedral chains whereas cations should, in addition, poison the growth of octahedral chains along their length since they share also edges with Fe octahedra. Such a poisoning effect has been identified for As-rich particles, which were shown to possess a much longer settling time than As-free particles.[43]

Effect of crystallinity on adsorption. The adsorption mechanisms of Cr(III) on goethite and hydrous ferric oxide have been compared.[12] On goethite, Cr(III) octahedra mainly share double corners with Fe octahedra and, possibly, a few edges. Conversely, on ferrihydrite, Cr(III) octahedra are principally engaged in edge linkages, and little double corner sharing is detected. It will be shown hereafter how the change in the relative proportion of binding sites is thought to reflect the relative density of surface sites. Goethite structure consists of infinite double chains, and the density of A-type sites located along the chains is high. Given the needle shape of goethite crystallites, the active unshared edges which are essentially located at the termination of the infinite double chains are seldom present on goethite. In contrast, hydrous ferric oxides have a disordered goethite-like structure, which means that octahedral chains are short in length (low crystallinity), and possibly part of them are single chains. This results in a tremendous increase in chain terminations and hence, in density of unshared edges available for cation adsorption. At the same time, the density of A-type OH sites, which are located along octahedral chains, are expected to be less affected in going from goethite to hydrous ferric oxides. This change of the relative density of surface sites with crystallinity, and more precisely the sharp decrease of the density of free edges upon ferrihydrite transformation is thought to explain, in addition to the variation of the specific surface area, the 500 times increase of enrichment factor for uranyl between goethite and hydrous ferric oxides.[49] XAS spectroscopy allows an interpretation at the molecular level of these specific changes measured macroscopically.

Natural samples. Isolated inner-sphere surface complexes have been found in a natural sample, such as chalcophanite, a Zn-rich phyllomanganate. As a phyllomanganate, it is composed, like birnessite, of layers of Mn(IV) octahedra (Figures 5a and 6) in which one out of seven Mn atoms is vacant. EXAFS data for chalcophanite[53] compares well to that obtained for a Cu(II)-saturated birnessite sample.[18] Zn- and Cu-RDFs have two peaks corresponding to sorbent O (d(Zn-O) = 1.90 Å and d(Cu-O) = 1.85 Å) and Mn (d(Zn-Mn) = 3.55 Å) and d(Cu-Mn) = 3.3 Å) neighbors (Figure 5a). These distances indicate that the Zn octahedra are bound by double corners to six basal Mn octahedra. Zn^{2+} and Cu^{2+} cations lie above and below Mn(IV) vacancies, thus compensating locally for the charge deficiency of the layer. The formation of these inner-sphere complexes between Zn or Cu and the birnessite-like structure explains the low ionic strength dependence of Zn and Cu adsorption onto δMnO_2.[55]

PATHWAYS OF SOLID SOLUTION FORMATION

Natural particles often include trace amounts of foreign metals in isomorphic substitution. Even though X-ray diffraction has proven fruitful for determining the extent of lattice substitutions (i.e., solid solutions), it is of little help for determining the site location of trace elements, and specifically when the host material is "amorphous." XAS is instead a unique structural tool for deciphering the actual structure of particles in the vicinity of substituting elements.

Natural samples. Case studies of isomorphic substitutions investigated by EXAFS include strontium substitution for calcium into coral calcite,[56] germanium (IV) substitution for Fe in goethite with appropriate local charge compensations,[57] and Ni, Cu, and Co substitution in lithiophorite.[58,59] Lithiophorite has a strong tendency to host 3d elements because it possesses a hybrid structure which consists of alternating MnO_2 and $Al(OH)_3$ layers.[60,61] Due to the high Z contrast between Al and Mn, it was possible to show by EXAFS spectroscopy that Ni and Cu are located within the dioctahedral $Al(OH)_3$ layers, whereas low-spin Co(III) either substitutes for Mn within the phyllomanganate layer or builds CoOOH domains (Figure 6).

Adsorption during "coprecipitation" processes. Coprecipitated Fe(III) and Cr(III) hydrous oxide gels are used in water treatment plants and in analytical chemistry to remove chromium from industrial effluent waters and other dilute solutions. The solubility of these gels is at least an order of magnitude lower than the solubility of chromium hydrous oxides obtained from homogeneous precipitation.[62] The Cr- and Fe-RDF of the coprecipitated gel are very similar to each other (Figure 11a).[12] Both display two metal-metal peaks beyond the first metal-oxygen one. The identity of their R position and of their relative intensity indicates that Cr(III) and Fe(III) possess a similar structure, that is, Cr(III) substitutes for Fe(III) in the αFeOOH-like framework. This isomorphic substitution is accounted for by the similarities of the electric charges and ionic radii of Cr^{3+} and Fe^{3+} and by the existence of isostructural Fe(III) and Cr(III) oxide polymorphs. On the other hand, the Cr hydrous oxide obtained from homogeneous or heterogeneous precipitation have a γCrOOH local structure (Figure 11b and 11c). The difference in efficiency of chromium extraction can be therefore attributed to the formation of different chromium phases. The solubility of the chromium coprecipitate (α(Fe,Cr)OOH) is clearly less than in the pure chromium oxyhydroxide (γCrOOH). The efficiency of the coprecipitation phenomenon explains, in turn, why in soils Cr mostly substitutes for Fe in iron oxyhydroxides, as evidenced by sequential chemical extraction.[63]

Fe(III) and Cr(III) are coprecipitated by raising the pH of a mixed, initially acid, solution. Due to the difference in their hydrolysis constant, precipitation of Fe(III) occurs at a much lower pH than that of Cr(III), and is characterized by the formation of stable polymers, the "16 Å precursor particles" (Figure 7) in conditions where Cr(III) is present as free ion ($pH < 2$). Therefore the "coprecipitation" phenomenon which occurs as the pH is raised is an adsorption of Cr(III) free ions onto the already formed Fe(III) colloids, rather than a coalescence of Fe(III) and Cr(III) isolated octahedra. This may involve up to 0.3 chromium atoms per Fe atom.[12]

Diffusion within a crystal lattice. Once adsorbed on a mineral surface, a cation may diffuse into the mineral lattice, provided its ionic radius is close enough to that of a lattice cation or to the size of a cavity. Two types of diffusion have been documented: diffusion within the phyllosilicate siloxane cavities and diffusion within the mineral lattice of phyllomanganate, calcite, and apatite.

The diffusion of cations and in particular that of K^+ and Cs^+ in the siloxane cavity of in phyllosilicates has been intensively studied by different spectroscopic methods.[64,65] Cs^+, for example, due to its radius (1.96 Å), can penetrate the siloxane cavity of smectites but not that of vermiculite whose diameter is 1.67 Å.[65]

Recently, the solid diffusion of Cr^{3+} within Mn(IV) vacancies of birnessite, a phyllomanganate, has been identified by EXAFS.[9] However, the product of the reaction is unstable. After electron transfer between Cr(III) and Mn(IV), the product of the reaction, the chromate oxyanion, is repelled away from the negatively charged birnessite surface (see Section 5.8.1).

Conversely, the adsorption and surface oxidation of Co(II) (the oxidation is attested both by XPS measurements[66] and by the large Mn(II) release[68]) may result in the diffusion of Co(III) within the phyllomanganate vacancies. This possible lattice diffusion is supported by the fact that in natural Co-containing Mn oxides (asbolane, lithiophorite), Co(III) and Mn(IV) always occupy equivalent lattice positions (Figure 6).[58] The Co for Mn substitution is sterically feasible owing to the equivalence of the ionic radii of Mn^{4+} (0.54 Å) and low spin Co^{3+} (0.53 Å) ions.[68] A similar oxydo-reduction process at the surface of Mn(IV) oxides has been reported recently in the case of Ce(III) and Tl(I).[10,69] The product of the reaction has been identified by XANES as Ce(IV) and Tl(III), respectively, which are, as is cobalt, much less soluble than their reduced form. These products remain therefore sorbed onto the manganese oxide surface.

Preliminary EXAFS results of Co(II) sorbed onto calcite indicate that solid diffusion may be involved in the removal of cobalt from solution, although dynamic dissolution/reprecipitation equilibrium and diffusion along microfractures

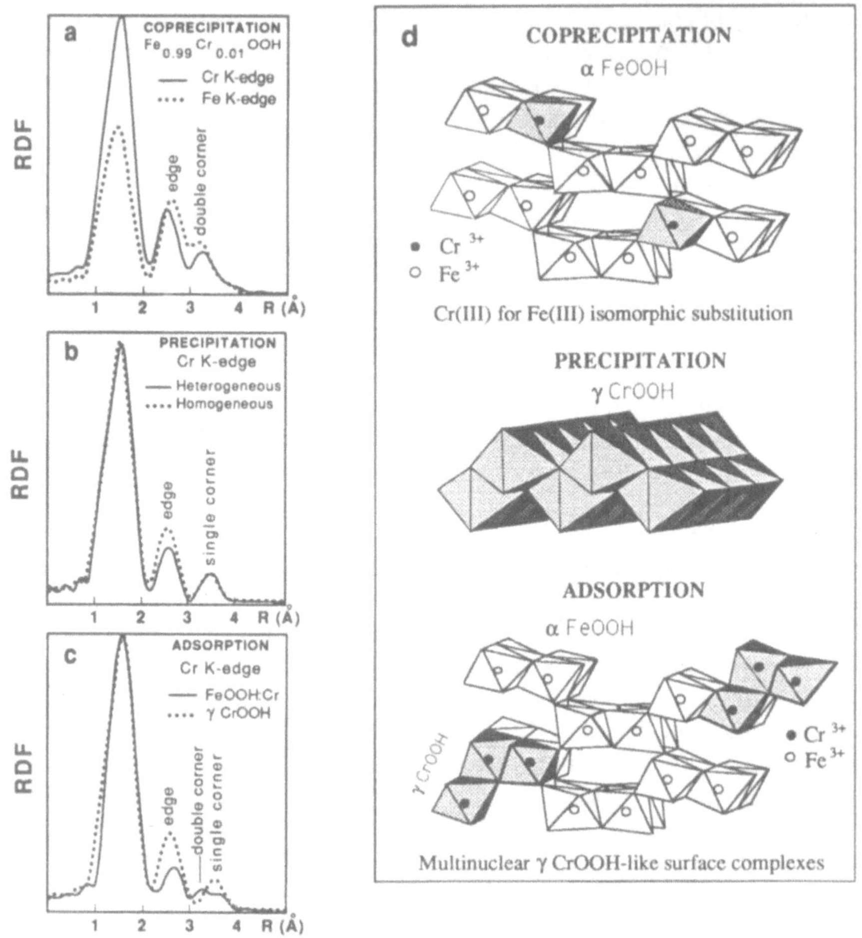

Figure 11. Fe(III)-Cr(III) coprecipitation versus Cr(III) adsorption onto hydrous ferric oxide. In the former mechanism, Fe and Cr atoms form an α(Fe,Cr)OOH solid solution, whereas in the latter, Cr builds multinuclear γCrOOH-like surface complexes. (Modified from Charlet and Manceau.[12])

could not be excluded in this study.[70] Solid state diffusion has been proven recently to occur within a few days by X-ray photoelectron spectroscopy (XPS) and time resolved laser fluorescence (TRLF) measurements. XPS spectra of Cd(II) sorbed at the surface of Iceland spar calcite crystals indicate that Cd disappears from the surface top 100 Å after only a month of storage of the crystal in vacuum.[71] Diffusion has been followed by TRLF in the case of Ce(III) sorbed onto apatite. The macroscopic removal of Ce(III) from solution is completed within a few minutes. However, the TRLF signal, indicative of Ce(III) present within the bulk of the mineral, takes days to reach a constant value.[10] In conclusion, since the kinetics of these mass transfer processes are much slower than those of adsorption, a single isotherm may no longer predict the behavior of the sorbate-sorbent system over a variety of time scales. This is a very important consideration when extrapolating experimental results of sorption studies to natural water-rock environments.

The phase formed by solid diffusion or adsorption on polymers cannot be distinguished, afterwards, from solids formed by true "coprecipitation" processes. Therefore, the distinction between the different modes of formation of solid solutions (namely solid diffusion, adsorption on polymers, and "true" coprecipitation) is only possible with laboratory systems. "True" coprecipitation process should seldom occur in nature since it would require very close precipitation edges for the two coprecipitated phases.

FROM ADSORPTION TO SURFACE POLYMERIZATION AND PRECIPITATION

In several XAS studies of the adsorption of hydrolyzable cations (Pb^{2+}, Co^{2+}, Cr^{3+}) precipitation has been reported. This precipitation had been induced by the mineral surface. Before reviewing these studies and the catalytic role of mineral surfaces in cation polymerization, it seems relevant to recall some macroscopic conditions which must be fulfilled in order for surface precipitation *not* to occur. As we will see, these conditions have been overlooked often in the experimental design of many XAS "adsorption" studies.

Conditions for Near Surface Precipitation

Two types of oversaturation may lead to surface polymerization and/or surface precipitation: (1) an oversaturation with respect to the surface sites (which occurs whenever the sorbate surface excess becomes larger than the density of surface site) and (2) a saturation with respect to the solubility of a new phase (which occurs when the ion activity product, IAP, for this phase becomes equal to its solubility product, K_{so}).

Oversaturation with respect to the surface sites. Two types of surface site occur at mineral-water interfaces, which differ by the origin of their charge and by the tools which can be used to determine their density.

The first type of surface site is a consequence of the charge created by vacancies (e.g., in birnessite) or isomorphic substitution by a heterovalent ion (e.g., Mg^{2+}/Al^{3+}, Fe^{2+}/Fe^{3+} in phyllosilicates, $Mn^{4+}/Mn^{3+}/Mn^{2+}$ in Mn oxides) within the lattice. Measurement of this surface lattice (or "permanent") charge density commonly involves that of the associated "cation (or anion) exchange capacity."[65] In phyllosilicates and phyllomanganates, these surface sites are a cavity bordered by six corner-sharing silica tetrahedra and six edge-sharing manganese octahedra, respectively. These cavities are bordered by six and three sets of lone-pair electron orbitals respectively emanating from the surrounding ring of oxygen atoms.

The second type of surface site originates from the breaking of a mineral lattice. Upon exposure to water, surface hydroxyl groups are formed by dissociative adsorption of water molecules, a process energetically favored because a better charge neutralization in the lattice is achieved.[72] The primitive surface species is denoted >Fe-OH in the case of an iron (oxyhydr)oxide surface, where > represents the bulk mineral lattice. Such a species undergoes protonation-deprotonation reactions, as well as complexation reactions with cations and anions. The measurement of the density of these sites has been reviewed by James and Parks[73] and by Davis and Kent.[74]

A close-packed layer of water molecules contains about 10 oxygen atoms per nm^2. This should be the upper limit of any surface site density. Any overestimation of this density, such as the use of 17 sites nm^{-2} (a value claimed to be derived for goethite from crystallographic considerations), leads to an underestimation of the degree of cation coverage. Furthermore, depending on the method used to get it (on the adsorbate used), the density of reactive surface sites varies, for instance for goethite, between 1 and 7 sites per nm^2.[2,73,74] Facing this intrinsic uncertainty, the criterion for surface site saturation is difficult to use as evidence for surface precipitation.

However, for a given sensitivity level of the XAS method, i.e., for a given adsorbate (Pb) to adsorbent (Fe) concentration ratio, the undersaturation criterion is more easily achieved on an amorphous phase than on the well-crystallized analogue, since the former develop a much larger specific surface area. For instance, the specific surface area of ferrihydrite is about 350 m^2g^{-1}, whereas that of goethite ranges between 10 and 40 m^2g^{-1}, the two phases being characterized by the same local structure, i.e., α-FeOOH. It has been shown that at similar [Pb]/[Fe] ratios, surface precipitation was occurring on goethite but not on ferrihydrite, where isolated adsorbed metal ions have been observed.[36,75]

Saturation with respect to the solubility of a pure phase. Adsorbed species may migrate along mineral surfaces and polymerize into surface clusters (Figure 8). These clusters form the link between adsorption phenomena and the heterogeneous nucleation of a new phase. Macroscopically, nucleation is restricted to supersaturated systems (i.e., to conditions where the ion activity product of the new phase (IAP) is greater than the solubility product (K_{so}) of this phase). As shown later, however, X-ray absorption studies provide evidence that polymeric surface clusters persist in systems undersaturated with respect to homogeneous precipitation ($IAP \leq K_{so}$). This apparent anomaly is accounted for by the physics of the different interfaces. The cluster-substrate energetics are more favorable compared to the cluster-solution interface: the surface reduces the free energy barrier for the mineral nucleation to occur.[76] Furthermore, the nucleation of a metastable phase has been shown to be kinetically favored over that of a less soluble analogue when the more soluble phase has a lower mineral-solution interfacial tension.[77] Although it will eventually transform to its most stable analogue via the "Ostwald Rule of Stages," the metastable phase may persist over geological time scales. Therefore, it is not surprising to observe in the lab the formation of such metastable phases, such as precipitates and hydroxy polymers of Cr (III) and Pb (III) on goethite and ferrihydrite, which shall be discussed now.

Surface Precipitation Case Studies

Laboratory studies. To date, only one EXAFS study has investigated a heterogeneous nucleation process, and it concerns the nucleation of hydrous chromium oxide onto hydrous ferric oxide.[12] Even at low surface coverage (<10% of the surface sites occupancy) Cr(III) was observed in a surface nucleated state. The polymeric surface clusters had a mixed α and γ CrOOH local structure, as indicated by the presence of three Cr-(Cr,Fe) peaks in the RDF (Figure 11c). They grew in epitaxy on an αFeOOH local structure adsorbent and persisted in undersaturated systems. With increasing surface coverage, nucleation became more extensive. The nucleated phase has a local structure representative of that of pure hydrous chromium oxide (γCrOOH), the two phases having identical solubility product. Thus Cr(III) sorbed as a polynuclear species and formed, even at relatively low coverage, a surface precipitate in conditions where solution precipitation did not occur. In the same manner, uranyl ions sorbed onto silica were found to be present as surface polymers in conditions where their polymerization in solution is negligible.[41]

As the IAP approches K_{so}, the presence of the adsorbent in the coordination sphere of the adsorbate atom is less and less detectable. The weakening of the substrate impact on the sorbate metal RDF has been observed in the study on Cr hydroxide heterogeneous nucleation discussed above[12] and also appeared in a series of studies on the heterogeneous precipitation of Co hydroxide at the surface of TiO_2 and Al_2O_3.[14,79] In the former, and although Co and Ti are barely distinguishable based on their small F_j contrast ($\Delta Z = 5$; see Section 2.2.3), the need to introduce a surface metal atom shell to fit the $\chi(k)$ function derived from the second Co K RDF peak may be justified since these experiments were run in undersaturated conditions (initial IAP = $10^{-17.7}$), and the solubility product of $Co(OH)_2$ is between $10^{-15.7}$ and $10^{-14.8}$.[78] On the contrary, in the Co-δAl_2O_3 experiments where the initial IAP was $10^{-16.8}$, the introduction of a surface metal atom shell was bearly substantiated in fitting the EXAFS data.[14,79]

Surface precipitation and epitaxial growth has also been described for sulfide minerals. MoS_2 crystallites have been shown to grow on δAl_2O_3.[80] These crystallites are bound to the alumina surface via their edges. Sites protruding from these edges have the highest catalytic power. As Co is mixed in small amounts with Mo before precipitation, Co is found, once precipitated, on the edges of the MoS_2 crystallites. At larger Co concentrations, mixed cobalt and molybdene sulfides, with a local millerite-type structure, have been shown to grow on the MoS_2 crystallite. EXAFS studies of multinuclear sulfido complexes have been restricted until now to catalytic research, but could be applied as well to the study of natural anoxic environments such as reducing sediments.

Natural systems. One of the most important contributions of EXAFS spectroscopy to the knowledge of the structural chemistry of low temperature materials is the recognition of their heterogeneous structure on a very fine scale. Recent EXAFS studies have shown that the actual structure at the 2 to 5 Å scale of low temperature minerals often differs from the average structure derived from diffraction-based techniques. Two types of structural situations have so far been identified: (1) a simple deviation from the random distribution of atoms along a solid-solution[81] (several examples now exist, and the nonrandom distribution of atoms in low-temperature materials seems to be the rule rather than the exception.); (2) existence of a discrete phase inter- or over-grown with the major one (one or both of the two phases are thought to result from a heterogeneous precipitation.). Three examples of such particles have been described and include MnO_2 phyllomanganate clusters in αFeOOH (Figure 12a), αFeOOH-like clusters in δMnO_2,[13] and αFe_2O_3-like clusters in αAlOOH[82] (Figure 12b). The three dimensional local structure of the latter inclusion has been described in detail, taking advantage of the polarized nature of the synchrotron beam. Hematite-like clusters have been found to develop within the channels of the diaspore lattice. The close analogy between the structure of these Fe clusters with the one described for γCrOOH multinuclear surface complexes sorbed onto hydrous ferric oxides (Figure 11d) is striking and strongly support their formation by a heterogeneous nucleation in conditions of undersaturation with respect to homogeneous nucleation. In other words, hematite clusters are seen as ancient multinuclear complexes that have formed at the time of the diaspore growth and which are now sealed within its bulk structure.

All these types of particle defects result from nonequilibrium crystallization processes in the thermodynamic and kinetic conditions which prevail at the Earth's surface. Deciphering the actual structure of these complex materials is a necessary step towards an advanced understanding of their conditions and mechanisms of formation.

FORMATION OF CLAY-LIKE STRUCTURES AT LOW TEMPERATURE

Phyllosilicate minerals are among the solid particles that exhibit the highest surface reactivity in soils and sediments. Their formation in low temperature environments has been the subject of ongoing studies.[83] XAS brings new light to this field as it reveals that the products of different removal processes may have a local structure similar to that of phyllosilicates, commonly referred to as "clays."

A Definition of a Clay-Like Structure

We shall consider as a clay-like local structure, any structure where octahedra complexes MeA^{m-6n}, comprising a metal cation Me^{m+} ($m = 2$ or 3) and six anions A^{n-} (A = O, OH), are polymerized to one another in a layer structure via the sharing of edges and linked through corners with silica sheet(s) (Figure 13). This

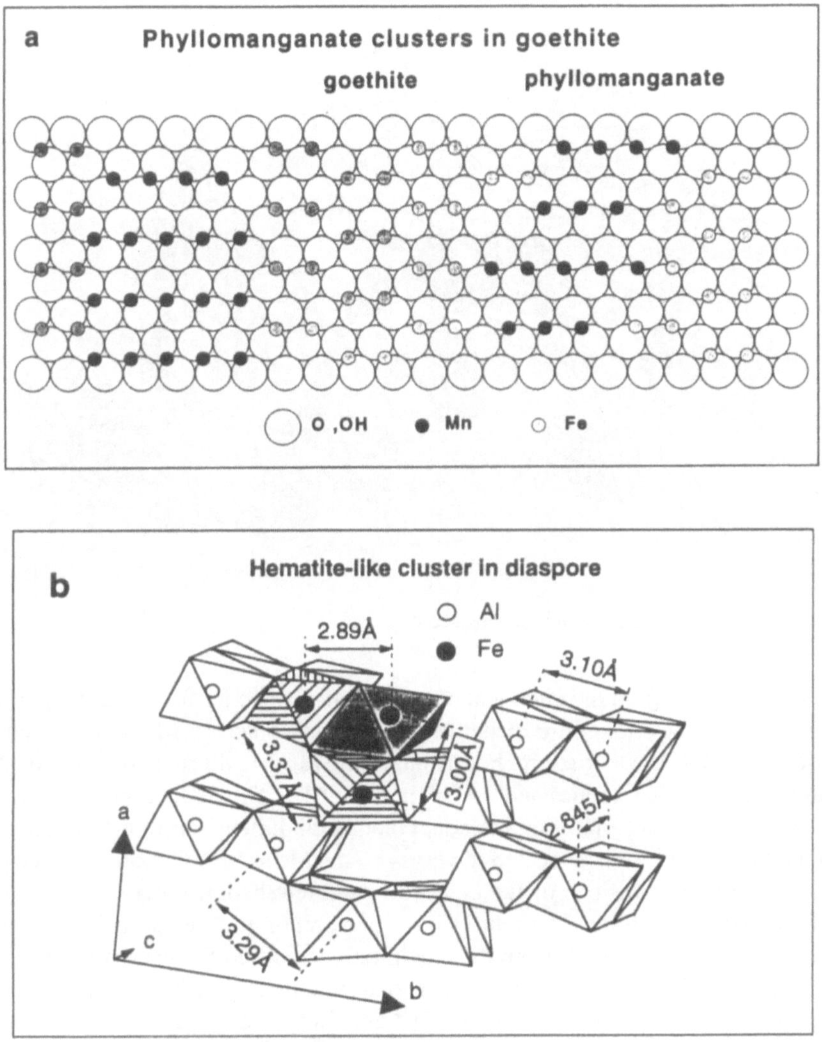

Figure 12. Examples of clusters embedded in a mineral matrix and viewed as relics from polynuclear surface complexes. (a) Phyllomanganate-like clusters in goethite. (Modified from Manceau et al.[8]) (b) Hematite-like clusters in channels of diaspore.[82]

definition does not require tetrahedral layers to be as large as octahedral ones. One may imagine islands of tetrahedral layers lying, for example, on octahedral sheets (or vice versa), building in that way a kind of disordered corrugated structure.[84] Let us analyze the criteria which can be used for identifying at the Me K-edge a trioctahedral clay-like structure.

In clays, Me atoms belonging to the octahedral sheet are surrounded by nearest Me atoms at about 3.0 to 3.12 Å and (Si,Al) atoms at 3.22 to 3.29 Å, the exact

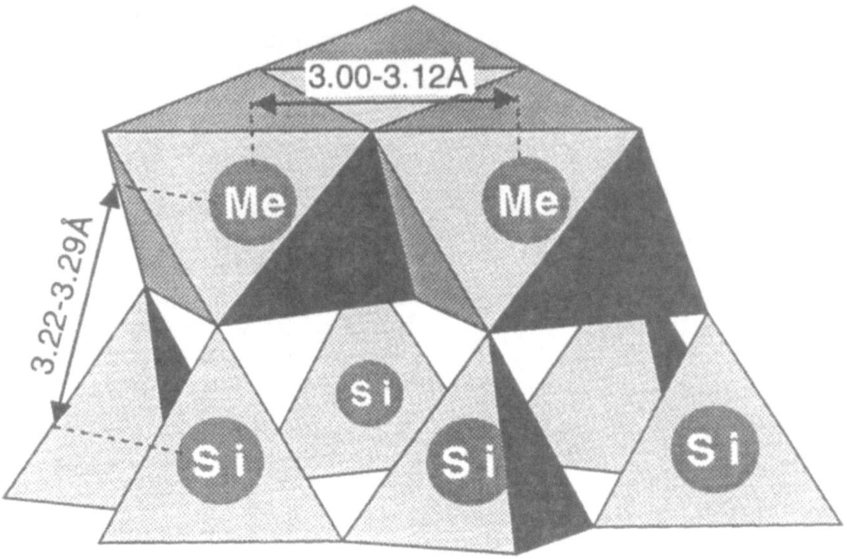

Figure 13. Local structure of a phyllosilicate.

distance depending on the chemical composition. A specific feature of this group of minerals is the shortening of the Me-Me distances across edges compared to those in the $Me(OH)_m$ structure. For instance, the Mg-Mg distance is equal to 3.06 Å in $Si_4Mg_3O_{10}(OH)_2$ (talc) and 3.12 Å in $Mg(OH)_2$. This reduction results from structural adjustments necessary to adapt the lateral dimensions of octahedral and tetrahedral sheets.[85] It is expected to vary with the Me to Si stoichiometry, i.e., the relative lateral dimensions of the octahedral and tetrahedral sheets.

Examples of such local structures have been reported to exist in XRD amorphous Fe-Si gels that result from the precipitation of silico-ferric complexes. Iron octahedra were found to be organized in a nontronite-type (ideally, $Si_4Fe_2O_{10}(OH)_2$) local structure.[86] The complexation of Fe^{3+} and SiO_4^{3-} takes place in solution, and these precursors already possess the local structure and overall stoichiometry of the well-crystallized mineral (i.e., nontronite) they will transform into upon aging.[86]

Cobalt and Nickel Sorption on Silicates

The sorption of Co(II) onto kaolinite occurs within the same pH range as the sorption onto SiO_2, when comparable cobalt and solid concentrations are used.[87] Sorption of Co(II) on SiO_2 particles has been further shown to be specific, as it induces a reversal of the particle electrophoretic mobility, and to involve the formation of surface Co(II) hydroxy polymers.[88,89]

Results of XAS studies conducted on such polymers formed on quartz,[90] kaolinite,[15,90] and loughlinite[91,92] (a sepiolite) have been compiled in Figure 14 and Table 1. These polymers were formed by a priori distinct surface phenomena such as "cation exchange" (Co-loughlinite) and "specific adsorption" (Co-kaolinite and Co-quartz). For comparison, data on the structure of synthetic $Co(OH)_2(s)$[15] and kerolite $[Si_4Co_3O_{10}(OH)_2]$,[92] a trioctahedral smectite formed by "coprecipitation" of Co with Si in basic conditions,[93] are shown as well. The two sixfold coordinated cations Co^{2+} and Ni^{2+} have similar ionic radii (Co^{2+}: 0.65 Å, and Ni^{2+}: 0.69 Å). Thus, data on Ni(II)-doped SiO_2 support catalysts, synthesized by adsorption of $[Ni(NH_3)_6]^{2+}$ (followed by desiccating and wetting of the solid[94]) are also included in Table 1.

In the Co-kaolinite, Co-loughlinite, and kerolite samples, the Co-Me and Co-Si distances (3.10 ± 0.04 Å and 3.27 ± 0.02 Å) are compromised between those existing in talc (d(Mg-Mg) = 3.06 Å, d(Mg-Si) = 3.25 Å)[95] and ferrous micas (d(Fe-Fe) = 3.11 Å, d(Fe-[Si,AI]) = 3.28–3.29 Å).[96] Since the ionic radius of Co(II) is between those of Mg^{2+} and Fe^{2+}, it can be concluded that the local structure around Co^{2+} is that of a clay. Larger Co-Co distances are observed in Co-quartz (3.13–3.16 Å) and Co-kaolinite (3.12–3.14 Å) atoms is correspondingly, where the number of nearest Si atoms is correspondingly low. This observation might support the idea of a nonstoichiometric clay-like structure in which the Si/Co molar ratio would be less than 1, that is less than the value of a 1:1 clay mineral. This hypothesis could then explain the intermediate range of the Co-Co distances found to be in between those of $Co(OH)_2$ and kerolite. However, a second explanation can be put forward for kaolinite. If one pays attention to the ratio of the number of Co and Si neighbors, it is striking that it is close to the value of 3 encountered in trioctahedral 1:1 clays, where each atom Me has six nearest Me neighbors across edges and two nearest Si across corners. This observation thus strongly suggests the formation of a stoichiometric $Si_2Co_3O_5(OH)_4$ structure. The reduced number of Co and Si neighbors (2.2 and 0.6, respectively, instead of 6 and 2) would then be explained readily by the small size and intrinsic disorder of particles, as recognized in nontronite-like silico-ferric gels.[86] The slight increase of the Co-Co distances is also compatible with this model, since in trioctahedral clays Me-Me distances are usually larger in 1:1 structures than in 2:1 structures.[97]

In conclusion, various a priori distinct surface phenomena, such as "cation exchange"[91] (Co-loughlinite), "specific adsorption"[15,90] (Co-kaolinite), or "precipitation"[86] (kerolite), lead to a unique local clay-like structure.[98] Further research along this line is needed to improve our understanding of the formation, at low temperature, of clay minerals.

RETURNING TO THE ISOLATED ION IN SOLUTION

Mineral Dissolution

Surface complex can destabilize surface metal ions relative to the bulk, and therefore lead to an enhanced dissolution of the mineral phase. Examples of such

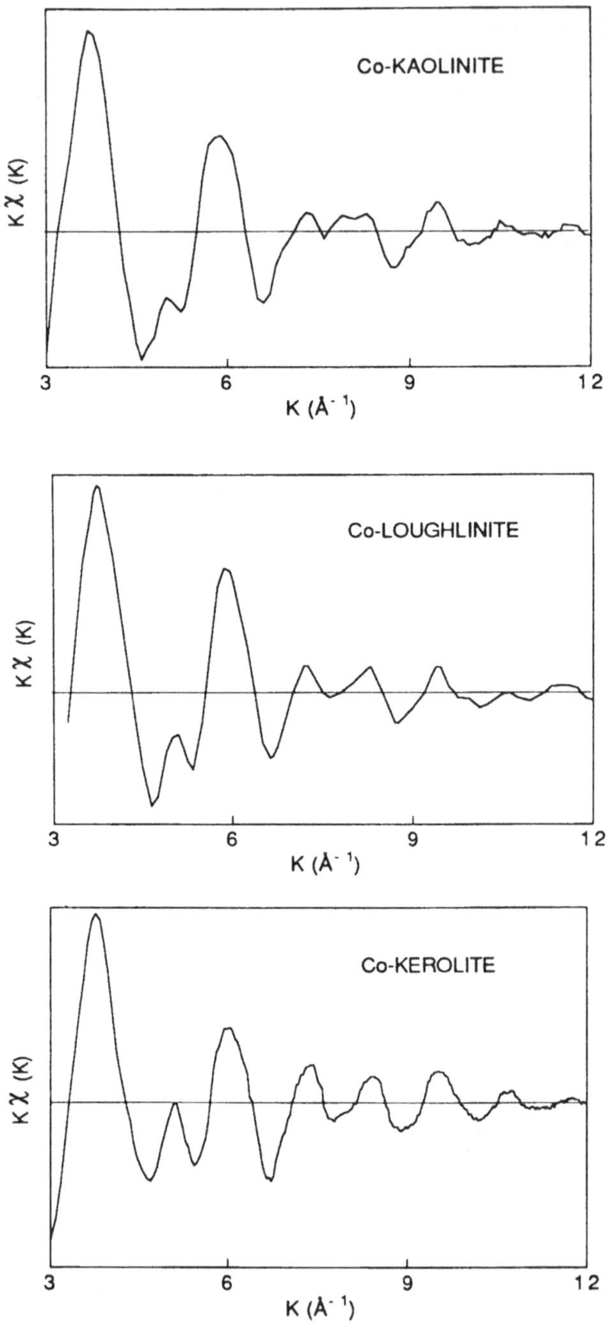

Figure 14. Formation by different mechanisms of clay-like local Co structures on or within different phyllosilicate minerals.[98] Notice the strong resemblance of Co K-edge EXAFS spectra of cobalt sorbed on kaolinite,[15] exchanged and/or substituted for Na and Mg at the surface of loughlinite[91] and in the lattice of kerolite.[92]

Table 1. Structural Information Derived from the Analysis of (Co,Ni)-Cation Contributions to EXAFS in Various Sorbed Samples and Minerals

	First Metal Shell			Second Metal Shell			
	Scatterer	R(Å)	N	Scatterer	R(Å)	N	Ref.
$Co(OH)_2(s)$	Co	3.17	6.0	—	—	—	79
$Co\text{-}SiO_2$	Co	3.13–3.16 ± 0.02	2.9 ± 0.4	Si	3.25–3.29 ± 0.02	n.d.	90
Co-kaolinite	Co	3.12–3.14 ± 0.02	2.2 to 5.7* ± 0.03	(Si,Al)	3.25±3.29 ±0.02	n.d.	90
Co-loughlinite	Co	3.07	6.0	(Si,Al)	3.25	4	92
Co-kerolite	Co	3.08	6.0	Si	3.25	4	92
$Ni\text{-}SiO_2$	Ni	3.10–3.11	6.0	Si	3.25	3.1	94
Talc	Mg	3.06	6.0	Si	3.25	4	95
Ferrous micas	Fe	3.11	6.0	(Si,Al)	3.28–3.29	4	96

Source: Charlet and Manceau.[98]

Note: * as a function of surface coverage;[90] n.d.: not determined.

complexes include proton, metal or ligand surface complexes, and this complexation can lead to electron transfer between the sorbate and the sorbant. The detachment of the surface metal has been shown to be the rate limiting step of these dissolution reactions.[99-101] XAS has in a few cases provided evidence for the formation of the surface activated complex which is assumed in these models.

The only EXAFS study of nonreductive dissolution concerns the corrosion of a U-containing borosilicate glass surface.[102] The leaching of these surfaces leads to the formation at the glass surface of hydrated uranyl units with incoherent U-O distances. The increasing splitting of the U-O distances has been interpreted as due to an ion exchange of sodium for protonated water. The uranyl surface complex is then detached from the glass surface together with the alkali ion, leading to a dissolution of the glass network and the formation of a gel layer. Prolongated corrosion results in the development of a surface structure resembling hydrated uranyl silicates.[103]

No XAS study of oxidative dissolution has been reported to date. Such a dissolution may involve, with Fe(II) sulfides and silicates, the reduction of a dissolved metal (Fe(III), Cr(VI)) at the mineral surface. As an example of the possibilities offered by the method, one may refer to the study of Cu^{2+} reduction at the surface of a conductive material.[21] An *in situ* quantitative measurement of the kinetics involved, and in particular of the fast reduction step of Cu^{2+} to Cu^{+}, was realized by dispersive XAS spectroscopy. The stabilization of Cu^{+} (otherwise an unstable ion) by the formation of a surface complex and the subsequent formation of surface metallic copper clusters was demonstrated.

The only XAS study of reductive dissolution concerns that of Mn(IV) and Mn(III) oxides by Cr(III).[8] Macroscopic studies indicate that this reaction proceeds at a rate independent of solution parameters, such as pH, oP_{O2}, and ionic strength, but dependent on, and even proportional to, the solid suspension concentration.[104,105] XANES spectroscopy allows an *in situ* measurement of the chromate ion (reaction product) concentration. A comparison of the rate of chromate ion formation by different Mn(IV) and Mn(III) polymorph oxides indicates that the reaction is faster on two-dimensional, versus three dimensional, substrates and when the concentration of active surface site, i.e., Mn(IV) vacancies increases. The reaction at the surface of birnessite ($Na_4Mn_{14}O_{27}.9H_2O$), a phyllomanganate, has been followed by EXAFS spectroscopy.[9] After 30 s equilibration, sorbed chromium atoms are still present in the +III oxidation state, as indicated by the XANES spectrum (Figure 15a). An intense second RDF peak is detected at the Cr K-edge, and its position falls at the emplacement of the Mn-Mn contribution of the sorbent (Figure 15b). Analysis shows 4 Mn neighbors at 2.90 Å.[9] Therefore, after a first electrostatic attraction due to the high negative charge present at the surface of birnessite,[106] the adsorbate ions (Cr(III)) can penetrate the Mn(IV) vacancies, due to a favorable steric fit with these reactive surface sites (Figure 15c). These vacant octahedral cages present in birnessite act as a molecular sieve for Cr(III) free ions, in the sense that polynuclear Cr(III) species are shown not to be oxidized, as they are too large to penetrate the reactive site.

After a 60 s equilibration time, a very sharp increase of the XANES peak (Figure 15a) and decrease of the first RDF peak (Cr-O contribution; Figure 15b) both indicate a dramatic increase of the Cr(VI) content, i.e., that electron transfer has occurred. At that time, Cr(VI), a very soluble anion, is released into solution together with Mn(II) ions.

Electrolyte Solutions

Once released into solution, the metal ion forms complexes with dissolved ligands. This phenomenon has also been intensively investigated by XAS spectroscopy, particularly in the case of chloro complexes.

The structure of chloro complexes was examined for a variety of metal ions by recording EXAFS and XANES spectra of the metal K-absorption edge. Some authors were further able to record Cl K-EXAFS and XANES spectra using fluorescence detection and a wiggler-magnet beam line.[107] Various metal concentrations (down to at least 10^{-3} mol dm^{-3}) and chloride to metal concentration ratios (2:1 to 100:1) were investigated. The pre-edge peak observed in Cl K-XANES spectra of such complexes provided evidence for a direct metal-chloride bonding, as it results from transitions to antibonding molecular orbitals derived from the metal 3d and the chlorine 3p orbitals. The distinction in the EXAFS spectra between O and Cl ligands as first neighbor atoms was feasible owing to phase shift and amplitude contrast between O and Cl neighbors.[11]

Fe(III) and Ni(II) complexes with chloride ions were shown to shift from an outer-sphere to an inner-sphere type as the concentration of stoichiometric solutions was increased above 0.4 mol dm^{-3}.[108] Cu(II) showed considerable initial bonding

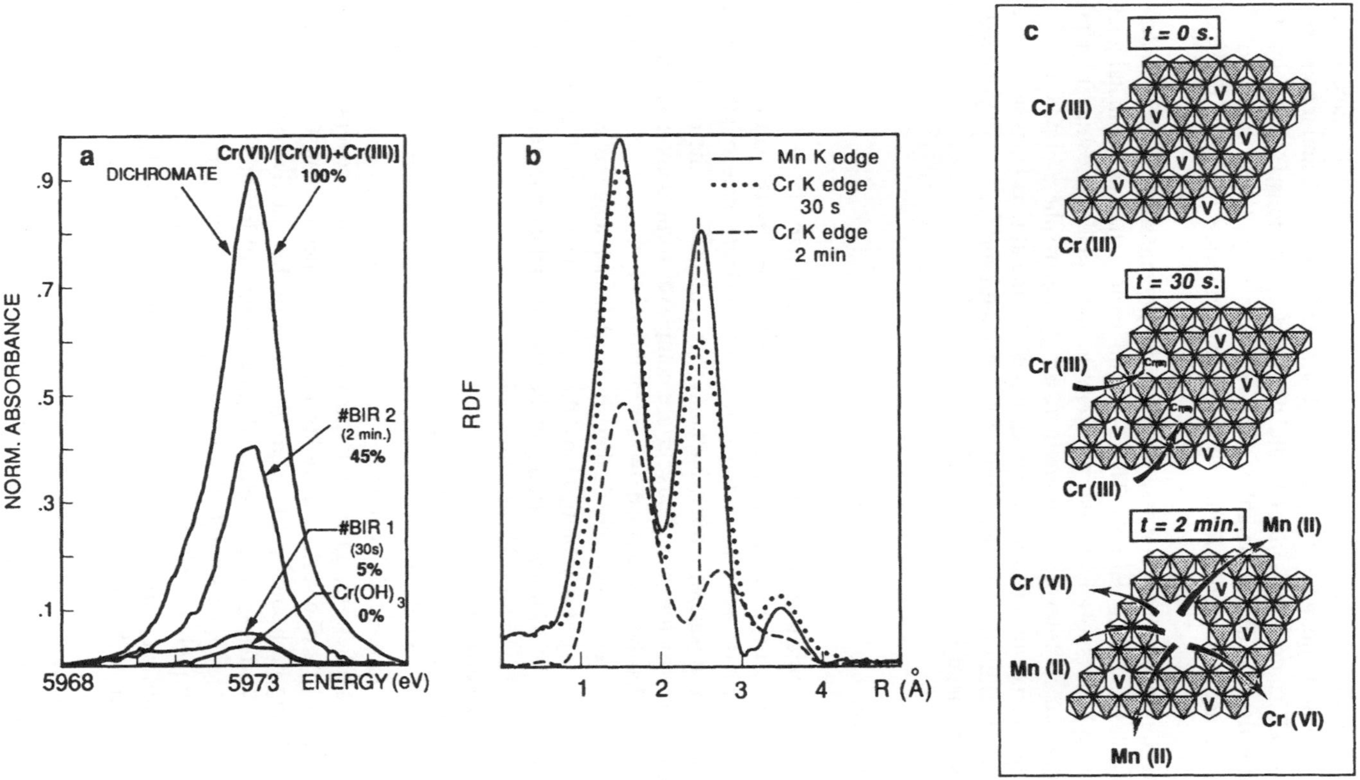

Figure 15. Cr(III) oxidation at the surface of birnessite. Pre-edge (a) and EXAFS (b) results, plus structural interpretation of the kinetics of oxidation (c). (Modified from Manceau and Charlet.[9])

to Cl^-, whereas Co(II) showed little or none.[109].Au(III)[110] and Zn(II)[111] complexations with chloride ions were also studied in detail in dilute solution. The number of chloride atoms in the Zn(II) inner shell increases from none to three as the $ZnCl_2$ concentration increases from 10^{-3} mol dm^{-3} to 1 mol dm^{-3}, and it increases also with the Cl:Zn ratio. In acidic 10^{-3} mol dm^{-3} $AuCl_3$ solutions, Cl was found to be present as first and second neighbor of Au. At increasing pH, the number of Cl atoms in the inner shell shifted from four to one as a consequence of the OH for Cl exchange.

Chloride ions have often been assumed in environmental studies to form outer-sphere complexes with metal ions. These examples show that, as the metal goes from a low to high chloride concentration environment (as in an estuary), metal ions form chloro complexes which are increasingly inner-sphere in nature. This has important consequences for the desorption of the metal ions from suspended particles, where chloride ions can compete directly with surface hydroxyl ligands for the system metal ions.

CONCLUSION

Although its application to the field of environmental particles is recent, XAS has already renewed our ideas on their structure and reactivity. Not only have hydrous oxide particles been shown to be ordered at the local scale, but they often possess a complex mosaic structure which may give indications of their conditions of formation. The structure of various surface complexes formed upon adsorption of trace elements has been described for several laboratory model systems, as well as for natural particles. It has been further shown that such adatoms would, even at low temperature, travel rapidly along the surface via lateral diffusion and polymerize with other adatoms, forming in the case of a silica surface, a unit with phyllosilicate local structure. This microscopic information finally allows, when coupled with aqueous chemical studies, a mechanistic understanding of dissolution reactions and particle growth poisoning.

These studies have only been possible following the availability of high flux synchrotron radiation and high counting-rate fluorescence X-ray detectors. Sorption phenomena studies are, however, limited to date to surface coverage higher than around 5 to 10% of a surface monolayer. A third generation of storage rings is currently being built and will be commissioned in the last five years of the twentieth century. These are, in chronological order, ESRF at Grenoble for Europe, APS at Argonne for the U.S. and SPring-8 near Osaka for Japan. These storage rings with their various insertion devices (wigglers and undulators) will provide the scientific community with X-ray beams of unprecedented properties. Gaining about ten orders of magnitude in brilliance with respect to a rotating anode X-ray tube, and three to four orders with respect to the most brilliant existing synchrotron X-ray sources, these seem *à peine croyable* (hardly believable) to the experienced experimenter. This will allow the study of much more diluted systems, such as natural samples, and the development of dispersive- and quick- XAS spectrometers will permit time-resolved XAS studies of the dynamics of many geochemical processes.

ACKNOWLEDGMENTS

The authors would like to thank the European Environmental Research Organization (EERO) for facilitating interactions between the Institut für Anorganische, Analytische und Physikalische Chemie and the Laboratoire de Minéralogie-Cristallographie. Laurent Charlet acknowledges the support by the Swiss National Science Foundation and Alain Manceau the support by the French Ministery of Environment, grants 90227 and 21125.

GLOSSARY

Angström (Å)	a non-SI unit equal to 10^{-10} m
DC linkage	linkage between octahedra in which a given octahedron shares two adjacent oxygen atoms with two different octahedra
E linkage	linkage between octahedra in which a given octahedron shares two oxygen atoms with an octahedron
EXAFS	extended X-ray absorption fine structure
F linkage	linkage between octahedra in which a given octahedron shares three oxygen atoms with a single octahedron
Ferrihydrite	an XRD "amorphous" hydrous ferric oxide
IAP	ion activity product, as defind in[78]
Mn K-edge	K-absorption discontinuity of X-rays for Mn (6539 eV)
RDF	radial distribution function
SC linkage	linkage between octahedra in which a given octahedron shares one oxygen atom with a single octahedron
XANES	X-ray absorption near-edge structure
XRD	X-ray diffraction
Z contrast	difference in atomic number (Z) between two elements

REFERENCES

1. Hochella, M.F., Jr. and A.F. White, Eds. *Mineral-Water Interface Geochemistry,* Reviews in Mineralogy 18 (Washington D.C.: Mineralogical Society of America, 1988), p. 603.
2. Van Griecken, R. "Single Particle Characterization Through Beam Technique," in *Sampling and Characterization of Environmental Particles, I*, J. Buffle and H. P. Van Leeuwen, Eds. Series of the Commission of Environ Anal. Chem. of IUPAC (Chelsea, MI: Lewis Publishers, Inc., 1992).

3. Brown, G.E., Jr., G. Calas, G.A. Waychunas, and J. Petiau. "X-ray Absorption Spectroscopy: Applications in Mineralogy and Geochemistry," in *Spectroscopic Methods in Mineralogy and Geology,* F.C. Hawthorne, Ed. Reviews in Mineralogy 18 (Washington, D.C.: Mineralogical Society of America, 1988), pp. 431–512.
4. Brown, G.E., Jr. "Spectroscopic Studies of Chemisorption Reaction Mechanisms at Oxide-Water Interfaces," in *Mineral-Water Interface Geochemistry*, M.F. Hochella and A.F. White, Eds., Reviews in Mineralogy 23 (Washington D.C.: Mineralogical Society of America, 1990), pp. 309–364.
5. Teo, B.K. *EXAFS: Basic Principles and Data Analysis* (Berlin: Springler-Verlag, 1986), p. 349.
6. Koningsberger, D.C. and R. Prins. *X-ray Absorption, Principles, Applications, Techniques of EXAFS, SEXAFS and XANES* (New York: John Wiley & Sons, 1988).
7. Manceau, A., J.M. Combes, and G. Calas. "New Data and a Revised Model for Ferrihydrite: a Comment on a Paper by R. A. Eggleton and R. W. Fitzpatrick," *Clays Clay Miner.* 38:331–334 (1990).
8. Manceau A., A.I. Gorshkov, and V. Drits "Structural Chemistry of Mn, Fe, Co, and Ni in Mn Hydrous Oxides. I. Information from XANES Spectroscopy," *Am. Mineral.* 77:1133–1143 (1992).
9. Manceau, A. and L. Charlet. "X-ray Absorption Spectroscopic Study of the Sorption of Cr(III) at the Oxide/Water Interface. I Molecular Mechanism of Cr(III) Oxidation on Mn Oxides," *J. Colloid. Interface Sci.* 148:425–442 (1992).
10. Bidoglio, G., P.N. Gibbson, E. Haltier, N. Omenetto, and M. Lipponen. "XANES and Laser Fluorescence Spectroscopy for Rare Earth Speciation at Mineral Water Interfaces," *Radiochim. Acta* 58–59, 191–197 (1992).
11. Combes, J.M., A. Manceau, and G. Calas. "Formation of Ferric Oxides from Aqueous Solutions: a Polyhedral Approach by X-ray Absorption Spectroscopy. I. Hydrolysis and Formation of Ferric Gels," *Geochim. Cosmochim. Acta* 54:583–594 (1989).
12. Charlet, L. and A. Manceau. "X-Ray Absorption Spectroscopic Study of the Sorption of Cr(III) at the Oxide/Water Interface. II. Adsorption, Coprecipitation and Surface Precipitation on Ferric Hydrous Oxides," *J. Colloid Interface Sci.* 148:443–458 (1992).
13. Manceau A., A.I. Gorshkov, and V. Drits. "Structural Chemistry of Mn, Fe, Co, and Ni in Mn Hydrous Oxides. II. Information from EXAFS Spectroscopy, Electron and X-ray Diffraction," *Am. Mineral.* 77:1144–1157 (1992).
14. Brown, G.E., Jr., G.A. Parks, and C.J. Chisholm-Brause. "In Situ X-Ray Absorption Spectroscopic Studies of Ions at Oxide-Water Interfaces," *Chimia* 43:248–256 (1989).
15. Chisholm-Brause, C.J., P.A. O'Day, G.E. Brown, Jr., and G.A, Parks ."Evidence for Multinuclear Metal-Ion Complexes at Solid/Water Interfaces from X-Ray Absorption Spectroscopy," *Nature* 348:528–530 (1990).
16. Manceau, A. and J.M. Combes. "Structure of Mn and Fe Oxides and Oxyhydroxides: a Topological Approach by EXAFS," *Phys. Chem. Miner.* 15(3): 283–295 (1988).
17. Combes, J.M., A. Manceau, and G. Calas, "Formation of Ferric Oxides from Aqueous Solutions: a Polyhedral Approach by X-ray Absorption Spectroscopy. II. Hematite Formation from Ferric Gels," *Geochim. Cosmochim. Acta* 54:1083–1091 (1990).

18. Arrhenius, G., K. Cheung, S. Crane, M. Fisk, J. Frazer, J. Korkisch, T. Mellin, S. Nakao, A. Tsai, and G. Wolf. "Counterions in Marine Manganates," in *La Genèse des Nodules de Manganäse*, Cl. Lalou, Ed. (Paris: Colloque International du CNRS, 1979), p. 333–256.
19. Chandesris D. "Thermal Disorder of Surface Atoms Studied by EXAFS," *J. Phys.* 47:479–486 (1986) .
20. Manceau, A. and V.A. Drits. "Local Structure of Ferrihydrite and Feroxyhite by EXAFS Spectroscopy," *Clay Miner.* (in press).
21. Tourillon, G., E. Dartyge, A. Fontaine, and A. Jucha. "Dispersive X-Ray Spectroscopy for Time-Resolved In Situ Observation of Electrochemical Inclusion of Metallic Clusters within Conducting Polymer," *Phys. Rev. Lett.* 57: 603–606 (1986).
22. Prieto, C., P. Lagarde, H. Dexpert, V. Briois, F. Villain, and M. Verdaguer. "X-Ray Absorption Spectroscopy of the Continuous Change from Ce(IV) to Ce(III)," *J. Phys. Chem. Solids* 53:233–237 (1992).
23. Bottero, J.Y., D. Tchoubar, A. Arnaud, and P. Quienne. "Partial Hydrolysis of Ferric Nitrate Salt. Structural Investigations by Photon-Correlation Spectroscopy and Small Angle X-Ray Scattering," *Langmuir* 7:1365–1369 (1991).
24. Tchoubar, D., J.Y. Bottero, P. Quienne, and M. Arnaud. "Partial Hydrolysis of Ferric Chloride Salt. Structural Investigation By Photon-Correlation Spectroscopy and Small-Angle X-Ray Scattering," *Langmuir* 7:398–402 (1991).
25. Drits, V.A. and C. Tchoubar. *X-ray Diffraction of Disordered Lamellar Structures. Theory and Application to Microdivided Silicates and Carbons* (NewYork: Springer-Verlag, 1990).
26. Combes, J.-M., A. Manceau, and G. Calas, "Study of the Local Structure in Poorly-Ordered Precursors of Iron Oxi-hydroxides", *J. Phys.* C8:697–701 (1986) .
27. Magini, M. "Structural Relationships Between Colloidal Solution and Hydroxide Gels of Iron (III) Nitrate," *J. Inorg. Nucl. Chem.* 39:409–412 (1977).
28. Parise, J.B., M.A.A. Schoonen, and G. Lamble. "An Extended X-Ray Absorption Fine Structure Spectroscopic Study of Amorphous FeS," in *Abstracts with Programs* (Washington, D.C.: Geological Society of America, 1991), p.123.
29. Morse, J.W., F.J. Millero, J.C. Cornwell, and D. Rickard. "The Chemistry of the Hydrogen Sulfide and Iron Sulfide Systems in Natural Waters," *Earth-Sci. Rev.* 24:1–42 (1987).
30. Eggleton, R.A. and R.W. Fitzpatrick. "New Data and a Revised Structural Model for Ferrihydrite," *Clays Clay Miner.* 36, 111–124 (1988).
31. Chukhrov, F.V., A. Manceau, B.A. Sakharov, J.M. Combes, A.I. Gorshkov, A.L. Salyn, and V.A. Drits. "Crystal Chemistry of Oceanic Fe-Vernadites," *Mineralogicheskii J.* 10:78–92 (1988).
32. Towe K.M. and W.F. Bradley. "Mineralogical Constitution of Colloidal Hydrous Ferric Oxides" *J. Colloid Interface Sci.* 24:384–392 (1967).
33. Harrison P.M., F.A. Fischbach, T.G. Hoy, and G.M. Haggis. "Ferric Oxyhydroxide Core of Ferritin," *Nature* 216:1188–1190 (1967).
34. Drits, V.A., B.A. Sakharov, A.I. Gorshkov, and A. Manceau. "New Structural Model for Ferrihydrite," *Clay Miner.* (in press).
35. Sposito, G. "Characterization of Particle Surface Charge," in *Sampling and Characterization of Environmental Particles I*, J. Buffle and H.P. Van Leeuwen, Eds., Series of the Commission of Env. Anal. Chem. of IUPAC (Chelsea, MI: Lewis Publ., Inc., 1992).

36. Manceau, A., L. Charlet, M.C. Boisset, B. Didier, and L. Spadini. "Sorption and Speciation of Heavy Metals on Hydrous Fe and Mn Oxides. From Microscopic to Macroscopic," *Appl. Clay Sci.* (in press).
37. Sposito, G. "Distinguishing Adsorption from Surface Precipitation," in *Geochemical Processes at Mineral Surfaces,* J.A. Davis and K.F. Hayes, Eds., ACS Symp. Ser. 323 (Washington, D.C.: American Chemical Society, 1986), pp. 217–228.
38. Bedzyk, M.J., G.M. Bommarito, M. Caffrey, and T.L. Penner. "Diffuse-Double Layer at a Membrane-Aqueous Interface Measured with X-Ray Standing Waves," *Science* 248: 52–56 (1990).
39. Hayes, K.F., A.L. Roe, G.E. Brown, K.O. Hodgson, J.O. Leckie, and G.A. Parks. "In Situ X-Ray Absorption Study of Surface Complexes: Selenium Oxyanions on αFeOOH," *Science* 238:783–786 (1987).
40. Hayes, K.F. and J.O. Leckie. "Modeling Ionic Strength Effects on Cation Adsorption at Hydrous Oxide/Solid Interfaces," *Colloid Interf. Sci.* 115:564–574 (1987).
41. Dent A.J., J.D.F. Ramsay, and S.W. Swanton. "An EXAFS study of uranyl ion in solution and sorbed onto silica and montmorillonite clay colloids," *J. Colloid Interf. Sci.* 150:45–60 (1992).
42. Sposito, G. and R. Prost. "Structure of Water Adsorbed on Smectites," *Chem. Rev.* 82: 553–572 (1982).
43. Waychunas, G.A., B.A. Rea, C.C. Fuller, and J.A. Davis."Fe and As K-edge EXAFS Study of Arsenate (AsO_4^{3-}) Adsorption on 'two-line' Ferrihydrite," *Geochim. Cosmochim. Acta* (submitted).
44. Combes, J.M. "Evolution de la structure locale des polymères et gels ferriques lors de la cristallisation des oxydes de fer. Application au piégeage de l'uranium," PhD Thesis, University of Paris, France (1988).
45. Parfitt, R.L., J.D. Russell, and V.C. Farmer. "Confirmation of the Surface Structures of Goethite (α-FeOOH) and Phosphated Goethite by Infrared Spectroscopy," *J. Chem. Soc. Faraday I* 72:1082–1087 (1976).
46. Parfitt R.L. and J.D. Russell. "Adsorption on Hydrous Oxides. IV Mechanisms of Adsorption of Various Ions on Goethite," *J. Soil Sci.* 28:297–305 (1977).
47. Buddemeier, R.W. and J.R. Hunt. "Transport of Colloidal Contaminants in Groundwater: Radionucleide Migration at the Nevada Test Site," *Appl. Geochem.* 3:535–548 (1988).
48. Paulson, A.J., R.A. Feely, H.C. Curl, E.A. Crecelcius, and Geiselman. "The Impact of Scavenging on Trace Metal Budgets in Puget Sound," *Geochim. Cosmochim. Acta* 52:1765–1779 (1988).
49. Szalay, A. "Cation Exchange Properties of Humic Acids and their Importance in the Geochemical Enrichment of UO_2^{2+} and Other Cations," *Geochim. Cosmochim. Acta* 28:1605–1614 (1964).
50. Leciejewicz, J. "Neutron-Diffraction Study of Orthorhombic Lead Monoxide," *Acta Crystallogr.* 14:66–81 (1961).
51. Tejedor-Tejedor, M.I. and M.A. Anderson, "In-Situ Attenuated Total Reflection Fourier Transform Infrared Studies of the Goethite (α-FeOOH)-Aqueous Solution Interface," *Langmuir* 2:203–210 (1986).
52. Lim-Nunez, R. and R.J. Gilkes. "Acid Dissolution of Synthetic Metal-Containing Goethites and Hematites," in *Proceedings of the Int. Clay Conf., Denver, 1985.* L.G. Schultz, H. van Olphen, and F.A. Mumpton, Eds. (Bloomington, IN: The Clay Mineral Society, 1987), pp. 197–204.

53. Manceau, A., J.M. Combes, and G. Calas. "Chemical and Structural Applications of X-ray Absorption Spectroscopy in Mineralogy," *J. Chim. Physi.* 86: 1533–1545 (1989).
54. Shannon, R.D. and C.T. Prewitt. "Effective Ionic Radii in Oxides and Fluorides," *Acta Crystallogr.* B25:925–946 (1969).
55. Catts, J.G. and D. Langmuir. "Adsorption of Cu, Pb and Zn by δMnO_2: Applicability of the Site Binding-Surface Complexation Model," *Appl. Geochem.* 1:255–264 (1986).
56. Pingitore, N.E., Jr., F.W. Lytle, B.M. Davies, M.P. Eastmann, P.G. Eller, and E.M. Larson. "Mode of Incorporation of Sr^{2+} in Calcite: Determination by X-ray Absorption Spectroscopy," *Geochim. Cosmochim. Acta* 56:1531–1538 (1992).
57. Bernstein, L.R. and G.A. Waychunas. "Germanium Crystal Chemistry in Hematite and Goethite from the Apex Mine, Utah, and Some New Data on Germanium in Aqueous Solution and in Stottite," *Geochim. Cosmochim. Acta* 51:623–630 (1987).
58. Manceau, A., S. Llorca, and G. Calas. "Crystal Chemistry of Cobalt and Nickel in Lithiophorite and Asbolane from New Caledonia," *Geochim. Cosmochim. Acta* 51:105–113 (1987).
59. Manceau, A., J. Rask, P.R. Busek, and D. Nahon. "Characterization of Copper in Lithiophorite from a Mn Banded Ore," *Am. Mineral.*, 75:490–494 (1990).
60. Wadsley, A.D. "The Structure of Lithiophorite. $(Al,Li)MnO_2(OH)_2$", *Acta Crystallogr.* 5:676–680 (1952).
61. Pauling, L. and B. Kamb. "The Crystal Structure of Lithiophorite," *Am. Mineral.* 67:817–821 (1982).
62. Sass, B.M. and D. Rai."Solubility of Amorphous Chromium(III)-Iron(III) Hydroxide Solid Solutions," *Inorg. Chem.* 26:228–2232 (1987).
63. Schwertmann, U. and M. Latham. "Properties of Iron Oxides in some New Caledonian Soils," *Geoderma* 39:105–123 (1986).
64. Sposito, G. *The Surface Chemistry of Soils* (London: Oxford University Press, 1984), p. 234.
65. Weiss, C.A., Jr., R.J. Kirkpatrick, and S.P. Altaner. "The Structural Environments of Cations Adsorbed onto Clays: ^{133}Cs Variable-Temperature MAS NMR Spectroscopic Study of Hectorite," *Geochim. Cosmochim. Acta* 54:1655–1669 (1990).
66. Crowther, D.L., J. Dillard, and J. Murray. "The Mechanism of Co(II) Oxidation on Synthetic Birnessite," *Geochim. Cosmochim. Acta* 47:1399–1403. (1983).
67. Loganathan, P. and R.G. Bureau. "Sorption of Heavy Metal Ions by Hydrous Manganese Oxide," *Geochim. Cosmochim. Acta* 37:1277–1293 (1973).
68. Burns, R.G. "The Uptake of Cobalt into Ferromanganese Nodules, Soils and Synthetic Manganese(IV) Oxides," *Geochim. Cosmochim. Acta* 40:95–102 (1976).
69. Bidoglio, G., P.N. Gibson, M. O'Gorman, and K.J. Roberts. "X-ray Absorption Spectroscopy Investigation of Surface Redox Transformations of Tl and Cr on Colloidal Mineral Surfaces," *Geochim. Cosmochim. Acta* (in press).
70. Xu, N., G.E. Brown, Jr., G.A. Parks, and M.F. Hochella, Jr. "Sorption Mechanism of Co^{2+} at the Calcite-Water Interface," *Abstracts with Programs* (Washington, D.C.: Geological Society of America 1990), p. 294.
71. Stipp, S.L., M.F. Hochella, Jr., G.A. Parks, and J.O. Leckie. "Cd^{2+} Uptake by Calcite Solid-State Diffusion and the Formation of Solid-Solution; Interface Processes Observed with Near-Surface Sensitive Techniques (XPS, LEED, AES)," *Geochim. Cosmochim. Acta* 56:1941–1954 (1992).

72. Boehm, P. "Acidic and Basic Properties of Hydroxylated Metal Oxide Surfaces," *Discuss. Faraday Soc.* 52:264–275 (1971).
73. James, R.O. and G.A. Parks. "Characterization of Aqueous Colloids by their Electric Double Layer and Intrinsic Surface Chemical Properties," *Surf. Colloid Sci.* 12:19–40 (1982).
74. Davis, J.A. and D.B. Kent. "Surface Complexation Modeling in Aqueous Geochemistry," in *Mineral-Water Interface Geochemistry,* M.F. Hochella and A.F. White, Eds., Reviews in Mineralogy 23 (Washington, D.C.: Mineralogical Society of America, 1990), pp.177–260.
75. Roe, A.L., K.F. Hayes, C.J. Chisholm, G.E. Brown, Jr., G.A. Parks, K.O. Hodgson, and J.O. Leckie. "In Situ X-Ray Absorption Study of Lead Ion Surface Complexes at the Goethite-Water Interface," *Langmuir* 7:367–373 (1991).
76. Stumm, W. and J.J. Morgan, *Aquatic Chemistry,* 2nd. ed. (New York, John Wiley & Sons, 1981), p. 780.
77. Van Cappellen, P. "The Formation of Marine Apatite. A Kinetic Study," PhD Thesis, Yale University, Yale, MA (1991).
78. Baes C.F. and R.E. Mesmer. *The Hydrolysis of Cations* (New York: John Wiley & Sons, 1976), p.458.
79. Chisholm-Brause, C.J., K.F. Hayes, A.L. Roe, G.E. Brown, Jr., G.A. Parks, and J.O. Leckie. "Spectroscopic Investigation of Pb(II) Complexes at the δ–Al_2O_3/water Interface," *Geochim. Cosmochim. Acta* 54:1897–1909 (1990).
80. Bouwens, S.M.A.M., J.A.R. van Veen, D.C. Koningsberger, V.H.J. de Beer, and R. Prins. "Extended X-Ray Absorption Fine Structure Determination of the Structure of Cobalt in Carbon-Supported Co and Co-Mo Sulfide Hydrodesulfurization Catalysts," *J. Phys. Chem.* 95:123–134 (1991).
81. Manceau, A. "Distribution of Cations Among the Octahedra of Phyllosilicates: Insight from EXAFS," *Can. Mineral.* 28:321–328 (1990).
82. Hazemann, J.L., A. Manceau, P. Sainctavit, and C. Malgrange. "Structure of the $\alpha Fe_xAl_{1-x}OOH$ Solid Solution. I. Evidence by Polarized EXAFS for an Epitaxial growth of Hematite-like Clusters in Diaspore," *Phys. Chem. Miner.* 19:25–38 (1992).
83. Van Olphen, H. *An Introduction to Clay Colloid Chemistry,* 2nd ed. (New York: John Wiley & Sons, 1977), p. 318.
84. Veblen, D. R. "Polysomatism and Polysomatic Series: A Review and Applications," *Am. Mineral.* 76: 801–826 (1991).
85. Bailey S.W. "Crystal Chemistry of the True Micas," in *Micas,* S.W. Bailey, Ed., Reviews in Mineralogy, Vol. 13, (Washington, D.C.: Mineralogical Society of America, 1984), pp.13–60.
86. Decarreau, A., D. Bonnin, A. Badaut-Trauth, R. Couty, and P. Kaiser. "Synthesis and Crystallogenesis of Ferric Smectite by Evolution of Si-Fe Coprecipitates in Oxidizing Conditions," *Clay Miner.* 22:207–223 (1987).
87. Leckie, J.O. Personal communication.
88. James, R.O. and T.W. Healy. "Adsorption of Hydrolyzable Metal Ions at the Oxide-Water Interface I. Co(II) Adsorption on SiO_2 and TiO_2 Model Systems," *J. Colloid Interface Sci.* 40: 42–52 (1972).
89. James, R.O. and T.W. Healy. "Adsorption of Hydrolyzable Metal Ions at the Oxide-Water Interface II. Charge Reversal of SiO_2 and TiO_2 Colloids by Adsorbed Co(II), La(III), and Th(IV) as Model Systems," *J. Colloid Interface Sci.* 40: 53–64 (1972).

90. O'Day, P.A., G.E. Brown, Jr., and G.A. Parks. "EXAFS Study of Aqueous Co(II) Sorption Complexes on Kaolinite and Quartz Surfaces," *Abstracts with Programs* (Washington, D.C.: Geological Society of America, 1990).
91. Fukushima, Y. and T. Okamoto. "Extended X-Ray Absorption Fine Structure Study of Cobalt-Exchanged Sepiolite," in *Proceedings of the International Clay Conference*, Denver, CO (Bloomington, IN: The Clay Minerals Society, 1985), pp. 9–16.
92. Manceau, A. and A. Decarreau. "Extended X-Ray Absorption Fine-Structure Study of Cobalt-Exchanged Sepiolite: Comment on a Paper by Y. Fukushima and T. Okamoto," *Clays Clay Miner.* 36:382–383 (1988).
93. Decarreau, A. "Partitioning of Divalent Transition Elements Between Octahedral Sheets of Trioctohedral Smectites and Water," *Geochim. Cosmochim. Acta* 49:1537–1544 (1985).
94. Bonneviot, L., O. Clause, M. Che, A. Manceau, and H. Dexpert. "EXAFS Characterization of the Adsorption Sites of Nickel Ammine and Ethylenediamine Complexes on a Silica Surface" *Catal. Today* 6:39–46 (1989).
95. Rayner, J.H. and G. Brown. "The Crystal Structure of Talc," *Clays Clay Miner.* 60: 1030–1040 (1973).
96. Takeda, H. and M. Ross. "Mica Polytypism, Dissimilarities in the Crystal Structures of Coexisting +M and 2M1 Biotite," *Am. Mineral.* 60:1030–1040 (1975).
97. Brindley G.W. and G. Brown. *Crystal Structures of Clay Minerals and Their X-ray Identification* (London: Mineralogical Society, 1980), p. 495.
98. Charlet, L. and A. Manceau. (in preparation).
99. Stumm, W. and R. Wollast. "Coordination Chemistry of Weathering: Kinetics of the Surface-Controlled Dissolution of Oxide Minerals," *Rev. Geophys.* 28:53–69 (1990).
100. Wieland, E., B. Wehrli, and W. Stumm. "The Coordination Chemistry of Weathering: III. A Generalization on the Dissolution Rates of Minerals," *Geochim. Cosmochim. Acta* 52: 1969–1981 (1988).
101. Hering J.G. and W. Stumm. "Oxidative and Reductive Dissolution of Minerals," in *Mineral-Water Interface Geochemistry*, M.F. Hochella and A.F. White, Eds., Reviews in Mineralogy 23 (Washington D.C.: Mineralogical Society of America, 1990), pp. 427–466.
102. Greaves, G.N., N.T. Barrett, G.M. Antonini, F.R. Thornley, B.T.M. Willis, and A. Steel. "Glancing-Angle X-Ray Absorption Spectroscopy of Corroded Borosilicate Glass Surfaces Containing Uranium," *J. Am. Chem. Soc.* 111:4313–4324 (1989).
103. Abrajano, T.A., J.K. Bates, A.B. Woodland, J.P. Bradley, and W.L. Bourcier. "Secondary Phase Formation During Nuclear-Waste Dissolution," *Clays Clay Miner.* 38:537–548 (1990).
104. Eary, L.E. and E. Rai. "Kinetics of Chromium (III) Oxidation to Chromium (VI) by Reaction with Manganese Dioxide," *Env. Sci. Tech.* 21:1187–1193 (1987).
105. Johnson, C.A. and A.G. Xyla. "The Oxidation of Cr(III) to Chromium (VI) on the Surface of Manganite (MnOOH)," *Geochim. Cosmochim. Acta* 55:2861–2866 (1991).
106. Golden, D.C., J.B. Dixon, and C.C. Chen."Ion Exchange, Thermal Transformations, and Oxidizing Properties of Birnessite," *Clays Clay Miner.* 34:511–520 (1986).
107. Sandstrom, D.R., B.R. Stults, and R.B. Greegor. "Structural Evidence for Solutions from EXAFS Measurements," in *EXAFS Spectroscopy: Techniques and Applications*, B.K. Teo and D.C. Joy, Eds. (New York: Plenum Publishing Corp., 1981), pp. 139–57.

108. Sandstrom, D.R. "Ni^{2+} Coordination in Aqueous $NiCl_2$ Solutions: Study of the Extended X-ray Absorption Fine Structure," *J. Chem. Phys.* 71:2381 (1979).
109. Sandstrom, D.R. "EXAFS Studies of Electrolyte Solutions," in *EXAFS and Near-Edge Structure III*, K.O. Hodgson et al., Eds. (New York:Springer-Verlag, 1984), p. 409–413.
110. Farges, F., J.A. Peck, and G.E. Brown, Jr. "Local Environment around Gold(III) in Aqueous Chloride Solutions: an EXAFS Spectroscopic Study," *Geochim. Cosmochim. Acta* (in press).
111. Brown, G.E., Jr., D.A. Parkhurst, and G.A. Parks. "Zinc Complexes in Aqueous Chloride Solutions: Structure and Thermodynamic Modelling," *Geochim. Cosmochim. Acta* (submitted).
112. McKale, A.G., B.W. Veal, A.P. Paulikas, S.K. Chan, and G.S. Knapp. "Improved ab initio Calculations for Extended X-Ray Absorption Fine Structure Spectroscopy," *J. Am. Chem. Soc.* 110:3763–3768 (1988).

CHAPTER 4

USE OF FIELD-FLOW FRACTIONATION TECHNIQUES TO CHARACTERIZE AQUATIC PARTICLES, COLLOIDS, AND MACROMOLECULES

Ronald Beckett and Barry T. Hart

TABLE OF CONTENTS

0-87371-895-X/93/$0.00+$.50

INTRODUCTION

The speciation of contaminants (e.g., nutrients, heavy metals, and organics) is very important in determining their transport, fate, and effects in aquatic systems, including rivers, lakes, oceans, and groundwaters.[1-3] Pollutants may occur as freely dissolved molecules or ions in the water or associated with larger macromolecules (e.g., humic substances), colloids, or particles, as depicted in Figure 1. The colloid phase, which is often defined as particles between 1 nm and 1 μm in size, is particularly important due to its high specific surface area and resulting potential adsorptive capacity, as well as the fact that such particles have little tendency to settle out of suspension. However, to date most work has been done using only two operationally defined fractions, "particulate" and "dissolved," separated generally on the basis of filtration through a 0.4 μm (or 0.45 μm) membrane filter, in which most of the important colloidal fraction is included in the "dissolved" fraction.

Hart and Hines[4] have recently pointed out that this simple two phase separation can hide the complexity of the interactions occurring and may provide incorrect information on the speciation and bioavailability of a particular contaminant. This can be illustrated by the results from a study of the processes controlling the behavior of copper and zinc in a river receiving input from a naturally oxidized ore body.[5] Water samples were first separated into "dissolved" and "particulate" fractions using tangential flow filtration (TFF) with a 0.2 μm membrane.[4] The "dissolved" fraction was then further separated into a colloidal fraction and a soluble fraction using a 10,000 relative molar mass filter (equivalent to approximately 3 nm diameter). Although the colloidal fraction made up only 4% of the mass of the "dissolved" fraction, it contained 62% of the "dissolved" Cu, 99.6% of the "dissolved" Fe and 99.9% of the "dissolved" Zn. This additional information, showing that in this polluted section of the Tambo River essentially all the Fe and Zn and a considerable proportion of the Cu are associated with colloidal matter and are not dissolved in the traditional sense, completely changes the interpretation of the possible toxic effects of the copper and zinc.

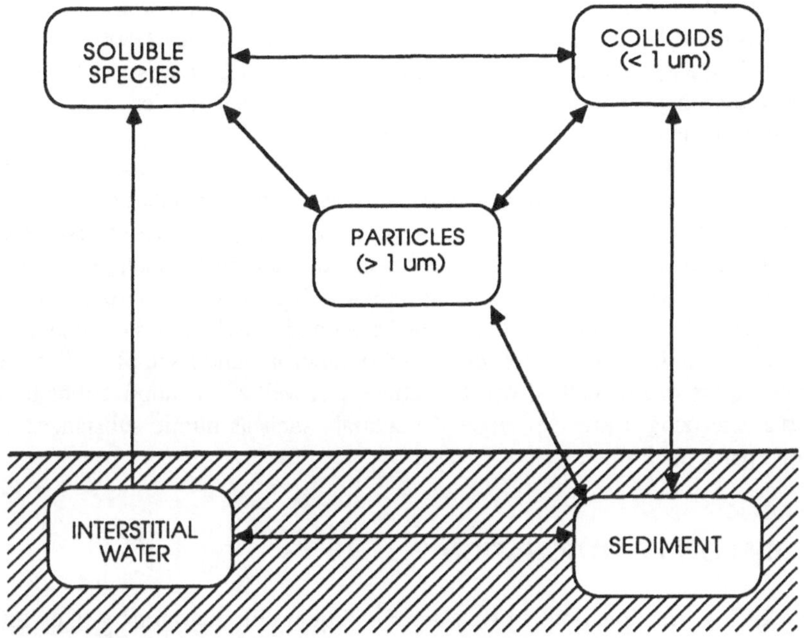

Figure 1. Major compartments for trace elements in aquatic systems.

Despite the considerable amount of evidence regarding the importance of suspended particulate matter (SPM) in determining the fate of contaminants in aquatic systems, and the growing evidence on the corresponding importance of aquatic colloidal matter, we still have a rather poor knowledge of the composition of these materials and of the way they associate with contaminants. This is in stark contrast with the very large volume of published work on bottom sediments and sediment-contaminant associations.[1,6]

Another very important component of natural waters is the dissolved organic matter. In particular, the humic substances fraction has been found to play a significant role in a number of processes.[7] The ubiquitous organic coatings on particles are responsible for establishing a negative surface charge[8,9] which arises predominantly from the ionizable carboxylic acid functional groups associated with the humic material. The actual charge and the corresponding electrophoretic mobility are greatly influenced by the concentration of Ca^{2+} and Mg^{2+} ions in the water.[9,10] An important implication of this negative charge is that these natural colloids are relatively stable in most freshwater systems, although they will tend to aggregate when they enter estuarine regions and the ocean.[11]

Humic substances, either dissolved or absorbed to particles, are capable of binding trace metals and organic pollutants, particularly those with low polarity. This strongly influences the partitioning of these contaminants between dissolved and colloidal or particulate phases, hence effecting their transport and toxicity in aquatic systems. Again, it can be said that there is insufficient knowledge about the chemical and physical nature of this important class of material. This point is well illustrated by the wide range of molecular sizes (at least 500 to 100,000 relative *M* mass) that have been reported for humic substances,[12] which could reflect either their inherent variability or perhaps the inadequacies of the methods currently available for such measurements.

The above discussion illustrates the need for better characterization methods for macromolecular, colloidal, and particulate species in natural waters. One approach that is often used to investigate very complex mixtures is to first separate the material into some simpler fractions based on parameters such as size, mass, density, or polarity. This chapter will review the status of a relatively new separation and sizing method, field-flow fractionation, which has the potential to yield very detailed information on complex environmental samples. The background theory of the method will be outlined, as well as the applications to date for characterizing a diverse range of materials such as humic substances, suspended particulate matter, and bacteria.

FIELD-FLOW FRACTIONATION

Field-flow fractionation (FFF) is a set of liquid chromatography-like elution methods which have been used to achieve high resolution separation and sizing of a wide range of particulate, colloidal, and macromolecular materials.[13-15] Environmental samples that have been studied to date include humic substances,[16,17] clays (Murphy, unpublished results), bacteria (Sharma, unpublished results), suspended particulate matter,[18-21] soils[22] and sediments.[23] The potential for FFF techniques to be applied to help understand complex environmental samples has scarcely been tapped, particularly when it is realized that they can be applied to samples ranging in size from molecules smaller than 1000 in relative *M* mass up to particles larger than 50 μm in diameter. This represents an enormous five orders of magnitude in size or perhaps 15 orders of magnitude in mass.

In this section we will outline the general principle of FFF as well as describing the different modes of operation, subtechniques, and run conditions that are used for analyzing various types of samples.

The FFF Principle

FFF separations are carried out within a flat open channel, usually having a rectangular cross section and possessing triangular end pieces where the sample and carrier fluid enters and leaves, as shown in Figure 2. This is in contrast to liquid chromatography, which most often employs a packed bed column or, in some cases, a fine capillary tube. The typical dimensions of an FFF channel are 30 to 100 cm long, about 1 to 3 cm wide, and only 0.05 to 0.5 mm thick.

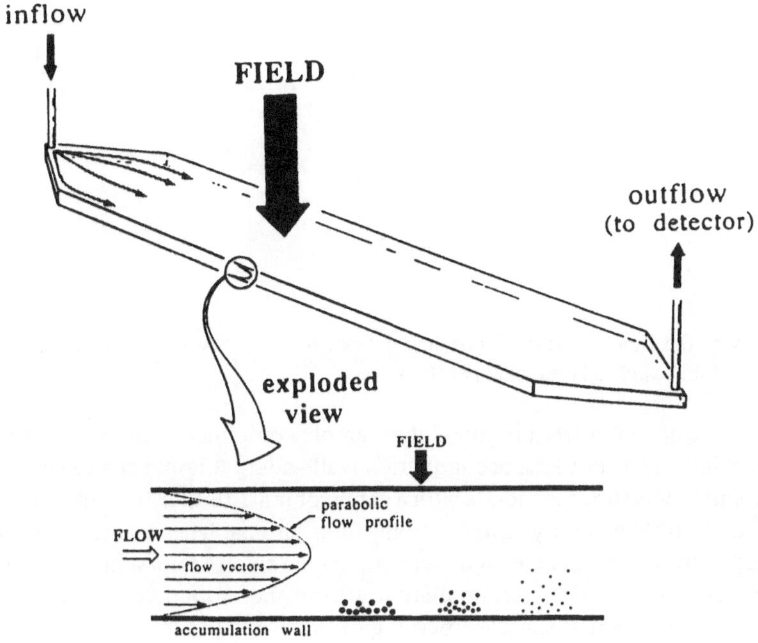

Figure 2. Schematic representation of an FFF channel with the separation mechanism for normal FFF shown in detail.

Thus, one of the obvious advantages of FFF is that there will be much less opportunity for either attractive or repulsive interactions with the stationary phase. Such unwanted interactions can give rise to anomalous retention behavior which will interfere with the size calibration. In the case of chromatography, sample adhesion or the presence of oversized particles may even cause column blockage, resulting in drastically reduced column lifetimes.

The separation mechanism employed in FFF is fundamentally different from chromatography and involves only physical interactions with no chemical interactions being present. The sample solution or suspension is first introduced into one end of the channel through a septum or injection valve, and then the flow is turned off. A field (such as gravitational, centrifugal, fluid crossflow, thermal gradient, electrical, or magnetic) is then applied at right angles to the flat face of the channel. This field drives the sample molecules or particles across the thin channel towards the accumulation wall. Each sample component forms an equilibrium cloud, whose average thickness or elevation above the wall depends on factors such as how strongly the particles interact with the field and the sample diffusivity. Following this relaxation period, which must be long enough for a particle situated at the top wall to be transported across the channel to the accumulation wall, flow of carrier liquid down the length of the channel is initiated, the run begins.

Laminar flow conditions (i.e., low Reynolds number) prevail in such an FFF channel, with the fluid velocity vectors having a typical parabolic profile across the thin dimension, as depicted in Figure 2. The linear fluid velocity is defined by the equation

$$v = 6 < v > \left[\frac{x}{w} - \left(\frac{x}{w} \right)^2 \right] \tag{1}$$

where

v is the fluid velocity at distance χ from the wall, <v> is the mean linear flow velocity and ω is the channel thickness

It will be apparent from Figure 2 that sample components forced to form more compact layers closer to the accumulation wall should migrate at a lower velocity than components forming clouds with a center of mass further from the wall where they can be influenced by faster moving fluid lamina. The degree of separation attained between two species will depend, among other factors, on the difference in the elevation above the accumulation wall of their cloud center of mass.

The concentration of sample emerging from the end of the channel is generally measured by passing the eluent through a sensitive detector. The plot of this detector response against elution volume V or time for elution t is called a fractogram (by analogy to chromatogram), and this contains important information about the sample, such as its particle size or molar mass distribution. It is common practice to record the relative elution behavior of a particular component by calculating the retention ratio, which is the ratio of the sample migration velocity down the channel v_r to the average linear carrier liquid velocity $<v>$. For runs conducted under constant experimental conditions (e.g., field strength, flow rate) the retention ratio may be calculated from

$$R = \frac{v_r}{< v >} = \frac{t^o}{t_r} = \frac{V^o}{V_r} \tag{2}$$

where

V_r and t_r are the sample elution volume and time respectively
V^o is the channel volume or void volume
t^o is the elution time for a totally unretarded highly diffusive molecule

Modes of FFF

The structure of the equilibrium sample clouds produced following relaxation depends on the relative magnitudes of the force on the particles generated by the applied field, the sample thermal motion (i.e., Brownian motion) which generates concentration diffusion transport, and any additional forces that the particles may experience under the particular experimental conditions that are set up in the channel. Three types of sample configuration are commonly encountered at the present time, which give rise to what are referred to as the different modes of FFF operation, namely normal FFF, steric FFF, and hyperlayer FFF.

Normal Mode of FFF

The applied field causes the particles to migrate at a terminal velocity U and they eventually concentrate at the accumulation wall. At this stage, back diffusion will tend to occur driven by the concentration gradient established adjacent to the accumulation wall. Equilibrium will be reached when the field induced flux of sample towards the wall is exactly balanced by the back diffusional flux at all positions across the channel. The resultant sample cloud has a concentration profile that decreases exponentially with distance from the wall according to the equation

$$c = c_o e^{-x/l} \tag{3}$$

where

c is the concentration at distance x from the accumulation wall
c_o is the concentration at the wall (when $x = 0$)

The characteristic mean cloud thickness l is determined by the relative magnitudes of the particles diffusion coefficient D, and their field induced transport velocity U according to the expression

$$l = \frac{D}{U} = \frac{kT}{F} \tag{4}$$

where

k is the Boltzman constant
T is the absolute temperature
F is the force on a particle due to the applied field

Alternatively, it can be seen that l is the ratio of the thermal energy kT to the driving force F exerted on the particles.

The net migration rate of a sample cloud down the channel will be governed by the integrated effect of the parabolic profile of fluid velocity vectors on the exponential distribution of sample particles across the channel (Figure 3a), as defined by Equations 1 and 3, respectively. It was originally shown by Giddings[24] that this gives rise to the fundamental equation for normal FFF, which relates the measured retention ratio R to the cloud thickness l as follows

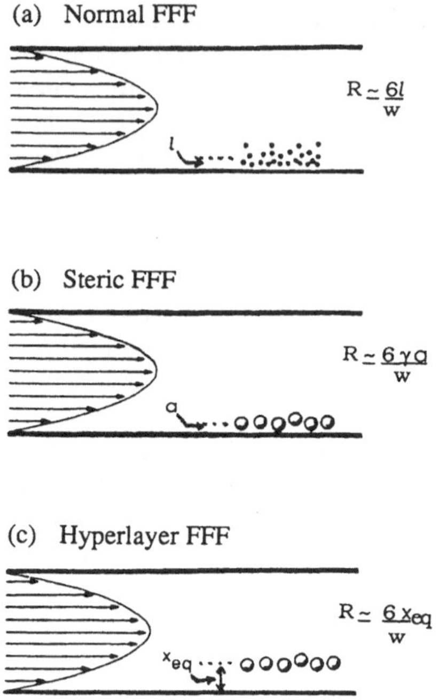

Figure 3. Schematic representation of the three major modes (or mechanisms) for FFF separations: (a) normal, (b) steric, and (c) hyperlayer.

$$R = 6\lambda\left[\coth\frac{1}{2\lambda} - 2\lambda\right] \simeq 6\lambda \tag{5}$$

where l is called the retention parameter and is the dimensionless cloud thickness relative to the channel thickness w

$$\lambda = \frac{\ell}{w} = \frac{kT}{Fw} \tag{6}$$

It should be noted that the sample cloud will only stay together provided that the rate of vertical mixing in the cloud due to Brownian motion is much greater than its horizontal velocity down the channel. For this reason, nonequilibrium peak broadening can result if the channel flow rate is too high, particularly in the case of poorly retained samples which have large values of λ and larger (hence less diffusive) sample components.

Having obtained l for a given sample component, by measuring its elution volume in an FFF experiment run in the normal mode, it is possible to calculate directly some fundamental parameters of the particles or molecules, such as their mass, diameter, diffusion coefficient, and so on. The particular parameters that can be obtained depend on the nature of the field that was employed in the FFF subtechnique used in the experiment. These relationships will be explored in the section on subtechniques of FFF which follows.

Steric Mode of FFF

When the field induced force on the particles is increased, a stage will be reached when the particle radius, a, becomes an appreciable fraction of the cloud thickness l. This will create perturbations to the exact form of Equation 5, which is based on the assumption that the particles are point masses and are thus unimpeded in their approach to the accumulation wall. Myers and Giddings[25] derived a simple modification to the approximate form of the general retention equation to take into account this steric exclusion effect:

$$R \simeq \frac{6l}{w} \quad (7)$$

where

a is the particle radius
γ is an empirical correction factor which takes into account several nonideal phenomena

The value of γ will be unity for a sample which behaves ideally.

In the extreme case, where the field driven flux of sample overwhelms thermal (Brownian) motion and hence concentration driven diffusion, the particles are all located against the accumulation wall, so that their center of mass is at a distance equivalent to one particle radius from the wall. Under these circumstances the larger particles, which extend further across the channel, will be pushed by the fluid flow at a higher velocity than smaller particles, which are only under the influence of slower moving liquid laminae. This situation is depicted in Figure 3b and results in an inversion of the elution order when compared to normal FFF, which, it will be recalled, results in the particles with lower effective mass being eluted first.

Extrapolation of Equation 7 to the steric limit, where $a >> l$, produces

$$R = \frac{6\gamma a}{w} \quad (8)$$

This condition is commonly encountered for particles with diameter greater than about 1 to 5 μm, although the position of the steric inversion point depends on a number of factors including the field strength, particle density, and flow rate.

The factor γ is added to take into account extraneous effects, such as particle-wall interactions (both attractive and repulsive) and hydrodynamic lift forces. These effects are complex; hence γ is usually unknown, and in general an accurate estimate of the particle size cannot be obtained directly from Equation 8. Thus, at this stage it is necessary to calibrate steric FFF retention data using particle size standards. A good range of polymer latex bead standards is commercially available, although it is generally necessary to compensate for the difference in density between the latex and sample particles. The appropriate calibration methods will be outlined in the discussion of the subtechniques below. With these calibration procedures, it should be possible to make steric FFF much more applicable to a range of samples. However, difficulties may still be encountered in dealing with heterogeneous samples containing particles of different density and nonspherical shape.

Hyperlayer Mode of FFF

If an additional force is present opposing the main external field, it may be possible to elevate the sample cloud above the accumulation wall, as illustrated in Figure 3c. The sample will take up a position in the channel where these two forces balance and will be swept down the channel by the corresponding flow vector, which can be obtained through Equation 1. This has obvious advantages in avoiding some of the uncharacterized wall interactions that can occur particularly with steric FFF. It is conceivable that such an opposing force could be provided by a second external field (e.g., electrical, gravitational, crossflow), repulsive wall interactions (e.g., electrostatic), creation of a density gradient across the channel, or hydrodynamic lift forces due to high carrier flow rates. The latter approach has proved to be particularly effective for separation of silt-sized particles and will be described in more detail.

It has been recognized for some time[26,27] that an increase in retention ratio can be obtained in steric FFF runs by increasing the channel flow rate. An increase in R for a given particle translates to an increase in γ in Equation 6. If γ becomes greater than unity, it implies that the particle center is elevated above its ideal value, which should correspond to the particle radius a . The force opposing the external field in this case is apparently due to the so-called hydrodynamic lift forces which are known to induce migration of particles away from the walls in tubular Poiseuille flow.[28]

This hydrodynamic lift force effect has been used in a practical sense, such as to prevent particle adhesion to the channel walls in cell separations[29] and to effect very high speed separations of large (> 1 μm) particles by steric FFF.[30] It should be pointed out that steric FFF can be considered as the limiting form of hyperlayer FFF obtained when the particles are forced to approach contact with the wall. The rather arbitrary distinction we will make between the two modes is to define hyperlayer FFF as being all cases where $\gamma > 2$, which occurs when the distance of the particle center above the wall is more than twice the particle radius.

Perhaps the first clear demonstration of the hyperlayer mode was reported by Giddings et al.,[31] who attained values for γ of up to 6, although undoubtedly some of the runs reported as steric FFF previously should strictly be classified as hyperlayer FFF by the current definition.

A cloud of particles positioned at a distance x_{eq} above the accumulation wall will be carried down the channel by the carrier at a velocity v_r as defined by Equation 1. Thus the retention ratio will be given by[32]

$$R = \frac{v_r}{<v>} = 6\left[\frac{x_{eq}}{w} - \left(\frac{x_{eq}}{w}\right)^2\right] \tag{9}$$

Since the value of x_{eq} depends on the balance of the force generated by the external field and the imperfectly characterized hydrodynamic lift force (both of which will normally be related in some manner to particle parameters such as diameter and mass), it is not possible at this stage to determine the particle size or mass from first principles. However, hyperlayer FFF runs are amenable to empirical procedures which, in some cases, lead to quite universal calibration curves. Hyperlayer FFF is a rapidly expanding technique which is challenging conventional methods used in the silt size range in terms of its speed, resolution, and accuracy, as will be illustrated in later sections.

Subtechniques of FFF

Different fields and gradients may be employed to drive the sample components across the thin FFF channel selectively . This gives rise to the various subtechniques of FFF, the most common being sedimentation FFF using gravitational or centrifugal fields, flow FFF using a fluid crossflow, and thermal FFF where a temperature gradient is employed. In addition, electrical and magnetic fields have been utilized with some success.

Each field will exert a force on the particles or molecules which will be related to some particular parameter. Hence, different subtechniques may give rise to distinct pieces of information about the sample. For example, sedimentation FFF yields the effective mass of the particles in the fluid, whereas flow FFF gives the diffusion coefficient.

Different subtechniques are better suited for the analysis of a specific range of sample types and sizes. For example, sedimentation gives exceptional resolution in the colloid size range, flow FFF is more universal, being suitable for a very wide range of water soluble macromolecules and particles up to 100 μm in diameter, and thermal FFF has been applied mainly for determining relative molar mass distributions of synthetic polymers dissolved in organic solvents. In this article we will concentrate on two subtechniques, namely sedimentation FFF and flow FFF, which have now been used at least to some extent for the characterization of environmental samples.

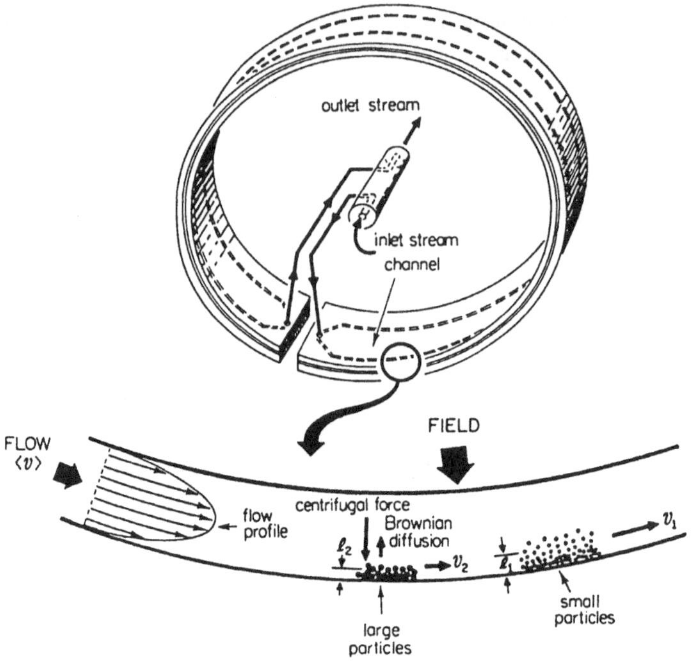

Figure 4. Diagram of a sedimentation FFF centrifuge with cross section of channel showing a normal mode separation.

Sedimentation FFF

Perhaps the simplest FFF experiment to imagine is to place the channel on a bench and to allow the Earth's gravitation to act as the field. The gravitational force is only sufficiently large to give appreciable retention for particles greater than a micron or so, and since such particles will have a low diffusion coefficient, this simple experiment is generally limited to steric FFF separations in the silt size range. It appears that increasing the flow rate to utilize the hyperlayer mode generally results in decreased resolution unless the field is also increased in order to keep the sample layers quite close to the wall.[27]

In order to achieve good retention in sedimentation FFF for submicron particles, it is necessary to increase the external field strength. This can be done conveniently by constructing a circular channel which is inserted inside a centrifuge basket, as shown schematically in Figure 4.

The force F exerted on a particle in a centrifugal field depends on the effective particle mass and the field strength G according to the expression

$$F = V_p \Delta\rho G \tag{10}$$

where

V_p is the particle volume
$\Delta\rho$ is the density difference between the particle and carrier liquid

Combining Equations 6 and 10 yields an expression for the retention parameter λ applicable for normal mode sedimentation FFF runs conducted at a constant field strength. On substituting $G = \omega^2 r$ and $V_p = \pi d^3/6$ this expression becomes

$$\lambda = \frac{6kT}{\pi\omega^2 rw\Delta\rho d^3} \tag{11}$$

where

ω is the angular speed
r is the centrifuge radius
k is the Boltzman constant
T is the absolute temperature
d is the equivalent spherical diameter of the particles

Thus, provided the particle density is known, it is possible to calculate, from first principles, the equivalent spherical diameter using the experimentally determined retention volume of the sample. This reliance on particle density may seem an impediment in some circumstances, but it also provides a means of calculating the particle density. Equation 11 can be transposed to give

$$\rho = \rho_p - \frac{6kT}{\pi\omega^2 rwd^3\lambda} \tag{12}$$

where

ρ_p is the particle density and ρ is the carrier liquid density

Thus, if separate runs are made with carriers of varying density, a plot of ρ against $1/\lambda$ yields both the diameter (from the gradient), and the particle density (from the y-intercept).

The resolution that can be attained in sedimentation FFF is excellent and stems from the cube dependence on d of the retention parameter λ (Equation 11) and hence elution volume V_r (Equations 2 and 5). This is formally expressed as the diameter based selectivity S_d which is defined as

$$S_d = \left|\frac{\delta V_r / V_r}{\delta d / d}\right| = \left|\frac{d\ln V_r}{d\ln d}\right| \tag{13}$$

where δV_r is the change in V_r caused by a very small increment in d denoted by

δd. For constant field sedimentation FFF runs, Equation 13 gives rise to the high selectivity value of 3. This high selectivity is offset somewhat by the fairly low column efficiencies obtained with FFF. The major contribution to peak broadening in normal FFF is the nonequilibrium effect caused when diffusion within the sample cloud is not rapid enough to prevent the differing fluid velocity vectors, which occur at various distances from the wall, spreading out the peak. This can be minimized by limiting the flow rate, but at the expense of increased run times. Despite the fact that the number of theoretical plates for FFF is often less than 500, quite low when compared to many chromatographic separations, the overall resolution of sample components with sedimentation FFF must still be considered excellent and is probably greater than any other particle sizing method commonly in use.

As mentioned previously, the upper limit for normal FFF is around 1 μm, above which steric or hyperlayer modes must be employed. Under most operating conditions generally used with sedimentation FFF for these larger particles, γ is < 2, thus falling within the definition of steric FFF. Although the diameter based selectivity for sedimentation/steric FFF[31] is < 1, some very impressive separations can be achieved. For example, Koch and Giddings[30] were able to separate a mixture of seven latex beads in the diameter range 2 to 45 μm to baseline resolution in < 4 min. Such analyses are done by carefully balancing the hydrodynamic lift force and centrifugal force to give the desired compromise between speed and resolution.

Since hydrodynamic lift forces depend on the particle size in an as yet unquantified manner and the sedimentation force depends on both size and density, it is not possible to determine the equivalent spherical diameter from first principle calculations. However, Giddings et al.[33] have recently reported a method for calibrating sedimentation/steric FFF runs using particle size standards, provided the density of both the sample and standards are known. This approach involves compensating for the difference in density of the standard and sample particles by altering the field strength, with all other experimental conditions being held constant, so that the force F exerted on like-sized particles is the same (see Equation 10).

Flow FFF

Dispersed molecules or particles can be swept towards the accumulation wall by a crossflow of carrier liquid in the subtechnique of flow FFF. To achieve this, the channel walls must be constructed of porous frits, as depicted in Figure 5, and generally a membrane is stretched over the accumulation wall to prevent loss of sample from the confines of the channel. All sample components will move with the linear crossflow fluid velocity U until they reach the wall. In this case the field force is induced by the frictional drag on a particle held stationary by the membrane and with the carrier liquid flowing past. This force is given by the simple law

$$F = fU \tag{14}$$

where f is the friction coefficient and U is obtained from the volumetric crossflow rate $\dot{V}_c$ and channel dimensions from

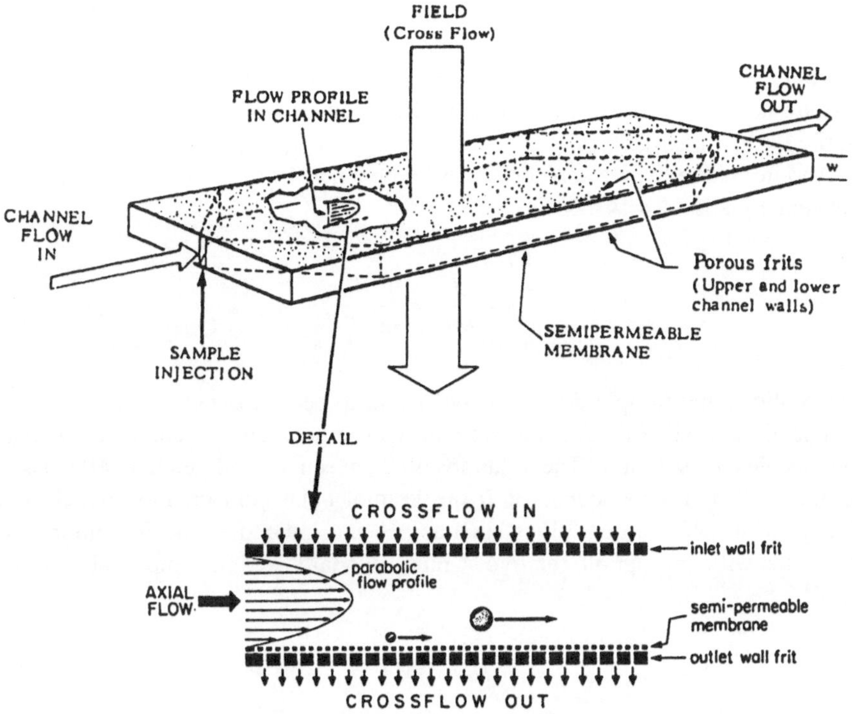

Figure 5. Diagram of a flow FFF channel showing the porous frits required for crossflow operation. Exploded view of the channel cross section shows a hyperlayer mode separation.

$$U = \frac{\omega \dot{V} c}{V^o} \tag{15}$$

Thus in the normal mode of flow FFF, an expression for λ is obtained by combining Equations 6, 14, and 15 to give

$$\lambda = \frac{kT}{\omega U f} = \frac{kTV^o}{\omega^2 \dot{V}_c f} \tag{16}$$

Using the Stokes equation (f = $3\pi\eta d$), which is applicable for spherical particles, this translates to

$$\lambda = \frac{kT}{3\pi\eta\omega U d} = \frac{kTV^o}{3\pi\eta\omega^2 \dot{V}_c d} \tag{17}$$

which enables the equivalent spherical diameter of the particles to be calculated.

Note that for flow/normal FFF the selectivity S_d is one; however, data reported in the literature so far show that the theoretical resolution is not being achieved, probably due to instrumental peak broadening effects. As better membranes and frits are found the performance of flow FFF, separations should improve considerably in this regard.

An alternative expression for λ can be obtained from Equation 16 using the Einstein equation ($f = kT/D$)

$$\lambda = \frac{D}{wU} = \frac{V^o D}{\omega^2 \dot{V}_c} \tag{18}$$

which allows the sample diffusion coefficient to be calculated.

The diffusion coefficient is a commonly used parameter to characterize macromolecules in solution. The relationship between D and relative M mass is complicated by a dependence of D on the molecular conformation which may change with solution conditions such as pH and ionic strength. Retention data from runs with appropriate relative M mass standards and the empirical relationship[34]

$$D = A' M^{-b} \tag{19}$$

are often used to establish a functional calibration procedure. This procedure will be illustrated in the section below on determining size distributions. Here A' and b are constants for a given polymer-solvent system and M is the relative molar mass.

A different experimental approach has been used recently which involves using only one liquid inlet flow at the start of the channel and two outlet flows as usual. Two configurations of this asymmetric flow FFF are currently in use. The normal flat channel can be modified by replacing the inlet frit (depletion wall) with a nonpermeable glass plate.[35] Alternatively, a fine porous cylinder, such as a hollow fiber filtration tube, can be employed.[36] This approach shows some promise but to date has only been used to separate proteins and some other biological materials.[37,38] As yet it has not been used in environmental studies and thus will not be discussed further in this paper.

Choice of FFF Conditions

Because of the inherent flexibility of FFF outlined above, the researcher is faced with a number of decisions in designing the best FFF experiment for a given type of sample. Although the factors which govern these decisions are quite complex, it is possible to offer some guidelines which, at the risk of oversimplifying the situation, we believe can act as a good starting point.

Table 1. Approximate Diameter Ranges (μm) over which Various Subtechniques and Modes of FFF are Applicable

Subtechnique Mode	Normal	Steric	Hyperlayer
Sedimentation	0.03–2	0.5–100	1–100
Flow	0.001–2	0.5–100	1–100

Subtechniques and Modes of FFF

The approximate diameter range over which each mode of operation and subtechnique is applicable is presented in Table 1. The transition between the normal mechanism and the steric/hyperlayer modes occurs at around 1 μm. However, it should be noted that this transition occurs over a diameter range (perhaps 1 to 2 μm) with the influence of the steric exclusion effect being first manifested as a perturbation in the normal mode of operation when Equation 11 is used to calculate the particle diameter. In general it is possible to manipulate the exact position of the transition from the normal mode by adjusting the field strength (increasing the field would decrease the diameter of the change over). In the case of hyperlayer FFF, the channel flow rate, and hence the hydrodynamic lift force, will also affect the diameter at which the change of mechanism (mode) occurs (Increasing the flow rate would increase the diameter of the change over position.). Such optimization of the run conditions would be desirable if the sample contains particles around 1 μm at either the upper or lower end of the size distribution, in which case it may be possible to ensure that the whole sample range is covered by the one operating mechanism.

The lower size end for sedimentation FFF is determined by the field strength that can be generated by the centrifuge and the particle density. For the instruments we have been using (e.g., FFFractionation Inc. Model S101) operating at a speed of 2500 RPM with particles of density 2.5 g cm^{-3}, the limit for reasonable retention to still be obtained will be about 0.03 μm, whereas the higher speed centrifuges which have been available through DuPont may decrease this limit to perhaps 0.01 μm.

Although the advantages of FFF generally become more apparent for larger macromolecules, separation of samples down to a relative *M* mass of 500 or so have been reported with flow FFF.[16] The limiting factor here would appear to be the stringent requirements for the membrane, which must reject the sample molecules and yet be permeable to other solutes in the carrier in order to avoid their accumulation at the membrane surface.

The largest particles that have been separated by steric FFF were 92 μm in diameter,[39] with hyperlayer FFF being restricted to less than 50 μm thus far.[32] The control gained through adjusting both the field strength and hydrodynamic lift forces should allow particles greater than this to be separated, although the size range that can be separated in a hyperlayer FFF run with constant parameters

Table 2. A Guide to the Preferred FFF Subtechniques and Modes for Different Size Ranges

Particle Diameter (μm)		
0.001–0.05	**0.05–2**	**2–100**
Flow FFF Normal Mode	Sedimentation FFF Normal Mode	Flow FFF Hyperlayer Mode

appears to be limited to a 10 to 20-fold change in diameter. Ultimately the upper limit will be determined by the channel thickness, which rarely exceeds 500 μm and certainly must not exceed a value where laminar flow conditions are degraded under the flow rates used. In addition, the judicial use of field and/or flow programming (which is discussed in the next section) should increase the range of sizes that can be resolved within a given run.

Table 2 indicates our preference for subtechniques and operating mechanisms recommended for use with various particle sizes. The ranges given should only be regarded as estimates, as the exact values depend on various factors as discussed above. Sedimentation FFF is preferred to flow FFF in the colloidal range when possible, due to its superior resolution. For silt size particles flow/hyperlayer FFF is recommended, mainly because the lack of any density dependence on retention allows for more reliable size calibration, which is particularly important for heterogeneous environmental samples.

Field Programming

Another feature that greatly enhances the flexibility of FFF is the ability to vary the field strength even during the course of a run. This is very useful in dealing with samples having a broad size distribution. We have found that with sedimentation/normal FFF with its exceptional selectivity ($S_d = 3$), it is almost mandatory to use field programming for natural colloidal materials. This is necessary not only to avoid unacceptably long run times (many hours) but also to prevent excessive dilution as the sample is separated, causing the detector signal to be indistinguishable above the baseline of the fractogram.[19] The fundamental strategy with normal FFF is to begin the run at a high field strength, in order to adequately retain the smallest particles of interest beyond the void peak, and then to decrease the field strength so that the larger particles elute in a reasonable time, taking care to maintain the resolution at a minimum desired level throughout. The need for programming is lessened in flow FFF due to its lower selectivity. Consequently, the majority of research into this aspect of FFF development has concentrated on sedimentation FFF.

Various field programming forms have been tested, including decay of the field strength by a step function or by using linear, parabolic, exponential, and power law equations.[40] The program that has been employed most commonly with sedimentation FFF in the past is a time delayed exponential decay form.[41] This involves starting with a constant field of G_o for a period t_1, before decreasing the field according to the equation

$$G = G_o e^{-(t-t_1)/\tau'} \tag{20}$$

where

τ' is the exponential decay constant
t is the time elapsed from the start of the run

An interesting feature of this program is that by choosing $t_1 = \tau'$ the linear time axis of the fractogram can be readily converted into a linear log d scale since t is proportional to log d.

More recently Williams and Giddings[42] introduced a new decay form referred to as the power program, in which the exponential decay equation above is replaced by the expression

$$G = G_o \left(\frac{t_1 - t_a}{t - t_a} \right)^p \tag{21}$$

where t_a and p are constants, which for sedimentation FFF are commonly set at $p = 8$ and $t_a = -8t_1$.

It can be seen that several parameters may be adjusted to optimize the separation for a given sample. These include the initial field strength G_o, the delay time t_1, the carrier flow rate $\dot{V}$ and the appropriate parameters which control the field decay rate for each program type (e.g., τ', t_a). These all combine to affect the retention and resolution in a fairly complex manner. Beckett et al.[19] have discussed in some detail the effect of program parameters in exponential decay runs. An obvious starting point is to set the initial field strength and constant field periods to provide sufficient retention of the smallest particles from the void peak.

The purpose of decreasing the field during a run is to shorten the run time, but it must be remembered that this will be achieved at the expense of some resolution. It was not until quite recently that the theoretical basis for making informed decisions about the effect of programming on resolution became available. Giddings et al.[43] introduced the concept of the fractionating power F_d as a more universal parameter for defining desired levels of separation in a continuous distribution. This is expressed as

$$F_d = \frac{R_s}{\delta d / d} = \frac{\delta t_r}{4\sigma_t} \cdot \frac{d}{\delta d} \tag{22}$$

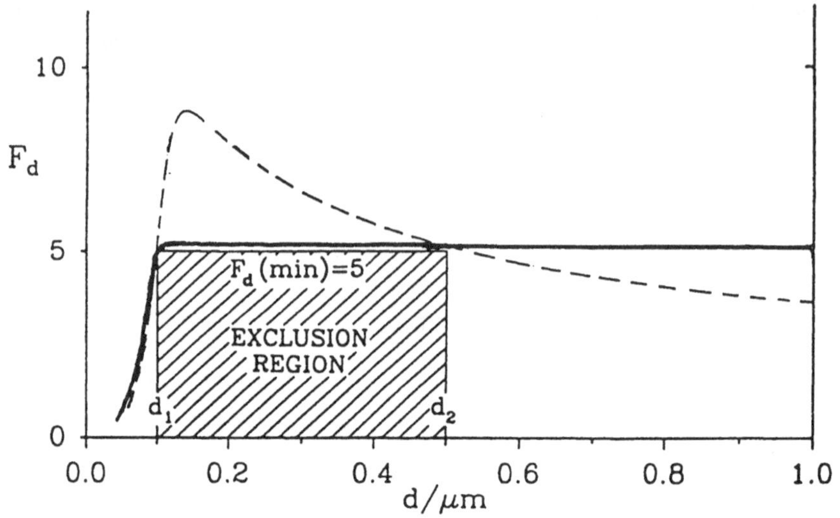

Figure 6. Fractionating power (F_d) plots for two runs designed to separate particles ($\Delta\rho$ = 1.5 g cm^{-3}) in the diameter range 0.1 to 0.5 μm with F_d = 5 (i.e., F_d curve must not enter the shaded exclusion region); solid line power program with p = 8, G_o = 137 g, t_1 = 17.8 min, t_a = 142 min and t^o = 5 min; dashed line exponential program with G_o = 78 g, t_1 = 31 min, τ' = 34 min and t^o = 5 min.

where

> R_s is the resolution, δt_r is the change in retention time due to an increment in diameter of δd and σ_t is the standard deviation in the retention time for particles of diameter d which is assumed to arise only from the nonequilibrium peak broadening effect

Thus F_d is the resolution per unit fractional change in diameter (i.e., when $\delta d = d$). For example, if it were desirable to separate particles differing by 10% in d with resolution 1 then F_d would need to be 10. If particles differing in d by 20% are desired at unit resolution, then F_d would be 5. The fractionating power can be used to monitor the degree of separation attained across the size range for a particular run as well as to test the effect of changing the various run parameters. Giddings et al.[43] developed analytical expressions for F_d and have outlined a systematic strategy that can be used to design exponential decay programs in order to achieve certain desired goals in the resolution over a specified size range.

Suppose that we wish to separate particles in the diameter range 0.1 to 0.5 μm ($\Delta\rho$ = 1.5 g cm^{-3}, η = 0.01 P, w = 0.0254 cm) with a fractionating power of 5. The restrictions that this will impose on the field program can be understood by referring to Figure 6 and realizing that the F_d versus d curve for the run must not enter the shaded exclusion region. First, choose the maximum flow rate possible (which then defines t^o since $t^o = V^o/\dot{V}$) since for a given set of exponential program conditions this will yield the maximum F_d.[19] Then, using the equations given by Giddings et al.,[43] set the minimum τ'/t^o value (Equation 55 in Reference 43) to give $F_d = 5$ at $d = 0.5$ μm and the G_o (Equation 56 in Reference 43) and t_l (Equation 57 in Reference 43) values to achieve $F_d = 5$ for the smallest particles (i.e., $d = 0.1$ μm). In our example, if we choose $t^o = 300$ s, the appropriate programming conditions would be $\tau' = 34$ min, $G_o = 78$ g, and $t_l = 31$ min, and the F_d curve obtained is as shown in Figure 6 (dashed line).

Not only can this approach be used to optimize run conditions for a particular program, but it can also be used to investigate the utility of different program forms.[40] Subsequently, Williams and Giddings[42] have devised a new program form (called the power program) that can result in improved uniformity of F_d across the sample size distribution. In this power program, the initial constant field period is followed by a field decay defined by Equation 21. It has been found that for sedimentation FFF, setting $p = 8$ results in an approximately constant F_d over a wide diameter range above a certain minimum value, a situation which is well suited for the characterization of broad particle size distributions. A suitable strategy for optimizing the power program parameters using the equations given in Giddings and Williams[42] is as follows. First, choose the desired level of fractionating power F_d* for the separation and the flow rate to be used in the run (i.e., t^o). For a series of likely initial field (G_o) values, use Equation 57 from Reference 42 and $t_a = -8t_l$ to calculate the required t_l and t_a values, respectively. The latter relationship is found to result in the sharpest break (knee) in the F_d versus d curve during the transition from constant field to decay conditions (i.e., at time t_l) without F_d exceeding F_d*. Now, plot the F_d versus d curves (using Equation 45 in Reference 42), and decide which G_o value will give the desired fractionating power (F_d*) for the smallest particles of interest in the sample.

Applying this methodology to the problem of separating a sample with $d =$ 0.1 to 0.5 μm (other parameters as given in the example above) to an F_d value of 5 suggests using the following parameters: $G_o = 137$ g, $t_l = 17.8$ min, and t_a $= 142$ min. Two advantages of the power program over the exponential program emerge from this example. First, beyond a certain minimum diameter the F_d versus d plot for the power program (solid line in Figure 6) is almost constant, which simplifies interpretation of the fractogram and the design of collection strategies if fractions are to be collected for subsequent testing. Second, for the 0.5 μm particles, the run time required with the power program was 168 min, which was somewhat less than for the exponential program of 193 min. This additional time results because, in order to obtain an F_d of 5 for the particles at each end of the required range (0.1 to 0.5 μm), particles in between are separated with greater than the specified resolution.

One aspect which should be borne in mind in designing programmed runs is the danger of inducing errors by decaying the field too fast. This secondary relaxation effect results if thermal motion is not rapid enough to allow the sample cloud to expand continuously to its equilibrium position as the field strength is decreased. In normal FFF this can cause a sample to elute later than expected, thus resulting in an overestimate in the particle diameter. Hansen et al.[44] have derived approximate expressions for the errors produced in exponential decay programs and have confirmed their applicability in experiments using sedimentation FFF. This theory can be used to correct for the perturbations caused by secondary relaxation, or it can give simple criteria for the run parameters τ' and t^o which will result in an acceptable level of error in a given analysis.

For samples with a broad distribution, the maximum discrepancy will be found for the largest particles which have the lowest diffusion coefficient. Thus, if we choose the maximum tolerable relative error in the diameter ($\delta d/d$) for these largest particles, then the decay constant (τ') should be greater than that given by the expression

$$\tau' = \sqrt[3]{\frac{\pi\omega^2 t^{o^2} d}{54kT(\delta d / d)}} \tag{23}$$

Similar work to evaluate secondary relaxation effects for the power program is currently in progress.

To date only a limited amount of work on programming has been conducted with flow FFF. This is because it is generally less urgently required than with sedimentation FFF due to its lower selectivity ($S_d < 1$). Additionally, with flow FFF there are considerable experimental difficulties involved in controlling both the field and flow fluid velocities to a given precision, while at the same time varying one or both of these parameters. A notable exception to this were the impressive separations of silt sized latex beads obtained by Ratanathanawongs and Giddings[32] using flow/hyperlayer FFF in which both the field and channel flow rates were programmed. In view of the potential that programming has to improve the performance of FFF separations and to strike the best compromise between resolution and run time, we foresee some rapid advances occurring in this aspect of FFF research in the next few years.

APPLICATIONS OF FIELD-FLOW FRACTIONATION

The remainder of this paper will be devoted to a description of the information that can be obtained using FFF analysis. Included will be some examples of how this may be used in studies of the environment as well as some predictions on future developments in the field.

Determination of Size Distributions

The fractogram in FFF is a plot of the detector response versus elution volume (or time). Using theoretical calculations as outlined previously or where necessary an empirical calibration, the abscissa (x-axis) of the fractogram can readily be converted into particle diameter or relative molar mass (M). The detector is generally used to monitor the mass concentration of sample in the eluent (dm'/dV) although, as is discussed below, this is often an imperfect measure.

A size distribution is usually depicted as a plot in which the area under the curve between specified size limits gives the fraction of the total sample in that size range. Thus, the ordinate in such a size distribution, which is often referred to as the frequency function by comparison with frequency histogram diagrams, should be $\mathrm{d}m'/\mathrm{d}d$ for a diameter distribution or $\mathrm{d}m'/\mathrm{d}M$ for a relative molar mass distribution, where d or M would be plotted on the abscissa respectively and m' represents the cumulative mass of eluted sample up to a specific point in the run. In the case of a particle size distribution, the frequency function would be given by

$$\frac{\mathrm{d}m'}{\mathrm{d}d} = \frac{\mathrm{d}m'}{\mathrm{d}V} \cdot \frac{\mathrm{d}V}{\mathrm{d}d} \tag{24}$$

The differential $\mathrm{d}V/\mathrm{d}d$ may be obtained directly if an analytical expression for V in terms of d is known. Alternatively, if d is obtained from V by numerical methods, as will be the case in more complicated programmed runs, then this scale transformation can be achieved by digitizing the fractogram and multiplying each ordinate value $(\mathrm{d}m'_i/\mathrm{d}V_i)$ by $\delta V_i/\delta d_i$, where δV_i is the difference between the elution volume for consecutive digitized points and δd_i is the corresponding difference in the particle diameter for these points. A similar approach can be used for obtaining relative molar mass distributions from the appropriate fractograms.

Relative Molar Mass Distribution of Humic Substances

Flow/normal FFF has been used to determine relative M mass distributions of fulvic and humic acids collected from various environments.[16,17] In this case a calibration line must be established through Equations 2, 5, 18, and 19. Different types of relative M mass standards can yield different calibration lines, as shown in Figure 7. In this case it is obvious that the sodium polystyrenesulphonate standards are more appropriate for humic substances, as the points corresponding to some reference materials (International Humic Substances Society) clearly fall on this line. These materials were extracted by an XAD8 resin adsorption method and had been studied by a number of different methods (e.g., vapor pressure osmometry, low angle X-ray scattering, ultracentrifugation), and thus a reasonably reliable average relative M mass could be assigned to each (Dr. R. Malcolm, USGS, Denver, CO, private communication). The samples were run in a 0.03 mol dm^{-3} ionic strength tris buffer at pH 7.9, and the concentration of humic substance injected was 0.25 mg cm^{-3}. It would seem that the solution molecular conformation of these humics resembles more closely the random coil polyelectrolytes than the globular proteins used for the other calibration line given in Figure 7.

Figure 8 gives an example of the conversion of the flow FFF fractograms into relative M mass distributions for a humic and fulvic acid. From the digitized relative M mass distribution data, it is a simple matter to calculate the number average ($\overline{M}_n$) and weight average ($\overline{M}_w$) relative M mass as well as the sample polydispersity ($\overline{M}_w/\overline{M}_n$). However, it is important to note the assumptions that are inherent in the calibration method used to arrive at the relative M mass numbers. One factor

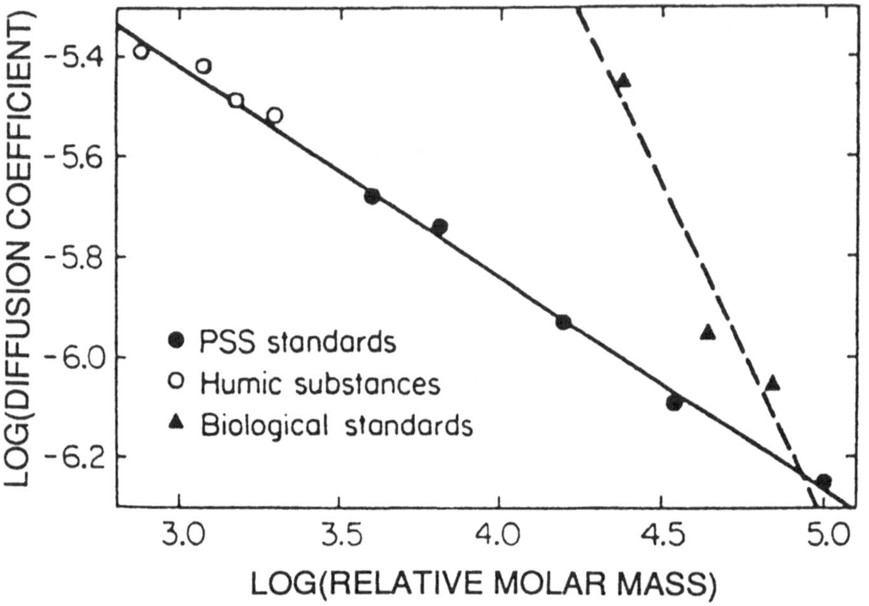

Figure 7. Calibration lines for flow FFF relative *M* mass determinations obtained using either sodium polystyrenesulphonate standards (● – ●) or biological standards (▲ – ▲). Also shown are data points for some reference humic substances obtained from IHSS (○).

limiting the accuracy of these average relative *M* mass measurements, particularly for the $\overline{M}_n$ values, is the lower resolution limit for flow FFF of around 500.

Two major assumptions are involved in the conversion from fractogram to relative *M* mass distribution. First, the calibration curve linking retention volume (or time) to relative *M* mass, which in this case was established using sodium polystyrenesulphonate standards, is taken to be applicable to all of the samples analyzed. A corollary to this point is that this relationship, which would be affected by changes in the molecular conformation, must also be maintained across the entire relative *M* mass range of each sample. Second, the solution absorbance at all elution volumes is taken to be proportional to the mass concentration of sample molecules in the fractionated sample (dm'/dV), which is equivalent to assuming that the mass absorbtivity of these samples is independent of relative molar mass. Similar assumptions are also common to other methods used for determination of relative *M* mass which require calibration and use light absorbance for sample detection (e.g., size exclusion chromatography). It may be that detectors based on other physical measurements (e.g., refractive index, light scattering) will prove to be more appropriate for humic substances, but this requires detailed evaluation.

We have investigated humic substances from various sites as illustrated in Table 3. A number of significant trends have emerged from these studies. First, the range of weight average relative *M* mass ($\overline{M}_w$) for all samples extended from

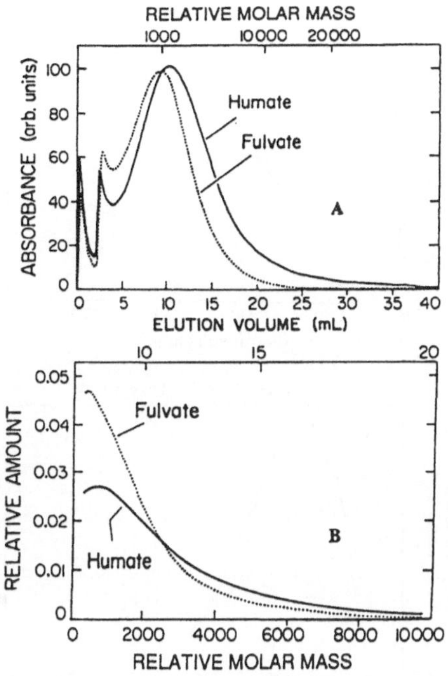

Figure 8. (a) Flow FFF fractograms of Suwannee River humic and fulvic acids; (b) relative molar mass distributions obtained from these fractograms using a polystyrenesulphonate calibration line similar to that shown in Figure 7.

Table 3. Number Average ($\overline{M}_n$) and Weight Average ($\overline{M}_w$) Relative Molar Mass and Polydispersity ($\overline{M}_w/\overline{M}_n$) Data for Some Humic Substances Measured by Flow FFF

Sample	$\overline{M}_n$	$\overline{M}_w$	$\overline{M}_w/\overline{M}_n$
Suwannee stream fulvate	1,150	1,910	1.66
Suwannee stream humate	1,580	4,390	2.78
Mattole soil fulvate	1,390	3,900	2.81
Mattole soil humate	1,940	6,140	3.16
Florida sand humate	2,250	7,960	3.54
Washington peat humate	3,020	17,800	5.89
Leonardite coal humate	3,730	18,700	5.01
Aldrich humate	3,070	14,500	4.72
Redwater Creek water	1,760	4,900	2.78
Inkpot Pond water	1,480	3,830	2.59

about 2000 to 20,000. Fulvic acids were lower in relative *M* mass than humic acids, and for the humates the relative *M* mass increased in the order *stream* < *soil* < *peat* < *coal*. Freshwater humic substances contained negligible material above 10,000 in relative *M* mass and had $\overline{M}_w$ values between 2000 and 5000. These values are much less than those reported by some studies in the past using various other methods,[45] but are consistent with recent work where care has

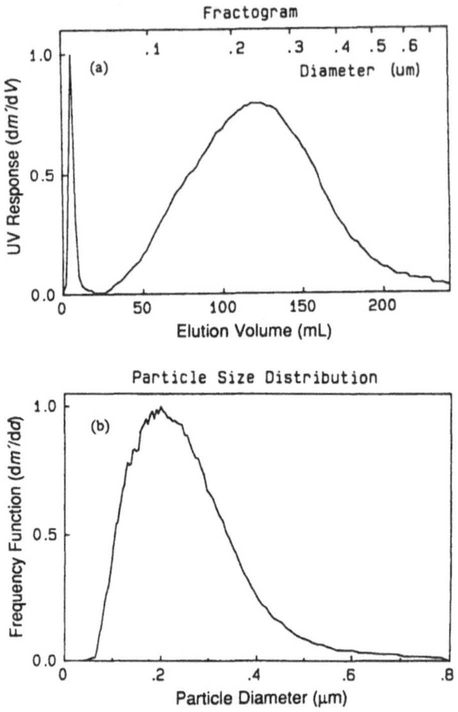

Figure 9. (a) Sedimentation FFF fractogram of a colloid sample collected from the Ovens River, Victoria; (b) particle size distribution calculated assuming a particle density of 2.5 g cm^{-3}.

been taken to avoid the many artifacts that can be encountered in these measurements.[46,47]

Flow FFF appears capable of producing reliable relative M mass measurements in times of 5 to 20 min, which should make it a useful analytical and separation method for natural organic matter. We envision that flow FFF should provide valuable information in studies of the nature, origin, and biogeochemical processes of humic substances in the environment.

Particle Size Distributions of Aquatic Colloids

The importance of colloidal particles in the biogeochemical processes occurring in natural waters is now widely accepted.[3] Although several techniques are currently available for separating and sizing submicron particles (centrifugation, filtration, light scattering) none of these has proven to be particularly adequate. The most important deficiencies result in a lack of resolution and/or accuracy in the resulting size distribution. Sedimentation FFF has the potential to fill this void with its high resolution and ability to supply size fractions for subsequent analysis. Figure 9 shows a fractogram and the calculated particle size distribution of a colloid sample collected from the Ovens River, Australia. The particle size axis

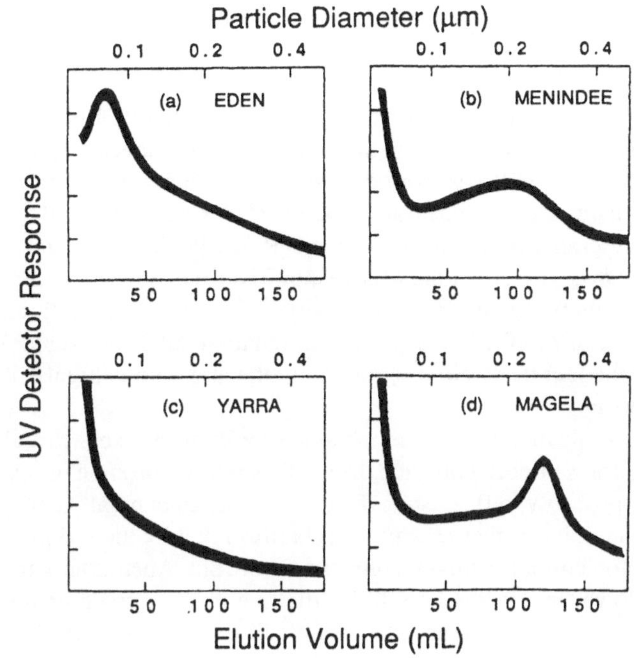

Figure 10. Sedimentation FFF fractograms of colloidal samples collected from different aquatic environments in Australia: (a) Eden catchment groundwater, N.S.W., (b) Menindee Lake, N.S.W.; (c) Yarra River, Vic.; and (d) Magela Creek, N.T.

is calculated utilizing equations analogous to Equations 2, 5, and 11 but taking into account the changing field strength during the run. Since retention in sedimentation FFF depends on the effective mass of the particles, the density of the particles must be known and is invariably assumed to be uniform for the whole sample. For mineral rich sediments, we generally assume a particle density of about 2.5 g cm^{-3} which is reasonable for silica and most silicates.

The other difficulty encountered arises from the use of a UV detector to monitor the concentration of particles in the eluent. The attenuation of the light intensity due to particles in the cell will be caused mainly by scattering rather than absorption. Hence, we would expect the absorbance reading on the detector to depend on both the mass concentration and the particle diameter as predicted by Mie theory.[48]

In the submicron range, this scattering mechanism would be expected to result in an underestimate in the sample concentration as the size decreased well below the wavelength of the radiation used (typically 254 nm). The corrections required for this are quite complex and will depend on the refractive index of the particles, which may not be known and in the case of natural samples will usually not even be uniform.

However, we have found that these errors can be tolerable, at least for distributions that are not too broad (such as sizes spanning less than a three- to fourfold range of diameters). The evidence for this is the close agreement between fractograms collected using the usual UV detector and an evaporative mass detector[19] or by inductively coupled plasma-mass spectrometry which gives elemental analyses.[49]

Karaiskakis et al.[18] were the first to report sedimentation FFF fractograms of colloidal suspended particulate matter collected from several Utah rivers. We have studied aquatic colloids from diverse environments in Australia, including rivers, lakes, dams and soil infiltration waters.[19-22] Reschiglian and Dondi[51] have reported fractograms of colloids from the River Po in Italy, and Taylor and Garbarino.[52] have worked on the Mississippi River, USA. Two common features encountered in these studies are the necessity to concentrate the colloids in natural waters by a factor of 10 to 100 in order to obtain a sufficient signal with a UV detector and the fact that field programming of some kind is highly desirable, as discussed above.

A review of all the fractograms obtained by these workers reveals that size distributions for samples collected from disparate environments can be quite distinct and probably reflect such factors as the catchment geology and the hydrological conditions that prevail. This is illustrated by the sedimentation FFF fractograms for colloid samples collected from four Australian sites shown in Figure 10. However, there is very little structure in these fractograms, and thus, except in some special circumstances, they will not provide enough information to enable the sample source to be identified.

It should be possible to utilize the steric or hyperlayer modes to size sediments in the silt range (2 to 63 μm). Sedimentation/steric FFF (using either gravity or a centrifuge) has been used to separate and size some minerals such as zirconia.[53]

A continuous separation device combining steric FFF and gravitational settling has been used for coal fly ash[54] and bottom sediments.[23] This method was developed mainly for the purpose of separating larger quantities (up to a few grams) of sediment, but with the current experimental device does not appear to give the high resolution of other FFF methods.

Flow/hyperlayer FFF is a rapidly emerging method for determining size distributions in the silt range which has several advantages as discussed above. Examples of the sample types that have been analyzed include chromatographic silica packing material,[32] coal, and limestone.[55] These promising preliminary results suggest that flow/hyperlayer FFF will prove very useful for sediment studies. However, one obstacle to be overcome is the as yet unknown influence of particle shape on these separations.

Characterization of Size Fractions

One of the main advantages of FFF sizing methods is that they are based on a separation procedure which greatly simplifies the problems associated with analyzing heterogeneous samples. In addition they provide the opportunity to collect fractions within given narrow size ranges for subsequent investigation by other techniques. This approach can provide valuable information on the nature of the sample as a function of particle size.

The scope of this aspect of FFF research is obviously very broad, although one limitation will be small sample size (generally < 1 mg) that is separated in a single FFF run; thus only quite sensitive analytical methods will be suitable. We will illustrate the utility of this methodology with work conducted over the past few years on the characterization of natural aquatic colloidal particles using Sedimentation FFF.

Electron Microscopy

Direct visual observation of the particles provides perhaps the best confirmation of the effectiveness and accuracy of a separation. A simple procedure for achieving this is to collect samples of eluent with a fraction collector, filter the particles onto a suitable membrane filter (e.g., 0.05 μm Nucleopore) so that the particles are not overcrowded (typically 5 to 10 ml of eluent), gold- or carbon-coat the specimen, and examine it using scanning electron microscopy (SEM).[50] Chittleborough et al.[22] have obtained even better quality pictures of fractionated soil colloids using transmission electron microscopy.

These electron micrographs have demonstrated that good separations are generally obtained with the elution order in accordance with the normal FFF mechanism (i.e., smallest particles eluting first). However, some discrepancies occur between the projected size measured from the scaled micrographs and that predicted from the FFF elution volume. Two reasons probably account for this — variations in the particle density which affects the FFF calculations and deviations of the particle shape from spherical geometry which influences the projected area of the particles. Provided the density of particles is known, combining the FFF and SEM data enables the average thickness and, hence, aspect ratios of the particles to be determined.[21]

Comparison of the electron micrographs obtained from different size fractions of a particular sample often shows that certain fractions are dominated by specific minerals (as reflected by their particle morphology). This type of information may well be useful in studies on the origin of suspended particulate material and its associated trace element or pollutant load.

X-ray Emission and Diffraction

Karaiskakis et al.[18] were able to obtain elemental analyses of river colloid size fractions separated by sedimentation FFF using an EDX surface scanning microprobe (EDX refers to energy dispersive analysis of emitted X-rays). Their data were for total concentration within the specified size range, as the sample was simply evaporated onto a carbon stub. Some trends in the element ratios of the fractions were apparent, implying a change in mineralogy.

We have attempted to use the EDX signal generated by the electron beam in an SEM, impinging on single colloidal particles isolated by filtering small aliquats of the sedimentation FFF eluent through membrane filters.[21,50] Although it is often possible to identify the mineral type, the process is somewhat tedious with such small particles, as 3 to 5 min may be required before the X-ray counts are significantly above the background noise even for the major elements. Barnard and co-workers have developed a similar approach by combining electron microprobe X-ray analysis and automatic image analysis, which makes it feasible to process hundreds of particles per sample. They have applied the technique to unfractionated silt particles[56] and samples fractionated by continuous flow centrifugation and tangential flow filtration.[57] Although this method will probably be of limited use for colloidal samples separated with sedimentation FFF, it should be readily applicable for the particle-by-particle analysis of silt sized sediments fractionated by steric or hyperlayer FFF.

Chittleborough et al.[22] have obtained X-ray diffractograms for fractions of soil colloids separated by sedimentation FFF and filtered onto low background silica wafers. Again the X-ray signal strength was a rather limiting factor; however, at least one mineral (kaolinite) could be detected in the sample. These preliminary results suggest that X-ray diffraction is a technique worth pursuing further for mineral identification within the size fractions, although attempts to increase the amount of material collected may be necessary. This could include combining samples collected from multiple runs or increasing the sample size injected. The latter can produce distortion of the peak shape due to overloading effects,[58] and this would need to be monitored in order to avoid excessive errors in the calculated particle size.

Inductively Coupled Plasma-Mass Spectrometry

One of the most sensitive instruments now available for elemental analysis is inductively coupled plasma-mass spectrometry (ICP-MS). This combines an induction furnace, which is used to atomize the sample particles and ultimately ionize the atoms within a hot gas plasma, with a quadruple mass spectrometer, which records the elemental composition via the various ion currents generated. The device is capable of rapid multielement analyses with the limit of detection for most elements being well below one part per billion. Eluent from an FFF containing the separated sample can either be introduced batchwise into the ICP-MS or the outlet tube from the FFF can be connected directly to the ICP torch, as the flow rate requirements for the two components are in most cases compatible. The latter gives rise to an integrated FFF-ICP-MS instrument, the output of which could be element based fractograms and size distributions.

If the concentration of element E in the eluent ($\mathrm{d}m'_E/\mathrm{d}V$) is monitored by the relevant ion current of the mass spectrometer (I_E), then the element based size distribution is obtained by plotting $\mathrm{d}m'_E/\mathrm{d}d$ versus d, where

$$\frac{\mathrm{d}m'_E}{\mathrm{d}d} = \frac{\mathrm{d}m'_E}{\mathrm{d}V} \cdot \frac{\mathrm{d}V}{\mathrm{d}d} \propto I_E \cdot \frac{\delta V}{\delta d} \tag{25}$$

The concentration of E in the particles ($dm_E'/\mathrm{d}m'$) for a given particle diameter can be obtained from

$$\frac{dm'_E}{dm'} = \frac{dm'_E}{dV} \cdot \frac{dV}{dm'} \propto \frac{I_E}{UV\ detector\ response} \tag{26}$$

This enables the elemental composition of the particles to be plotted as a function of particle size.

Taylor et al.[49] have recently demonstrated the feasibility of this ICP-MS based methodology by analyzing a number of minerals and river borne colloids separated by sedimentation FFF. They have used successfully both the batchwise fraction collection and introduction method as well as the first directly interfaced FFF-ICP-MS instrument, which they constructed (Murphy et al., paper in preparation).

An example of such an element based fractogram and the calculated size distribution and particle composition distribution are given in Figure 11a, b, and c, respectively. In this case the Al content of Darling River suspended particulate matter is fairly constant, indicating a uniform mineral composition over the particle size range covered (0.05 to 0.6 μm). The ICP-MS instrument is so sensitive that even trace elements (e.g., Zn, Cu, Rb) are capable of being detected, making this an extremely powerful tool for assessing the pollutant distributions and perhaps for providing the elusive fingerprint to help establish the source of suspended particulate matter or sediments collected at a given location.

Chittleborough et al.[22] have used ICP-MS to obtain Al, Fe, and Mg data on sedimentation FFF generated fractions of several soil colloid samples. In addition, K levels were monitored using inductively coupled plasma-atomic emission spectrometry (ICP-AES). The variations in the element ratios could be interpreted in terms of varying mineralogy. For example, the increasing K:Al ratio with particle size indicated that the proportion of micas increased with the increase in particle size. This result was supported by the morphological information obtained with transmission electron microscopy.

The successful analysis of K in sedimentation FFF effluent samples obtained by Chittleborough is a very encouraging result. It indicates that the more widely available ICP-AES instrumentation may be coupled with FFF instruments to provide compositional size distributions, at least for some of the major elements present.

Other Characterization Methods

Many other physical and analytical measurements may be performed on the separated sample to better characterize the particles contained within specified size ranges. The major criterion is that the technique be sensitive enough to detect the very low concentration contained in the eluent.

Photodiode array detectors may be used to obtain spectral information on separated mixtures.[17] It would seem likely that rapid scanning UV and electrochemical detectors, as well as fluorescence and Fourier transform infrared spectrometers, could also be used, particularly for the identification of polymers and

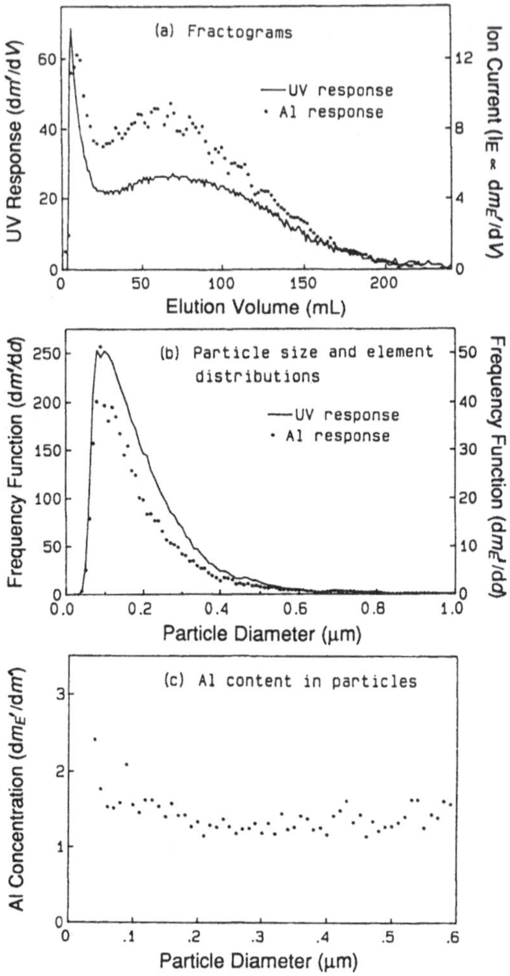

Figure 11. (a) Normal particle based fractogram (UV detector response – solid line) and element based fractogram (ICP-MS aluminum ion current – points) for Darling River suspended colloid sample, (b) size distributions, and (c) element composition distribution calculated from data shown in fractograms.

other macromolecules. Currently we are using a fluorescence detector to measure the fractograms of stained bacteria in complex mixtures of suspended particles, as outlined in the section below.

We have shown above that the elemental composition of particles may be determined by ICP-MS and ICP-AES. Provided the sample species are in a dissolved form or are amenable to chemical digestion, atomic absorption spectrometry (AAS) should also be feasible. Flame AAS may be suitable for some major element, and it may be possible to link this instrument directly to form an FFF-AAS hybrid. It is more likely that furnace AAS will be required to attain the sensitivities necessary, and this would involve batchwise introduction of collected fractions, aided perhaps by automatic sample injection.

The use of radioisotope labeled samples provides an extremely sensitive detection method. We have been using this in some of the adsorption experiments described below. To date we have collected fractions for measurement in a scintillation counter; however, continuous flow radioactive detectors are commercially available.

Caldwell and her co-workers have combined the results obtained from sedimentation FFF and photon correlation spectroscopy (dynamic light scattering) of the separated fractions to measure both the size and density of emulsion droplets[59] and latex beads.[60] Dynamic light scattering provides an independent measure of the hydrodynamic radius of the particle, which can be used together with the FFF retention volume (through Equations 2, 5, and 11) to yield the particle density. With this approach, Caldwell has also been able to obtain estimates of the thickness of polymer layers adsorbed to latex beads.[60]

The preliminary studies with various additional techniques that have been mentioned above serve to illustrate the general point that characterization of heterogeneous materials such as those found in the environment is a very complex problem. Generally, each measurement technique yields only a single parameter. Thus, to fully characterize a particle for its size, density, shape, and composition requires that much information must be collected. In order to take full advantage of the high resolution separations achieved with FFF methods, as many different detection and analysis techniques as possible should be employed.

Adsorption/Complexation Studies

The fate and behavior of contaminants in natural waters is often controlled predominantly by adsorption or complexation interactions with particles or macromolecules. Many studies have investigated the equilibrium reactions between pollutants and the whole particulate or dissolved organic matter sample. Beckett et al.[20] have pointed out how FFF techniques can be used to obtain detailed adsorption information as a function of adsorbent size. The experimental procedure employed in such studies involves three steps:

1. Adsorption of the pollutant to the whole adsorbent sample.
2. Separation of the sample using the appropriate FFF method and if necessary collection of eluent fractions across the fractogram.
3. Analysis of the eluent or eluent fractions for the amount of pollutant adsorbate present.

The adsorbent concentration at any point along the fractogram (dm'/dV) is usually measured by a detector placed at the outlet of the FFF channel. This is most commonly a UV detector monitoring either turbidity or absorbance of the adsorbent in the eluent. The pollutant concentration (dP/dV) is monitored by the response of a second detector or within the collected fractions using a suitable analytical technique (e.g., ICP-MS, radioactivity). The amount of pollutant adsorbed per unit mass of adsorbent (dP/dm') at any point along the volume axis of the fractogram is given by

$$\frac{dP}{dm'} = \frac{dP}{dV} \cdot \frac{dV}{dm'} \propto \frac{\textit{pollutant response}}{\textit{UV detector response}} \tag{27}$$

Here the pollutant response refers to any measured quantity that is proportional to the pollutant concentration in the eluent. With a particulate adsorbent, the surface adsorption density or amount of pollutant adsorbed per unit adsorbent surface area (dP/dA) is given by

$$\frac{dP}{dA} = \frac{dP}{dV} \cdot \frac{dV}{dm'} \cdot \frac{dm'}{dA} \propto \frac{\textit{pollutant response}}{\textit{UV detector response}} \cdot d \tag{28}$$

since at a given eluent volume it can be shown that $ddm'/dA = \rho d/6$, provided we assume the particles are spherical.

This analysis is illustrated in Figure 12 for the adsorption of the herbicide glyphosate (14_c labeled) onto Darling River suspended particulate matter separated by sedimentation/normal FFF. Figure 12a gives the adsorbent (UV detector response) and pollutant (scintillation counter response) based fractograms. The adsorption distribution (a plot of dP/ddm' versus d) is shown in Figure 12b, and the adsorption density distribution (a plot of dP/dA versus d) is shown in Figure 12c. For a homogeneous sample, we would expect dP/ddm' to decrease with increasing particle size, but the adsorption density dP/dA should remain constant. This is often found to be the case, but with some systems the adsorption density increases significantly across the size range. This could be caused by a number of trends within the particle distribution, such as changes in mineralogy, texture, or shape.

To date we have confined our studies to pollutant-colloid interactions, but the methods should be equally applicable to the investigation of complexation of pollutants to natural organic matter, such as humic substances using flow/normal FFF or the adsorption by larger silt size particles using flow/hyperlayer FFF. The results obtained so far with sedimentation FFF suggest that such techniques will yield very useful information on the speciation of pollutants in aquatic systems.

Characterization of Aquatic Biota

Microscopic organisms such as phytoplankton and bacteria play an important but sometimes underestimated role in the biogeochemical cycling of elements in the aquatic environment. There is a need for better characterization and enumeration methods to assist in studies aimed at a more complete understanding of this role.

Aquatic bacteria are generally somewhat smaller than cultured bacteria, and they occur in a complex matrix of organic and inorganic particles. These factors

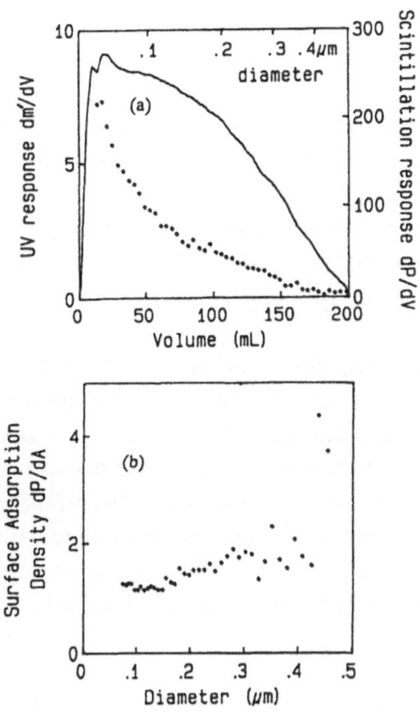

Figure 12. (a) Fractogram of Darling River suspended particulate matter (UV detector response – solid line) and adsorbed glyphosate based fractogram (scintillation counts – points), (b) surface adsorption density distribution for glyphosate adsorbed onto Darling River suspended colloids.

add to the difficulty of sizing and counting bacteria by the common but tedious method of epifluorescence microscopy. The recently developed automated method of flow cytometry for cell counting and sorting also has not been particularly successful with natural bacteria due to their small size (< 1 μm). Single particle counters (e.g., Coulter counter, HIAC particle size analyzer) can be used to count cells, but the absolute lower size limit is about 0.5 μm, and these methods do not distinguish between cells and the high concentrations of nonbiotic particles found in natural samples.

Sedimentation FFF should be a useful method for determining the bacterial biomass of natural bacteria in a water or biofilm sample. In addition, it can be used to characterize the bacteria population for parameters such as cell size, density, and shape. The method currently being developed in our laboratory involves staining the bacteria with a fluorescent dye designed to bind to bacterial DNA (e.g., DAPPI) and using a fluorescence detector to record the fractogram (Sharma et al., paper in preparation).

Two advantages result from the use of the fluorescence detector compared to a UV detector in this application. First, fluorescence is more sensitive, thus lowering the detection limit and reducing the concentration factor required to run natural water samples. Second, since the fluorescent dye specifically binds to the bacteria DNA, the signal should reflect only the amount (probably cell number concentration) of bacteria present even in the presence of other particles that would normally contribute to the turbidity signal of a UV detector and may mask the contribution of bacteria. However, one difficulty that has been only partially overcome to date is the occurrence of strong autofluorescence in natural waters, sediments, and biofilm samples, particularly at emission wavelengths around 500 nm. This does not seem to be an insurmountable problem, as it should be possible to avoid these wavelengths by choice of more suitable fluorescent dyes and experimental conditions or by using a purification step to isolate the bacteria before FFF analysis.

The specific gravity of bacteria cells can be estimated using sedimentation FFF in two ways. First, it may be determined by performing runs with different carrier density and employing Equation 12 and the method described in the section above on subtechniques of FFF. Low molecular weight density modifiers may decrease the diameter of cells by osmosis, as has been observed for red blood cells using flow/hyperlayer FFF.[61] This effect may be minimized by fixing the cells with formaldehyde and by using high molecular weight density modifiers.

The second method for determining the cell density is to collect fractions of the FFF eluent and measure the cell diameter by an independent method, such as epifluorescent microscopy on dynamic light scattering. The retention parameter λ, corresponding to the mean elution volume for a given fraction (Equations 2 and 5), and the mean cell diameter d are then substituted into Equation 12 to calculate the density of the cells.

Our results on cultured bacteria indicate that the cell density is relatively constant over the entire size range of the sample but can vary significantly between bacteria types. In addition, we have found that the fluorescence signal for a given culture is related to the cell number concentration and is also independent of cell size. This would be expected if the amount of cell DNA is constant for all cells in the population. Once the average cell density has been determined by one of the above methods, it is then possible to convert the fractogram into a particle size distribution. This size distribution could be expressed in terms of the absolute number of cells rather than simply the fraction of cells present in a given size range, provided the fluorescence signal per cell calibration is performed.

Microbial ecologists are interested in measuring the cell biomass (dry cell mass) concentration in various aquatic environments. Again, FFF provides two approaches to this rather difficult and tedious task. First, if the dry cell mass per cell volume has been determined (there are some estimates in the literature that could be used), then the fractogram data (which, if the cell density is determined, yields cell number versus cell volume) can be used to compute the amount of biomass injected. Alternatively, we can use the fact that the sedimentation FFF elution volume is directly related to the buoyant cell mass, and since the cell water has zero buoyant density, this is simply the buoyant mass of the organic matter in the cell. If we take an estimate of the average density of this cell organic matter (1.5 g cm^{-3}, for example) then the biomass per cell at any point along the fractogram can be estimated. Combining this with the number of cells present, which is obtained from the fluorescence detector response, provides a quite direct method for measuring the bacterial biomass in a water sample.

Larger single-celled organisms, such as algae and cyanobacteria (so-called blue-green algae), should be amenable to characterization and enumeration by analogous methods using flow/hyperlayer FFF. As discussed earlier, in this case the cell shape will have some effect on the elution volume and hence the calculated equivalent spherical diameter. Although only very preliminary results are available[37] (Beckett et al., unpublished data), there is great incentive to develop these methods due to increasing concerns about eutrophication and, in particular, the serious effects of toxic blue-green algae.

CONCLUSIONS

Field-flow fractionation is a versatile method which can be used to separate and size a wide range of environmental materials. The size range covered extends from about 1 nm to 100 μm, and not only can size distributions be generated, but fractions within a known size range can be collected for subsequent analysis by other techniques. In favorable cases, this can enable the chemical composition as a function of particle size to be determined. However, since FFF is limited to handling sample sizes of less than a milligram, very sensitive analytical methods are needed.

Various subtechniques (fields) and modes of operation (separation mechanisms) can be used in FFF. The most promising of these for environmental samples appear to be flow/normal FFF for macromolecules and colloids (< 2 μm), sedimentation/normal FFF for colloids (~ 0.05 to 2 μm), and flow/hyperlayer FFF for particles (> 2 μm).

The most extensive studies in the environmental field to date have been carried out on suspended colloids from rivers using sedimentation FFF.[19-21] The relative *M* mass distributions for humic substances and other related organic compounds can be measured using flow FFF, which fills a real need since the other methods available all suffer from certain deficiencies. Many other applications can be suggested, some of which have been the subject of preliminary testing. Examples include bacteria, algae, soil colloids, silt sized sediments, and various industrial effluents.

FFF has been known for over 25 years, and in that period the theory of the method as well as many experimental strategies for analyzing different types of samples have been established.[13] However, it was not until quite recently that the instrumentation has become commercially available (DuPont Co., DE; FFFractionation Inc., UT).

With this sound background we predict that FFF should now emerge quite rapidly to become a familiar technique in the analytical laboratory. It will be of particular use for very heterogeneous samples where a high resolution preseparation is most advantageous in a characterization procedure. Thus, FFF techniques represent a significant addition to the instrumental methods available to help with the difficult task of understanding aquatic processes.

ACKNOWLEDGMENTS

The FFF research in our laboratory has been supported by the Australian Research Council and the Land and Water Resources Research and Development Corporation. We acknowledge the efforts of our former and present graduate students, Geoff Nicholson, Marcia Hansen (University of Utah), Deirdre Hotchin, Reshmi Sharma, Myhuong Nguyen, and Rick Wood, who have performed most of the experiments. Finally, we thank Professor J. Calvin Giddings, who invented FFF and who has given us much assistance.

REFERENCES

1. Salomons, W. and U. Forstner. *Metals in the Hydrosphere* (Heidelberg: Springer-Verlag, 1984).
2. Hart, B.T. "Uptake of Trace Metals by Sediments and Suspended Particulates: A Review," *Hydrobiologia* 91:299–313 (1982).
3. Hart, B.T., Ed. *The Role of Particulate Matter in the Transport and Fate of Contaminants* (Melbourne: Water Studies Center, Chisholm Institute of Technology/Monash University, 1986).
4. Hart, B.T. and T. Hines. "Trace Elements in Rivers," in *Trace Elements in Natural Waters*, B. Salbu and E. Steinnes, Eds. (New York: CRC Press, 1992).
5. Hart, B.T., S. Sdraulig, and M.J. Jones. "Behaviour of Copper and Zinc Added to the Tambo River, Australia by a Metal-Enriched Spring," *Aust. J. Mar. Freshwater Res.* 43:457–489 (1992).
6. Forstner, U., and G.T.W. Wittmann. *Metal Pollution in the Aquatic Environment*, 2nd ed. (Heidelberg: Springer-Verlag, 1981).
7. Beckett, R., Ed. *Surface and Colloid Chemistry in Natural Waters and Water Treatment* (New York: Plenum Press, 1990).
8. Hunter, K.A. and P.S. Liss. "Organic Matter and the Surface Charge of Suspended Particles in Estuarine Waters," *Limnol. Oceanogr.* 27:322–335 (1982).
9. Beckett, R. and N.P. Le. "The Role of Organic Matter and Ionic Composition in Determining the Surface Charge of Suspended Particles in Natural Waters," *Colloids Surf.* 44:35–49 (1990).
10. Tipping, E. and D. Cooke. "The Effects of Adsorbed Humic Substances on the Surface Charge of Goethite in Freshwaters," *Geochim. Cosmochim. Acta* 46:73–80 (1982).
11. Beckett, R. "The Composition and Surface Properties of Suspended Particulate Matter," in *The Role of Particulate Matter in the Transport and Fate of Pollutants*, B. T. Hart, Ed. (Melbourne: Water Studies Center, Chisholm Institute of Technology/Monash University, 1986), pp.113–142.

12. Aiken, G.R., D.M. McKnight, R.L. Wershaw, and P. MacCarthy, Eds. *Humic Substances in Soil, Sediment and Water: Geochemistry, Isolation and Characterization* (New York: John Wiley & Sons, Inc., 1985).
13. Giddings, J.C. "Field-Flow Fractionation," *C & E News* 66:34–45 (1988).
14. Caldwell, K.D. "Field-Flow Fractionation," *Anal. Chem.* 60:959A–971A (1988).
15. Janca, J. *Field-Flow Fractionation: Analysis of Macromolecules and Particles.* Chromatographic Science Series, Vol. 39 (New York: Marcel Dekker, Inc., 1988).
16. Beckett, R., Zhang Jue, and J.C. Giddings. "Determination of Molecular Weight Distributions of Fulvic and Humic Acids Using Flow Field-Flow Fractionation," *Environ. Sci. Technol.* 21:289–295 (1987).
17. Beckett, R., J.C. Bigelow, Zhang Jue, and J.C. Giddings. "Analysis of Humic Substances Using Flow Field-Flow Fractionation" in *The Influence of Aquatic Humic Substances on Fate and Treatment of Pollutants*, P. MacCarthy and I. H. Suffett, Eds., ACS Advances in Chemistry Series No. 219 (Washington, D.C.: American Chemical Society, 1989), pp. 65–80.
18. Karaiskakis, G., K.A. Graff, K.D. Caldwell, and J.C. Giddings. "Sedimentation Field-Flow Fractionation of Colloidal Particles in River Water," *Int. J. Environ. Anal. Chem.* 12:1–15 (1982).
19. Beckett, R., G. Nicholson, B.T. Hart, M. Hansen, and J.C. Giddings. "Separation and Size Characterization of Colloidal Particles in River Water by Sedimentation Field-Flow Fractionation," *Wat. Res.* 22:1535–1545 (1988).
20. Beckett, R., D.M. Hotchin, and B.T. Hart. "Use of Field-Flow Fractionation to Study Pollutant-Colloid Interactions," *J. Chromatogr.* 517:435–447 (1990).
21. Beckett, R., G. Nicholson, D.M. Hotchin, and B.T. Hart. "The Use of Sedimentation Field-Flow Fractionation to Study Suspended Particulate Matter," *Hydrobiologia* in press (1992).
22. Chittleborough, D.J., D.M. Hotchin, and R. Beckett. "Sedimentation Field-Flow Fractionation: A New Technique for the Fractionation of Soil Colloids," *Soil Sci.* in press (1992).
23. Beckett, R. "The Application of Field-Flow Fractionation Techniques to the Characterization of Complex Environmental Samples," *Environ. Technol. Lett.* 8:339–354 (1987).
24. Giddings, J.C. "The Conceptual Basis of Field-Flow Fractionation," *J. Chem. Educ.* 50:667–669 (1973).
25. Myers, M.N. and J.C. Giddings. "Properties of the Transition from Normal to Steric Field-Flow Fractionation," *Anal. Chem.* 54:2284–2289 (1982).
26. Giddings, J.C., M.N. Myers, K.D. Caldwell, and P.J. Pav. "Steric FFF as a Tool for the Size Characterization of Chromatographic Supports," *J. Chromatogr.* 185:261–271 (1979).
27. Caldwell, K.D., T.T. Nguyen, M.N. Myers, and J.C. Giddings. "Observations on Anomalous Retention in Steric Field-Flow Fractionation," *Sep. Sci. Technol.* 14:935–946 (1979).
28. Segr, G. and A. Silberberg. "Radial Particle Displacements in Poiseuille Flow of Suspensions," *Nature* 189:209–210 (1961).
29. Caldwell, K.D., Z.Q. Chen, P. Hradecky, and J.C. Giddings. "Separation of Human and Animal Cells by Steric Field-Flow Fractionation," *Cell Biophys.* 6:233–251 (1984).

30. Koch, T. and J.C. Giddings. "High-Speed Separation of Large (>1 μm) Particles by Steric Field-Flow Fractionation," *Anal. Chem.* 58:994–997 (1986).
31. Giddings, J.C., X. Chen, K.G. Wahlund, and M.N. Myers. "Fast Particle Separation by Flow/Steric Field-Flow Fractionation," *Anal. Chem.* 59:1957–1962 (1987).
32. Ratanathanawongs, S.K. and J.C. Giddings. "High-Speed Size Characterization of Chromatographic Silica by Flow/Hyperlayer Field-Flow Fractionation," *J. Chromatogr.* 467:341–356 (1989).
33. Giddings, J.C., M.H. Moon, P.S. Williams, and M. N. Myers. "Particle Size Distribution by Sedimentation/Steric Field-Flow Fractionation: Development of a Calibration Procedure Based on Density Compensation," *Anal. Chem.* 63:1366–1372 (1991).
34. Tanford, C. *Physical Chemistry of Macromolecules* (New York, John Wiley & Sons, Inc., 1961), Ch. 6.
35. Wahlund, K.-G. and J.C. Giddings. "Properties of an Asymmetric Flow Field-Flow Fractionation Channel Having One Permeable Wall," *Anal. Chem.* 59:1332–1339 (1987).
36. Jönsson, J. and A. Carlshaf. "Flow Field-Flow Fractionation in Hollow Cylindrical Fibres," *Anal. Chem.* 61:11–18 (1989).
37. Wahlund, K.-G. and A. Litzén. "Applications of an Asymmetric Flow Field-Flow Fractionation Channel to the Separation and Characterization of Proteins Plasmid Fragments Polysaccharides and Unicellular Algae," *J. Chromatogr.* 461:73–87 (1989).
38. Litzén, A. and K.-G. Wahlund. "Improved Separation Speed and Efficiency for Proteins, Nucleic Acids and Viruses in Asymmetrical Flow Field-Flow Fractionation," *J. Chromatogr.* 476:413–421 (1989).
39. Giddings, J.C. and M.N. Myers. "Steric Field-Flow Fractionation: A New Method for Separating 1–100 μm Particles," *Sep. Sci. Technol.* 13:637–645 (1978).
40. Williams, P.S., J.C. Giddings, and R. Beckett. "Fractionating Power in Sedimentation Field-Flow Fractionation with Linear and Parabolic Field Decay Programming," *J. Liq. Chromatogr.* 10:1961–1998 (1987).
41. Yau, W.W. and J.J. Kirkland. "Retention Characteristics of Time-Delayed Exponential Field-Programmed Sedimentation Field-Flow Fractionation," *Sep. Sci. Technol.* 16:577 (1981).
42. Williams, P.S. and J.C. Giddings. "Power Programmed Field-Flow Fractionation: A New Program Form for Improved Uniformity of Fractionating Power," *Anal. Chem.* 59:2038–2044 (1987).
43. Giddings, J.C., P.S. Williams, and R. Beckett. "Fractionating Power in Programmed Field-Flow Fractionation: Exponential Sedimentation Field Decay," *Anal. Chem.* 59:28–37 (1987).
44. Hansen, M.E., J.C. Giddings, M. R. Schure, and R. Beckett. "Corrections for Secondary Relaxation in Exponentially Programmed Field-Flow Fractionation," *Anal. Chem.* 60:1434–1442 (1988).
45. Thurman, E.M. *Organic Geochemistry of Natural Waters* (Dordrecht, The Netherlands: Nijhoff/Junk, 1985).
46. Wershaw, R.L. and G.R. Aiken. "Molecular Size and Weight Measurements of Humic Substances," in *Humic Substances in Soil, Sediment and Water: Geochemistry Isolation and Characterization*, G.R Aiken, D.M. McKnight, R.L. Wershaw, and P.MacCarthy, Eds. (New York: Interscience, 1985), Chapter 19, pp. 477–492.

47. Reid, P.M., A.E. Wilkinson, E. Tipping, and M.N. Jones. "Determination of Molecular Weights of Humic Substances by Analytical (UV) Scanning Ultracentrifugation," *Geochim. Cosmochim. Acta* 54:131–138 (1990).
48. Kerker, M. *The Scattering of Light* (New York, Academic Press, Inc., 1969).
49. Taylor, H.E., J.R. Garbarino, D.M. Hotchin, and R. Beckett. "Inductively Coupled Plasma-Mass Spectrometry for Use as an Element Specific Detector for Sedimentation Field-Flow Fractionation Particle Separation," *Anal. Chem.* in press (1991).
50. Nicholson, G.J. "Studies on Sedimentation Field-Flow Fractionation and Coagulation of Fluvial Colloidal Matter," M. Appl. Sci. Thesis, Chisholm Institute of Technology/Monash University, Melbourne, Australia (1987).
51. Reschiglian, P. and F. Dondi. "Characterization of Colloidal Particles in River Water and Turbidity Control by Power Field-Decay in Sedimentation Field-Flow Fractionation," FFF-89 First International Symposium on Field-Flow Fractionation, Salt Lake City, UT (1989), p. 54.
52. Taylor, H.E. and J.R. Garbarino. "Utilization of Inductively Coupled Plasma-Mass Spectrometry as a Detector for Sedimentation Field-Flow Fractionation," FFF-89 First International Symposium on Field-Flow Fractionation, Salt Lake City, UT (1989), p. 50.
53. Giddings, J.C., M.N. Myers, M.H. Moon, and B.N. Barman. "Particle Size Distribution II: Assessment and Characterization." in *Particle Separation and Size Characterization by Sedimentation Field-Flow Fractionation*, T. Provder, Ed., ACS Symposium Series No. 472 (Washington, D.C., American Chemical Society, 1991), pp.198–216.
54. Schure, M.R., M.N. Myers, K.D. Caldwell, C. Byron, K.P. Chan, and J.C. Giddings. "Separation of Coal Fly Ash Using Continuous Steric Field-Flow Fractionation," *Environ. Sci. Technol.* 29:686–689 (1985).
55. Barman, B.N., M.N. Myers, and J.C. Giddings. "Rapid Particle Size Analysis of Ground Minerals by Flow/Hyperlayer Field-Flow Fractionation," *Powder Technol.* 59:53–63 (1989).
56. Barnard, P.C., R.E. Van Grieken, and D. Eisma. "Classification of Estuarine Particles Using Automated Electron Microprobe Analysis and Multivariate Techniques," *Environ. Sci. Technol.* 20:467–473 (1986).
57. Hart, B.T., G. Douglas, R. Beckett, A. Van Put, and R.E. Van Grieken. "Characterization of Colloidal and Particulate Matter Transported by the Magela Creek System, Northern Australia," *Hydrolog. Process.* in press (1992).
58. Hansen, M.E., J.C. Giddings, and R. Beckett. "Colloidal Characterization by Sedimentation Field-Flow Fractionation. VI. Peturbations due to Overloading and Electrostatic Repulsion," *J. Colloid. Interface Sci.* 132:300–312 (1989).
59. Caldwell, K.D. and J.T. Li. "Emulsion Characterization by the Combined Sedimentation Field-Flow Fractionation-Photon Correlation Spectroscopy Methods," *J. Colloid Interface Sci.* 132:256–268 (1989).
60. Li, J.T. and K.D. Caldwell. "Sedimentation Field-Flow Fractionation in the Determination of Surface Concentration of Adsorbed Materials," *Langmuir* 7:2034–2039 (1991).
61. Giddings, J.C., B.N. Barman, and M.-K. Li. "Separation of Cells by Field-Flow Fractionation and Related Methods," in *Cell Separation and Technology*, ACS Symposium Series, D. Kompala and P. Todd, Eds. (Washington, D.C.: American Chemical Society, 1990), pp.128–144.

CHAPTER 5

ENVIRONMENTAL AEROSOL CHARACTERIZATION BY SINGLE PARTICLE ANALYSIS TECHNIQUES

C. Xhoffer and R. Van Grieken

TABLE OF CONTENTS

0-87371-895-X/93/$0.00+$.50

INTRODUCTION

Micrometer-size particles are very important in all aspects of life: soils, sands, sediments, pigments, plant materials, solid insecticides, as well as living cells. In addition, human activities cause the production of various potentially unfavorable substances, as, for example, there are fly-ash, asbestos fibers, fallout particles and particulate pollutants in water. Among all these particles, atmospheric aerosols form one of the major constituents within the Earth's environment. They originate from a variety of sources, both natural and anthropogenic, and some have toxic or carcinogenic effects on human life.

The most important characteristics of aerosols are their shape, size, optical and electrical properties, structure, heterogeneity, and chemical composition. The last-mentioned concerns the elemental, isotopic, and molecular composition and is often the most difficult and crucial information needed for the characterization of particles.

For many years, the primary information on chemical aerosol composition was given by bulk analysis of filter samples. However, in the last two decades, more refined and powerful analytical instruments have been developed that are capable of analyzing aerosols at the single particle level. These techniques opened new directions and possibilities in aerosol research, especially where source apportionment of the constituents of atmospheric aerosols are concerned.

For a general overview on single particle analysis techniques and their applications in various environmental domains, we refer to a recent review publication.[1]

Throughout the years, our laboratory has gained much experience in the analysis of individual aerosol particles from various environments: coastal, marine, urban, and forestal. The combination of results from sensitive and specific analytical techniques, such as EPMA, LAMMA, SIMS, EELS, and STEM, available at the Micro- and Trace Analysis Centre, are necessary to study and learn as much as possible about aerosols. This chapter will outline this work in the context of characterizing and apportioning elemental concentrations to their respective sources and of elucidating formation mechanisms and chemical transformation reactions of environmentally important particle types.

MICROBEAM ANALYSIS TECHNIQUES

Electron Probe X-ray Microanalysis (EPXMA)

At the University of Antwerp, a JEOL Superprobe JXA-733 (Tokyo, Japan) is used for single particle analysis. In the automated PRC routine for particle recognition and characterization, a very narrow electron beam is raster scanned over a preset sample area of a Nuclepore filter. Each particle on the filter is localized, sized, and chemically analyzed by measuring the energy of the X-rays generated within that specific particle. Typical working conditions are a beam current of 1 nA at an acceleration voltage of 20 to 25 keV. X-ray spectra are accumulated for about 20 s in the energy-dispersive (EDX) mode using a Si(Li) detector. The entire data are stored on floppy disk and can be processed at any time. This technique is very fast and efficient for the characterization of individual particles. However, it is not straightforward to interpret the huge amount of information available from EPXMA measurements. Therefore, multivariate clustering procedures are used to classify each particle according to a most representative particle type. Most often, hierarchical and nonhierarchical clustering procedures are used, occasionally in combination with principal component analysis (PCA). The purpose of the PCA method is to represent the variation present in the data in such a way that, without losing significant information, the dimensionality is reduced. The particles are thus classified according to their matrix composition by multivariate methods, resulting in the assessment of the abundances for the different particle types that can be correlated with their source of emission or their formation mechanism. General information about these classification procedures can be found in the textbook of Massart and Kaufman.[2]

Laser Microprobe Mass Analysis (LAMMA)

The commercial LAMMA-500 of Leybold-Heraus (Cologne, Germany) makes use of an Nd-Yag laser to generate very short and intense light pulses for the vaporization and ionization of a microvolume of the sample. The positive and negative ions are accelerated and detected in a mass spectrometer. The obtained mass spectra are also stored on a magnetic device for off-line processing. The interpretation of the LAMMA-data is based on an attentive inspection of the mass spectra and the comparison with reference spectra of standards. LAMMA measurements on impactor samples reveal additional data about the low Z-elements ($Z < 11$) and about the speciation of elements as S and N. Because of the favorable detection limits of LAMMA, trace elements are more often detected compared to EPXMA. Also, organic and inorganic molecular information can be obtained when the instrument is used under mild operating conditions, usually defined as laser desorption excitation.

Electron Energy-Loss Spectroscopy (EELS)

Our laboratory is equipped with a ZEISS 902 CEM microscope (Oberkochen, Germany). Accelerated electrons of one specific energy (80 keV) are focused on a thin foil supporting the particles. Only the transmission electrons are selected and dispersed by a magnetic prism spectrometer according to the energy lost during the inelastic event. By scanning the dispersion plane over a detector and plotting the intensity I(E) of the transmitted signal versus the energy-loss, an EEL spectrum is collected. Element specific images (ESI) result in element specific maps and can be obtained using the spectrometer as an imaging forming system.

EELS is a high spatial resolution technique suitable for the determination of low-Z elements. It offers the possibility of analyzing and mapping element specific features down to the submicron level. The detection limit is comparable to EPXMA and is in the range of 1000 ppm.

Secondary Ion Mass Spectrometry (SIMS)

SIMS analysis at our institute is performed with a CAMECA IMS-300 (Paris, France) instrument. The samples are bombarded with 14 kV O^- primary ions rastered over a small sample area. The charged secondary ions are then accelerated over a potential and analyzed by a magnetic sector according to their mass/charge ratio. The generated secondary ions can thus be collected as mass spectra, as in-depth profiles, or as distribution images of the sputtered surface.

SIMS is capable of generation surface specific signals from a depth of a few nm. For a microanalysis approach, only minimum sampling volumes (0.01 mm^3) are required. In principle, all elements can be detected, and isotope ratio measurements can be obtained.

APPLICATIONS TO ENVIRONMENTAL AEROSOL RESEARCH

Marine Aerosols

North Sea Environment

It is known that the atmospheric input of heavy metals in the North Sea by wet and dry deposition processes is of importance as a pathway for the pollution emitted by the surrounding industrialized countries.[3] In Table 1 the different input routes (riverine, direct discharges, dumping, and the atmosphere) of trace metals to the North Sea are compared. The atmospheric input constitutes a major pollution source for metals such as Zn, Pb, Cu, and Cd but also for organic material such as PAH, hexachlorohexanes, and PCB's. Most of this North Sea data are obtained by bulk chemical analysis, and microanalytical techniques have rarely been invoked. Today, the development of suitable software for automated particle analysis and data processing routines by multivariate analysis highly reduces the analysis time necessary to characterize an aerosol sample based on individual particle analysis measurements.

Ship based measurements are very well suited for atmospheric environmental aerosol research purposes. Large sampling times over a very wide region can be achieved with relatively simple equipment. Hitherto, this strategy has often been applied and proven to be very succesful for the sampling of aerosols over the oceans. Ship based measurements generally reflect the result of interaction processes between the sea surface and the atmosphere. However, such sampling

Table 1. Total Input of Heavy Metals into the North Sea and Relative Contribution of the Different Sources

	Cd	Cu	Pb	Zn	Ref.
Rivers (%)	19	34	14	31	103
Discharges (%)	6	6	2	5	103
Dumping (%)	9	29	35	34	104
Atmosphere (%)	66	31	49	30	105

strategies, as well as sampling from coastal stations, can sometimes be biased by contamination due to sea spray, rust particles, and paint chips from the vessel. Aerosol sampling with the aid of an aircraft has the technical possibility of sampling both at very low altitudes, down to 10 to 15 m above sea level, and at much higher altitudes. Therefore, an evaluation can be made of the contribution of the sea itself as a source for trace metals in the air. Therefore, the combined use of a research vessel and an aircraft for aerosol sampling offers the possibility to cover a large area, to obtain vertical profiles, and to avoid any contamination by local sources of pollutants.

For the ship based measurements, series of aerosol samples were collected on board the research vessel "R/V Belgica" during various cruises over the North Sea, the English Channel, and the Celtic Sea in a time range of 4 years (1984–1987). The sampling for EPXMA was done on 47 mm diameter Nuclepore filters with a pore size of 0.4 μm. At selected times, additional samples were collected for LAMMA measurements with a low volume single orifice impactor for periods of 10 min. For this, the aerosol particles were directly impacted on a 100 nm thick Formvar foil supported by 300 mesh electron microscope grids. The sampling times of both filtration and impactor samples were optimized to obtain an appropriate loading on the collection surface to allow single particle analysis by the microanalytical techniques.

Another series of airborne particulate matter was collected with the aid of an aircraft over the Southern Bight of the North Sea. The sampling campaign started in September, 1988, and ended in October, 1989. Using a twin engine Piper Chieftain PA 31-350, 108 samples were collected. Detailed information on the sampling instrumentation and sampling strategy can be found elsewhere.[4] Briefly, after localizing the inversion layer, six tracks of ca. 110 km were flown at six different heights and equally spaced between the sea surface (level 1) and the inversion layer (level 6). Particulate matter was sampled and deposited on 0.4 μm pore-size Nuclepore membrane filters using an isokinetic inlet.[5] Also, giant aerosol particles (particles with radius > 1 μm and especially those larger than 10 μm) were collected by a specially developed sampling device. The giant aerosol collector consists of a circular impaction surface (diameter = 1 cm) covered with a particle sticky tape and supported by a vertical bar. The hole is exposed outside the airplane perpendicularly to the streamlines of the air around the plane and directed upwind. These giant aerosols are of importance since they have a high impact on the total deposition process and on cloud and precipitation processes.[6,7] Calculations by Dedeurwaerder[8] showed that deposition of giant particles explains 94% of the total dry deposition above the North Sea for Cd, 96% for Cu, 85% for Pb, and 88% for Zn. For both the normal and giant aerosol mode, data reduction and interpretation was based on hierarchical and nonhierarchical cluster algorithms combined with principal factor analysis (PFA) with orthogonal Varimax rotation.[9,10]

The EPXMA results after hierarchical cluster analysis of either the ship- or the aircraft-based measurements generally reveal 10 major particle types. Differences in particle type abundances are observed for the various sampling campaigns since the chemical composition of the airborne particulate matter strongly depends not only on the meteorological conditions and the back trajectories of the air masses, but also on the sampling location and altitude of sampling. We shall discuss each identified particle type separately with special reference to their formation mechanisms and chemical changes during their lifetime in the atmosphere and to their specific emission sources.

Sea salt. The presence of sea salt in the atmospheric aerosol is attributed to a pure marine source. The breaking of waves is the main process for the generation of fresh sea salt into the atmosphere. These contributions of NaCl in the aerosol are more pronounced under more remote sampling conditions or when the back trajectories of the sampled air come from far over the Atlantic Ocean without continental interferences.

Transformed sea salt. This particle type is rich in S and Cl, but also mixtures of $NaNO_3$, Na_2SO_4, and NaCl are possible.[11] These S- and Cl-rich particles are formed by the conversion of NaCl into Na_2SO_4 by SO_2 or H_2SO_4 implying halogen displacement reactions in sea salt particles.[12,13] The contribution of S enrichment in sea salt aerosols is greater and clearly pronounced in samples for which an important anthropogenic influence on the marine atmosphere is expected. In the aerosol samples collected during prevailing continental influence, pure sea salt particles are no longer detected. The formation of mixed sea salt/mineral aerosols can be explained by cloud coalescence processes between mineral and sea salt containing particles.[14] Sometimes nitrate enrichment has been observed in LAMMA spectra. Laboratory experiments have shown that the formation of $NaNO_3$ on micrometer size NaCl particle samples by impaction can be enhanced by the uptake of gaseous HNO_3 from the passing airstream.[15] Such nitrate coatings on environmental sea salt particles have also been reported for LAMMA studies on aerosols from the Bahamas and the Sargasso Sea.[16] Besides typical positive and negative mode LAMMA spectra of untransformed and highly transformed sea salt particles, also mixtures of Na-Cl-Pb have been observed. These could be Pb-rich particles from automobile exhausts which have coagulated with atmospheric sea salt particles. Up to 49% of the total giant particles detected in marine air masses above the Southern Bight of the North Sea have been identified as sea salt and aged sea salt. Principal factor analysis (PFA) performed on this giant particle data set[10] have shown that the marine source (high loadings observed for Na, Mg, S, and Cl) is slightly anticorrelated with altitude. This could be expected from the mechanism by which major and trace elements are injected from the sea into the atmosphere.[17]

Sulfur-rich particles. For this particle type, no associations to other detectable elements with $Z > 11$ are present. High particle type abundances have been observed during sampling campaigns for which the air masses have had long residence times over industrial regions (e.g., south of England or from above Eastern European countries). Consequently, emissions must be related to industrial and automotive sources all over Europe. Most of these S-rich particles have a diameter in the submicrometer range. A number of secondary reactions can take place on particulate sulfur.[18,19] Neutralization reactions can occur between NH_4^+ (often present in continental aerosols) and H_2SO_4 with the formation of various ammonium salts such as $(NH_4)_2SO_4$, $(NH_4)HSO_4$, and $(NH_4)_3H(SO_4)_2$.[20] Also biological particles and carbonaceous material (composed of low-Z elements such as H, C, N, and O), with their complex morphology and wet surfaces, provide an attractive nucleation surface for SO_2 absorption and conversion to sulfate.[21] LAMMA spectra of such mixed carbonaceous and S-rich particles have been detected abundantly in the 0.5 to 1 μm size range of the impactor samples. Also Fe and Mn that catalyze the SO_2 oxidation processes[22] can frequently be detected in these particles. A smaller fraction of the S-rich particles have a marine origin, as is the case for dimethyl sulfoxide (DMSO) and derived compounds. Low particle number concentrations have been detected at the most westerly sampling locations, where only pure marine conditions are encountered. In open ocean waters, the predominant volatile sulfur compound is dimethyl sulfide (DMS), representing almost 90% of the marine sulfur emissions.[16] Methanesulfonic acid (MSA), one of the oxidation products of DMS, has been found to be mainly associated with the smallest aerosol particles, whose LAMMA spectra match the reference fingerprint spectra of MSA salts, sodium methanesulfonate, and ammonium methanesulfonate. Analogue LAMMA results have been observed for Antarctic aerosols,[23] but this will be discussed in more detail later in this chapter.

Calcium sulfate particles. These particles are very often found in both marine and continental aerosols. Andreae et al.[14] postulated some possible marine formation mechanisms, such as fractional crystallization of marine aerosols or interaction processes between marine or terrestrial airborne $CaCO_3$ with atmospheric SO_2 and/or H_2SO_4, e.g., within clouds. However, for the North Sea atmospheric environment, these mechanisms are of minor importance since the major abundances for $CaSO_4$ particles have been present in the samples influenced by the continent. In these cases, NaCl has been virtually absent. It is remarkable that all $CaSO_4$-rich filters sampled above the North Sea have been influenced by continental air masses travelling from over the South of England. Thus, important sources for $CaSO_4$ particles above the North Sea are combustion processes and eolian transport from the continent. Note that large quantities of $CaCO_3$ and $Ca(OH)_2$ are used in thermal power plants, as a desulfurization agent, to enfavor the oxidation of SO_2 and neutralize the SO_3; this leads to sulfates. From X-ray spectra and LAMMA-spectra, it has been observed that, for some particles, higher S contributions are present, with relative S-intensities much higher than normally detected for $CaSO_4$. Hence, this particle type could partly be identified as $CaSO_4$ enriched with S. Also, Harrison and Sturges[18] explained the formation of $CaSO_4.(NH_4)_2SO_4$ aerosols by the coagulation of $CaSO_4$ with submicrometer sulfate aerosols.

Ca-rich particles. This particle type shows only Ca as a detectable element in the X-ray spectrum. The particles are assigned to $CaCO_3$ (C and O cannot be detected in conventional EPXMA.), and they can, just as $CaSO_4$, originate from both marine and continental sources. In some Ca-rich particles, sulfur has been detected. Yet, these have not been classified into the $CaSO_4$ group because of too low S-characteristic X-ray intensities. This particle type may represent the initiation of Ca-S-rich particle formation from $CaCO_3$ and/or derived components with atmospheric SO_2 or H_2SO_4. According to the obtained data set, Ca-rich particles above the North Sea cannot unambiguously be apportioned to one specific source type.

Aluminosilicate-rich particles. This mineral type of particles finds its origin on the continent. The major elements are Al, Si, S, K, Ca, and Fe, and the minor ones, Ti, Cr, Mn, Ni, and Zn. From morphology studies, smooth and nearly spherical particles could be identified as fly-ash and therefore be differentiated from soil dust. These typically shaped particles frequently have been observed on filters when the sampled air masses were mainly influenced by Eastern European emissions. Quantitative EDX-analyses have been performed by Storms[24] and Rojas[9] and Van Grieken on a dozen fly-ash particles by applying a spherical ZAF-correction procedure to the X-ray intensities.[25] The following average composition was calculated for this group: 25% Si, 18% Al, 5.5% K, 3.0% Fe, 0.9% S, 0.4% Ca, and 47% O. These data are in close agreement with the results reported by Kaufherr and Lichtmann[26] for fly-ash particles sampled in the U.S. Positive LAMMA spectra of these spherical aluminosilicates also show the presence of trace elements, such as V, Ga, Rb, Cs, Ba, and Pb. However, one must realize that the distinction between soil dust and fly-ash is still dubious.

During long periods of fog, high relative humidities, and inversion layer conditions as encountered during some sampling campaigns, aluminosilicate particles can undergo secondary transformation reaction by SO_2 and/or H_2SO_4. Mineral particles that are initially hydrophobic can probably be wetted with acidified water droplets and left behind as cloud condensation nuclei (CCN) for further reactions. These CCN can, in an initial state, be composed of an aluminosilicate nucleus covered by a S-rich surface coating. But further transformation, element-specific extraction processes are also possible. Such reactions can be responsible for the further breakdown of silicate mineral particles, forming silicon-rich clusters. In the negative ion mode, LAMMA spectra from such particles point to the presence of mass peaks (m/z = 97 from HSO_4^-) of sulfate species such as $(NH_4)_2SO_4$ or NH_4HSO_4. Other mass peaks in the negative ion mode of such spectra are related to the matrix of the aluminosilicate.

The giant aluminosilicate particle fraction, consisting of both wind blown dust and fly-ash, is positively correlated with altitude, which could be explained by the presence of a long-range transported mineral aerosol at high altitudes. The size distribution of the giant aluminosilicates shows a bimodal character, with maxima of the average diameters centered around 4 μm and 15 μm. This suggests two completely different sources, and morphology studies have confirmed the existence of smaller spherical particles (fly-ash derived) and a larger, irregularly shaped fraction.

Silicon-rich particles. These mineral quartz particles are irregularly shaped. They can often be derived from soil dust or emitted into the atmosphere during the combustion of coal in power plants. Therefore, they are frequently found in the presence of aluminosilicates. The fact that these particles, just as the Fe-rich particles, are found in the smallest size range (below 1 μm) could strengthen the hypothesis that they are formed during combustion processes.

Titanium-rich particles. For some North Sea samples, rather high Ti-rich particle abundances have been observed. The main source for Ti-release into the atmosphere is pigment spray, but minor pollution processes and sources, such as soil dispersion, asphalt production, and coal-fired boilers and power plants have also been recognized.[27] Sometimes, contributions of Cr, Si, Zn, Pb, and Ba are observed in this particle type.

Iron-rich particles. These particles show a spherical shape with a mean diameter of 0.6 μm, and they are quite common at all heights over the Southern Bight airshed. They are mostly associated with siderurgical activities in Northern France.[28] Within the Fe-rich cluster, three different Fe-rich fractions can be recognized. The first and second subgroup, characterized by pure Fe and Fe-Zn-Mn respectively, are mainly produced by ferrous metallurgy processes. Very often, S is associated with Fe-rich particles, and this represents the third subgroup. These are pyrite and iron sulfate, probably formed by reaction between iron oxide and sulfuric acid during or after their release in ferrous metallurgy-related combustion processes.

Miscellaneous particle types. Each of the miscellaneous particle types observed accounted for less than 10%. These particle types are not always present in air masses above the North Sea. They are often characterized by the presence of some heavy metals. Variations within these groups are mainly due to different meteorological conditions and seasons during which the sampling took place. Although their abundances are quite low compared to the other groups, the presence of these rather rare particles can sometimes be apportioned to one specific source. For example, higher abundances of heavier elements like Pb and Br together with Cl have been detected. These Pb-rich particles originate from automobile exhaust emissions. Indeed, Pb is added to the gasoline together with ethylene dihalide compounds (Br, Cl). The emitted lead halides can readily be converted to lead sulfates by reaction with SO_2, H_2SO_4, or $(NH_4)_2SO_4$ with the loss of HBr.[29] However, none of the transformed lead halide particles have shown associations with S, in spite of the very high S concentrations present in the corresponding sampled air masses.

Characteristic X-ray spectra have shown that P is sometimes associated with S and Cl or with an organic fraction. The organic phosphorus fraction is formed primarily through biological activities. Bubble bursting in the sea can cause enrichment of P in the sea salt aerosols by fractionation out of the sea surface microlayer.[30] Chen et al.[31] found that the major sources for particulate P in marine aerosols of New Zealand are soil particles containing both naturally occurring and fertilizer-derived P, as well as sea salt particles and industrial emissions. Na, Al, and V can be associated with P as markers for a marine source and an anthropogenic pollution source (burning of biological material), respectively. However, for the North Sea samples, no associations to these markers have been observed, and hence these P-rich particles probably have a marine origin. Alternatively, Ca and P are often present in the same particles. This particulate fraction can be fertilizer-derived or is a residual from biological material (e.g., pollen) and can be transported over the North Sea by wind action. Ca- and P-rich particles, although with different elemental X-ray intensities, have been identified in aerosol samples taken above the equatorial Pacific Ocean where only marine influences are expected.[32]

Heavy metal concentrations of Cu and Zn emitted by metallurgical processes have been confirmed to reach maxima with East-Southeast winds[3] as previously reported by Kretzschmar and Cosemans.[33] Also, burning processes of either organic material or of municipal waste in incinerators are responsible for the emission of K- and Zn-chlorides.

Particles with no major X-ray intensities have been denoted as organic. They have seemed anticorrelated with altitude and could therefore be related to biogenic material such as pollen, spores, and bacteria. Sometimes, V and Pb are associated with organic matter, and these particles are probably pollution-derived, especially near combustion sources where their contents are highest. The chemical complexity of these particles indicates that several atmospheric interaction processes can be involved in the formation.

Besides hierarchical cluster analysis, other multivariate techniques, like nonhierarchical clustering or principal component analysis, can be useful to gain extra information from automated individual particle analysis. As an example we

Table 2. Cumulative Eigenvalues and the Loadings for the First Three Principal Components Derived from the Covariance Matrix for Two PCA Performed on the North Sea Data

	Principal Components					
	For all Samples (51)			For the Third Cluster (29)		
	1	2	3	1	2	3
Cum. Percentage	54	75	91	61	80	88
Variables	**Loadings**					
Sea salt	0.99	0.01	−0.09			
Aged sea salt	0.42	−0.10	0.22			
Marine Fraction (MF)				−0.50	0.86	−0.10
Sulfur	−0.66	−0.40	−0.62	0.97	−0.04	−0.21
$CaSO_4$	−0.41	0.91	−0.01	0.25	−0.13	0.06
Ca-rich	0.24	−0.19	0.00	−0.08	0.29	0.39
Aluminosilicates	−0.42	−0.39	0.80	−0.90	−0.37	−0.22
Quartz	−0.39	−0.18	0.26	−0.21	−0.46	0.01
Tri-rich	−0.23	−0.14	−0.23	0.29	−0.06	0.02
Fe-rich	−0.33	−0.21	0.00	0.11	−0.22	0.90

show here the application of a principal component analysis (PCA) performed on the above described North Sea data set.[34] The use of PCA allows the study of differences between all samples on the basis of their abundance variations between the identified particle types. The relative abundances of nine particle types (nine variables) for 51 samples were used as input data for the so-called Data Processing Program (DPP).[35] The loadings of the first three principal components, listed in Table 2, are plotted in Figure 1a, while the scores are represented in Figure 1b. The first three principal components explained 91% of the total variance present in the original data set. Three main clusters were recognized. The first group (14 samples out of a total of 51) had a high score on the first principal component determined by high abundances of NaCl and/or transformed NaCl. These samples thus showed a more pronounced marine character. A cluster of eight samples was separated by a second principal component and related to high $CaSO_4$ abundances. Because of a low negative score on the first principal component, the emission of $CaSO_4$ must be related to continental anthropogenic sources. A third cluster (29 samples) was elongated in the direction of the first and third principal components. The third component was related to high sulfur and aluminosilicate abundance variations. The results of these 29 samples were subjected to a second principal component analysis in order to enhance the differences between a marine and continental character. A new variable, the marine fraction, had to defined as

$$\textit{Marine fraction}\ (\%) = \frac{100 < ss + aged\ ss >}{< ss + aged\ ss >_{marine\ cluster}}$$

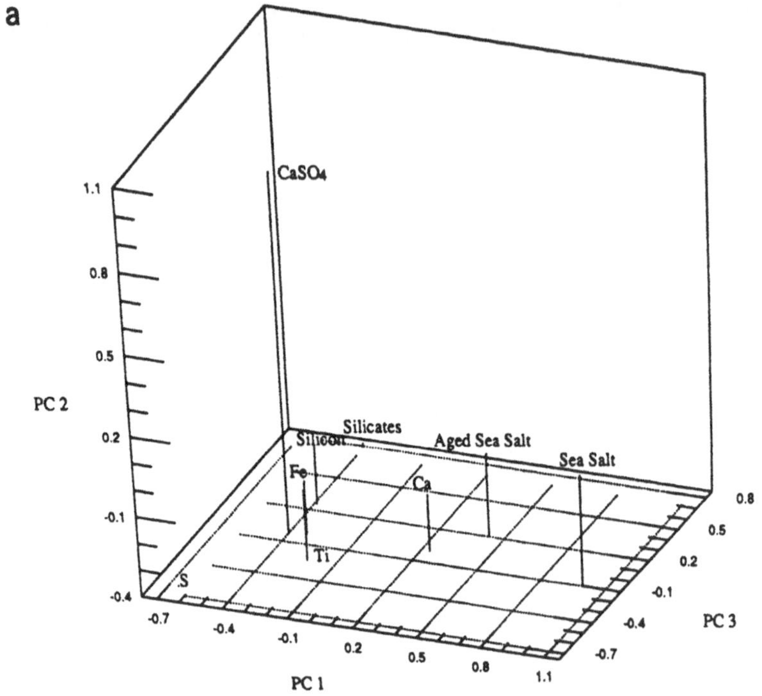

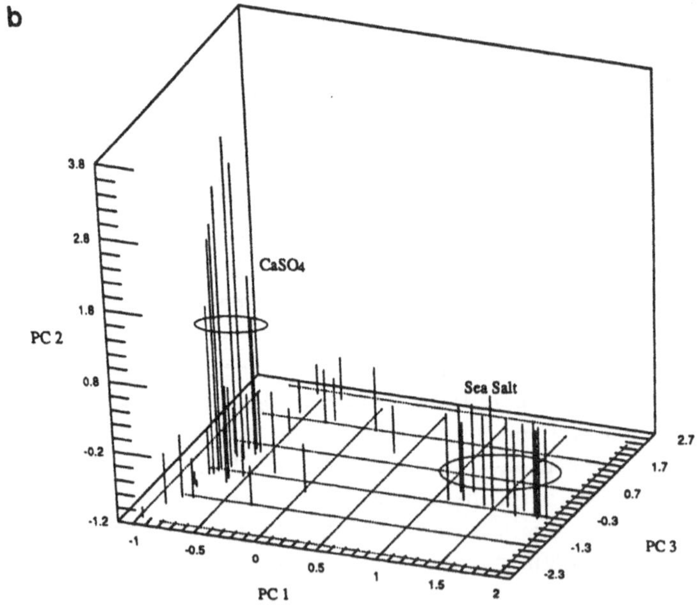

Figure 1. Loadings (a) and component scores (b) of the first three principal components (51 samples) obtained by PCA of North Sea aerosol results. (Reprinted from Xhoffer, C. et al. *Environ. Sci. Technol.* 25:1470–1478 (1991). With permission.)

where

> ss and aged ss represent respectively the sea salt and aged-sea salt particle type for each sample, and $<ss + aged\ ss>_{marine\ cluster}$ is the mean of the sum of the sea salt and aged-sea salt particle-type abundance for the marine cluster

The calculated value for the latter was 89%.

The results after a dual PCA performed on the EPXMA data set of the North Sea are also presented in Table 2. The loadings and scores are represented in Figure 2a and 2b, respectively. Seven samples showed a high correlation with the marine fraction, which means that most of the analyzed particles from these samples originate from mixed marine/continental air masses. The other identified cluster contains nearly 100% continentally derived particles (22 samples). Variations within this last group were mainly due to different silicate/sulphur ratios. The determination of the silicate/sulphur ratio reflects the contribution ratio of their anthropogenic sources. Table 3 summarizes the relative particle abundances for each of the four cluster groups, as well as the mean abundance for all particles analyzed. From the mean marine fraction value, we can conclude that the mean of all samples taken during the various North Sea campaigns is also representative for mixed sampling conditions.

Transect from Marine to Industrial Environment

The LAMMA technique has been applied to a set of aerosol samples collected on a transect from a beach site toward and through a heavily polluted industrialized area. The whole sampling strategy was performed under constant influence of steady clean onshore tradewinds. The aim was to find answers to the questions concerning sources, transport processes and physico-chemical changes of airborne particulate matter.

The samples were taken near the Brazilian coast, in the state of Bahia, north of the city of Salvador.[36] The sites were (1) a beach location about 20 m from the waterline of the Atlantic Ocean, (2) a hill with low vegetation, 10 km from the coast, and (3) a strongly polluted site 20 km from the coast in the center of a large industrial complex, including petrochemical, fertilizer, and metallurgical plants.

Atmospheric aerosols were sampled by means of a five-stage single-orifice cascade impactor with specified cut-off diameters of 4, 4 to 2, 2 to 1, 1 to 0.5, and 0.5 to 0.25 μm, respectively.

LAMMA spectra were classified for different aerosol particle types according to the major elemental and cluster ion peaks observed for a given sampling site. The positive and negative LAMMA spectra of nearly all particles could be classified according to one of six representative particle types.

Positive mode LAMMA spectra of sea salt aerosols are identified as Na- and K-chlorides. When applying the low laser energy desorption mode on these particles, it becomes clear that a significant fraction of these particles are coated

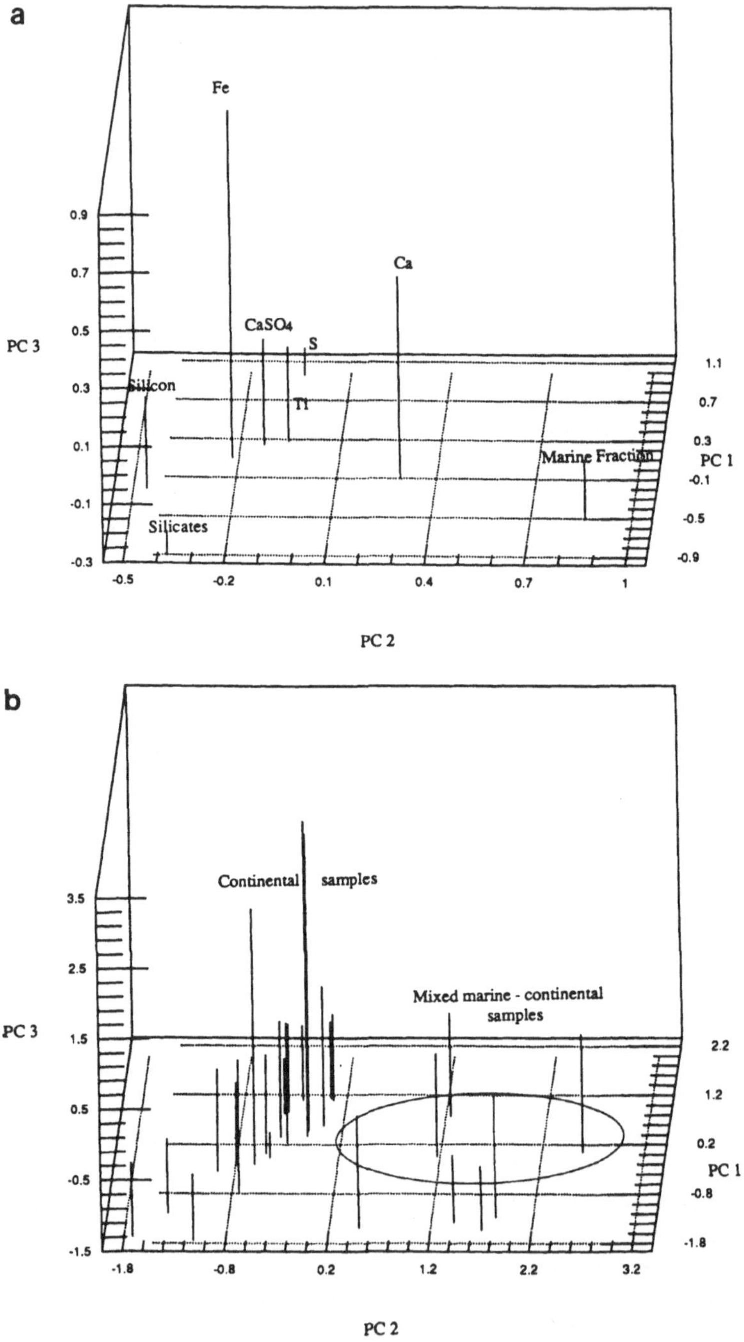

Figure 2. Loadings (a) and component scores (b) of the first three principal components, obtained by a second PCA on the third cluster (29 samples) resulting from the first PCA of North Sea aerosol results. (Reprinted from Xhoffer, C. et al. *Environ. Sci. Technol.* 25:1470–1478 (1991). With permission.)

Table 3. Mean Relative Abundances of Each Particle Type Calculated for the Four Clusters that were Formed by Principal Component Analysis

Type	Marine	$CaSO_4$	Mixed	S + Silicates	Mean
Number	14	8	7	22	51
Sea salt	76	1	16	1	23
Aged sea salt	13		17		6
S-rich	2	16	21	41	23
$CaSO_4$	2	67	7	9	16
Ca-rich	4		5	2	3
Aluminosilicates	2	9	30	27	18
Si-rich	1	4	3	6	4
Ti-rich				5	2
Fe-rich	1	4	3	9	4
Marine fraction	100	1	37	1	37

with a layer of $NaNO_3$ devoid of chloride. These chemical transformation reactions can be the result of differential crystallization of liquid seawater containing nitrate as a result of a heterogenous reaction between HNO_3 vapor and the particles.[18] Only a minor fraction of the 1 to 2 µm sea salt particles have been found to be rich in $NaNO_3$ in the samples taken at the beach location. Inland, the $NaCl$-$NaNO_3$ particles become predominant in the 1 to 4 µm size range, and in the industrial area, pure sea salt particles are not observed at all. This drastic abundance difference of $NaNO_3$–rich particles on the transect gives direct indication of a very fast and efficient gas-to-particle conversion mechanism.

Positive LAMMA spectra of Ca-rich particles result from various Ca-compounds, such as $CaSO_4$, $CaCO_3$, and CaO. In some cases, negative mode spectra have revealed the presence of the sulfate ion. EPXMA have shown the presence of $CaCO_3$ or CaO by the intense Ca signal in the X-ray spectrum and the absence of other elements with $Z < 11$.

The fourth characteristic particle type represents typical crust-derived aerosols. However, their size distribution does not plead for a local soil dispersion source. EPXMA measurements on the 0.5 to 0.25 µm stage of sample 1 have shown particles rich in Al, Si, S and Fe, and predominantly particles rich in Al and Si and less S and Fe in the 1 to 0.5 µm size range. Part of these particles have been identified as fly-ash particles associated with carbonaceous material and sulfate, most probably derived from the common charcoal fires on the beach.

A fifth heterogeneous group gathers diverse spectra which exhibit mass peaks from a combination of the following elements: Al, Si, Ca, Ti, V, Mn, Fe, Co, Sr, Ba, and NH_4^+. No evidence has been found for soot or sulfur compounds. On the contrary, phosphorus and nitrogen oxy-salts have been observed. Because of the high variability in composition and the unusual metal enrichments, the corresponding particles are classified as pollution particles from the industrial complex and the local metallurgy.

The final group refers to particles which, in addition to crustal elements or V, yield positive mass peaks corresponding to organic N-compounds such as amines. These particles are observed in the smallest size-range, and their relative abundance increases from the seaside station towards the industrial area. Amines can be released into the atmosphere by decomposition of proteins of biogenic origin, by the combustion of polyamides or by the emissions from local amine-producing industry.

In conclusion, LAMMA has shown that the formation of particulate nitrate is actually occurring in coastal areas with their high atmospheric sea salt loads, and that pollution contributes to this production mechanism. In the accumulation mode particles, sulfur is present in compounds such as $(NH_4)_2SO_4$ or NH_4HSO_4.

Remote Aerosols

The Antarctic Peninsula

Due to growing industrialization and the long-range transport of pollutants (linked to high stack technology), it has become increasingly difficult to define the composition of baseline natural aerosols in the absence of any pollution. The Antarctic continent is the most distant area from the world's predominant pollution sources, and the Antarctic aerosol has therefore been recognized as a "background aerosol." Also, the ocean surrounding the Antarctic continent is very productive from the biological point of view. So, properties of marine aerosol particles can also be studied.[37]

We will now report EPXMA and LAMMA results from five years of continuous aerosol sampling in the Antarctic Peninsula.[23,24,38]

The sampling station is located at the Brazilian Antarctica Station, "Commandante Ferraz" (62.1° S, 58.4° W), on King George Island, Admiral Bay, Antarctic Peninsula. For EPXMA, stacked filter units were used to separate the coarse and fine aerosol fraction. The coarse mode aerosols ($2.0 < d_p < 15$ μm) were collected on an 8 μm pore-size 47 mm Nuclepore filter, while the fine fraction ($d_p < 2.0$ μm) was collected on a 0.4 μm pore-size Nuclepore membrane. Sampling for LAMMA was performed with a single-orifice 6-stage Batelle-type cascade impactor.[39] For six samples, roughly 100 particles from each of the impactor stages with cut-off aerodynamic diameters of 2, 1, and 0.5 μm were analyzed by LAMMA.

Table 4 shows the various particle types present on the different impactor stages. No significant differences between samples were observed, and they were all averaged out for the whole sampling period. The given particle type abundances should be considered as a slightly biased rough estimate of the actual situation,since particles other than the dominating sea salt were sometimes subjectively selected on the basis of their appearance.

Coarse mode aerosol. The major component of this aerosol fraction is sea salt with a mean diameter of 3 μm. It contributes for about 70% of the total particles and is recognized by its cubic shape. Also, five other marine components have been detected in a large number of particles. They represent 25% overall and are characterized by the presence of Mg, S, Cl, K, and Ca and usually exist as

Table 4. Particle Types Detected with LAMMA and Approximated Particle Type Abundances for Three Different Impactor Stages, Sampled in the Antarctic Peninsula

0.5 μm	1 μm	2 μm
Sea salt (>94%)	Sea salt (98%)	Sea salt (97%)
SO_4^{2-} rich (5%)	SO_4^{2-} rich (1%)	Aluminosilicates (2%)
Organic (<1%)	Zn-rich (<1%)	Zn-rich (<1%)
	K-rich (<1%)	K-rich (<1%)
		Fe-rich (<1%)

mixtures of Mg, K, Ca chlorides and sulfates. These are typical marine components that are formed during the crystallization processes while seawater evaporates.[40] About half of this particle fraction has a diameter in the submicrometer range. $CaSO_4$ particles are present as well, and they can be identified by either a needle-like or spherically shaped structure depending on the way the particles were formed.[14] Positive mode LAMMA spectra of mixed S-rich marine particles show mainly clusters of Na/S/O, K/S/O, and Na/K/S/O, and for the negative mode; S^-, HS^-, and SO_n^- mass peaks. A possible explanation for this sulfate enrichment is agglomeration of sea salt particles with very small sulfate particles which are very abundant in the South Pole aerosol.[41,42] More than 95% of the particles detected in the coarse mode aerosol are produced by a marine source. The remaining 5% of particles are identified as soil dust, organic material, or a mixture of a soil dust with a marine component.

Fine mode aerosol. It is important to indicate that particles smaller than 0.1 μm escape the detection for EPXMA. A total of seven representative particle types were discriminated by a cluster procedure for the fine mode aerosol particles. NaCl still makes a predominant contribution of more than 75%. Sometimes, sea salt mass spectra from LAMMA include C_n-clusters. Comparison with standard spectra reveal that they originate from compounds having carbon chains within their structures (e.g., black carbon and/or fatty acids). Most likely, the surface microlayer provides such organic material that can be concentrated in sea spray droplets. Pure $CaSO_4$ and $CaSO_4$ mixed with NaCl and $MgCl_2$ account for almost 10% of the total particle population. This type of complex particles was already observed in other Antarctic studies.[43,44] It is most striking that sulfur appears predominantly as $CaSO_4$ instead of $(NH_4)_2SO_4$ or NH_4HSO_4. About 70% of these latter two exist in the size range of $0.03 < r < 0.1$ μm.[45] It has been noticed that there exists seasonal variability of the $CaSO_4$ population. A maximum is observed during the summertime experiments. Various authors agree that, especially in summer, sulfate tends to dominate the Antarctic aerosol by number and usually by mass.[42,46] This would suggest a biogenic origin for the sulfur in these particles. However, other formation mechanisms, like bubble bursting,[47,48] in-cloud processes,[14] or sequential evaporation processes of seawater droplets[49] cannot be excluded. Wouters et al.[23] found evidence by LAMMA for the presence of methane sulfonic acid (MSA) as an S-rich component in the Antarctic aerosol. The m/z = 141 ($Na_2CH_3SO_3^+$) peak in the positive, and the m/z = 95 ($CH_3SO_3^-$) one in the negative spectra showed up as rather dominant features in the mass spectrum. Comparison of these spectra with data from several standards such as the Na-salt of MSA ($NaCH_3SO_3$) and Na_2SO_4 were indicative of the presence of MSA compounds. From the LAMMA spectra, a significant part of all counterions for sulfate seems to be Na^+. This could mean that the MSA-rich particles are in fact highly transformed sea salt.[50] MSA is a reaction product of the photo-oxidation of dimethylsulfide (DMS), which originates from processes such as the metabolism of certain marine algae and decay processes of dead krill. The emission of the oxidized form of DMS, namely dimethylsulfoxide (DMSO), derived from the ocean surface, contributes significantly to the global sulfur cycle. MSA is produced in the gaseous state; it dissolves into aqueous aerosol droplets and accretes on aerosol particles in the atmosphere.[51] The organo-sulfur compounds are eventually converted to nonsea salt sulfate (nss-SO_4^{2-}).[52] According to Legrand et al.,[53] the H_2SO_4 present in Antarctic ice is most probably due to biogenic activity of the surrounding waters. Berresheim[54] investigated atmospheric distributions of DMS, SO_2, MSA, and nss-SO_4^{2-} in the surroundings of the Antarctic Peninsula. The local MSA/nss-SO_4^{2-} ratio appeared high compared to values over other oceans.

Other minor particle types observed for the fine mode filters are soil dust, pure S-rich, Si + NaCl, $MgCl_2$, organic matter, and pure Si. They contribute to less than 15% of the total population, and neither of them have shown any seasonal variability. Various metallic particles, such as Fe-rich and Zn-rich particles both containing Pb, were only observed by LAMMA because of the higher sensitivity compared to EPXMA. Parungo et al.[41] stated that the Fe-Pb-containing particles could be meteoritic dust. The origin of the Zn-Pb-rich particles is not clear but might be contamination of the galvanized material extensively used in the base station. Also, intense K-peaks associated with much less intense but nevertheless obvious Pb-isotope patterns have been detected in some particles. They could originate from biogenic emissions. According to their size (< 1 μm), however, long range transport does not seem so obvious. Another possibility, of course, is local contamination. Similar particles were detected by Sheridan and Musselman,[55] who attributed them to wood smoke.

The Amazon Basin

The Amazon Basin is the world's largest tropical forest region. It occupies roughly 3/8 of the total Brazilian area. The large scale study outlined in the following paragraph provided a unique opportunity to assess the impact of this ecosystem on the atmosphere. The Amazon Boundary Layer Experiment refers to a long-term study of the chemistry of the atmospheric boundary layer supported within the Global Tropospheric Experiment (GTE) of the NASA Tropospheric Chemistry Program.

It has long been known that vegetation can act as an aerosol source during burning activities,[14] by mechanical abrasion of plant leaves,[56] but also by natural release. Beauford et al.[57,58] showed that plants are capable of generating particulate Zn, Pb, and Cu. Emitted plant wax particles contain trace amounts of K, P, Ca, Mg, Na, and Cl.[59,60] According to Nemeruyk,[61] plant transpiration causes migration of Ca^{2+}, SO_4^{2-}, Cl^-, K^+, Mg^{2+} and Na^+. Of course, the nature and frequency of the emissions will depend largely on the plant itself and on meteorological factors, but generally, a forest ecosystem definitely influences the atmosphere.

Only a few comprehensive studies on the composition and origin of aerosols in the tropical forest have been published.[62-64] We will outline some major EPXMA and LAMMA results of aerosols taken during various sampling strategies, spread out over a time of about 4 years.[24,28,64,65]

Sampling was carried out both from two ground-based stations and by aircraft flights during the ABLE-2A experiment in July through August, 1985. One sampling site, named Duke Reserve, is located 25 km from Manaus, in the state of Amazonas, Brazil. Samples were taken under the canopy at 1.5 m above ground level. The other sampling station, Bacia Modelo Tower, is situated at about 70 km north of Manaus. Here samples were taken on a 45-m high tower near ground level (1.5 m), in the middle of the tower (20 m) and at the top about 15 m above the canopy. Both geographical locations are upwind from Manaus, so that no anthropogenic contributions are to be expected. Aerosol samples were also collected during a flight of the NASA Lockheed Electron aircraft at an altitude of 150 m and higher above ground level. They are considered as being representative of the boundary layer. Detailed descriptions of these sampling strategies can be found elsewhere in literature.[62]

The coarse and fine mode aerosols for EPXMA were collected in the same way as described for the Antarctic samples. The LAMMA samples were collected by a 10-stage Batelle-type cascade impactor with cut-off diameters of 16, 8, 4, 2, 1, 0.5, 0.25, 0.12, and 0.06 μm and a backup filter. Dry and wet season LAMMA samples were taken. Satellite photographs assured the absence of plumes from vegetation burning during the wet season weather conditions.

Morphological and elemental EPXMA data of all sampling locations were treated by hierarchical cluster analysis. Careful examination of the obtained particle types showed two main particle groups, namely aerosols rich in Al, Si, Ti, and Fe denoted as mineral dust and one group containing aerosols emitted by plants. The morphological structure of the latter particles was diversified and some particles were clearly pollen grains, fungi, algae, and plant fragments. Leaf fragments were enriched in the coarse fraction compared to the fine fraction, suggesting that part of the coarse-mode aerosols are produced as a result of leaf abrasion by wind action in the canopy. Biogenic emission accounts for almost 80% of the total number of particles. The most frequently observed particles by EPXMA are those containing high P, S, and K concentrations. There is also a significant contribution of Cl among the different particle types. Sodium is present together with S, Cl, and K, and is detected in 8% of the particles. Most of the particle types have a large contribution of low-Z elements (C, H, N, O); this can

Table 5. Hierarchical Cluster Analysis of the EPXMA-Data for the Coarse (a) and Fine (b) Filter Mode Sampled at the Duke Reserve Site in the Amazon Basin

(a) Group	Percentage			
	Tower[(a)]	Duke (Ground)[(a)]	All Samples[(b)]	Elements Present
1	22.5	9.8	16.5	Al,Si,Fe
2	21.7	46.7	34.4	S,K,(P)
3	17.4	2.3	9.8	S
4	14.1	6.7	10.3	Ca
5	10.6	17.2	13.7	K,(P)
6	5.7	1.2	3.5	Fe
7	5.7	11.3	8.4	Na,S,Cl,K
8	2.3	4.8	3.4	Cl,(K)
(b) Group				
1	43.7	35.1	39.1	S
2	25.1	14.3	23.5	Al,Si,Fe
3	12.6	34.8	22.1	S,K
4	6.1	7.2	7.1	Si
5	5.7	2.9	1.0	ME^{n+}
6	2.6	2.8	2.7	S,Ca,K
7	2.3	0.4	1.5	Ca
8	1.7	2.5	2.1	K
9	0.2	—	0.9	Cl

[a] Results obtained after a nearest centroid sorting perfomed on the individual samples of either the tower (top) or the Duke (ground) location.

[b] Results obtained after a hierarchical cluster analysis performed on all samples of either the coarse or fine mode aerosols.

be deduced from the high spectral background present in the corresponding X-ray spectra. These particles are assumed to be generated by vegetation.

The mineral fraction of the fine mode aerosol is determined by two different particle types, identified as soil dust and quartz. The silicon rich particles have been frequently observed in the fine fraction filters. They might originate from both soil dust or vegetation sources. The most important contribution is made by the particle type containing mainly S, K, Ca, and P. These aerosol particles originate from biogenic emissions and were released by the vegetation; indeed, their composition agreed with the one expected for plant material. This large group may be subdivided into more specific particle types, each probably originating from a different biogenic source or process. The most abundant element is sulfur, which is present in a considerably large amount in almost all particle types related to the vegetation. Table 5 gives the hierarchical cluster results of both the coarse and fine fraction for the same stacked filter unit samples from the Ducke Reserve. Some low abundance particle types were present in the fine mode aerosol which are more difficult to identify, namely Zn–S-, Na-S-, Mn-, and Cl-rich particle types. They are all believed to originate from vegetation.

LAMMA of the organic material results in mass peaks that are indicative of the fragmentation of hydrocarbons and terpenes, while oxygen containing ions are formed by the fragmentation of oxygenated compounds such as alcohols, esters, etc. Since the LAMMA spectra are a superposition of many different organic compounds and because of the fragmentation behavior of long chain hydrocarbons, a detailed structural interpretation of the spectra is not possible. Negative mode spectra show the frequent and very intense appearance of small organic fragments such as CN^-, CNO^-, $HCNO^-$, $HCOO^-$, CH_3COO^-, $C_xH_y^-$ ($x < 3$, $y < 7$), etc. These fragments have, however, little or no decisive value for compound determination. They appear in all particle types detected but most often in MSA-, K-, P-, and salt-rich particles.

The conjugated base of methane sulfonic acid (MSA) was frequently detected in these aerosols, especially among the smallest ones. Until recently, this species was believed to originate exclusively from the photochemical oxidation of organic sulfur products (DMS, MeSH) that are emitted by oceanic and continental ecosystems.[14,66] Recent laboratory studies, however, suggest that MSA is also released during biomass burning. In this particular case, both sources are possible; biomass-burning plumes were observed during flights over the southern periphery of the Amazon Basin. MSA can be detected in both positive and negative mode spectra as $CH_3SO_3^-$ and $Na_2(CH_3SO_3)^+$. Heterogeneous nucleation reactions and/or adsorption effects are responsible for the presence of MSA in the particulate fraction. MSA occurs together with $(H)SO_x^-$ ions in the negative mode, and with NH_4^+ or sometimes "salt" ions in the positive mode spectra. So possible counterions are NH_4^+, Na^+, K^+, Mg^{2+}, or Ca^+.

The ammonium ion is often detected among the smallest particles. The substantial aerosol NH_4^+ concentrations arise from chemical transformation of atmospheric NH_3. Sources of NH_3 are emissions from the forest and wetlands. From the negative mode spectra, plausible candidates for counterions appear to be SO_4^{2-} or maybe organics. However, Andreae and Andreae[67] believe that ammonium salts of organic acids are too volatile to form stable aerosols.

Nitrate does not appear too often in the spectra. This might be a consequence of prolonged storage: nitrates are not stable. Nitrate was found to be associated with salt particles and with aluminosilicates. It is mostly detected in the coarse mode, as is consistent with both experimental findings[68] and theoretical considerations.[69] Just as ammonium, nitrate is believed to originate from emissions of N-containing gases from the forest ecosystem. Incorporation in the aerosol occurs by the reaction of atmospheric nitric acid with particulates. As it is sometimes detected in salt spectra, direct emission from vegetation is another possible source for part of the nitrate.

Other spectral patterns often encountered in the tropical forest aerosol are attributed to K-rich particles. Their positive and negative mode fingerprint spectra compare very well with reference spectra of K_3PO_4, K_2HPO_4, and KH_2PO_4, Although the relative intensities of the clusters most closely correlate with the KH_2PO_4 fingerprint, the spectra are not identical. This leads to the assumption that the phosphate group in the unknown sample is probably linked to an organic chain (e.g., in phospholipids).

Another frequently appearing particle type has been identified as being a mixture of several salts. "Salt," in this context, refers to particles whose positive mode spectra look like mixtures of several sulfates, carbonates, chlorides, and sometimes nitrates and phosphates. At first sight, they are very similar to transformed sea salt spectra, since elements such as Na, Mg, and K are abundantly present. The only differences lie in their significantly higher Ca content, and the appearance of Ca_xO_y clusters and occasionally of PO^+ in the LAMMA spectra. The negative mode spectra often show quite intense small organic fragments. The appearance of all these nutrient elements (Na, Mg, K, Ca, Cl, and SO_4^{2-}) is attributed to the emission of aerosols during plant transpiration, and they could be emitted simultaneously with the plant wax.

LAMMA results also suggest that the aluminosilicate particles can be divided into different types according to their surface coating. Low laser energy excitation shots revealed desorption of organics, Cl^- and PO_n^- -ions from the mineral core.

The metal-rich group contains predominantly Cr/Fe/Pb-rich, Cu-rich, and Ti-rich particles. Beauford et al.[58] showed that plants are capable of emitting heavy elements such as Zn and Pb. But then, roughly 60% of these particles show no peaks other than the metal(oxide) related ones, and there is no reason to assume that heavy metals should be emitted separately from the biologically important elements like Mg, Ca, K, and Na. This, and the fact that their relative abundances are rather low, points to an anthropogenic origin.

Some minor differences were observed by LAMMA for samples taken during the dry versus wet season. K-rich and P-rich particles were totally absent during the wet periods. As biomass burning activities do not take place during this time of the year, this might imply a vegetation burning source, although season related production of vegetative aerosols may not be excluded. Andreae et al.[70] do report elevated K-concentrations (up to a factor of 3) in biomass burning haze, but they do not report data on phosphorus. Furthermore, the MSA related compounds had a much lower incidence in the wet season samples. EPXMA research of samples collected during the wet season at the Bacia Modelo Tower showed a strong increase of the soil dust component. So, one must keep in mind that the relatively high percent contributions of the soil dust component masked the contribution of the vegetation. This led to results that differed significantly from what was found for the other samples taken during the dry season. For the other particle types and/or compounds, there were neither marked differences in relative abundance nor in size distribution for both seasons.

Differentiation between the samples taken at the tower and at ground level was done by applying a nearest centroid sorting (nonhierarchical clustering according to Bernard et al.[71]). The presence of aluminosilicates is a factor of two more pronounced in the tower samples. One should expect more soil dust near the ground surface, but since the surface is covered with biological material and the upward transportation is shielded by the canopy, the aluminosilicates are likely to be transported over large distances away from the sampling location. Also the S-rich particle type is nearly absent at the ground level station (5%), while they are detected for up to 28% in the tower samples. The samples collected at the ground level showed larger contributions of aerosols related to the vegetation. Thus. the P-S-K-rich particles are more abundant at the ground level station. The abundances of organic particles is variable over all samples, and no straightforward correlation with height could be revealed. No large differences were observed for the samples collected halfway up the tower and at the top of the tower.

The composition of the particle types present in the boundary layer (altitude above 150 m) is very similar to the composition of those collected at the top of the tower. However, at higher altitudes, lower filter loadings are observed. Compared to the samples collected at lower altitudes, chlorine is present in rather high concentrations (sometimes up to 4%). This element is more often associated with S, P, and K. Even Na is present in 75% of these particles within this Cl-rich group. These large Cl-concentrations can be explained by a marine influence. Sea salt can be transported together with soil dust to the Amazon Basin, where they agglomerate with other particle types. Generally, it can be stated that there exists a homogeneous mixing of particles within the boundary layer. For the aerosols collected at higher altitudes (between 500 and 5000 m), more than 90% of all particles can be related to soil dust; less than 4% are characterized as Ti-rich, plant related, and Fe-rich particles. The remaining part can be attributed to a sea salt derived fraction.

Remote Mountain Site (Bolivia)

The Chacaltaya sampling location is situated near the top of a 5245-m high Chacaltaya Mountain, about 25 km east of La Paz (Bolivia). The soil in the neighborhood is stony, partly snow covered year round, without vegetation, and with little dust. Fractionated aerosol samples were obtained with a single orifice Batelle type cascade impactor, separating six fractions (16, 16 to 8, 8 to 4, 4 to 2, 2 to 1, and 1 to 0.5 μm). Each fraction was submitted to EPXMA and LAMMA. Particle classification was performed by a hierarchical cluster analysis.[72] Silicates, a group in which quartz and aluminosilicates are classified, originate from the suspension of soil dust. The highest weight fractions were found on the larger impactor stages, so that the soil dust should originate mainly from short distance sources. Minor fractions of the silicate group contain Ti, Zn, or Pb, implying a crustal source for these elements. Modified silicates were present and are characterized by a silicate nucleus on which sulfates, phosphates, and chlorides are deposited. They show highest abundance for the smallest stages. Since no moist damage was observed in the impaction spots, sulfur coatings on the surface of these particles is mainly due to atmospheric processes that occurred before impaction. Another type of silicate particles are agglomerated, with whiskers that contain almost stoichiometric amounts of Ca and S, so that their composition is most probably $CaSO_4$. Some Chacaltaya samples contained high relative weight fractions of silicates, with equal atomic fractions of K and Cl, suggesting the presence of KCl on the particles. No significant sources of Cl exists in the vicinity of Chacaltaya except volcanic emissions. Indeed, the sampling site is located nearby volcanos like Guallatiri and Tacora. Shaw[42] suggests that volcanic activity can explain the high Cl concentrations and its correlation with Si. Most probably, Cl enters the environment in a gaseous form, as supposed by Adams et al.,[77] who found also strong correlation between these two elements.

Sulfate particles only constitute a minor weight fraction among the present groups. LAMMA results showed that the negative mass spectra of these particles closely correspond to those obtained for $(NH_4)_2SO_4$ standards.[74] Some of the S-rich particles had an exceptionally high K/Na concentration ratio. Their source in the continental South American fine mode aerosol involves gaseous precursors, possibly in forest burning activities.[75] Also, the expected carbonaceous particles in these samples have been identified with the typical C_n-fragmentation pattern. These particles also contained a high concentration of K. Gypsum particles ($CaSO_4.2H_2O$) tend to be present in the fine aerosol fraction. Its source is probably anhydride deposits in the arid Altiplano, immediately west of the Chacaltaya Mountain, or $CaSO_4$ fibers released from sulfate-bearing silicate particles in the air or during the impaction process. LAMMA showed much more complex spectra than the pure $CaSO_4$ fingerprint. From the stoichiometry of $CaSO_4$, it follows that only part of the SO_4^{2-} can be due to this material. The earlier discussed condensation sulfate aerosol is indeed abundantly present in the sample. The consistent presence of Pb in the LAMMA spectra of the fine particulate sulfate suggests that the element has a gaseous precursor. Other atmospherically enriched elements (As, Cu, Zn, ...) were not detected in the same particles, probably as a result of inferior detection sensitivity.

Also, miscellaneous particle types were observed. Calcium phosphate particles most probably originate from the Amazon Forest. Small weight fractions contained high concentrations (more than 50 wt %) of metals like Ti, Cu, Zn, Fe, Pb and rarely elements like Mo, Ce, La, and Bi. For Cu, Zn, and Pb, it was proven by Beauford et al.[57] that plants growing on a soil containing trace amounts of heavy elements can emit a submicrometer metal containing aerosol, of which the chemical and morphological characteristics are not yet described. Only very few Ti-rich particles were detected. Ti is a typical crustal element, and pure Ti-spectra can probably be assigned to the mineral rutile (TiO_2), whereas particles containing equal amounts of Fe and Ti can be associated with ilmenite ($FeTiO_3$).

Urban Aerosols

The Antwerp (Belgian) Environment

Automated EPXMA was used for the identification of about 8000 individual particles sampled at an urban site near the city of Antwerp, the second largest city in Belgium, with a population of ca. 600,000 inhabitants within a metropolitan area of 210 km^2. The city is located within a very dense traffic system and

Table 6. **Mean Cluster Analysis Results for 16 Samples of Air Particulate Matter, Sampled at an Urban Site in Antwerp, Belgium**

Particle Type	Abundance (%)	Elements
Soil dust	30.0	Al, Si, Fe, (S, Ti, Zn, Ca)
Auto exhaust	19.0	Pb, (S, Cl, Br)
Sulfates	18.9	S, (Ca)
Sea salt	3.2	Cl, (Na, K, S)
Biological	2.3	P, S, K, Ca, (Si, Cl)
Miscellaneous	26.6	Fe, Cu, Zn, Sb, Cr, Ba, Al, Cd, Mn, As

surrounded by large nonferrous industries and chemical plants. It is also located within a few hundred km of several other large industrial concentrations, such as the German Ruhrgebiet and Rotterdam and only 100 km away from the North Sea.

Aerosol filter samples for EPXMA were taken at the top of a 15-m high building located 5 km south of the city center. The site is dominated by grassland, and major highways are within 1 km distance. A large nonferrous plant and waste incinerator are located at 5 and 2 km, respectively. About 400 particles of the urban Antwerp environment were analyzed by LAMMA.

From a cluster analysis, six general denominators (particle classes) were retained[19] and labeled according to their presumed source, i.e. soil dust, auto exhausts, sulfates, sea salt, biological particles, and miscellaneous particle types (Table 6). The different particle classes follow.

Sulfur-rich particles. These particles were identified by their high elemental S-concentration. Many studies of sulfate-containing aerosols, collected at urban sites, have shown that particulate sulfur is present in the submicrometer size fraction (0.01 to 0.5 µm), with a maximum number concentration around 0.04 µm diameter.[45,76] In each sample, two size modes were observed: fine sub-µm (mean diameter of 0.25 µm) and coarse particles (mean diameter of 2.0 µm). The possible sources for sulfate are numerous, but most often anthropogenic emissions, arising from combustion of fossil fuels, constitute the main source within this environment. Energy dispersive EPXMA provides no means of distinction between H_2SO_4 and the different $(NH_4)_2SO_4$ compounds formed throughout neutralization reactions. However, several authors have reported morphological criteria for their differentiation. H_2SO_4 particles are dome-shaped and very often surrounded by concentric rings of smaller droplets.[77] They are electron translucent and quite unstable upon electron beam irradiation. On the other hand, pure $(NH_4)_2SO_4$ particles appear as wart-shaped, with few or no rings of smaller droplets, and they are stable upon irradiation.[78] In this study, all samples showed an absence of droplet rings, leading to the conclusion that all of the sulfate was present as ammonium sulfate. From LAMMA, it seemed that about 50% of the particles were rich in ammonium or at least associated with ammonium. Possible counter ions are NO_3^-, SO_4^{2-}, and Cl^-. In the positive mode spectra, sulfate and nitrate were always found to occur in clusters with Na or, less frequently, with K. This does not mean that they cannot be coupled with NH_4^+. Negative mode spectra of the smallest particles often showed very intense HSO_4^- ion peaks, which are indicative for $(NH_4)2SO_4$.

Calcium sulfate. A Gaussian size distribution was observed with a maximum around 0.9 μm. Most of the $CaSO_4$ particles are formed of aggregations of smaller fragments with several crystalline forms. The average S/Ca mass ratio for this particle type is equal to 0.72, which is close to the theoretical value of 0.80. No $CaCO_3$ particles were observed, probably because of quite efficient secondary reactions with SO_2.

Auto exhaust particles. This particle type was frequently observed in almost all samples. A more detailed description of this aerosol particle group is given later in this chapter.

Sea salt particles. A minor component of the air particulate material (a few percent of the total abundance) consisted of sea salt particles, identified by their high Na, K, and Cl concentrations and by their distinctly cubic shape. Also, transformed sea salt particles were observed. They are derived from a marine source rather than from industrial processes.

Biological particles. Particles with X-ray spectra showing high background intensities due to Brehmsstrahlung from light elements such as H, C, N, and O were identified as biological particles originating from pollen, spores, plant and leaf fragments, and insects. Mamane and Noll[21] observed similar particles sampled at rural sites. Most of these were visually identified in the secondary electron image mode but could not be detected with the automated analysis mode of our instrument as their backscattered electron signal fell generally under the image threshold setting. Nevertheless, some biological particles were detected because they contained elements with higher atomic mass, e.g., K, S, and P. Note that biological particles are often 15 μm and are excluded from the analysis.

Soil-derived particles. Resuspension of soil material by the action of the wind is the main source of the aluminosilicate aerosols in the Antwerp environment. They make up nearly 30% of all particles analyzed. Their composition is similar to that of clay minerals such as kaolinite and montmorillonite, feldspars, and pure quartz. No discrimination was made between soil-dust particles and fly-ash particles, which have similar chemical composition. This implies that the soil contribution is somewhat biased by the contribution of fly-ash particles.

Miscellaneous particles. A wealth of particles is present with a composition that does not lead immediately to a unique source identification. Most of these particles could, however, be attributed to anthropogenic emissions. The most abundant particle group in the "miscellaneous" class consists of pure Fe-oxides. They can be divided into spherical particles and nonspherical types. The former are characteristic for formation mechanisms at high temperatures, while the latter are

Table 7. EPXMA Results of the Santiago Fine and Coarse Mode Aerosol after a Hierarchical Cluster Analysis

	Abundance (%)		
Particle Type	**Fine Mode**	**Coarse Mode**	**Major Elements**
Soil dust	28	58	Si, Ca, Fe, Al
Sea salt	19	11	Cl, S, K, Na
Sulfate	15		S, Cu, Pb
Aluminosilicate	12		Al, Si
Metallurgical	10	6	S, Cu, Zn, Pb
Gypsum	9	6	Ca, S
Si-Fe-rich	4	3	Si, Fe, S
Biomass burning	3	1	K, Cl, S
Organic		8	none
Ca-Si-rich		7	Ca, Si

probably naturally occurring oxyironhydroxides (ferrite, hematite, goethite, and magnetite). Oxides of metals such as Zn, Cu, Sn, Cr, Fe, and Pb are originating from various abrasion processes of metal objects (Zn and Cr), from the emissions of incinerators (Zn and Pb), from the lead-producing industry (Zn, Pb, Sb), and from oil burning processes (V and Ni).

The Santiago (Chile) Environment

Fifty-one pairs of aerosol samples were collected at the Universidad de Santiago de Chile Planetarium from January, 1987 until March, 1987. The fixed sampling site was 4 km west of the city center of Santiago, in a valley at an altitude of 600 m, with a population of four million.

Individual aerosol particles were studied by EPXMA and LAMMA.[79,80] About 2500 particles in both the fine and coarse mode fraction were subjected to EPXMA. Table 7 represents the number percentage of each particle type, the major elements detected, and the possible identification.

The most characteristic particle clusters recognized in the fine mode aerosol are soil dust, (transformed) sea salt, S-rich, aluminosilicates, metal-rich gypsum, Si-Fe-rich and particles derived from biomass burning. Only the two former particle types are produced by natural sources and account for about 50% of the total particle number concentration. The others are all generated by anthropogenic activities. The particles associated with seaspray, sulfate, metallurgical emissions and wood-burning processes have sub-μm diameters, whereas those related to the earth's crust range from 1 to 1.6 μm diameter.

The most abundant natural particle types for the coarse mode also correspond to soil dust and sea salt derivation, but their contribution are much more pronounced (more than 70% of the total fraction). In addition, an organic fraction (8%) was noticed, and morphological inspection indicated that some of these particles refer to biological material. Particles associated with anthropogenic activities appeared less frequently. They presented the same elemental composition as those in the fine mode, characterizing metallurgical, gypsum, Si-Fe-rich and wood-burning release particles. The average particle diameters for the different coarse mode classes ranged between 1.4 and 4.2 μm.

The results of EPXMA pointed toward the presence of a marine aerosol forming part of the Santiago de Chile atmospheric particulate matter. In fact, Prèndez and Ortiz[81] reported that the city of Santiago had nonmarine impact mainly due to its topographical configuration and geographical location. This assumption was based on enrichment factor calculations performed on total suspended matter elemental concentrations. Hence, another microanalytical technique, namely LAMMA, was invoked for confirming marine influences. Indeed, among the particles in the size range of 0.25 to 0.5 μm, positive ion LAMMA spectra were mainly characterized by Na and Na-derived molecular fragments. There was also S-enrichment present in most of the spectra pertaining to sea salt derived aerosol particles. These transformed sea salt particles (Na, Cl, and S) seemed to be most abundant in the stages with smaller cut-off diameters. This is evidence that marine aerosols are indeed transported into the city of Santiago, suffering a S-enrichment generated by anthropogenic sources. The negative LAMMA ion mode had abundant sulfates and nitrates, and in many cases it was uniquely composed of the carbon series. With relatively low frequency, phosphates also formed part of the LAMMA spectra.

Industrial Aerosols

Asbestos Particles

Asbestos is used in over 3000 fields of application and ca. 70% of the use is through transferred products (e.g., asbestos cement) in which the composition is often drastically changed compared to the original product.

The risk to health, occupational and environmental, rising from the use of asbestos has been widely publicized.[82] From the biochemical and environmental standpoint, fine respirable asbestos fibers of high surface area have considerable capacity to absorb other substances. They may therefore selectively absorb the constituents of living cells and thereby upset the delicate balance of cell reproducibility, but they may also absorb pollutants from the atmosphere, to be transported to the lung and deposited in the tissue.[83] So, the toxicity of asbestos fibers is believed to be associated with the surface absorption power and reactivity of the minerals. Because of this, our laboratory performed many microanalytical and surface sensitive analyses for the characterization of the surface of asbestos fibers.

In order to study the capabilities of LAMMA to detect compounds present at the surface of individual fibers, several species such as benzo(a)pyrene (BaP), benzidine, and dimethylaniline (DMA) were absorbed onto the different types of asbestos.[84] Both of these surface-added compounds can easily be detected under specific laser desorption conditions. Sometimes, organic contaminants were detected at the asbestos surface, probably antioxidants from polyethylene used for storing the samples.[85] Mass spectra taken at high laser energy shots showed considerable fragmentation of both the organic compounds and the asbestos substrate. However, for the latter, mass spectra are significantly different after treatment compared to those of the raw fibers, suggesting leaching of magnesium and iron. For DMA, LAMMA could corroborate at the microscopic level earlier findings in which the catalytic oxidation of DMA at the surface of kaolinite, sepiolite, and other zeolites was proven.[86]

It was possible to obtain depth-profiles of inorganic constituents using SIMS. Measuring the $^{24}Mg/^{28}Si$ ratio for HCl treated samples indicated Mg diffusion to the surface. The results of leaching experiments of chrysotile with oxalic acid are quite different. Depth profile observations are consistent with a diffusion controlled process in which Mg diffuses through the external silica layer.

Chrysophosphate could become a viable alternative for untreated chrysotile with a significantly reduced absorption capacity for organic compounds. Positive and negative LAMMA spectra recorded in low laser desorption conditions revealed interesting information on the surface composition alterations.[87] Magnesium and other elements are easily leached from the surface layers of asbestos fibers. For chrysotile, the disappearance of magnesium has been related to changes in biological activity. It has been shown that a reduction in hemolytic activity of chrysotile fibers can be achieved by a superficial coating of organo-silane derivatives.[88] The chemical surface treatment of chrysotile fibers with organic reactants makes use of a mild acid etching in order to dissolve the outer magnesium hydroxide monolayer, thus bringing free silanol groups to the surface of the fibers. These can then react through condensation with an organosilane compound. A systematic decrease in mass peak intensity was observed for the treated fibers compared to standards, indicating the removal of the outer brucite ($Mg(OH)_2$). Most attention has been paid to the acid attack of the chrysotile fibers, which leaves intact the fibrous morphology. For this, electron energy-loss spectroscopy (EELS) was used, since it offers the possibility of detecting differences in elemental composition and performing element specific images (ESI) with sufficient resolution for mapping the reaction products on the treated chrysotile surfaces. The conventional transmission electron microscope mode has proven that indeed, after silanation, the structural integrity of the fibers is preserved. Element specific imaging also revealed that the surface modification is only partially effective and that the coating is not distributed homogeneously.[89] The validity of the obtained EELS and ESI results was compared to the results of parallel analysis on standard untreated chrysotile asbestos fibers.

Another attempt to modify the surface properties of asbestos involves the use of $TiCl_3$ as a reagent. Biological investigations have demonstrated a reduction in the surface reactivity as measured by the hemolytic effect on red blood cells.[90] EELS and ESI were used to confirm the presence of titanium at the surface.[91] Titanium is fairly evenly distributed on the chrysotile surface. High-magnification ESI images ($\times$ 140,000) of cross sectioned treated fibers demonstrated unambiguously the presence of titanium inside the fiber as well as on the fiber's external surface. The elemental map inside the fiber tubes follows the spiral curvature of the asbestos fibrils. This could be an indication that the fibers unroll due to an acid lixivation prior to the chemical Ti treatment or that a Ti diffusion process has taken place.

Fly-Ash Particles

The optimization of oil burning processes in power stations has been studied intensively during the last fifteen years. Generally, the reduction of emissions of sulfur oxides is an important aspect of the struggle against acid rain. It was found that oil fly-ash particles are able to carry high concentrations of sulfur oxides absorbed to particles of a large area.[92] Since most of these particles have a very porous and spongy structure and are very hygroscopic,[93] they can absorb large amounts of atmospheric water in their pores. After emission by the stack, the absorption of water increases the weight and the density. Consequently, these particles fall out at a relatively short distance from the power plant and transport their acidic content to the place of fall-out. Particles with typical fly-ash characteristics were found on Belgian historical monuments that are presently attacked by aggressive acid depositions.[94,95] On the contrary, when emitted in a dry state and under certain meteorological weather conditions, fly ash particles can have long residence times in the air. Eastern European fly-ash particles have been identified over more remote regions of the North Sea.[34]

Fly-ash samples from a Belgian thermal power station have been obtained at isokinetic sampling conditions by a cyclone with a backup filter.[72] Some characteristic oil fly-ash particle morphologies were observed. The first type, called the spongy type, is approximately spherical, with numerous craters on the surface suggesting a hollow structure. The surface itself is smooth but sometimes deformed under the influence of the high temperatures. Some particles show a foamy structure instead of a smooth surface. Others have an intermediate outlook between the first and the second type, indicating a possible transition from the first type into the second during the burning process. Element mapping over a single particle, using a wavelength dispersive system (WDS), has revealed that both the elements sulfur and vanadium are homogeneously dispersed over the particle. LAMMA as carried out by Denoyer et al.,[96] have yielded evidence for the carbonaceous nature of these particles and for the presence of oxides of sulfur and vanadium. Also, morphology differences have been observed for the smaller particles (1 to 5 μm) on the filter fraction. The spherical type mostly contains V, Fe, Ni, and some S, but also silicate particles have been found. Polyhedral particles contain mostly sulfates of alkali elements. Furthermore, all particles seem to be massive without apparent voids.

In general, the main elements present in both the micrometer and submicrometer particles are Si, Al, K, Mg, S, Ca, Fe, and Ti. Depending on the burning technology and the coal type, elements such as P, Na, Cl, V, Ni, As, Rb, Sr, Pb, and many others may be present as minor constituents.[26,97]

Zinc Smelter Particles

The EPXMA technique in combination with cluster analysis has been used for the chemical and morphological characterization of resuspended soil particles, sampled in the vicinity of a zinc smelter.[98] Furthermore, the contribution of each particle type as a function of receptor site has been estimated. It is apparent that there is a large difference between chemical composition of the fine and coarse atmospheric particles.

For the fine mode particles, most of the calculated clusters were composed of non-Zn containing particles, as for example, (a) an aluminosilicate fraction (29%) with high concentrations of crustal elements such as Al and Si that are representative of the resuspended soil fraction of the aerosol, (b) an aggregate fraction of $CaSO_4$ with aluminosilicates, (c) a Cl-rich fraction containing $(NH_4)_2SO_4$ particles, and (d) naturally occurring organic particles.

The Zn-rich particles can also be classified into different groups according to their associations with various elements. Zn was found (a) in the presence of typical soil derived elements such as Al, Si, and Fe, thus representing zinc-contaminated soil particles (20%) and (b) to contain ZnS derived from the storage yard and formed as an aggregate of platelets (7%); (c) ZnO particles are derived from the roast ovens where the ZnS is oxidized to ZnO showing a smooth surface morphology, and (d) Fe-Zn-rich particles have been identified as Zn-contaminated goethite. All the Pb-containing particles were clustered together (11%). Because of the high but variable concentrations for Si, Fe, and Zn, this cluster is most likely to contain particles of different composition and origin.

Nearly identical cluster compositions have been classified for the coarse particle mode. The major differences have been observed for the Pb-containing particle type that seems to be twice as much abundant in the coarse fraction (27%). The soil related particles still represent one third of all analyzed particles (31%), whereas the Zn-contaminated soil cluster contributes close to 15% which is comparable with the fine mode fraction. The nonheavy metal containing compounds (5%) are substantially less abundant in the coarse fraction. One significant difference is the abundance of the Cl cluster. This cluster makes little or no contribution to the coarse fraction, while its abundance equals nearly 8% in the fine fraction, an indication that they are possibly formed as fine NH_4Cl condensates from the gas phase. The factory is also known as an emitter of Cd, but except for a few exceptional particles, no Cd was detectable, apparently due to the insufficient sensitivity of the energy dispersive EPXMA analysis.

Car Exhaust Particles

A characterization of individual particles present in the Antwerp aerosol was performed by Van Borm et al.[99] Attention was focused on the Pb particles originating from car exhaust, as identified from a set of Pb-containing particle types. Some 1500 Pb-containing particles were detected among a total of 7200 particles analyzed with EPXMA, allowing their classification into five distinct particle types. LAMMA was especially used to detect ammonium compounds in relation to Pb. The principal source of particulate Pb in the urban atmosphere is combustion of leaded petrol.

All lead-containing particles found could be divided into five main particle types: lead sulfates, lead halides, lead associated with medium-Z elements, soil related lead, and lead associated with heavy metals such as Ni, Zn, As, Se, Ag, Cd, Sn, and Sb. About 67% of all Pb-containing particles have high concentrations of sulfur. On the average, it seems that the individual particles do not undergo massive and complete sulfation and are only partly converted to pure lead sulfate species. Chlorine and bromine could not be detected by either EPXMA and LAMMA within this particle type. This strongly indicates the absence of halides in the Pb-S-rich aerosol particles. The lead halides form the second most abundant cluster-containing Cl and Br (22% by number).

A number of metals were found to be associated with Pb (6%) and are believed to originate from anthropogenic emissions such as refuse burning or nonferrous emissions.

Pb associations with medium-Z elements such as P, K, and Ca represent 3% of all Pb-rich particles. It has been found primarily associated with both K and S, and presumably originates from refuse burning, a documented source type in the Antwerp urban aerosol.[100] Soil-related Pb is classified as particles found associated with Si (2% abundance) and originating from resuspended soil contaminated with Pb from auto exhaust.

As already mentioned before, unlike EPXMA, LAMMA can provide information about the presence of ammonium compounds. Therefore, more attention has been focused on the relation between NH_4^+ and Pb-rich particles. About 90% of the Pb-containing particles have been found to be associated with NH_4^+. Both positive and negative mode spectra closely resemble those of a $Pb(NH_4)_2(SO_4)_2$ standard. The positive mode spectra of this compound are dominated by the Pb isotope peaks. The presence of ammonium is indicated by the mass peak at m/z = 18 (NH_4^+). The negative mode spectra are dominated by the typical $(H)SO_n^-$ mass peaks, HSO_4^- being the most intensive. For the urban particles, the average $NH_4^+/^{204}Pb^+$ intensity ratio is extremely variable, ranging from 0.9 to 150. Usually high ratios, compared with those of the standard, could result from ammonium sulfate surface coatings. The fact that many other particles of different types also exhibit NH_4^+ in their spectra supports this suggestion. Therefore, one cannot state that all particles investigated are in fact pure $Pb(NH_4)_2(SO_4)_2$ particles; only those with a ratio near 0.9 might be. Others may be PbO or $PbSO_4$, coated with $(NH_4)_2SO_4$ present in the urban aerosol.

A last remarkable detail that emerges from these analyses is that, with EPXMA and LAMMA, lead halides and lead sulfates are found, but never clear "intermediate" particles (i.e., Pb-particles containing both Cl/Br and sulfate). Halides seem to be almost completely removed from the lead sulfate particles. Post and Buseck [101] tentatively identified lead halide particles as a-2PbBrCl.NH_4Cl. Biggins and Harrison[102] identified various Pb-sulfate compounds using X-ray diffraction (XRD), the most important being $Pb(NH_4)_2(SO_4)_2$. But then, XRD is limited to crystalline compounds, while EPXMA and LAMMA are not.

CONCLUSIONS

The combination of automated EPXMA and cluster algorithms, in combination with microanalytical techniques such as LAMMA, SIMS, and EELS, appeared in many field studies to be very useful for the characterization of individual particulate matter. Besides the chemical composition, molecular and morphological information provide considerable extra knowledge.

Cluster analysis is very often applied on an individual particle data set for creating more order out of chaos and for reducing the large and complex data into more characteristic and interpretable particle groups.

The basic idea of receptor modelling is measuring concentrations of various species (elements) at ambient sampling sites (the receptors) and using that information to identify the number of major particle sources, to determine the source profiles, and to obtain the total mass source apportionment.[103]

Specific to these environmental studies, we can state that the composition of the North Sea aerosol particles (giant particles included) appears to be diverse and largely determined by the air mass trajectories. Aerosol concentrations near sea level and at distinct altitudes above the surface vary greatly, depending on proximity to natural and man-made sources. The influence from both the U.K. and the European mainland on the North Sea environment is often so predominant that hardly any marine contribution is observed.

The aerosol in the Antarctic Peninsula is dominated by sea salt particles. Analyzing a large number of individual aerosol particles, it was observed that sulfur for particles in the sub-μm size range appears preferentially as $CaSO_4$ instead of the expected $(NH_4)_2SO_4$ or NH_4HSO_4. These $CaSO_4$ particles show a clear seasonal variability, with higher population during summertime. Most particles analyzed have a marine origin: they are generated by "bubble bursting" and subsequently transformed in the atmosphere. Associations of excess sulfate and methane sulfonate are the most striking alterations.

It can be stated that the vegetation in the Amazon Basin is an important source of aerosol particles. The most abundant particle types are produced by biological sources and can be described as a mixture of different salts and organic fragments. Their importance on a global scale and their impact on the biogeochemical cycle is not yet fully studied and might be much greater than so far supposed.

The composition of an urban aerosol will of course largely depend on its geographical location, the activities performed locally and the industries surrounding the sampling site. Still, some trends can be defined. Soil dust particles are usually abundant. Other particle types often found are sulfates, car exhaust particles, and various particles derived from diverse sources such as oil burning processes, abrasion processes, and emissions from incinerations.

REFERENCES

1. Xhoffer, C., L. Wouters, P. Artaxo, A. Van Put, and R. Van Grieken. "Characterization of Individual Particles by Beam Techniques," in *Environmental Particles, Vol I. IUPAC Monography*, J. Buffle and H.P. van Leeuwen, Eds. (Chelsea, MI: Lewis Publishers, 1992).
2. Massart, D.L. and L. Kaufman. *The Interpretation Of Analytical Data by the Use of Cluster Analysis* (New York: John Wiley and Sons, Inc., 1983).
3. Injuk, J., Ph. Otten, C. Rojas, L. Wouters, and R. Van Grieken. "Atmospheric Deposition of Heavy Metals (Cd, Cu, Pb and Zn) into the North Sea," Final Report to Rijkswaterstaat, Dienst Getijdewateren, s'Gravenhage, The Netherlands, Project NOMIVE*2 Task No. DGW-920, University of Antwerp, Belgium (1990).
4. Otten, Ph., C. Rojas, L. Wouters, and R. Van Grieken. "Atmospheric Deposition of Heavy Metals (Cd, Cu, Pb and Zn) into the North Sea," Second Report to Rijkswaterstaat, Dienst Getijdewateren, s'Gravenhage, The Netherlands, Project NOMIVE* Task No. DGW-920, University of Antwerp, Belgium (1989).
5. Pena, J.A., J.M. Norman, and D.W. Thomson. "Isokinetic Sampler for Continuous Airborne Measurements," *J. Air Pollut. Control Assoc.* 27:337–340 (1977).
6. Noll, K.E. and M.J. Pilat. "Size Distribution of Atmospheric Giant Particles," *J. Atmos. Environ.* 5:527–540 (1971).
7. Whitby, K.T., R.B. Husar, and Y.H. Liu. "Aerosol Size Distribution of Los-Angeles Smog," *J. Colloid Interface. Sci.* 39:177–204 (1972).
8. Dedeurwaerder, H.L. "Study of Dynamic Transport and of the Fall-out of Some Ecotoxicological Heavy Metals in the Troposphere of the Southern Bight of the North Sea," PhD Dissertation, University of Brussels (VIB), Belgium (1988).
9. Rojas C.M. and R. Van Grieken. "Electron Microprobe Characterization of Individual Aerosol Particles Collected by Aircraft above the Southern Bight of the North Sea," *Atmos. Environ.* in press (1991).
10. Van Malderen, H., R. Rojas, and R. Van Grieken. "Characterization of Individual Giant Aerosol Particles above the North Sea," *Environ. Sci. Technol.* in press (1991).
11. Bruynseels, F., H. Storms, R. Van Grieken, and L. Van Der Auwera. "Characterization of North Sea Aerosols by Individual Particle Analysis," *Atmos. Environ.* 22:2593–2602 (1988).
12. Hitchcock, D.R., L.L. Spiller, and W.E. Wilson. "Sulfuric Acid Aerosols and HCl Release in Coastal Atmospheres: Evidence of Rapid Formation of Sulfuric Acid Particulates," *Atmos. Environ.* 14:165–182 (1980).
13. Clegg, S.L. and P. Brimblecombe. "Potential Degassing of Hydrogen Chloride from Acidified Sodium Chloride Droplets," *Atmos. Environ.* 19:465–470 (1985).
14. Andreae, M.O., R.J. Charlson, F. Bruynseels, H. Storms, R. Van Grieken, and W. Maenhaut. "Internal Mixture of Sea Salt, Silicates, and Excess Sulphate in Marine Aerosols," *Science* 232:1620–1623 (1986).
15. Otten, P., F. Bruynseels, and R. Van Grieken. "Nitric Acid Interaction with Marine Aerosols Sampled by Impaction," *Bull. Soc. Chim. Belg.* 95:447–453 (1986).
16. Kolaitis, L.N., F. Bruynseels, and R. Van Grieken. "Determination of Methanesulfonic Acid and Non-sea-salt Sulfate in Single Marine Aerosol Particles," *Env. Sci. Technol.* 23:236–240 (1989).
17. Blanchard D.C. *Air-Sea Exchange of Gases and Particles*, 1st ed., Nato ASI Series, P.S. Liss and W.G.N. Slinn, Eds. (Dordrecht, The Netherlands: D. Reidel Publishing Company, 1983), pp. 407–444.

18. Harrison, R.H. and W.T. Sturges. "Physico-chemical Speciation and Transformation Reactions of Particulate Atmospheric Nitrogen and Sulfur Compounds," *Atmos. Environ.* 18:1829–1833 (1984).
19. Van Borm, W., F. Adams, and W. Maenhaut. "Characterization of Individual Particles in the Antwerp Aerosol," *Atmos. Environ.* 23:1139–1151 (1989).
20. Charlson, R.J., D.S. Covert, T.V. Larson, and A.P. Waggoner. "Chemical Properties of Tropospheric Sulfur Aerosols," *Atmos. Environ.* 12:39–53 (1978).
21. Mamane Y. and K.E. Noll. "Characterization of Large Particles at Rural Site in the Eastern United States: Mass Distribution and Individual Particle Analysis," *Atmos. Environ.* 19:611–622 (1985).
22. Kleinman, M.T., R.F. Phalen, R. Mannix, M. Azizian, and K. Walters. "Influence of Fe and Mn Ions on the Incorporation of Radioactive (SO_2)-S-35 by Sulphate Aerosols," *Atmos. Environ.* 19:607–610 (1985).
23. Wouters, L., P. Artaxo, and R. Van Grieken. "Laser Microprobe Mass Analysis of Individual Antarctic Aerosol Particles," *Int. J. Environ. Anal. Chem.* 38:427–438 (1989).
24. Storms, H. "Quantification of Automated Electron Microprobe X-ray Analysis and Application in Aerosol Research," PhD Dissertation, University of Antwerp, Belgium (1988).
25. Armstrong, J.T. and P.R. Buseck. "Quantitative Chemical Analysis of Individual Microparticles using the Electron Microprobe: Theoretical," *Anal. Chem.* 47:2178–2192 (1975).
26. Kaufherr, N. and D. Lichtmann. "Comparison of Micron and Submicron Fly Ash Particles using Scanning Electron Microscopy and X-ray Elemental Analysis," *Environ. Sci. Technol.* 18:544–547 (1984).
27. Hopke, P.K. *Receptor Modelling in Environmental Chemistry* (New York: John Wiley and Sons, Inc., 1985).
28. Bruynseels, F. "Application of Laser Microprobe Mass Analysis in Aerosol Research," PhD Dissertation, University of Antwerp, Belgium (1987).
29. Sturges, W.T. and R.M. Harrison. "Bromine in Marine Aerosols and the Origin, Nature and Quantity of Natural Atmospheric Bromine," *Atmos. Environ.* 20:1485–1496 (1986).
30. Graham, W., S.R. Piotrowicz, and R.A. Duce. "The Sea as a Source of Atmospheric Phosphorus," *Mar. Chem.* 7:325–342 (1979).
31. Chen, L., R. Arimoto, and R.A. Duce. "The Sources and Forms of Phosphorus in Marine Aerosol Particles and Rain from the Northern New Zealand," *Atmos. Environ.* 19:779–787 (1985).
32. Xhoffer, C. "Electronenprobe Microanalyse Van Vlieg-as en Partikels uit het Marine Milieu," MSc Thesis, University of Antwerp, Belgium (1987).
33. Kretzschmar, J.G. and G. Cosemans. "A Five Year Survey of Some Heavy Metal Levels in Air at the Belgian North Sea Coast," *Atmos. Environ.* 13:267–277 (1979).
34. Xhoffer, C., P. Bernard, and R. Van Grieken. "Chemical Characterization and Source Apportionment of Individual Aerosol Particles over the North Sea and the English Channel using Multivariate Techniques," *Environ. Sci. Technol.* 25:1470–1478 (1991).
35. Van Espen, P. "A Program for the Processing of Analytical Data (DPP)," *Anal. Chim. Acta* 165:31–49 (1984).

36. Bruynseels, F., H. Storms, T. Tavares, and R. Van Grieken. "Characterization of Individual Particle Types in Coastal Air by Laser Microprobe Mass Analysis," *Int. J. Environ. Anal. Chem.* 23:1–14 (1985).
37. Fitzgerald, J. "Marine Aerosols: A Review," *Atmos. Environ.* 25A, 3/4:533–545 (1991).
38. Artaxo, P., M.L.C. Rabello, W. Maenhaut, and R. Van Grieken. "Trace Elements and Individual Particle Analysis of Aerosol Particles from the Antarctic Peninsula," *Tellus*, submitted (1991).
39. Mitchell, R. and J. Pilcher. *Ind. Eng. Chem.*, 51:1039 (1959).
40. Borchert, B. in *Chemical Oceanography*, Vol. 2, G. Skirrow and J.P. Riley, Eds. (London: Academic Press, Inc., 1965).
41. Parungo, F., B. Bodhaine, and J. Bortniak. "Seasonal Variation in Antarctic Aerosol," *J. Aerosol Sci.* 12:491–504 (1981).
42. Shaw, G. "X-ray Spectrometry of Polar Aerosols," *Atmos. Environ.* 17:329–332 (1983).
43. Parungo, F., E. Ackerman, W. Caldwell, and H.K. Weickmann. "Individual Particle Analysis of Antarctic Aerosol," *Tellus* 31:521–529 (1979).
44. Harvey, M.J., G.W. Fisher, I.S. Lechber, P. Isaac, N.E. Flower, and A.L. Dick. "Summertime Aerosol Measurements in the Ross Sea Region of Antarctica," *Atmos. Environ.* 25A, 3/4:569–580 (1991).
45. Mészáros, E. "The Atmospheric Aerosol" in *Atmospheric Particles and Nuclei*, G. Gotz, E. Mészáros, and G. Vali, Eds. (Budapest: Akademiai Kiaido, 1991), pp. 17–77.
46. Cunningham, W. and W. Zoller. "The Chemical Composition of Remote Area Aerosols," *J. Aerosol Sci.* 12:367–384 (1981).
47. Blanchard, D.C. and A.H. Woodcock. "Bubble Formation and Modification in the Sea and its Meteorological Significance," *Tellus* 9(2):145–158 (1957).
48. Stramska M., R. Marks, and E. Monaham. "Bubble-Mediated Aerosol Production as a Consequence of Wave Breaking in Supersaturated (Hyperoxic) Seawater," *J. Geophys. Res.* 95(C10):18281–18288 (1990).
49. Harvie, C.E., J.H. Weare, L.A. Hardie, and H.P. Eugster. "Evaporation of Seawater: Calculated Mineral Sequences," *Science* 208:498–500 (1980).
50. Mallant, R., G. Kos, and A. Van Westen. Proceedings of the 2nd Int. Aerosol Conf., Berlin, (September 1986), pp. 49–52.
51. Brimblecombe, P. and S. Clegg. *J. Atm. Chem.* 22:437 (1985).
52. Saltzman, E., D. Savoie, J. Prospero, and R. Zika. "Methanesulphonic-acid and Non-sea salt Sulphate at Flanning and American Samoa," *Geophys. Res. Lett.* 12:437–440 (1985).
53. Legrand, M., C. Lorius, N. Barkov, and V. Petrov. "Vostok (Antarctica) Ice Core: Atmospheric Chemistry Changes over the Last Climate Cycle (160,000 years)," *Atmos. Environ.* 22:317 (1988).
54. Berresheim, H. "Biogenic Sulphur Emissions from Sub-Antarctic and Antarctic Oceans," *J. Geophys. Res.* 92:3245–3262 (1987).
55. Sheridan, P. and I. Musselman. "Characterization of Aircraft-Collected Particles Present in the Arctic Aerosol; Alaskan Arctic, Spring 1983," *Atmos. Environ.* 19:2159–2166 (1985).
56. Bigg, E.K. and D.E. Turvey. "Sources of Atmospheric Particles over Australia," *Atmos. Environ.* 12:1643–1655 (1978).
57. Beauford, W., J. Barber, and A.R. Barringer. "Heavy Metal Release from Plants into the Atmosphere," *Nature* 256:35–36 (1975).

58. Beauford, W., J. Barber, and A.R. Barringer. "Release of Particles Containing Metals from the Vegetation into the Atmosphere," *Science* 195:571–573 (1977).
59. Wils, E.R.J., A.G. Hulst, and J.C. Hartog. "The Occurrence of Plant Wax Constituents in Airborne Particulate Matter in an Urbanized Area," *Chemosphere* 11:1087–1096 (1982).
60. Simoneit, B.R.T. "Application of Molecular Marker Analysis to Reconcile Sources of Carbonaceous Particulates in Tropospheric Aerosols," *Sci. Total. Environ.* 36:61–72 (1984).
61. Nemeruyk, G.E. "Migration of Salts into the Atmosphere during Transpiration," *Sov. Plant Physiol., Engl. Transl.* 17:560–566 (1970).
62. Artaxo, P. and C. Orsini. "Emission of Aerosol by Plants Revealed by Three Receptor Models," in *Aerosols: Formation and Reactivity*, G. Israel, Ed. (Great Britain: Pergamon Journals, Ltd., 1986), pp. 148–151.
63. Maenhaut, W. and K. Akilimali. "Study of the Atmospheric Aerosol Composition in Equatorial Africa using PIXE as Analytical Technique," *Nucl. Instrum. Methods* B22:254–258 (1987).
64. Artaxo, P., H. Storms, F. Bruynseels, R. Van Grieken, and W. Maenhaut. "Composition and Sources of Aerosols from the Amazon Basin," *J. Geophys. Res.* 93 (D2):1605–1615 (1988).
65. Wouters, L. "Laser Microprobe Mass Analysis of Individual Environmental Particles," PhD Dissertation, University of Antwerp, Belgium (1991).
66. Lovelock, J.E., R.J. Maggs, and R.A. Rasmussen. "Atmospheric Dimethyl Sulfide and the Natural Sulphur Cycle," *Nature* 237:452–453 (1972).
67. Andreae, M.O. and T.W. Andreae. "The Cycle of Biogenic Sulfur Compounds over the Amazon Basin. I. Dry Season," *J. Geophys. Res.* 93:1487–1497 (1988).
68. Talbot, R., M.O. Andreae, T.W. Andreae, and R. Harris. "Regional Aerosol Chemistry of the Amazon Basin during the Dry Season," *J. Geophys. Res.* 93:1499–1508 (1988).
69. Basset, M. and S. Seinfield. "Atmospheric Equilibrium Model of Sulfate and Nitrate Aerosols — II. Particle Size Analysis," *Atmos. Environ.* 18:1163–1170 (1984).
70. Andreae, M.O., E.V. Bowell, M. Garstang, G.L. Harris, G.F. Hill, D.J. Jacob, M.C. Pereira, G.W. Sachse, A.W. Setzen, P.L. Silva Dias, R.W. Talbot, A.L. Torres, and S.C. Wolfsy. "Biomass-Burning Emissions and Associated Haze Layers over Amazonia," *J. Geophys. Res.* 93:1509–1527 (1988).
71. Bernard, P., R. Van Grieken, and D. Eisma. "Classification of Estuarine Particles using Automated Electron Microprobe Analysis and Multivariate Techniques," *Environ. Sci. Technol.* 20:467–473 (1986).
72. Raeymaekers, B. "Characterization of Particles by Automated Electron Probe Microanalysis," PhD Dissertation, University of Antwerp, Belgium (1986).
73. Adams, F., P. Van Espen, and W. Maenhaut. "Aerosol Composition at Chacaltaya, Bolivia, as Determined by Size-Fractionated Sampling," *Atmos. Environ.* 17:1521–1536 (1983).
74. Surkyn, P., J. De Waele, and F. Adams. "Laser Microprobe Mass Analysis for Source Identification of Air Particulate Matter," *Int. J. Anal. Chem.* 13:257–274 (1983).
75. Lawson, D.R. and J.W. Winchester. "Sulfur and Trace Element Concentration Relationships in Aerosols from the South American Continent," *Geophys. Res. Lett.* 5:195–198 (1978).

76. Okada, K. "Number-Size Distribution and Formation Process of Submicrometer Sulfate-Containing Particles in the Urban Atmosphere of Nagoya," *Atmos. Environ.* 19:743–757 (1985).
77. Grass, J.L. and G.P. Ayers. "On Sizing Impacted Sulfuric Acid Aerosol Particles," *J. Appl. Met.* 18:634–638 (1979).
78. Webber, J.S. "Using Microparticle Characterization to Identify Sources of Acid-Rain Precursors reaching Whiteface Mountain, New York," in *Electron Microscopy in Forensic, Occupational and Environmental Health Sciences*, S. Bassu and J.R. Milette, Eds., (New York: Plenum Press, Inc., 1986).
79. Rojas, C.M., P. Artaxo, and, R. Van Grieken. "Aerosols in Santiago de Chile: A Study using Receptor Modelling with X-ray Fluorescence and Single Particle Analysis," *Atmos. Environ.* 2:227–241 (1990).
80. Kolaitis, L. "Applications of Laser Ionization/Desorption in Mass Spectrometry," PhD Dissertation, University of Antwerp, Belgium (1988).
81. Prèndez, M. and J. Ortiz. "Elemental Composition of Airborne Particulate Matter from Santiago City, Chile," *J. Air Pollut. Control Assoc.* 34:54–56 (1984).
82. Elmes, P.C. "Current Information on the Health Risk of Asbestos," *R. Soc. Health J.* 96:248–252 (1976).
83. Hudson, A. *Phil. Trans. R. Soc. London, Ser. A* 286:611–624 (1977).
84. Van Espen, P., J. De Waele, E. Vansant, and F. Adams. "Study of Asbestos using SIMS and LAMMA," *Int. J. Mass Spectrom. and Ion Phys.* 46:515–518 (1983).
85. De Waele, J. K., J.J. Gijbels, E.F. Vansant, and F.C. Adams. "Laser Microprobe Mass Analysis of Plastic-Contaminated Asbestos Fiber Surfaces," *Anal. Chem.* 55:2255–2260 (1983).
86. Vansant, E.F. and S. Yariv. "Adsorption and Oxidation of N,N-dimethylaniline by Laptonite," *J. Chem. Soc., Faraday. Trans. I* 73:1815–1824 (1977).
87. De Waele, J. K. and F.C. Adams. "Applications of Laser Microprobe Mass Analysis for Characterization of Asbestos," *Scanning Electron Microsc. III*:935–946 (1985).
88. Feuerbacher, D.G. and G.T. Dimataris. "Comparative Cytotoxicity and Mutagenicity of Organosilane-reacted Chrysotile Asbestos," Clay Minerals Society Annual Meeting 1980.
89. Xhoffer C., P. Berghmans, I. Muir, W. Jacob, R. Van Grieken, and F. Adams. "A Method for the Characterization of Surface-modified Asbestos Fibres by Electron Energy-loss Spectroscopy and Electron Spectroscopic Imaging," *J. Microsc.* 162:179–184 (1990).
90. Cozak, D., C., Barbeau, F. Gauvin, J-P. Barry, C. DeBlois, R. De Wolf, and F. Kimmerle. *Can. J. Chem.* 61:2753–2760 (1983).
91. Berghmans P. and F.C. Adams. "Electron Energy Loss Spectroscopy (EELS) and Electron Spectroscopic Imaging (ESI) for the Localization of Titanium in Chrysotile Asbestos," *Surf. Interface Anal.* submitted (1991).
92. Tartarelli, R., P. Davini, F. Morelli, and P. Corsi. "Interactions between SO_2 and Carbonaceous Particulates," *Atmos. Environ.* 12:289–293 (1978).
93. Cheng, R.J., V.A. Mohen, T.T. Shen, M. Current, and J.B. Hudson. *JAPCA* 26:787–790 (1976).
94. Roekens, E., L. Leysen, Z. Komy, and R. Van Grieken. Proc. of 2nd Int. Colloq., Materials Science and Restoration, (1986), pp. 487–489.
95. Leysen, L.A., J.K. De Waele, E.J. Roekens, and R. Van Grieken. "Electron Probe Micro-analysis and Laser Microprobe Mass Analysis of Material Leached from a Limestone Cathedral," *Scanning Microsc.* 1:1617–1630 (1987).

96. Denoyer, E., R. Van Grieken, F. Adams, and D. Natusch. "Laser Microprobe Mass Spectrometry. I. Basic Principles and Performance Characteristics," *Anal. Chem.* 54, 26A–41A (1982).
97. Sándor, S., Sz. Török, C. Xhoffer, and R. Van Grieken. "Individual Coal Flyash Particles Analysis by EPMA," *Proceedings of the XIIth International Congress for Electron Microscopy* (San Francisco: San Francisco Press, Inc., 1990), pp. 254–255.
98. Van Borm, W. and F. Adams. "Cluster Analysis of Electron Microprobe Analysis Data of Individual Particles for Source Apportionment of Air Particulate Matter," *Atmos. Environ.* 22:2297–2307 (1988).
99. Van Borm, W., L. Wouters, R. Van Grieken, and F. Adams. "Lead Particles in an Urban Atmosphere: An Individual Particle Approach," *Sci. Tot. Environ.* 90:55–66 (1990).
100. Van Borm, W. "Source Apportionment of Atmospheric Particles by Electron Probe X-ray Micro Analysis and Receptor Models," PhD Dissertation, University of Antwerp, Belgium (1989).
101. Post, J.E. and P.R. Buseck. "Quantitative Energy-Dispersive Analysis of Lead Halide Particles from the Phoenix Urban Aerosol," *Environ. Sci. Technol.* 19:682–685 (1985).
102. Biggins, P.D.E. and R.M. Harrison. "Atmospheric Chemistry of Automotive Lead," *Environ. Sci. Technol.* 13:558–565 (1979).
103. Gordon, G.E. "Receptor Models," *Environ. Sci. Technol.* 14:792–800 (1980).
104. PARCOM. Secretariat of the Paris Commission. Report on Land-based Inputs of Contaminants to the Waters of the Paris Convention in 1987, PARCOM 11/6/1–E (1989).
105. STWG: Scientific and Technological Working Group of the Department of the Environment of the European Community. "Quality Status of the North Sea," Second International Conference on the Protection of the North Sea, London (1987).
106. Otten, Ph. "Transformation, Concentrations and Deposition of North Sea Aerosols," PhD Dissertation, University of Antwerp, Belgium (1991).

CHAPTER 6

SAMPLING AND CHARACTERIZATION OF COLLOIDS AND PARTICLES IN GROUNDWATER FOR STUDYING THEIR ROLE IN CONTAMINANT TRANSPORT

John F. McCarthy and Claude Degueldre

TABLE OF CONTENTS

0-87371-895-X/93/$0.00+$.50

INTRODUCTION

Contaminant transport models generally treat the subsurface environment as a two-phase system in which contaminants are distributed between a mobile aqueous phase and an immobile solid phase (rock or soil constituents). Contaminants with a high affinity for sorbing to rock or aquifer media are assumed to be retarded, relative to the rate of groundwater flow (Figure 1). However, an increasing body of evidence indicates that under some subsurface conditions, components of the solid phase may exist as colloids and particles suspended in and transported with the flowing groundwater. Association of contaminants with this additional mobile phase may enhance the rate of contaminant transport; conversely, deposition of the colloidal particles may reduce the permeability of a formation and thereby decrease transport.[1,2] Accurate assessment of current contamination problems, engineering and safety assessment of containment and disposal options, and development of cost effective remediation strategies all require fundamental understanding of the potential role of colloidal particles in enhancing or diminishing contaminant transport in the subsurface.

A variety of inorganic and organic materials exist as colloids and small particles in groundwater, including mineral precipitates (notably iron, aluminium, calcium and manganese oxides, hydroxides, carbonates, silicates, and phosphates, but also including oxides and hydroxides of actinide elements such as uranium, neptunium, plutonium, and americium), rock and mineral fragments (including layer silicates, oxides, and other weathering products of mineral phases), "biocolloids" (including viruses and bacteria), microemulsions of nonaqueous phase liquids, and macromolecular components of natural organic matter (NOM, including some components of humic substances and other organic polymers such as exocellular biopolymeric material secreted by microorganisms). This chapter will focus primarily on inorganic colloids and particles, but will also refer to data on biocolloid transport. The distinction between "colloidal" and "dissolved" NOM is related to the aggregation properties of NOM molecules and is poorly understood. NOM (as well as some anthropogenic organics such as surfactants) may play an important role in sorbing and facilitating the transport of contami-

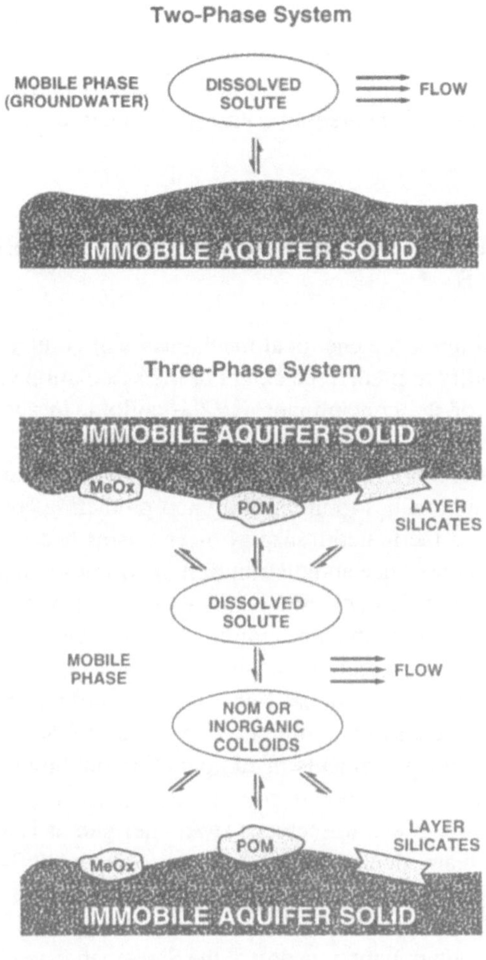

Figure 1. The potential role of mobile colloids in enhancing the transport of highly adsorbing contaminants is illustrated. When colloids are present, consideration must be given to the distribution of contaminants between the aquifer, colloid, and groundwater, as well as the interactions of the colloids with each other (agglomeration) or with the aquifer surfaces (filtration). POM and MeOx refer to particulate organic matter and metal oxide surfaces, respectively. (Reprinted from McCarthy, J. F. and J. M. Zachara. *Environ. Sci. Technol.* 23(5):496–503 (1989). With permission.)

nants,[2] but the processes controlling the transport of NOM may be mechanistically different from those relevant to the stability and transport of inorganic colloids. NOM will be discussed, however, in the context of its specific chemical effect on the colloidal stability of inorganic colloids and particles and of its effect on the association of contaminants with organically coated particles.

The objective of this chapter is to highlight the sampling and analytical considerations that need to be addressed in studies of groundwater colloids. Current information on the nature and abundance of groundwater colloids will be discussed, as will some of the mechanisms by which colloids may be generated and stabilized in natural subsurface environments. However, we will first place this information in an environmental context by briefly reviewing available studies on colloid transport and potential cotransport of contaminants adsorbed to the colloids.

ROLE OF COLLOIDS IN THE SUBSURFACE TRANSPORT OF CONTAMINANTS

Our understanding of the chemical mechanisms of contaminant-colloid interactions and our ability to predict the extent of the association vary widely depending on the nature of the contaminant and the colloid. In addition, much of the available information is drawn from studies of colloids from surface waters; information is much more limited for colloids recovered from groundwater systems. In general, approaches to monitoring and predicting contaminant transport have ignored colloid-facilitated transport mechanisms because little information is available on the abundance and distribution of colloids and particles in groundwater, their affinity to bind contaminants, or their mobility in subsurface systems.

Measurements of contaminant migration at waste sites clearly demonstrate the inadequacy of existing predictive capabilities. For example, at a Defense Programs site at Los Alamos National Laboratory, plutonium and americium disposed of at a liquid seepage site migrated up to 30 m;[3] predictions, which were based on laboratory measurements of radionuclide binding to immobile subsurface materials and which ignored colloids and particles, estimated migration would be limited to a few millimeters. At another site at Los Alamos, not only were plutonium and americium detected in monitoring wells 3400 m from a liquid waste outfall, but the transported radionuclides were shown by ultrafiltration to be present as colloids and particles (25 to 450 nm in diameter).[4]

An analysis of contaminant transport at the Savannah River Laboratory showed that observed transport exceeded predictions for a number of sites, including hazardous chemical and mixed waste sites present as seepage basins, shallow land burial and rubble pits, and solvent burning pits.[5,6] Computer models based on predicted values for contaminant sorption to immobile aquifer material (and ignoring colloids) predicted "zero" concentrations in groundwater; however, field data showed measurable concentrations of waste migrating from the sites. When models were modified to include an empirically fitted "facilitated transport" component, predictions improved significantly.[6] Column studies with soils from this site demonstrated that significant amounts of a mixture of metals pulsed on the columns migrated rapidly and coeluted with a peak in turbidity of the column effluent; furthermore, the metals were detected on the surfaces of the turbidity-causing colloids in the eluent.[7]

Similarly, Coles and Ramspott[8] found much more rapid migration of ruthenium in the field than expected from batch adsorption studies. A well 91 m from a nuclear detonation cavity at the Nevada Test Site (NTS) was pumped to induce flow into the well from the cavity. No radioactivity was observed for 2 years, but, with further pumping, the concentration of both ^{3}H (a nonreactive tracer of water flow) and ^{106}Ru increased at the same rates, suggesting that both travelled at the same velocity from the detonation cavity. Laboratory measurements of the equilibrium sorption coefficient (K_d) for Ru with rock and water from the site predicted that the Ru should have travelled 30,000-fold slower than the ^{3}H-water (i.e., the Ru should have moved only 3 cm during the time the ^{3}H-water migrated 91 m).[8] Although that study did not look for colloids in the groundwater, Buddemeier and Hunt[9] did demonstrate the presence of inorganic colloidal particles with sorbed metals and lanthanide radionuclides in groundwater at another detonation cavity at NTS.

In contaminant plumes at the Chalk River Nuclear Laboratory in Canada, K_d-based models of radionuclide mobility were inadequate to predict the transport of ^{137}Cs migrating from a glass block;[10] the glass contained mixed fission products and actinides and had been placed below the water table 20 years earlier. Sampling of the extended plume of low-level ^{137}Cs demonstrated that most of the radionuclide in the groundwater was present within the particulate phase.[11]

Contaminants can become associated with inorganic colloids by a number of mechanisms. Some transition metals and transuranic elements are known to form aqueous colloids by condensation or homogeneous nucleation from a solution when supersaturation with respect to the metal mineral occurs.[12-14] Metal contaminants can also adsorb to existing inorganic colloids such as clay minerals or hydrated oxides. Metallic and organic cations can be associated with layer silicate clays by ion exchange.[15] Surfaces of Fe-, Mn-, Al-, and Si-oxide particles strongly adsorb certain metallic cations and organic and inorganic acid anions.[16] Metals and organic acids may adsorb to the surface of calcium carbonate by complexation or exchange with structural ions.[17] Coatings of NOM alter the surface properties of layer silicate clays and metal oxide colloids, making them more or less reactive with contaminant ions.[18] The affinity with which hydrophobic organic contaminants bind to sediments and aquifer material is directly related to the organic carbon content of the particles;[19] however, in low-organic-carbon sediments (<0.001 g C/g sediment), the dependence of sorption on organic carbon diminishes (especially for less hydrophobic or ionized organic contaminants), and interactions with mineral surfaces begin to dominate as a contaminant retention mechanism.[20-22] Contaminant binding to all the substrates noted above is influenced to varying degrees by other solutes that may compete for common adsorption sites on the colloid surface. The importance of these entities in facilitating contaminant transport in groundwater depends on the colloid surface area, the number of reactive sites per unit of surface, the preconditioning of surfaces by strongly bound cosorbates, and the strength and reversibility of the contaminant-surface reaction. Unfortunately, few data of this type exist for groundwater colloids.

Two lines of evidence suggest that colloids may influence the transport of subsurface contaminants: laboratory column studies demonstrate cotransport of contaminants sorbed to mobile colloids, and field studies demonstrate the association of contaminants with natural groundwater colloids. For example, the pesticides DDT and paraquat were shown to be cotransported with montmorillonite through soil columns under saturating conditions.[23,24] Champ et al.[25] demonstrated the rapid transport of Pu in undisturbed aquifer cores eluted with groundwater for 8 months; half of the eluted Pu was particulate (>450 nm) and almost 25% was colloidal (3 to 450 nm). Newman[7] introduced a pulse input of metals (Pb, Cr, Cu, and Ni) on a column of aquifer sediments; the metals eluted from a soil column as two peaks. The first peak coincided with a peak in turbidity, and surface analysis of eluted colloids demonstrated that the metals were adsorbed to the surface of the turbidity-causing colloids.[7]

Several field studies have demonstrated the association of contaminants with colloidal material from groundwater. At the Nevada test site, transition metals (manganese and cobalt) and lanthanide (cerium, europium) radionuclides were associated with inorganic colloids (3 to 50 nm by ultrafiltration) recovered from groundwater inside a nuclear detonation cavity and from a well in a permeable fractured lava and tuff formation 300 m from the cavity.[9] Plutonium and americium were associated with siliceous colloids (25 to 450 nm by ultrafiltration) in a shallow alluvial aquifer at Los Alamos National Laboratory.[4] Filterable particles (>400 nm) containing radionuclides of Co, Zr, Ru, Cs, and Ce were recovered from contaminant plumes in groundwater at the Chalk River Nuclear Laboratory.[11] Uranium and daughter species such as thorium were found associated with iron- and silicon-rich colloids (18 to 1000 nm by ultrafiltration) downgradient from a uranium deposit in Australia.[26]

Although contaminants may sorb to natural colloidal material and be transported through small laboratory columns; the significance of this process to field-scale migration of contaminants is dependent on the distances the colloidal material will migrate through vadose and saturated zones. Colloids are mobile in some subsurface environments, and that mobility is controlled by the stability of the colloids in groundwater and the chemical interactions between colloids and immobile matrix surfaces, as well as by hydrological factors that control flow rates and flow paths of groundwater.

To be mobile over significant distances, a colloidal particle must be stable [i.e., resistent to either dissolution (chemical stability) or aggregation with other particles (colloidal stability)] and must avoid filtration (i.e., avoid being physically trapped by the immobile aquifer media). The stability of colloids in groundwater depends on the interaction energy between similar particles and between particles and aquifer solids.[1,27,28] Colloidal particles and immobile materials in aquifers possess surface charges controlled by ionizable surface groups. Changes in aqueous chemistry may cause aggregation of colloidal particles or may favor the attachment to the aquifer media. Destabilization can occur due to compression of the double layer surrounding particles by increased counterion concentrations,[7,29,30] by pH-induced changes in surface charge,[31] or by the presence of strongly binding ions that decrease the net surface charge.[32] Conversely, transport of colloids through porous media columns is enhanced by conditions promoting the colloidal stability of particles. For example, low ionic strength of the eluting solution increases the mobility of latex microspheres,[33-35] montmorillonite,[24] hematite,[36] and natural soil colloids.[7] Likewise, increased pH (above the pH_{zpc} of the colloid), or the presence of sorbing anions such as arsenate and phosphate at lower pHs can electrostatically stabilize colloids and promote their transport through negatively charged porous media.[7,33,36-38] Increases in temperature can also reduce the repulsive energy between surfaces and result in dispersion of clay particles and subsequent plugging of pores in the formation in the downgradient direction.[39] Thermal effects on particle stability may be particularly important around high-level radioactive storage sites, but no data are available that address this issue.

Changes in aqueous chemistry can disperse and remobilize trapped particles. For example, changing to low-ionic-strength water promoted the mobilization of latex microspheres previously deposited on a column.[33] In a field study at an artificial recharge site in California, clay colloids were dispersed from surface soil during recharge by low-ionic-strength water; however, increasing the ionic strength of the recharge water by addition of gypsum (calcium sulfate) at the recharge area prevented the release of clays from soil aggregates.[40]

If particle-particle and/or particle-media attachment are energetically favorable, then deposited particles will accumulate on media surfaces. This deposit can cause a decrease in media permeability as porosity decreases, resulting in decrease in the mean groundwater flow rate. For example, laboratory studies demonstrated that the permeability of porous media was reduced due to mechanical straining of clay particles by the media.[24,41,42] Large changes in permeability of aquifers have been observed when the electrolyte concentration of the pore water was altered.[43-45] This coupling between particle deposition and associated permeability reduction is important to transport of both colloid-associated and dissolved contaminants; however, the significance of these processes may be reduced in heterogeneous natural systems in which preferential flow through larger openings predominates.

There are but a few field studies that have attempted to test and model the transport of colloids and particles in groundwater experimentally. Harvey et al.[46] investigated the transport of latex microspheres and indigenous bacteria through a sandy aquifer in Cape Cod, Massachusetts, using both forced- and natural-gradient experiments. O'Melia[28] attempted to apply filtration theory to describe the breakthrough of six types of latex microspheres that differed in size and surface characteristics. The passage of the microspheres under the natural aquifer flow regime was monitored with samples collected 6.9 m from the injection well. The concentration histories of the microspheres are shown in Figure 2. In general, the removal of particles decreased with increasing particle size. The kinetics of filtration in porous media was described by Yao et al.:[47]

$$\frac{dC}{dL} = -\frac{3}{4}\alpha_{(p,c)}\eta_{(p,c)}\frac{(1-\varepsilon)}{a_c}C \qquad (1)$$

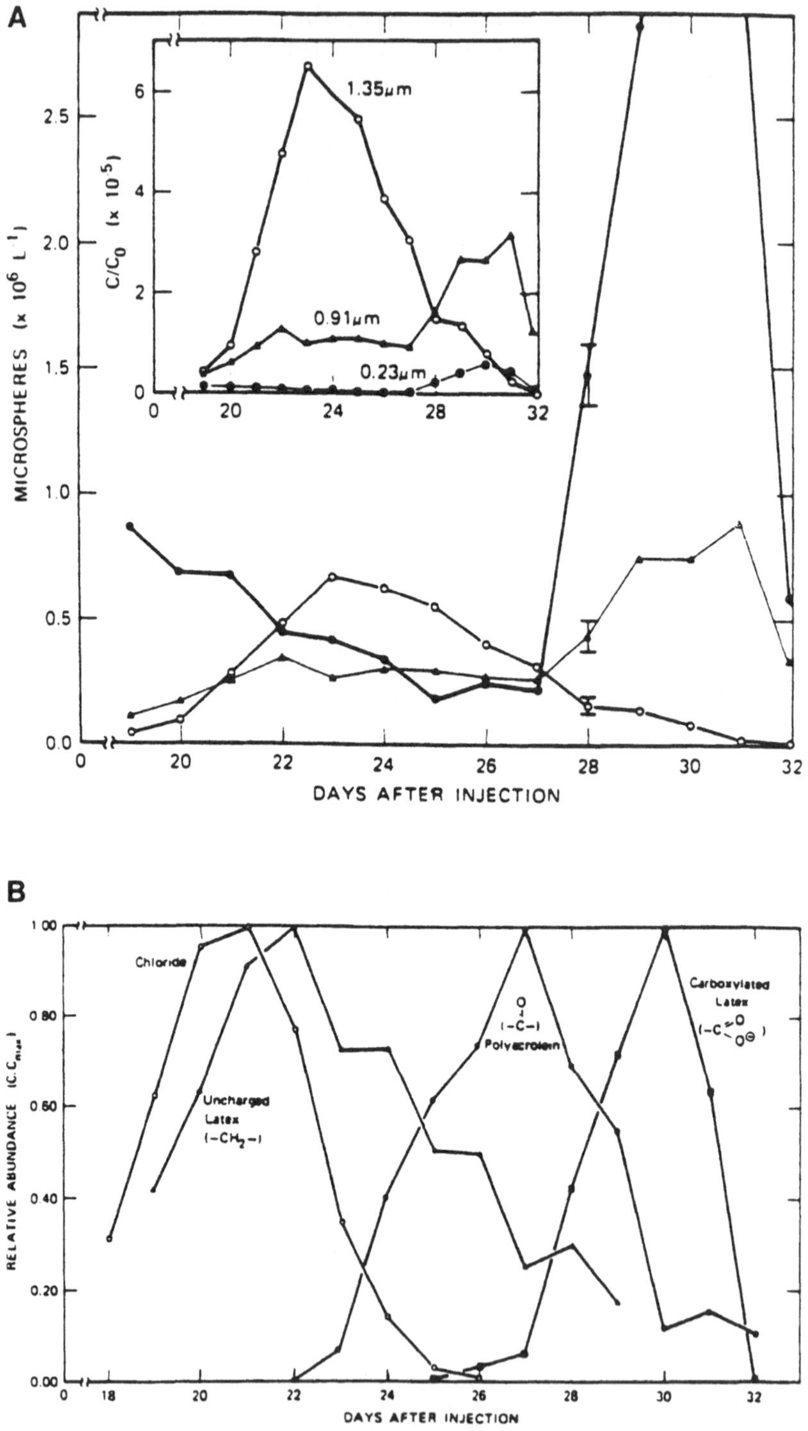
A
MICROSPHERES (x 10^6 L^-1)
DAYS AFTER INJECTION
C/C0 (x 10^-5)
1.35μm
0.91μm
0.23μm
B
RELATIVE ABUNDANCE
DAYS AFTER INJECTION
Chloride
Uncharged Latex
(-CH2-)
Polyacrolein
Carboxylated Latex

Figure 2.

where

C is the number concentration of particles of a specific size
L is the length of travel for the particles
a_c is the radius of the sand grains
ε is the porosity of the porous media
$\alpha_{(p,c)}$ and $\eta_{(p,c)}$ describe the kinetics of the deposition of suspended particles (p) on the collector or media (c)

The single-collector efficiency, η, is the rate at which particles strike a surface of a single porous media grain divided by the rate at which particles move toward the grain. The term (η) represents the sum of physical factors determining particle collision, including Brownian diffusion (η_D), fluid flow (interception; η_I), and gravity (η_G). The effects of diffusion, fluid flow, gravity, hydrodynamic retardation, electrostatic effects, and London-van der Waals forces on particle deposition have been theoretically evaluated,[48] and the value of η can be estimated based on parameters such as the particle diameter and density and the fluid density and viscosity.[47]

The attachment coefficient, or sticking probability, $\alpha(c,p)$ is the rate at which particles attach to the collector surface divided by the rate at which particles strike the collector surface. Ideally, $\alpha = 1$ for a completely destabilized system. The value of α can be experimentally obtained from measurement of particle removal by integrating Equation 1 with boundary conditions $C = C_o$ at $L = 0$ and $C = C_L$ at $L = L$:

$$\alpha_{(p,c)_{exp}} = -\left(\frac{4a_c}{3(1-\varepsilon)\eta_{(p,c)_{theor}} L}\right)\ln\left(\frac{C_L}{C_o}\right) \tag{2}$$

Calculations of particle transport and attachment in the field experiment of Harvey et al.[46] are presented (see C_L/C_o) in Table 1.[28] The high removal rates observed in the field experiment (only 0.01 to 3% of the injected particles were recovered at the sampling wells; Table 1) are due to the long travel distance and

Figure 2. Concentration histories for microspheres recovered at a sampling well 6.9 m downgradient in a sandy, freshwater aquifer in Cape Cod, Massachusetts are shown. The mean grain size, average porosity, and hydraulic conductivity are 0.5 mm, 0.38, and 0.1 cm/sec, respectively. The ambient flow under the natural gradient conditions under which the experiment was conducted was 0.3 to 0.5 m/d. (A) Concentration histories for 0.23, 0.91, and 1.35 μm (diameter) carboxylated microspheres. Dimensionless concentration history is depicted in the inset. (B) Concentration histories for chloride (a nonreactive tracer of groundwater flow), uncharged (0.53 μm diameter), polyacrolein (0.84 μm diameter), and carboxylated latex (0.53 μm diameter) microspheres. Data have been normalized to maximum concentration. (Reprinted from Harvey, R. W., L. H. George, R. L. Smith, and D. R. LeBlanc. *Environ. Sci. Technol.* 23:51–56 (1989). With permission.)

Table 1. Particle Transport and Attachment in the Cape Cod Aquifer[a]

Particle Diameter (nm)[b]	Type[b]	C_L/C_0[b]	$\eta_{(p,c)\text{theor}}$	$\alpha_{(p,c)\text{exp}}$
230	Carboxyl	0.0001	0.215	0.0033
530	Carboxyl	0.0004	0.123	0.0049
930	Carboxyl	0.0006	0.086	0.0067
1350	Carboxyl	0.0012	0.069	0.0076
600	Uncharged	0.0005	0.114	0.0052
850	Polyacrolein	0.031	0.090	0.0030

Source: O'Melia.[28]

[a] Parameters: Approach velocity of flow = 0.125 m/d; a_c = 0.025 cm; T = 11°C; L = 6.9 m; ε = 0.38; particle density assumed as 1.05 g/cm^{-3}.
[b] Data from Harvey et al.[46]

high mass transport rate ($\eta(_{p,c})_{theor}$) resulting from the low flow velocity in the aquifer. The attachment probabilities ($\alpha_{(c,p)}$; Table 1) for all six particles are remarkably uniform and suggest that the surface properties of the latex particles that control attachment are dependent primarily on the solution chemistry of the groundwater; O'Melia[28] suggests that NOM in the groundwater enhances the colloidal stability of the particles and reduces the value of $\alpha_{(c,p)\text{exp}}$. In spite of the similarities in the attachment probabilities, the different latex particles differ by a factor of 300 with respect to retention. O'Melia[28] attributes this to the differences in mass transport by convective Brownian diffusion (η_D), which is inversely related to particle size.[47] A similar attempt to use filtration theory to model movement of bacteria was reported by Harvey and Garabedian,[49] using data from another bacterial injection experiment in the same aquifer. The model reasonably accounted for the concentration histories of bacteria in sampling wells. Estimated attachment efficiencies were similar to those reported by O'Melia.[28] In general, the results of these studies appear to indicate that theories of the kinetics of particle deposition in porous media may be useful to describe particle transport and deposition in aquifers.

The success of filtration theory in these studies in the relatively well-sorted porous media of the Cape Cod sandy aquifer is encouraging; however, it is not clear that this approach is useful in all aquifers. In fact, there is considerable field data suggesting that transport of colloids, including bacteria and viruses, can occur very rapidly and over long distances in at least some locations. The mechanism of transport in these cases may involve travel through highly transmissive zones of preferred flow (examples of preferred flow paths can include macropores caused by root channels in soils, layered deposits of coarse gravels between more finely structured horizons, or fractures in consolidated formations; Figure 3). For example, biocolloids have been reported to migrate large distances in groundwater. Bacteria injected into groundwater have been reported to travel up to 920 m at rates of 200 to 350 m/d, and viruses injected into a well migrated 680 m downgradient at a rate estimated between 36 and 180 m/d.[50] Yeast cells injected into a well moved 7 m in a sand and gravel aquifer in less than 48 h, and moved faster than nonreactive tracers of groundwater flow such as iodide and bromide.[51] Inorganic colloids have also been documented to migrate hundreds of meters. Asbestos fibers were detected in an aquifer recharged from a reservoir containing high levels of fibers.[52] Rapid movement of layer silicates and organic matter through the vadose zone has also been observed.[53]

Migration documented in some reports may be associated with flow-through channels and secondary pore structure rather than through the intergranular pore space. In laboratory columns, the organic colloid, blue dextran (2,000,000 dalton relative *M* mass) eluted from the column faster than ^{3}H-water, presumably by being forced through larger pores and excluded from the smaller pores.[54,55] Likewise, Harvey et al.[46] observed that bacteria injected into the Cape Cod aquifer eluted slightly ahead of the nonreactive tracer. Preferential flow paths such as fractures, solution channels, or soil macropores can greatly enhance transport and reduce retention of particles (Figure 3). Smith et al.[56] recovered 22 to 79% of bacteria injected onto intact soil columns, but only 0.2 to 7% of the bacteria were recovered if the columns were prepared from mixed, repacked soil; the results suggested that flow through macropores present in the intact columns bypassed the adsorptive surfaces of the porous media. In the field, Pilgrim and Huff[53] observed the rapid movement of layer silicates and organic matter through the vadose zone during storm events and attributed transport to flow-through macropores. The role of fractures on particle transport was examined by Toran and Palumbo,[35] who simulated fractures by inserting small (0.2 or 1 mm in diameter) tubes into a sand column. The "fractures" significantly increased the transport (decreased retention) of latex microspheres and bacteria (Figure 4).

In summary, there appears to be a preponderance of evidence that groundwater colloids are able to bind contaminants and may, in at least some circumstances, be mobile in aquifers. Unfortunately, much of the information specific to groundwater colloids is more anecdotal than systematic and is insufficient to evaluate reliably the significance of colloids as a transport vector or to develop a predictive capability. Nevertheless, recent attempts to assess the long-term significance of groundwater colloids to contaminant transport have highlighted the importance of kinetic assumptions within the model framework. If the contaminant-colloid interaction is assumed to involve a rapid and reversible linear association, the transport of the contaminant is only slightly faster in the presence of colloids because the contaminant is stripped from the mobile colloids as the advancing front of colloids arrives at "clean" portions of the aquifer, due to desorption and rapid reequilibration with sorption sites on the aquifer matrix. Continued migration of the "stripped" colloids has little effect on contaminant mobility.[57,58] However, in a more complex model incorporating rapid but *irreversible* association, mobile colloids could significantly accelerate contaminant migration; the transport of the contaminant would then be strongly dependent on the extent of colloid-aquifer media interaction.[58] Unfortunately, little is known directly about

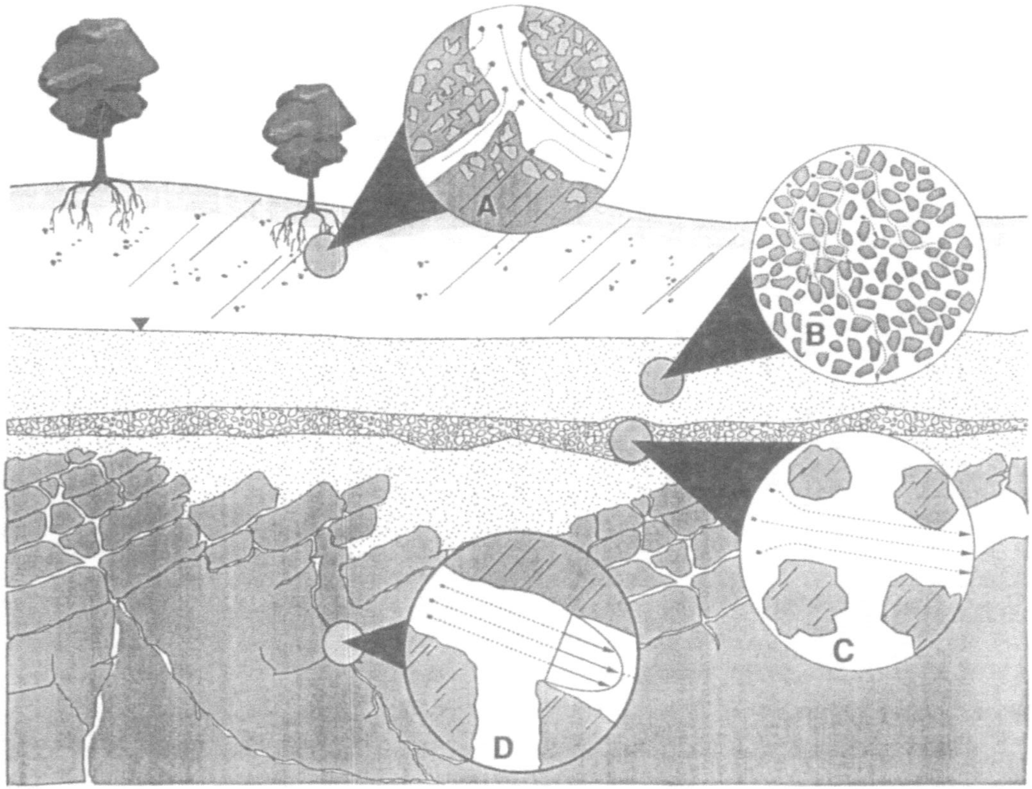

Figure 3. Diagrammatic illustration of the potential role of physical heterogeneity and preferential flow on the transport of contaminants. (A) In the unsaturated soil zone, macropores can form from decayed roots, etc., and provide zones of rapid movement during storm events when

the macropores become saturated with water. Some of the soil colloids and particles can be mobilized and transported during storm flow. (B) In well-sorted, saturated porous media, such as sand, colloids have opportunities to contact the immobile media and be retained by diffusion, interception, or settling. Colloid transport in such zones may be well described by filtration theory.[47] (C and D) Colloid transport can be greatly enhanced in high permeability zones where contact with immobile surfaces is limited, such as in very coarse, gravelly lenses (C) or in fratures (D) within consolidated formations. In such cases, the rates of water flow are fast (relative to flow within intergranular pores). In fractures, the velocity gradient across such flow paths tends to keep the colloids entrained in regions away from the fracture surfaces.

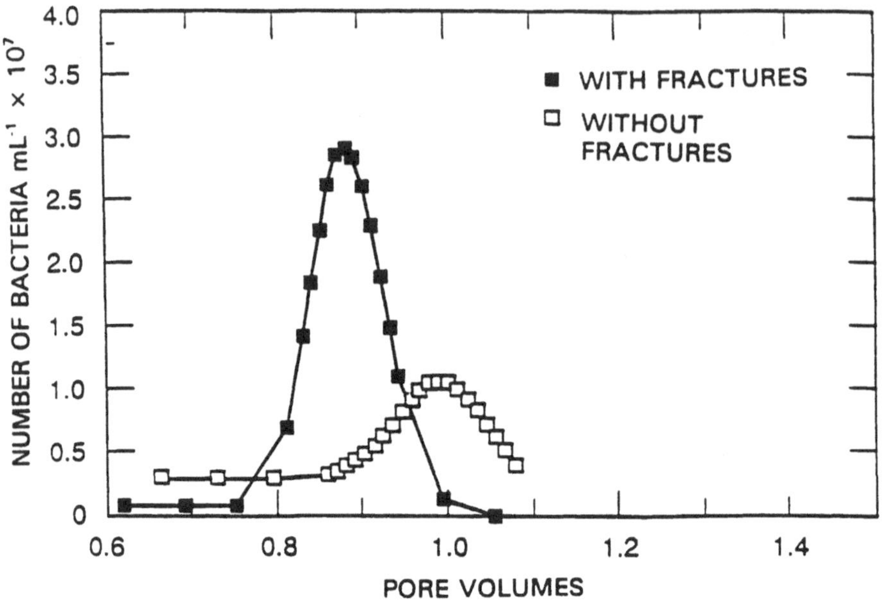

Figure 4. Concentration histories are shown for a methanotrophic bacteria eluted from sand columns (5 cm diameter by 65 cm length; average linear flow velocity of 4.2 cm/h). A 30 ml pulse (5% of a pore volume) of bacteria (10^8 cells/ml) were injected on the columns. The open symbols represent the breakthrough of bacteria in columns without "fractures." The solid symbols represent the breakthrough of bacteria in columns containing 15 randomly distributed "fractures" composed of 12 cm long tubes (1 mm diameter). The bacteria were transported more rapidly (retardation factor of 0.86 versus 0.98) and with less retention (73% retained on the column versus 93% retained) in columns that contained the "fractures" compared to similar columns without tubes. (From Toran, L. and A. V. Palumbo. *J. Contam. Hydrol.* 9:289–303. With permission.)

the reversibility of contaminants sorbed to groundwater colloids. While it has been noted that the uranium activity ratios ($^{234}U/^{238}U$) in aquifers in England,[59] Australia,[26] Germany,[60] and Canada[61,62] are different for colloid, rock, and water phases, this should not be used as evidence to directly conclude that colloids are not in equilibrium with the water. The exchangeable uranium (e.g., at the colloid-solution interface) may be in equilibrium with the truly dissolved uranium in the water, while the nonexchangeable components of uranium (e.g., the bulk mineralized portion of the rock or colloid) may contain uranium with a different isotopic ratio. This makes it difficult to arrive at firm conclusions about the equilibrium state of uranium between the colloid and water based on the isotopic ratios of the whole colloidal phase. Actually, the reversibility of contaminant sorption can be obstructed if the sorbed contaminant is "trapped" inside colloid aggregates.[63] Further progress in reducing uncertainties about the role of colloids will require development of reliable and systematic data on the nature and abundance of colloids in groundwater, including information about their formation and transport as well as about the extent and kinetics of both their association with and dissociation from contaminants.

SOURCES OF GROUNDWATER COLLOIDS

Although a detailed discussion on the nature and abundance of groundwater colloids will be presented later in this chapter, it is useful now to introduce the general hydrogeochemical processes thought to be responsible for colloid generation in subsurface environments. Subsequent discussion of the issues in sampling and characterization of groundwater colloids can then be better understood in the context of the types of geochemical alterations that can influence colloid generation in groundwater.

Colloids in groundwater can arise from a number of sources and be removed or stabilized by a number of hydrogeochemical mechanisms. For soils and sediments, colloids may be detrital (incorporation of preexisting colloids during formation of sedimentary deposits) or may be formed *in situ* as the result of geochemical alteration of thermodynamically unstable primary mineral phases or by generation of inorganic solid phases. The nature of the detrital colloidal material depends on the depositional environment and the mineral composition of the geological deposit, but can include a variety of layer silicates as well as iron- and aluminum-oxides. The presence of clay-sized material in many subsurface deposits suggests that the material is not innately mobile. The immobilization may result from aggregation of fines to form larger particles or from attachment of the colloids to the aquifer media either by electrostatic forces or by chemical cementation resulting from deposition of secondary phases such as iron oxides, carbonates, or silica. Mechanisms postulated for the generation of colloids include the following processes.

Dispersion

Colloidal particles can be dispersed and become mobile in aquifers as a result of changes in the groundwater chemistry such as a decrease in ionic strength or changes in ionic composition from calcium- to sodium-dominated chemistry. For example, introduction of low-ionic-strength water into an artificial recharge basin in a sandy aquifer in California dispersed poorly-crystallized, submicron-sized particles and caused turbidity in wells several hundred meters downgradient. The high turbidity levels were well correlated with low specific conductance of the groundwater, and treatment of the recharge area with gypsum relieved the turbidity because the increased levels of Ca^{2+} coagulated the clay colloids.[40]

Decementation of Secondary Mineral Phases

Colloidal particles may be immobilized within aquifers by secondary mineral phases that cement the colloids to each other and to the larger mineral grains. Geochemical or microbiological changes that result in dissolution of these cementing phases can result in release of colloids. For example, Gschwend et al.[64] observed 10 to 100 mg/l of silica colloids in groundwater receiving recharge from evaporation ponds and a fly-ash basin. The infiltrate was enriched in carbon dioxide and dissolved the soil-cementing carbonate mineral, thus releasing the silica colloids. Similarly, reducing conditions promoted by microbial oxidation of organic matter from overlying swamps appeared to cause mobilization of silicate clays by dissolving ferric hydroxide coatings binding the clay particles to aquifer solids.[65]

Geochemical Alteration of Primary Minerals

In crystalline formations, groundwater infiltrates through fractured systems. The fonnation (granite, gneiss, etc.) may be fractured due to tectonic stresses, formation of cooling joints (igneous rocks), or metamorphic structural changes. Alteration zones (i.e., alteration products along fracture walls) become a pathway for water in which colloids are generated by two mechanisms: microerosion of primary minerals and secondary phase production.[66] Erosion in granite begins with the resuspension within the fracture of nonbonded material, which is formed primarily by crushing of minerals due to tectonic activity within the formation. Particles can also be produced to a much lesser extent by the mechanical action of infiltrating water, as well as by the chemical dissolution of primary minerals between less soluble mineral grains. As an example, feldspar in granite is more rapidly weathered than quartz or biotite; thus, a biotite particle in contact with a feldspar or plagioclase phase may be freed when the feldspar or plagioclase is altered to clay and subsequently removed.

In addition to these classical mechanisms of primary colloid phase generation, colloids may be generated during production and microerosion of secondary mineral phases. The influx of water that is not in equilibrium with the minerals of the rock may lead to dissolution of the primary minerals and result in supersaturation of the fluid phase with respect to secondary minerals. As the fluid moves along the flow path, compositional changes occur, leading ultimately toward equilibrium of the water with the primary minerals, while a series of secondary minerals precipitate as crystalline phases (e.g., gibbsite, kaolinite, and muscovite) or amorphous phases (e.g., allopane and halloysite). Prior to precipitation, particles are produced by kinetically controlled processes including precursor nucleation and Ostwald ripening.[67] This model has been used to describe colloid generation onto the rock phase; however, it should be noted that this model can be extended to heteronucleation onto preexisting colloids within the water phase in order to describe colloid generation within the water. It must be recognized that the composition and size distribution of colloidal species is the result of complex processes. Nevertheless, characterization of groundwater colloids (size distribution and composition) may yield information about colloid stability as well as about their generation processes, indicating, for example, whether colloids are similar mineralogically to the parent rock or appear to be secondary mineral phases.

Homogeneous Precipitation

Colloidal particles can also be formed *in situ* by several mechanisms. Changes in groundwater geochemical conditions such as pH, major element composition, redox potential, or partial pressures of CO_2 can induce supersaturation and coprecipitation of colloidal particles. The precipitates can include major elements such as oxides of iron and manganese, calcium carbonates, and iron sulfides, as well as minor elements such as carbonates and sulfides of metals and radionuclides. Transient changes in groundwater chemistry associated with formation of such precipitates can occur in shallow systems influenced by variations in chemistry of recharge water, but can also occur due to microbiological activity or anthropogenic influences.

Contaminant plumes or geochemical gradients in pH, ionic composition, or organic material resulting from waste disposal activities can provide conditions that can result in formation of colloidal precipitates.[68] As examples, Gschwend and Reynolds[13] observed precipitation of ferrous phosphate colloids (1 to 10 mg/l of 100 nm-sized particles) downgradient of a sewage infiltration site. Solubility calculations suggested that the dissolved phosphate ions from the sewage and reduced iron in the aquifer exceeded the solubility product of ferrous phosphate, resulting in the formation of insoluble colloids. Other studies have documented the formation of iron oxide colloids in groundwaters as a result of changes in pH and oxygenation which caused the solubility limit of Fe(III) to be exceeded.[69,70] Many strongly hydrolyzing radionuclides also form submicron-sized particles. For example, colloidal particles of hydrated uranium oxide were formed by reducing a solution U(VI) to the less soluble U(IV) oxidation state.[12]

NOM Effects on Stability

Natural organic matter may be a critical factor in maintaining the colloidal stability of newly formed or dispersed particles, especially for colloids expected to have a net positive charge and to be inherently unstable in (generally negatively charged) aquifers. This principle of increased colloidal stability due to association of NOM with particles was demonstrated in surface waters.[71-73] Ryan and Gschwend[65] postulated that colloidal hydrous oxides of Fe, Al, and Ti in a coastal sedimentary aquifer were stabilized as suspensions in groundwater by coatings of organic carbon on the inorganic particles. Liang et al.[70] observed formation of high concentrations of stable suspensions of 200 nm amorphous iron oxide colloids following injection of oxygen into a suboxic, Fe(II)-rich aquifer. The surface potential of the colloids was negative at neutral pH, and organic carbon was detected on the particles, suggesting that it was the coatings of NOM on the particles that were responsible for the particles remaining dispersed in the groundwater.

SAMPLING AND CHARACTERIZATION OF GROUNDWATER COLLOIDS

Our understanding of the subsurface environment is limited by the techniques we use to characterize it. In no aspect of geochemical investigations is this limitation of greater consequence than in describing the nature and abundance of colloidal particles in groundwater.[74] Effective studies of groundwater colloids require correct sampling of the *mobile* material in groundwater so that suspended colloids are included, but particles that would be immobile in groundwater are excluded. Crucial factors to be considered in sampling colloids include well construction and development, pump selection and rate of groundwater sampling, purging of standing water, selection of sampling tubing, and possible alteration of samples during collection and preservation of the samples prior to analysis. Inadequate attention to these factors can lead to erroneous conclusions concerning the presence and chemical nature of colloidal particles.

Water Access Concept

Access to groundwater can be achieved in a variety of ways. Most often, a vertical bore hole is created by drilling. A solid piece of pipe is used to keep the well open in either unconsolidated materials or unstable rock. The well can be fully penetrating (constructed in such a way that it withdraws water from the entire thickness of the formation), or partially penetrating (which draws water from only a limited portion of the total thickness of the formation). The water enters the well through a well screen, a tubular device at the end of the well casing with either slots, holes, gauze, or continuous wire wrap.

Recent advances in drilling technology make it possible to change the angle of the rod during drilling so that a horizontal bore hole is created at a desired depth. The orientation of the well (vertical versus horizontal) may have significant implications to colloid sampling in fractured rock. Fractures may be important conduits for transport of colloids and small particles and are generally oriented in a vertical direction within the formation. Thus, horizontal wells have a higher probability of intercepting these zones of preferential flow than do vertical wells. There has been no systematic comparison of recovery of colloids in vertical versus horizontal wells.

Another method of accessing groundwater for colloid investigations has involved tunneling into consolidated formations in, for example, the Swiss Alps.[75,76] Tunnels involve a much larger scale of disruption to the formation and to the natural hydrology of the site. Furthermore, unlike wells that can be isolated from the atmosphere by inserting inflatable packers within the bore hole, tunnels remain open to the air. This large opening alters the total piezometric potential of the water, and thus the distribution of the water flow within the formation. In addition, the partial pressure of gases in the groundwater is altered. Introduction of oxygen can affect the redox status of the groundwater and can lead to formation of mineral oxide colloids, while degassing of carbon dioxide can disrupt the normal *in situ* pH and alter chemical equilibria of carbonate-forming species. However, a well-designed underground laboratory can have advantages in producing samples that may be more representative of the formation than those recovered from a bore hole. The geochemical and hydrological properties of subsurface environments are inherently heterogeneous, and the size of the sampling devices (e.g., a screened portion of a well) may be much smaller than the size of the natural heterogeneity (e.g., bedding planes, fractures, clay lenses, etc.) at a field site (e.g., Figure 3). A single well, therefore, cannot accurately represent the spatial variability in the abundance and distribution of colloids within natural formations; larger-scale sampling structures, such as tunnels, may have advantages in this respect. In addition, the presence of a bore hole may also modify the *in situ* hydrology by either blocking or connecting fractures or porous zones, which may in turn either produce colloids (erosion artifacts) or reduce the abundance of colloids reaching a well. Regardless of the technique used to access the subsurface environment, the limitations of the groundwater sampling method must be recognized with respect to the potential effects of those perturbations on collection and characterization of colloids.

Well Construction and Development

Drilling redistributes material, creates fine particles, and may introduce materials into the bore hole; however, some techniques create more sampling problems than others. Figure 5 illustrates some well drilling systems. Drilling techniques that utilize drilling muds (usually organic-rich slurries of bentonite clays) to lubricate the drill bit and carry away cuttings can create a zone around the well that is contaminated with colloids and fine particles which can form a barrier to flow. Water-rotary and cable tool methods are cleaner, but the large amounts of water introduced by these techniques can affect subsurface hydrogeochemistry. Augering (with casing for deeper unconsolidated formations) is less disruptive because foreign materials are not introduced and less fine material is produced than by high-speed drilling or hammering. In more consolidated sediments, air-rotary drilling carries cuttings away using compressed air and the formation groundwater, but may alter the redox conditions around the well. Well packing used around the well screen in consolidated formations may contribute colloids or may filter them out. Some of these problems can be minimized by adequate well development, which should be designed to remove particles from the vicinity of the well through relatively rapid pumping, along with occasional back pressure on the formation to dislodge trapped particles. Rates of water removal during purging of stagnant water in the well or during sampling should be *much* lower than during well development and ideally should approach the rate at which water would enter the well under natural groundwater flow conditions.[77,78]

Representative sampling of groundwater requires use of materials that retain their integrity over the entire length of the casing. Metal corrosion of carbon or galvanized steel casing can be encountered in both oxidizing and reducing conditions and can be aggravated by contact with highly saline groundwater. Stainless steel casing minimizes this concern, but even this material is sensitive to pitting in the presence of chloride ions at low pH. Polyvinyl chloride (PVC), and especially Teflon™ construction, minimizes these problems, but are not structurally adequate for deep wells.[77,78]

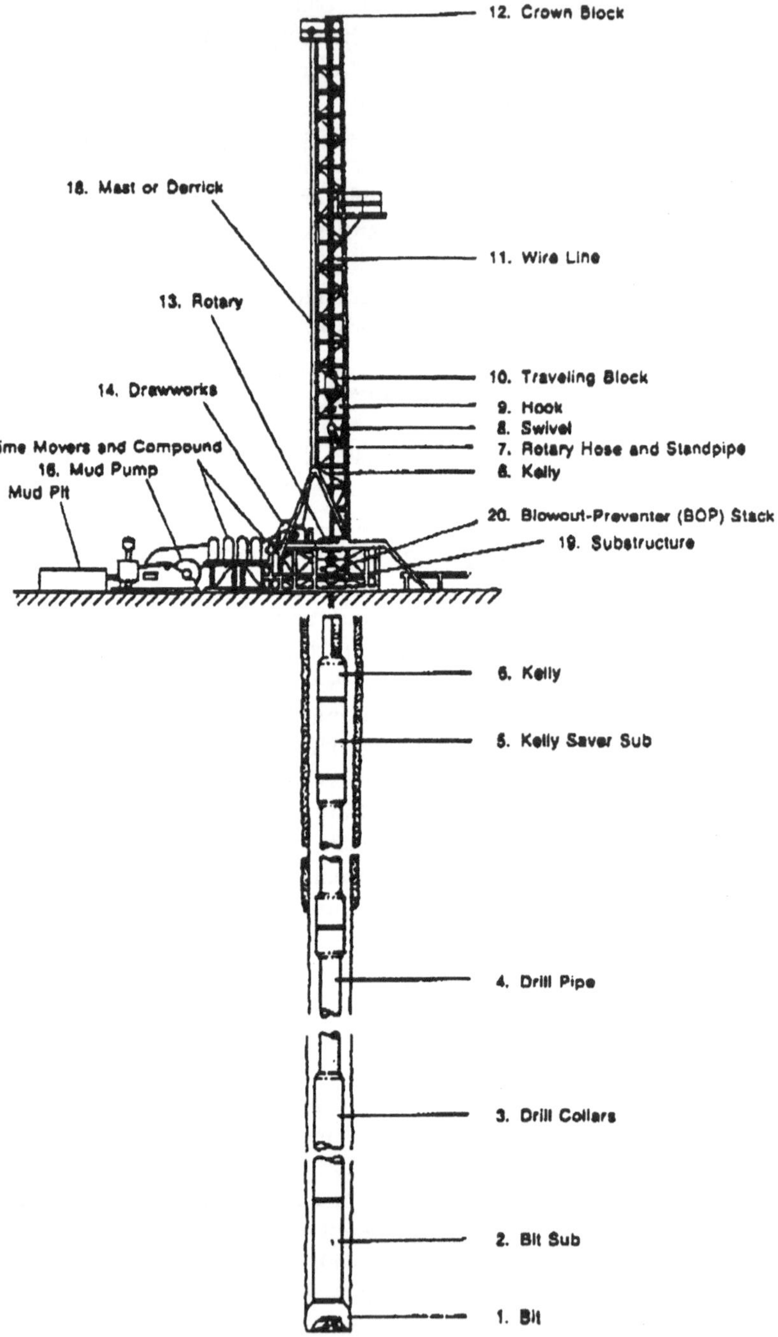

Figure 5A.

Figure 5. Examples of well drilling systems are illustrated. (A) The systems and components of a typical rotary drilling are illustrated. The typical rotary drilling method involves

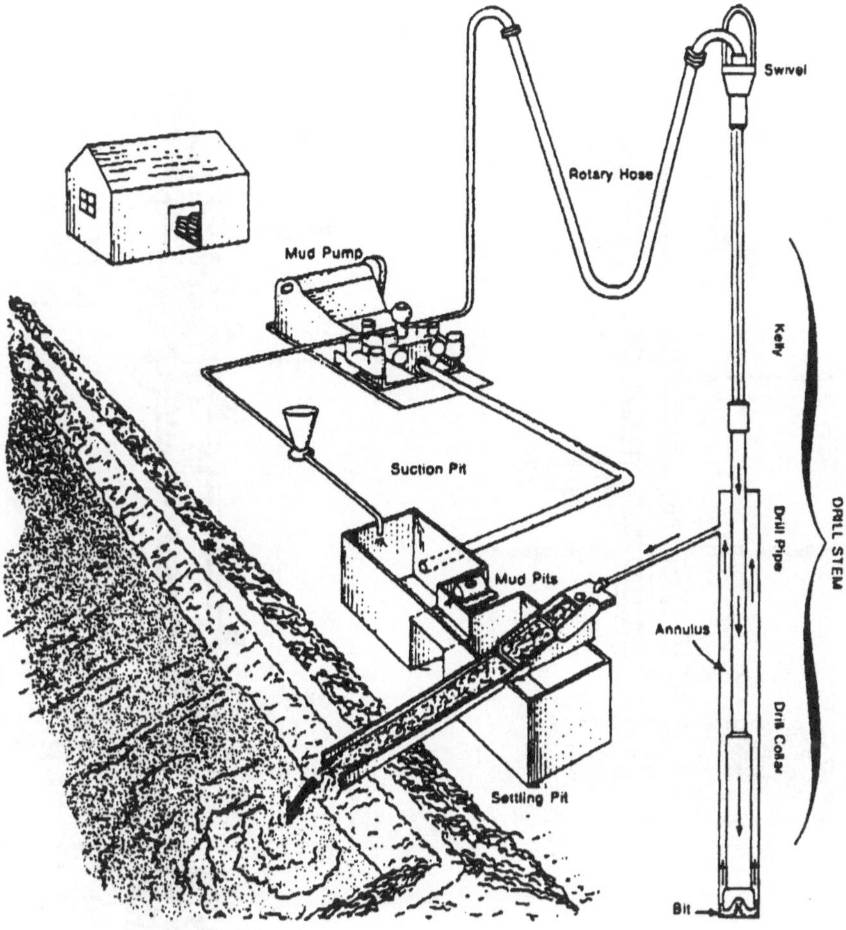

Figure 5B.

rotating a bit, and cuttings are removed by continuous circulation of a drilling fluid as the bit penetrates the formation. The bit is attached to the lower end of a drill pipe, which transmits the rotating action from the rig to the bit. In the direct rotary system, the drilling fluid is pumped down the drill pipe and out through ports or jets in the bit; the fluid then flows upward in the annular space between the hole and drill pipe, carrying the cuttings in suspension to the surface. (B) The components of a typical drilling mud circulation system are illustrated. The drilling mud is pumped down the center annulus, and the muds and cuttings are removed through the outer annulus. The cuttings are removed, and the suspended muds are pumped back down the drill pipe. (C) A dual-walled reverse circulation system is illustrated. This method is useful in loosely consolidated materials and reduces intrusion of the drilling fluids to the formation. Fluid or air is pumped down the outer annulus of a dual-walled pipe, and cuttings are removed through the inner pipe. (D) The hollow-stem auger consists of flights welded onto a larger diameter pipe with a cutter head mounted at the bottom. The drill rod passes through the center of the auger sections. The hollow-stem auger can be used as a temporary casing to prevent caving and sloughing of the bore hole wall, and permits installation of screens without using casing or drilling muds. (From Driscoll, F.G. Johnson Filtration Systems, St. Paul, MN (1986), pp. 286–333. With permission.)

Fig. 5C

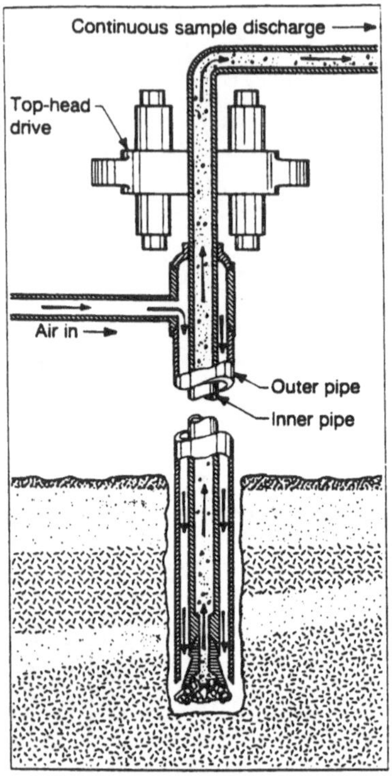

Fig. 5D

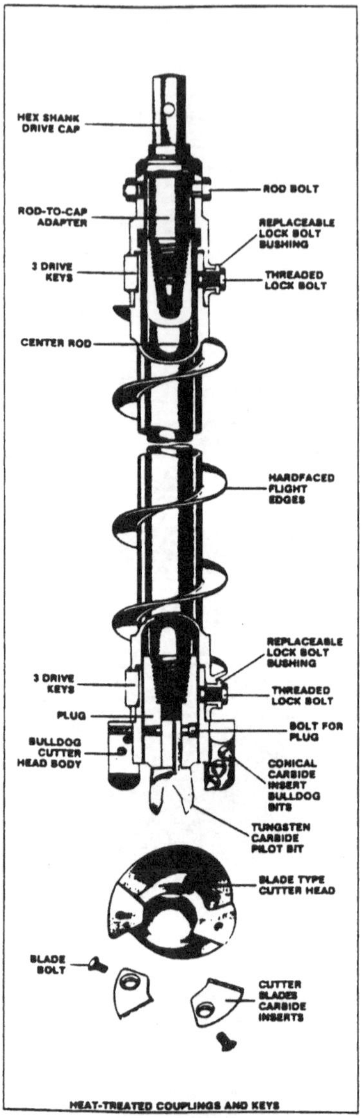

Figure 5C,D.

The well should be allowed to reequilibrate with the formation for a time sufficient to permit stabilization of chemical parameters and particle abundance. The length of time will depend on the drilling method and permeability of the formation. The rebuilding of natural *in situ* conditions by local advective flushing may take months or more to remove any bore hole fluid contamination, as well as to restore the *in situ* colloid population disrupted during well emplacement and development. For example, at the Leuggern Site in Switzerland, a deep well (1688 m below the surface) equilibrated for 3 months under natural artesian flow (i.e., groundwater flows without pumping because the pressure on the groundwater exceeds that at the opening of the sampling well) of 20 ml/min; over this period, the redox potential dropped from 0 to –200 mV and the colloid concentration dropped three orders of magnitude (10 μg/l to 10 μg/l).[79]

Fluid Sample Collection (Rate of Groundwater Purging and Sampling)

Water remaining in the well casing between samplings is likely to be unrepresentative of the formation water except when groundwater velocities in the zone of interest are relatively high. The fluid in the well bore must be removed prior to sample collection. For both well purging and for sample collection, fluid collection rates should be slow, ideally no faster than the ambient groundwater flow velocity.[80,81] The sampling zone should be isolated with packers to minimize the volume that needs to be purged and to avoid air contact with the formation water. Several methods of sample recovery are possible. Artesian flow offers the greatest potential for recovery with the closest approximation of *in situ* hydrological conditions and avoids concerns about pump selection and pumping rates. However, rapid geochemical changes can occur (such as oxidation and precipitation of metal oxides) unless the seep, fracture, or bore hole is isolated from the atmosphere. In the tunnel at the Grimsel test site in Switzerland, for example, the water flowing locally from a fracture of the migration zone was collected in a Teflon™ cup pressed into a hollow in the tunnel wall. The cup was sealed from the atmosphere with a series of Teflon™ O-rings secured with a stainless steel cross screwed into the rock.[75]

Bailing is a poor method for sample recovery because settled particles are resuspended within the wells; bailed samples consistently contained orders-of-magnitude higher levels of colloids than were recovered by slow pumping (100 ml/min) using a bladder pump within the same well and after the same purging times.[80,81]

Groundwater samples are most often recovered by pumping, but the recovery of colloids is affected by the type of pump and by the rate of pumping. Increased shear in the flow velocities around the well screen may resuspend settled particles and detach submicron particles.[82] Ryan[83] calculated the shear rate produced by pumping at 1 and 4 l/min (assuming a 5 cm in diameter well screened over 1 m length in an aquifer with a porosity of 0.4) exceeds that required to detach TiO_2 particles from glass[84,85] or to break up clay aggregates.[86] Pumping at 0.1 l/min does not produce such shear rates and should, therefore, produce samples that contain only naturally suspended colloids. Puls et al.[87] compared pump types (bladder and low- and high-speed submersible pumps) and pumping rates (0.6 to 92 l/min) to the abundance of colloids in groundwater samples recovered from a series of wells in an unconsolidated alluvium from a granite porphyry in Arizona.

Colloid concentrations took longer than other field parameters to stabilize during purging of the wells: about 50% longer than dissolved oxygen or redox potential and about twice as long as specific conductance, pH, or temperature (Figure 6). Colloid concentration (estimated by light scattering) stabilized at 0.1 mg/l during pumping at 1.1 l/min, but increased to 0.7 mg/l when the pumping rate was increased to 3.8 l/min before stabilizing again at 0.1 mg/l. When the pumping rate was increased to 30 l/min, colloid concentrations jumped to 4.4 mg/l before stabilizing at 0.2 mg/l; the highest pumping rate produced colloids of larger size than were recovered at the lower discharge rates (Figure 6).

Sampling Tubes

Sample tubing has intimate contact with the groundwater, and the slow pumping rates recommended for colloid sampling can increase this contact time even further. In addition to concerns about sample alteration due to corrosion, sorption, or leaching, artifacts resulting from diffusion of gas across the tubing must be considered, especially for deeper formations and slow pumping rates. For example, while Teflon™ is valued for its low adsorptive capacity and inertness,[88] Holm et al.[89] confirmed the transfer of oxygen and carbon dioxide across Teflon™ sample tubing and calculated that at low flow rates, changes in Fe speciation (and, for example, possible precipitation of iron oxide colloids) could occur as a result of oxygen diffusion through 16 m of sample tubing used to sample initially anoxic or suboxic waters. The amount of gas transferred depended on the tubing length and was inversely related to the pumping rate.[89] The effect may be minimized if the water transfer line is encased in a nitrogen-purged outer tube, encased in a less permeable outer jacket (such as latex rubber) or by trading the advantages of Teflon™ for a less permeable material such as PVC or nylon.[89]

ANALYSES OF GROUNDWATER COLLOIDS

Analytical Situation

The location where any analyses are performed is an important consideration because it affects several factors that control the stability of a colloidal phase. Three main sampling/characterization sites are defined: *in situ*, on-site, and in the laboratory. The initial *in situ* state of the groundwater is described by thermodynamic data (temperature, pH, total pressure on the fluid phase, partial pressure of specific gases such as oxygen and carbon dioxide, redox potential, and the molar concentrations of dissolved species) and hydrologic data (such as the piezometric gradient, porosity, water flow velocity), as well as the mass of colloids in a given volume of groundwater and the specific size distribution of colloids. It is the goal of colloid sampling to minimize changes in those parameters which may affect the abundance or properties of colloidal particles.

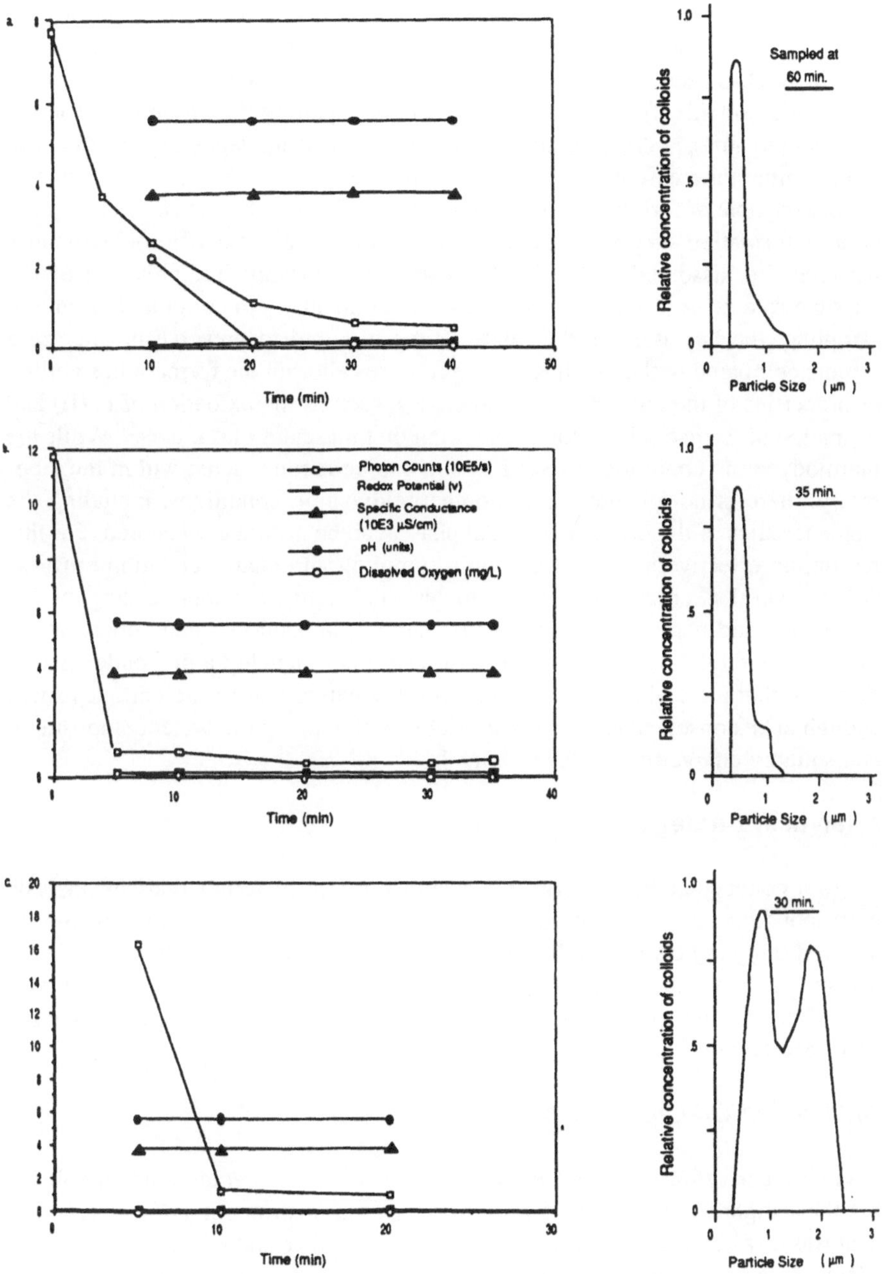

Figure 6. Changes in water quality indicators during purging of Well 503, Pinal Creek, Arizona (Table 6) are shown for: (a) bladder pump (1.1 l/min); (b) low-speed submersible pump (3.8 l/min); (c) high-speed submersible pump (30 l/min). Particle size distribution was estimated using photon correlation spectroscopy. (From Puls, R. W., J. H. Eychaner, and R. M. Powell. *Haz. Waste Haz. Mater.* 9(2): in press (1992). With permission.)

In situ thermodynamic conditions can, as discussed above, be significantly disrupted by sampling. Introduction of oxygen alters the redox chemistry of the system and has been shown to result in the rapid oxygenation and precipitation of iron oxide colloids.[69,70,79] In deeper formations, opening the groundwater in the bore hole to atmospheric pressure can result in significant degassing of CO_2, with a concomitant increase in pH as well as a disruption of the chemical speciation of cations capable of forming carbonate complexes. Both of these changes can cause colloid formation and alter the distribution of some metals or radionuclides between the dissolved and colloidal phase, thus biasing interpretation of the possible role of *in situ* sorption of these contaminants onto colloids. For colloid sampling, the key issue is the relative time required to preserve or analyze a sample compared to the kinetics of the processes altering the formation, stability, or properties of the colloids; some processes, such as the oxidation of Fe(II) and formation of ferric oxide colloids, can occur on timescales of minutes.[90] While the thermodynamic conditions existing *in situ* can be reconstructed within the laboratory, there is no guarantee that nonthermodynamic conditions, including the characteristics of the original colloidal phase, can be accurately restored. For this reason, an effective strategy for colloid sampling and characterization must involve several alternate approaches to test and confirm understanding of the abundance and nature of colloids under natural, undisturbed conditions. While this ideal is rarely achieved, it is at least possible to develop a preponderance of evidence that the colloid observations are reasonable; that is, are particles small enough to be nonsetteable, suitably insoluble within the *in situ* water composition, and suitably charged to confer stability?

Analytical Strategy

As a conceptual framework, it is useful to identify four modes of analysis corresponding to where the sample is collected and analyzed. In a well-constructed sampling campaign, it is advantageous to collect and analyze samples in several ways to improve and confirm interpretation of the data. The four modes of analysis (at-line, off-line, on-line, and *in situ*) are illustrated in Figure 7, and are described below.

At-Line and Off-Line Analysis

At-line and *off-line* approaches are distinguished by the requirement of collecting the sample and transporting it to the measuring instrument. *Off-line* refers to analyses of a sample at a centralized laboratory facility and offers advantages of efficiency and economy as well as the analytical benefits associated with the use of sensitive and sophisticated instrumentation. The obvious disadvantage is the delay between sampling and analysis, which can have a devastating effect on the colloidal phase (e.g., aggregation and growth of colloids). It is these disadvantages that encourage *at-line* analysis using dedicated instruments installed in close

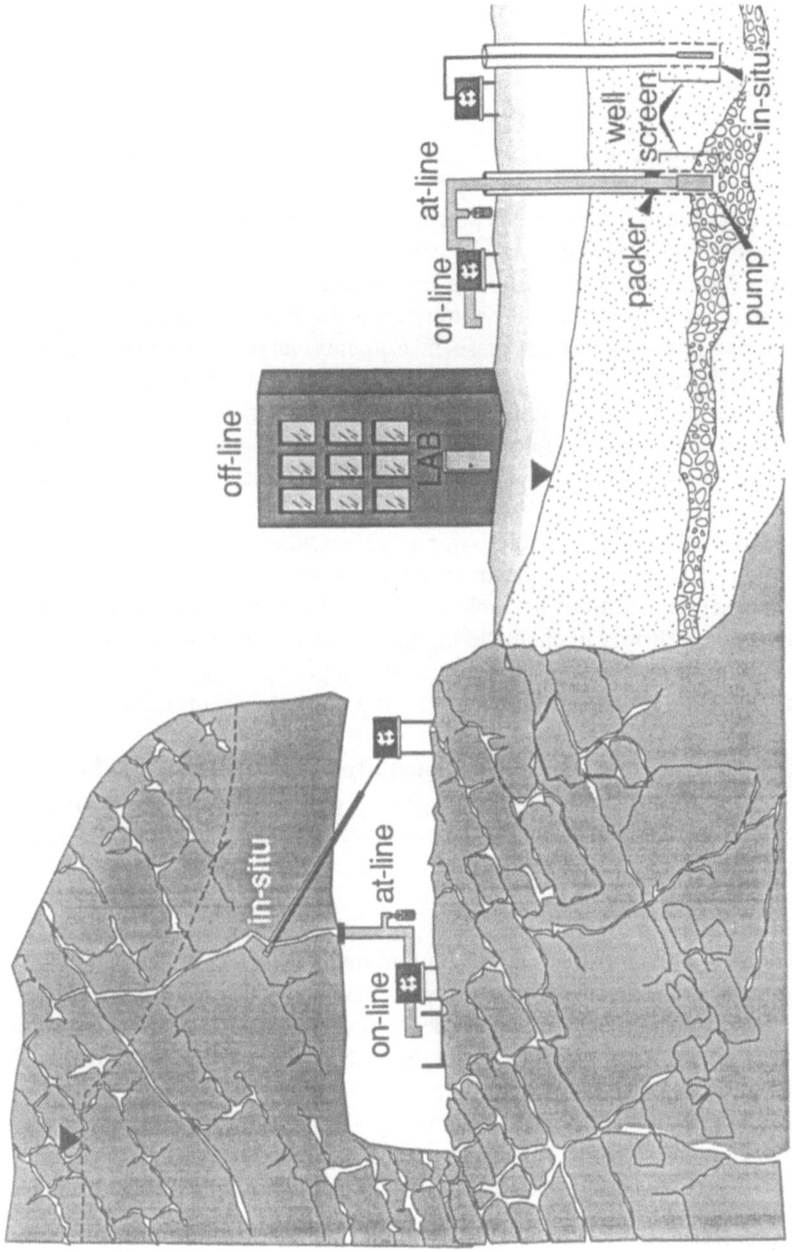

Figure 7. Different modes of sampling groundwater for colloidal particles are illustrated. The situation shown on the left involves sampling deep groundwater within a tunnel. Sampling from wells is illustrated on the right. Note that the screened interval of one of the wells intercepts a coarse gravel lens which may be a zone that preferentially transports colloids. The other well, illustrated with an *in situ* downhole detection device, is screened at the same depth but does not intercept the same gravel lens. Even in wells located in proximity to each other, physical heterogeneities in the subsurface may affect the nature and abundance of recovered colloids.

proximity to the groundwater sampling line. Advantages of this mode of analysis include rapid analysis following collection of the sample and closer control of the analysis by the field researchers. In addition, immediate analysis of real-time data makes it possible to adjust the sampling protocols and improve the quality and reliability of the data.

On-Line Analysis

The *on-line* approach involves analyses of the sample before it exits the groundwater sampling tube. Flow-through cells can measure basic thermodynamic parameters such as pH, electrode potential, dissolved oxygen, and temperature. For some analyses, the sampling tube can be attached directly to the analytical instrument. For example, the sampling line from the well can be attached to the flow-through cuvette of a spectrophotometer or fluorometer, or the cuvettes for light-scattering or microelectrophoresis instruments can be modified to accept a sample directly from the sampling tube. Particles can be collected using on-line filters for determining their mass and composition or for visualization and surface analysis using scanning electron microscopy with chemical analysis. Automated sampling systems have been devised to extract the sample, condition it, and transfer it to the analytical instrument for measurement. For example, Degueldre et al.[75,91] used a small (8 ml) filter cell to minimize possible sorption surfaces and developed a light-triggered relay system to automatically inject and repressurize the cell with sequential additions of 5 ml volumes of fresh groundwater. The gas for pressurizing the cell was in chemical equilibrium with the groundwater, and a final rinse of ultrapure water was used to prevent salt precipitation during drying. Volumes ranging from 0.5 to 1000 ml were filtered through a 1 cm^2 filter surface using this method.

On-line analysis can be conducted continuously by monitoring groundwater flowing through an instrument (e.g., on-line particle counting), or on an intermittent basis with occasional injection of the sample stream into the instrument. This sampling mode has the advantage of minimizing changes in the physical and chemical state of the sample prior to analysis.

In general, on-line and off-line analyses differ in their purposes and endpoints. Analyses at central laboratories are quite appropriate for elucidating the chemical composition of the groundwater or of a colloidal mixture (collected, for example, on a filter). On-site analyses are directed at describing phenomena that are subject to alteration due to collection and storage. Rapid analysis at the field site is much preferred for measurements of thermodynamic parameters (including temperature, dissolved oxygen, alkalinity, pH, electrode potential, and further quantification of redox status through field measurement of key redox couples) as well as for analysis of the abundance and nature of colloidal particles (using field-portable instrumentation and techniques such as light scattering or filtration). Recently, new approaches promise even greater improvement in measurement of the natural state of a groundwater system by permitting observation without the need to remove the sample from the well. This constitutes the fourth mode of colloid characterization, which follows.

In Situ Analysis

In situ analysis involves measurements obtained directly inside the bore hole using probes or fiber optic sensors for measuring particle size by light scattering, or detecting the presence and concentrations of chemicals (particularly contaminants) using absorption or fluorescent spectra or Raman emission. Most work in instrumentation for *in situ* analysis has been focused on detection of contaminants in groundwater monitoring wells but could be adapted readily for detection and characterization of NOM or organically coated particles in groundwater.[92,93]

Sample Processing and Analyses at the Field Site

Tables 2 and 3 summarize the various techniques available for colloid characterization. The analytical framework for their application is described in this and the following section. At least partial characterization of the groundwater colloids can be performed at the field site using noninvasive techniques (i.e., without removal of water) such as light scattering (static or dynamic) or single-particle counting (Tables 2 and 3). More often, however, classic separation techniques (micro- or ultrafiltration) are used to concentrate the colloidal particles in the liquid phase for additional analyses or to collect the colloidal particles on the filter membranes or other sample carrier for microscopic examination and chemical analysis in the laboratory. Ideally, the filtration apparatus would be connected directly with the sampling line, carefully rinsed with ultrapure water, and, if the groundwater is not oxic, purged with inert gas to prevent contact with oxygen.

For many groundwater sampling situations, the groundwater is not oxic, and care must be exercised to prevent alteration of the *in situ* redox state of the sample. Liquid samples can be collected anoxically for on-site or laboratory analyses using argon- or nitrogen-filled glass vials or bottles fitted with a Teflon™-coated silicone rubber septum seal (thick butyl rubber plugs are more effective barriers to oxygen than the septa, but contact of the sample with the rubber may alter the sample). The groundwater sampling line can be attached to a hypodermic needle to inject the sample through the septum, and the inert gas can be vented through a second needle. To further minimize opportunities for diffusion of oxygen into the sample, the vial should be filled to eliminate all headspace and then submerged in cool water until analyzed to reduce diffusion of air through the septum. Because oxygen diffuses more readily through a gaseous phase, these precautions impose a barrier of water and the rubber septum to protect the sample from air. The absence of a gaseous headspace further limits diffusion of air through the septum and into the sample.[83,94]

Aseptic techniques may be required to prevent introduction of nonindigenous bacteria during sample collection, especially if the samples are not refrigerated. Not only will the bacteria produce artifacts detected as increasing populations of "biocolloids," but they may alter the redox status of the sample and consume, or at least alter, any natural organic matter present in the sample.[95] Modification of the sample by native groundwater bacteria can also be a concern but can be controlled by refrigeration or by poisoning the sample (although the possible effects of the poison on the sample must be considered).

Treatment of anoxic samples can be conducted in glove boxes, but this cumbersome apparatus can often be eliminated by sample manipulations using gas-tight glass syringes and argon-filled dilution or reagent bottles with septum seals. For example, rapid analyses of redox couples in the field are important data for confirming and understanding the electrode potential. This can be particularly important in natural systems since the redox reactions in groundwater are generally in disequilibrium, and the apparent redox potential measured by the electrode is represented by the dominant redox couple. Field portable reagent kits are available which can permit rapid and simple colorimetric analysis of important couples [e.g., Fe(II) and Fe(III)].[70,96] These measurements can be performed without a glove bag by using syringes for transfer and dilution of the sample and by assuring that a stream of nitrogen or argon is passed over the vials when the reagent powder is introduced into the sample.[94] In a shallow, suboxic, ferrous-iron-rich aquifer in South Carolina, the measured redox potential agreed well with that calculated from the Fe(II)/Fe(III) couple measured using these field kits as described above.[70] However, measurements of redox potential may take months before constant potential and equilibrium are established at the electrode when the redox-sensitive species are present at very low concentrations.[76,97]

In general, evaluation of the precautions required to prevent modification of the sample requires consideration of the mass of potentially unstable constituent in the sample [e.g., Fe(II)], compared to the quantity of contaminant (e.g., oxygen) that can be introduced into the sample between the time of collection and analysis.

Colloid Characterization

Many of the methods available for analysis of environmental colloids are outlined in Tables 2 and 3. Unfortunately, there is no "Colloidograph" which produces in-line simultaneous information on the size distribution, concentration, and composition of the original colloidal phase. All that can be done is to adapt the on-, at-, and off-line analytical systems so that a reasonable facsimile of the *in situ* colloidal phase is assembled from the combination of techniques for analyzing discrete particles (in solution or on membranes; Table 2) or for analyzing the bulk composition of either the unaltered groundwater, liquid concentrates produced by cross-flow or tangential-flow filtration, or dried particles retained on micro/ultrafilters or other sample carriers (Table 3).

Table 2A. Methods Used To Characterize Groundwater Colloids By Discrete Particle Analysis: Single-Particle Analysis in the Fluid Phase

Method	Analytical Principle	Information Provided	Mode of Analysis	Size Analyzed (nm)	Minimum Concentration[a] (Particles/ml)	Advantages/ Limitations	Refs.
Optical microscopy	Imaging of drop of water in light beam	Particle number and morphology	At-line, off-line	≥200	$\geq 10^4$	Easy/Limited resolution at small sizes	
Single-particle counting/ electrical impedance	Change in impedance due to particle occluding an aperture within an electrolyte solution	Number and size distribution of particles	Off-line	≥500	10^7	Good for polydispersed suspension/Possible coagulation in electrolyte solution; limited to larger particles	116
Single-particle counting/flow cytometry	Single-particle analysis using light scattering, absorptiontion, or fluorescence	Particle number, size, and spectral properties (including concentration of adsorption sites for fluorescent substances on particles)	Off-line	≥1000	10^7–10^8	Size analysis and sorting of particles for additional analyses/ Needs high levels of large particles	117–120
Single-particle counting/light scattering	Light scattering from a single particle in a laser beam	Number and size distribution	On-line; possibilities for *in situ*	≥50	10 to $\geq 10^5$	Good for polydispersed systems, including small colloids/May require dilution of sample	76, 121
Inductively coupled plasma-atomic emission spectroscopy	Emission spectroscopy of the light flash from a single atomized particle	Elemental analysis; particle concentration; relative size	Off-line	≥500	10^3	Sensitivity varies between elements	98, 121

Table 2A. (continued) Methods Used To Characterize Groundwater Colloids By Discrete Particle Analysis: Single-Particle Analysis in the Fluid Phase

Method	Analytical Principle	Information Provided	Mode of Analysis	Size Analyzed (nm)	Minimum Concentration[a] (Particles/ml)	Advantages/ Limitations	Refs.
Inductively coupled plasma-mass spectroscopy	Mass spectroscopy of ion flash from single atomized particle	Elemental and isotopic analysis; concentration; relative size	Off-line		10^3	Sensitivity varies between elements	99
Micro-electrophoresis	Optical microscopic measurement of rate and direction of colloid movement in an electrical field	Average charge of surface of colloid	At-line; off-line	>200 (Depends on refractive index particle)	$\geq 10^4$	critical for evaluating colloid stability/No information on micro-heterogeneity of surface charge	105

[a] The relationship between particle number and the mass concentration of colloids (μg/l) is given in Table 4.

Table 2B. Methods Used To Characterize Groundwater Colloids By Discrete Particle Analysis: Analysis of Single Particles on a Filter or Sample Carrier

Method	Analytical Principle	Information Provided	Mode of Analysis	Size Analyzed (nm)	Minimum Concentration[a] (Particles/ml)	Advantages/ Limitations	Refs.
Optical microscopy	Imaging in light beam	Particle number and morphology; fluorescence microscopy for enumerating stained bacteria	At-line, off-line	≥200	$\geq 10^4$	Easy; microbial enumeration on formalin-preservedsamples/ Limited resolution at small sizes	123, 124
Scanning electron microscopy (SEM)	Electron-beam surface imaging of particles retained on membrane filter using detection of emitted secondary electrons	Particle number, size, morphology; possible to link with microchemical analyses by EDS	Off-line	20–300,000	10^5	Visualize and count heterogeneous particles/Possible artifacts due to vacuum-coating	107, 125
Transmission electron microscopy (TEM)	Electron-beam imaging through sample on a grid film	Evaluation of particle number and accurate determination of size; possible to link with microchemical analyses	Off-line	≥1	10^8	Excellent resolution at small sizes/Images only electron opaque structures in 2-dimensions; difficult to count absolute number of particles; potential for artifacts in sample preparation	75, 91
Energy dispersive spectroscopy (EDS) (with SEM or TEM)	Analysis of X-rays generated from electron interactions with sample surface	Elemental analysis of particle imaged in SEM or TEM	Off-line	>1000 for SEM; >10 for TEM	Like SEM; Like TEM	Detects only major (not trace) constituents	13, 65

Table 2B. (continued) Methods Used To Characterize Groundwater Colloids By Discrete Particle Analysis: Analysis of Single Particles on a Filter or Sample Carrier

Method	Analytical Principle	Information Provided	Mode of Analysis	Size Analyzed (nm)	Minimum Concentration[a] (Particles/ml)	Advantages/ Limitations	Refs.
Electron energy loss spectroscopy (EELS) (with TEM)	Detect change in energy of electrons	Elemental analysis of particle	Off-line	≥5	Like TEM	Difficult to interpret	91
Electron microprobe analysis (EMPA)	Electron-beam surface imaging with detection of the wavelengths of emitted X-rays	Quantitative mapping of elemental composition	Off-line	≥1,000–2,000		Elemental mapping; quantitative analysis/Poor resolution compared to SEM	
Laser microprobe mass analysis (LAMMA) (transmission mode on TEM sample carrier)	Photoionization and time-of-flight mass spectral analysis	Chemical and isotopic analysis of single particle	Off-line	>500	Like TEM	Very sensitive; better size resolution than LAMMA in reflectance mode/Ionization is a function of energy of incident beam and must be standardized; interference from ionization of sample carrier	91
LAMMA (reflectance mode on membrane filter)	Laser photoionization with back-reflected ion and time-of-flight mass spectral analysis	Chemical and isotopic analysis of single particle	Off-line	>5,000	Like SEM	Very sensitive/Ionization is a function of energy of incident beam and must be standardized; less size resolution than LAMMA in transmission mode	91

Laser ablation ICP-MS	Photoionization of sample surface with molecular analysis	Elemental and isotopic	Off-line	>10,000	Like SEM	Ionization is a function of energy of incident beam and must be standardized	
Scanning tunneling microscopy	Tip of probe "floats" over sample; motion detected piezoelectrically	3-Dimensional topography of electrically conducting surface; number and size of particles	Off-line	0.1–1,000	Function of sample prepara-tion	Atomic resolution/ Limited to electrically conductive surfaces	126, 127
Atomic force microscopy	Vertical motion of probwe measured as tip "rides" on surface of sample	3-Dimensional topography of even nonconducting surface; number and size of particles	Off-line	0.11–1,000	Function of sample prepara-tion	Can image nonconduc-ing surfaces/Probe can move colloid	128, 129

[a] The relationship between particle number and the mass concentration of colloids (μg/l) is given in Table 4. Estimates of the colloid population required for microscopic investigations are indicated in Table 5.

Discrete particle analysis provides information on the number, size, and composition of heterogeneous populations of colloidal particles (Table 2) and can be used to estimate mass concentration of colloids (Table 4). The unaltered groundwater, or a liquid concentrate produced by filtration, can be observed directly on a glass slide using optical microscopy, but the resolution of light microscopy limits its utility to larger-sized particles. Several methods of single-particle counting are available but may be limited to larger-sized particles (flow cytometry and electrical impedance methods) or require counting in an electrolyte solution that may promote aggregation of particles (electrical impedance method). For natural systems (i.e., generally polydispersed suspensions that can include small colloids), the number and size distribution of particles can be determined more accurately using single-particle counting methods based on light scattered by individual particles in a stream of groundwater passing through a light beam. The procedure can be conducted on-line, but requires dilution in ultrapure water (e.g., produced fresh on-site)[76] in order to achieve a concentration in which only a single particle at a time is present in the stream passing through the laser beam. Dilution to low ionic strength also helps to stabilize particles.

The diluted groundwater can be returned to the laboratory for ICP analysis of single particles (Table 2A). For single-particle ICP, the nebulized fluid sample in the plasma phase produces a flash of light or ions for each particle. The light flash is analyzed as a function of time for a single element; for a given emission line, the total number of photons is a function of the amount of that element in the individual particle, and the frequency of a given photon flash is a function of the particle concentration. Model $MnCO_3$ colloids of 500 nm have been successfully analyzed in this way.[98] Alternately, ions yielded from a single particle in the plasma can be analyzed in a mass spectrometer for determination of the number and isotopic composition of individual particles; this method has been tested for detection of 300 to 3000 nm quartz colloids, but it is reasonable to expect a better detection limit for more readily ionizable and detectable elements such as silica.[99]

Particles collected on filters (or other sample carriers) can be visualized and analyzed individually using optical or scanning electron microscopy (Tables 2 and 5) for determining the size distribution of particles. Optical microscopy is useful for observing and enumerating bacteria, especially when the viable microbiota are stained, for example, with acridine orange which causes living cells to fluoresce in a background of other nonliving particles. The chemical composition of discrete particles can be determined using micro- or nanobeam analyses, such as energy dispersive X-ray analysis, X-ray photoelectron and auger electron spectroscopy, laser ablation ICP-MS, or laser microprobe mass analysis (in either transmission mode if the sample is mounted on a TEM sample carrier, or in the reflection mode if the sample is on a membrane). Advanced scanning probe microscopes (scanning tunnelling or atomic force microscopes) provide information on the three-dimensional topography of individual particles at the atomic scale (Table 2B).

Additional information about the concentration and chemical composition of groundwater colloids is available from bulk analysis of particles in liquid concen-

Table 3A. Methods Used for Bulk Analyses of Particle Population: Analyses of Fluid Phase (either unaltered groundwater or concentrates produced by cross-flow or tangential filtration)

Method	Analytical Principle	Information Provided	Mode of Analysis	Size Analyzed (nm)	Minimum Concentration[a] (mass (μg) or particles/ml)	Advantages/ Limitations	Refs.
Ultracentrifugation	Gravitational sedimentation with detection by adsorption or light scattering	Estimate of size distribution of particles and macromolecules based on sedimentation coefficient	Off-line	Molecules to 10,000	Depends on scattering or UV absorption of sample	Eliminates artifacts due to adsorption on membranes/Results results can vary with patticle density	75, 130
Turbidity	Light scattering	Estimate of particle concentration	At-line, on-line	80–800	10^9 (Assuming 300 nm colloids)	Inexpensive; easily run in flow-through mode/No size distribution	80, 83, 87
Static light scattering	Average scattering at different angles	Average particle concentration	Off-line, at-line, on-line, possibly *in situ*	3–600	10^9 (Assuming 600 nm colloids)	No size distribution without filtration and calibration	75
Photon correlation spectroscopy (PCS)	Dynamic light scattering in fluid phase	Size distribution of particles; may require micro/ultra-filtration reduce heterodispersity	Off-line, at-line, on-line, possibly *in situ*	3–10,000	$>10^7$ (Assuming 100 nm colloids)	Field-portable/ Inherent limitation in data analysis of ploydisperse or non-spherical particles	8, 70, 75, 83, 87
Microelectrophoresis	Rate and direction of colloid movement in an electrical field, based on laser light scattering	Average charge of surface of colloid populations	At-line; off-line	>40 (Depends on refractive index of particle)	$\geq 10^4$	Critical for evaluating colloid stability/ No information on charge of individual particles	65, 70, 75, 105

Table 3A. (continued) Methods Used for Bulk Analyses of Particle Population: Analyses of Fluid Phase (either unaltered groundwater or concentrates produced by cross-flow or tangential filtration)

Method	Analytical Principle	Information Provided	Mode of Analysis	Size Analyzed (nm)	Minimum Concentration[a] (mass (μg) or particles/ml)	Advantages/ Limitations	Refs.
Diffraction spectroscopy	Low angle light scattering	Size distribution	At-line	4,000–400,000	10^4 (Assuming 10,000 nm particles)	Rapid/Limited resolution	131
Laser-induced photoacoustic spectroscopy (LIPAS)	Energy absorption with acoustic detection	Average particle concentration	Off-line	≥1	10^7 (Assuming 100 nm colloids)	Can provide information on speciation/No size distribution without filtration and calibration	132, 133
Total organic carbon (TOC) analysis	Oxidation of organic carbon to CO_2, usually with infrared detection of CO_2	Concentration of dissolved, colloidal, or particulate carbon	On-line; off-line	Depends on pore size of filter membrane	≥0.01 mg carbon/l (minimum concentration limited primarily by possibilities of contamination)	Identify NOM as potential sorbent, cation complexant, or electrostatic stabilizer/Potential underestimation of TOC unless high-temperture combustion technology is used	76, 91, 94

[a] The relationship between particle number and the mass concentration of colloids (μg/l) is given in Table 4.

Table 3B. Methods Used for Bulk Analyses of Particle Population: Analysis of Filter Membranes of Sample Carrier Containing Deposited Particles

Method	Analytical Principle	Information Provided	Mode of Analysis	Size Analyzed (nm)	Minimum Concentration[a] (mass (μg) or particles/ml)	Advantages/ Limitations	Refs.
Gravimetry	Mass	Mass of particles deposited on filter membranes	Filtering on–line or off-line; analysis off-line	Determined by pore size of filter membrane	10 μg of filtered solids	Direct determination of mass/Subject to filtration artifacts	62, 75
Neutron activation analysis	Nuclear activation and detection	Elemental and isotopic analysis	Filtering on-line or off-line; analysis off-line	Determined by pore size of filter membrane	Varies between isotopes	Sensitive, specific	135, 136
Pyrolysis gas chromato-graphy	Partial combustion and chromato-graphic separation of products	Semiquantitative information on chemical composition of organic particles or coatings	Filtering on-line or off-line; analysis off-line	Determined by pore size of filter membrane	10 μg	Structural information on organic matter; sensitive/Interpreta-tion is complex	136
X-ray diffraction	X-ray scattering	Mineralogy of crystalline colloids on filter	Filtering on-line or off-line; analysis off-line	Filterable particles ≥100	100 μg of filtered solids	Identify crystalline colloids	62, 65, 70
Infrared spectroscopy	Absorption of infrared light	Mineralogy of crystalline and amorphous colloids on filter; chemical compostion of organic particles and coatings	Filtering on-line or off-line; analysis off-line	Filterable particles	<1 μg of filtered solids	Difficult to calibrate for quantitative analysis; moderately sensitive/Information on chemical structure	136, 144

Table 3B. (continued) Methods Used for Bulk Analyses of Particle Population: Analysis of Filter Membranes of Sample Carrier Containing Deposited Particles

Method	Analytical Principle	Information Provided	Mode of Analysis	Size Analyzed (nm)	Minimum Concentration[a] (mass (μg) or particles/ml)	Advantages/ Limitations	Refs.
X-ray photo-electron and auger electron spectroscopy (XPS and AES)	Electron spectroscopy	Semiquantitative analysis of surface chemical composition and redox	Filtering on-line or off-line; analysis off-line		Like gravimetry	Chemical analysis of individual particles	91, 137

[a] The relationship between particle number and the mass concentration of colloids (μg/l) is given in Table 4.

Table 3C. Methods Used for Bulk Analyses of Particle Population: Analyses of *Either* Fluid Phase *or* Dissolved Filter Membranes with Deposited Particles

Method	Analytical Principle	Information Provided	Mode of Analysis	Size Analyzed (nm)	Minimum Concentration[a] (mass (μg) or particles/ml)	Advantages/ Limitations	Refs.
Inductively coupled plasma-atomic emission spectroscopy (ICP-AES)	Aerosol atoms detection in plasma torch by emission spectroscopy	Elemental analysis of concentrate from filtration	Filtering on-line or off-line; analysis off-line	Depends on pore size of filter membrane	μg/l level (in general)	Sensitive and specific multielement analysis	103
Inductively plasma-mass spectroscopy (ICP-MS)	Aerosol atoms detection in plasma torch by mass spectroscopy	Elemental and isotopic analysis of concentrate from filtration	Filtering on-line or off-line; analysis off-line	Depends on pore size of filter membrane	ng/l^{-1} level (in general)	Very sensitive and specific/Limited experience in groundwater colloid studies	104
Nuclear spectroscopy (α or γ counting)	Passive detection of spectrum of energy emitted upon decay of a radioelement	Activity of radioactive elements associated with the colloidal phase	Like ICP	Like ICP	Function of the radioelement	Specific activities of radionuclide; implications to contaminant transport; can detect natural fall-out radionuclide tracers/May require sample concentration or long acquisition times	4, 59

[a] The relationship between particle number and the mass concentration of colloids (μg/l) is given in Table 4. Estimates of the colloid population required for microscopic investigations is indicated in Table 5.

Table 4. Relationship Between Mass Concentration and Number Concentration for Populations of Different-Sized Colloids

Colloid Diameter (nm)	Colloid Number (pt/l)	Colloid Mass (µg/l)
10	10^{12}	1
10	10^{11}	0.1
10	10^{10}	0.01
100	10^{10}	10
100	10^{9}	1
1000	10^{9}	1000
1000	10^{8}	100

Note: The mass of colloids (c_m; µg colloid/l) is related to the number of particles (c_n; particles/l) as a function of particle density (ρ)and particle diameter (d):

$$c_m = c_n \frac{\pi \rho d^3}{6}$$

The average particle density is assumed to be 2.0, taken as an average for clay, hydrated silica, and iron hydroxide.

trates or deposited on membranes or other sample carriers (Table 3). The classic protocol includes concentration of the colloidal phase at the field site using a series of filters with different pore sizes to produce a series of either liquid concentrates or filters on which the colloids have been deposited. It must be recognized that this approach is based on several assumptions,[75,100] including colloids larger than the nominal pore size of the filter are retained totally and no smaller colloids are retained; no reactions other than straining filtration occur between the colloid and the membranes (i.e., collection due to electrostatic or chemical interactions is ignored); the colloids retained on the filter adhere strongly enough to prevent losses before analysis; and in the case of SEM analyses, sample preparation does not alter the membrane or the colloids. Clearly, these assumptions are not always valid, and data must be evaluated and interpreted with a cautionary awareness of these caveats. Size distribution derived solely from filtration data must often be viewed as semiquantitative. Nevertheless, micro- and ultrafiltration and analysis of the filtrate and retained phases remain as one of the principal techniques for determining the concentration, size distribution, and chemical composition of natural (i.e., polydispersed) colloids.

If liquid concentrates are produced or if colloids are collected on filter membranes for analysis, it must be recognized that the composition of the concentrates or deposited material can be affected by ion retention for filters of even rather large size cut-offs.[101] This is mostly due to electrostatic interaction between ions and the charged surface of the loaded membrane, which will be affected by the salt concentration in the water. To recognize and account for this potential artifact, liquid concentrates should be prepared at several concentration factors, and colloids should be collected on membranes from different volumes of groundwater.

Table 5. Particle/Colloid Population Required for Microscopic Investigation

Grid for Transmission Electron Microscopy[a]			Membrane for Scanning Electron Microscopy[b]			Membrane for Optical Microscopy[c]		
Volume Filtered (μl)	Colloid Diameter (nm)	Colloid (pt/l^{-1})	Volume Filtered (ml)	Colloid Diameter (nm)	Colloid (pt/l^{-1})	Volume Filtered (ml)	Colloid Diameter (nm)	Colloid (pt/l^{-1})
2	10	2×10^{15}	10	100	8×10^{10}	**10**	500	4×10^{10}
	100	2×10^{13}		1000	8×10^{8}		**3000**	$\mathbf{9 \times 10^{8}}$
20	10	2×10^{14}	**100**	**100**	$\mathbf{8 \times 10^{9}}$	100	500	4×10^{9}
	100	$\mathbf{2 \times 10^{12}}$		1000	8×10^{7}		3000	9×10^{7}
			1000	100	8×10^{8}	1000	500	4×10^{8}
				1000	8×10^{6}		3000	9×10^{6}

Note: The population of particles or colloids required for observation on different sample carriers is indicated as a function of the diameter of the colloid and the colloid concentration, in particles (pt) per liter. Highlighted entries indicate typical conditions for preparation of groundwater samples. The concentration of colloids shown in the table represents the optimal detection, corresponding to 5% coverage of the sample carrier.

[a] Carrier diameter: 3mm; surface area: $7mm^2$.

[b] Filter diameter: 1cm; surface area: 1.2 cm^2.

[c] Filter diameter: 4.5 cm; surface area: $15cm^2$.

The elemental composition of the samples can then be analyzed as a function of the concentration factor or volume filtered. Nonlinearities in the correlation between the mass of individual elements and the concentration factor can be corrected in some cases by testing filtration using synthetic groundwater to quantify ion retention.[75]

In groundwater systems with relatively high concentrations of colloids or colloid-associated contaminants, it may not be necessary to concentrate the groundwater in order to detect colloids. The presence and composition of particles can be determined by analysis of filtrates (unretained material) from tangential-flow ultrafiltration at different pore sizes. The concentration and composition of the retained colloidal material are calculated by differences in the composition of the filtrates produced by the different pore sizes of filters.[70,94] It should be noted that the composition of cations in the filtrates from cross-flow filtration is affected by sorption onto colloids deposited during filtration; Puls et al.[87] reported 10 to 50% error in concentration of several cations in filtrates of groundwater with high concentrations of colloids.

Samples can be analyzed for their bulk chemical composition (Table 3) by atomic absorption spectroscopy (AAS)[102] or, preferably, by inductively coupled plasma techniques using atomic emission spectroscopy (ICP-AES),[103] which is subject to fewer interferences than AAS. These results may be confirmed or complemented by ICP with mass spectroscopic detection (ICP-MS), with detection limits in the ng/l range.[104] Membranes can be dissolved and analyzed similarly to determine the elemental composition of retained material. The membranes must be handled carefully since the mass of groundwater colloids loaded onto the membrane is small and the polymer membranes can electrostatically attract particles and aerosols from air. Therefore, the membranes (both at the site or in the laboratory) should be handled under clean conditions using either a glove bag[75] or a clean atmosphere system.[105] Membranes can first be analyzed gravimetrically to determine the mass of particles, then by X-ray diffraction for mineralogical composition prior to dissolution for characterization by AAS, ICP-AES, or ICP-MS. It may be useful to check the homogeneity of a sample by cutting the filter and analyzing the parts separately. The elemental analyses can be correlated with mineral composition[26,106] such as:

- Si with SiO_2
- Al with $Si_4Al_2O_{11}$ (clay)
- Fe with $Fe(OH)_3$
- likewise for TiO_2, ThO_2, and ZrO_2

It must be recognized that these correlations, without confirmation by single-particle analysis, are provisional at best. For example, the correlation between Al and clay may be affected by the presence of $Al(OH)_3$ colloids. In all analyses, quality assurance is critical and includes analyses of reagents and blanks (including blanks for dissolved, clean membranes) and synthetic groundwater. Data quality can be cross-checked by comparing chemical, gravimetric, and particle counting results.[75]

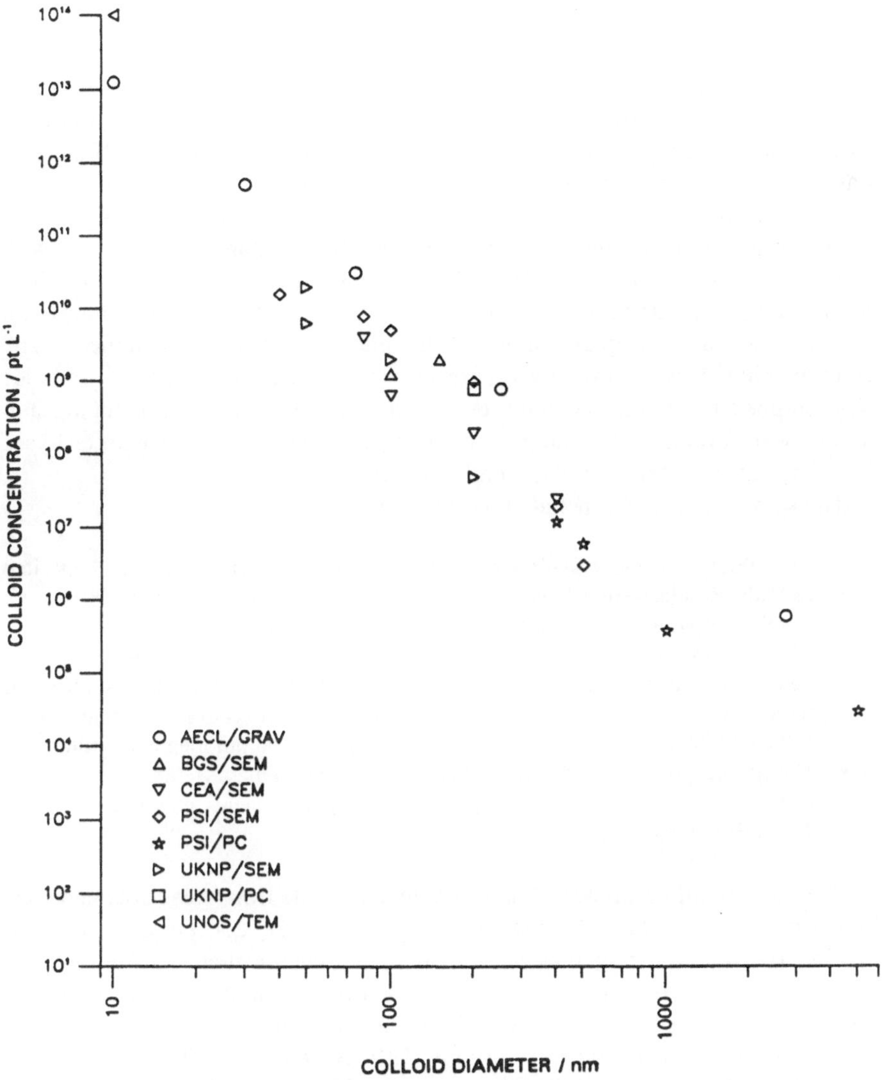

Figure 8. The main result of the Grimsel Colloid Exercise interlaboratory comparison was the cumulative size distribution of colloidal particles recovered at the Grimsel Test Site.[75] Laboratories that contributed to this exercise included: Atomic Energy of Canada, Limited (AECL); British Geological Survey (BGS); Commissariat de l'Energie Atomique (CEA); Paul Scherrer Institute (PSI); Harwell Laboratory Nuclear Physics Division (UKNP); and the University of Norway (UNOS). Techniques used to characterize colloids include gravimetry (GRAV), scanning electron microscopy (SEM), single-particle counting (PC), and transmission electron microscopy (TEM). The cumulative contribution of different sized particles to the total colloid concentration in Grimsel groundwater (particles [pt]/l) is indicated. The cumulative size distribution follows Pareto's power law: $\log C_n = 15.8(\pm 0.4) - 3.2 (\pm 0.2) \log (d)$, where C_n is the cumulative particle concentration for sizes ranging from 25 μm to d. (From Degueldre, C., et al. Paul Scherrer Institute, Report TM-39, Würenlingen and Villigen, Switzerland (1989). With permission.)

Model Case Study: The Grimsel Colloid Exercise

In order to test the reliability, accuracy, and reproducibility of the analytical strategies described earlier, a benchmark test was performed using natural deep groundwater. The intercomparison exercise involving 12 laboratories (Figure 8) was conducted at the Grimsel Test Site, Switzerland.[75] The exercise focused on the water which flowed in a well-characterized fracture zone and was sampled from a Teflon™-encapsulated source at constant temperature, flow rate, and with a constant chemistry and colloid content (no daily or seasonal variation). The colloid sampling was based on both direct water sampling on-site, along with various concentration/separation on-/at-line schemes performed in duplicate. The sampling phase was followed by a characterization phase conducted after samples were shipped to off-line laboratories. In addition, the effects of shipping and storage were evaluated by comparison of results obtained on-site with analyses performed in the laboratory by similar techniques.

The sampling scheme included the following:[75]

- Ultrafiltration using cross-flow membranes, producing membranes with deposited colloids. Results were obtained using different filtration volumes (20 to 200 ml) and membranes with different pore size cut-offs (3 to 450 nm).
- Ultrafiltration using tangential-flow filtration systems, producing liquid concentrates of the colloids (concentration factors of 20 to 300). Both a hollow fiber unit (pore size cut-off of 2.1 nm) and a flat membrane unit (cut-off of 1.5 nm) were compared, with and without prefiltration through a 1000 nm membrane filter.
- The filtrates for each of these separations were analyzed.
- Unfiltered groundwater was collected in glass, polyethylene, polypropylene, and Teflon™ bottles for analysis.

The exercise differentiated samples produced on-site from those obtained after fluid samples were returned to the laboratory. The colloid concentration and size distribution were determined by SEM, gravimetry, chemical analysis of the fluid sample after micro/ultrafiltration, and single-particle counting. The colloid concentrations were also evaluated by TEM, static and dynamic light scattering and laser-induced photoacoustic spectroscopy (LIPAS; see Tables 2 and 3).

Cumulative particle-size distribution results are shown in Figure 8.[75] Particle concentrations were estimated using SEM, gravimetry using membranes prepared on-line at the site, and single-particle counting. The concentration of colloids greater than 10 nm in diameter in the Grimsel water was approximately 10^{14} particles/l, with about 10^{10} particles/l in a size range greater than 50 nm and 2×10^7 particles/l larger than 450 nm. On a mass basis, the total concentration of particles and colloids larger than 10 nm was 200 ± 100 μg/l, with approximately 50 μg/l of colloidal material between 40 and 450 nm in diameter, and about 100 μg/l larger than 450 nm. The estimates of colloid concentration ranged over an order of magnitude among the 12 analytical groups and among the different characterization techniques. The enumeration of colloids is complicated by difficulties in distinguishing single colloids from aggregates of those particles. In addition, comparisons of particle concentration by gravimetry of dry cakes on filters following sequential filtration is based on assumptions about the "average" density of the colloidal phase; a density of 2 g/cm^3 was used since the colloids were mostly clays and silica. Gravimetry suffers the additional difficulty of being unable to distinguish the mass of colloids from any artifacts (precipitates, contamination), while these can potentially be observed and ignored on filters visualized by SEM. These considerations make single-particle counting more powerful than bulk particle analysis.

The Grimsel colloids consisted of silica, illite/muscovite, biotite, calcium silicate, and very small amounts of NOM. Some bacteria were also observed. Because of the relatively high pH (9.6), the particles were negatively charged. This reduced the risks of aggregation and adsorption onto the walls of the container. Examples of SEM photomicrographs of colloids from Grimsel and from other sites are shown in Figure 9.

The main recommendations from this exercise are the need for *in situ*/on-line sampling, tracking of sources of artifacts (e.g., ion retention leading to precipitation or aggregation), and identifying possible sources of contamination during sampling, storage, and transport. A combination of different techniques is recommended to assure reliability of data on natural groundwater colloids.

NATURE AND ABUNDANCE OF GROUNDWATER COLLOIDS

Although the data base on the nature and abundance of groundwater colloids has increased greatly over the last few years (Table 6; Figure 9), it is not yet possible to predict reliably the nature and abundance of colloidal particles as a function of the chemical, hydrological, and mineralogical properties in a range of subsurface environments. Nevertheless, some general relationships are beginning to develop.

- Colloids appear to be ubiquitous. Although concentrations are extremely low in some systems (as low as 1 to 25 μg/l), there is no study which demonstrates the absence of colloidal particles (Table 6).
- In general, colloid concentrations are lowest in deep, geochemically stable subsurface environments; shallow aquifer systems generally appear to have higher colloid concentrations even in the absence of geochemical instabilities.
- Regardless of the geology or depth of the geological system, higher levels of colloids are routinely associated with some hydrogeochemical perturbation. For example, in studies of a series of fractured granitic systems, colloid concentrations are 20- to 1000-fold higher in groundwater zones affected by inputs of surface water or in hydrothermal zones with large temperature and pressure gradients, compared to stable hydrogeochemical systems (Table 6).[75,76,107,108] In fractured rhyolitic tuffs, colloid concentrations were several-fold higher in locations affected by underground nuclear testing, compared to locations with similar geology but remote from the test

Table 6. Nature and Abundance of Colloidal Particles in Various Subsurface Systems

Site Location	Geology (depth from surface/m)	pH	Eh /mv	t /°C	Colloid Diameter (nm)	Colloid Concentration (mg/l)	Colloid Composition	Comments	Refs.
Grimsel Test Site (CH)	Granite/mylonite (500)	9.6	−250	12	10–1,000	0.1	Clay/silica	Stable hydrogeochemical; system stable groundwater flow rate	75, 91
Zurzach (CH)	Granitic/fractured; 2 sources (470)	8	−180	37	10–1,000	0.025	Silica/clay	Stable hydrogeochemistry; steady flow from artesian system	97
Leuggern (CH)	Granitic/fractured (1680)	8	−150	66	10–450	0.025	Silica/clay	Stable hydrogeochemistry; steady flow from artesian system	79
Bad sackingen (D)	Granitic/fractured (82–201)	6.5	340–400	30	10–450	≈0.025	Silica/clay	Stable hydrogeochemical system; water pumped	107
Menzenschwand (D)	Granitic/fractured (240)	6.5	+ (O_2 detected)	12	10–1,000	0.4	Clay/silica	Mixed groundwaters (introduced oxic surface waters)	114
Transitgas tunnels, Grimsel (CH)									76, 110
Sources 1–8	Granitic/fractured; 8 sources (200)	8–9	0	8	100–10,000	0.1–2	Clay/silica	Mixed groundwaters	
Sources 9–13	Granitic/fractured; 5 sources (200)	8–9	−80 to 0	20–30	100–10,000	5–20	Clay/silica	Hydrothermal fracture zone	
Sources 14–16	Granitic/fractured; 5 sources (200)	8–9	+200 to +300	5	100–10,000	0.5	Clay/silica	Direct transit of surface water	

Fanay-Augéres, Massif-Central (F)	Granitic formation (280)	6.0	398	16	100–1000	>0.2	Silica, NOM	Mixed water (groundwater and surface water with high levels of NOM [≈2 mg C/l])	139
Wellenberg SB6 (CH)	Marlstone	9.0 (360–420)	–500	12	100–1000	2–3	Clay	≈2 mg/l of Fe(II) presents the potential for artifacts resulting from oxidation; low Ca^{2+} ≈ 1–2 mg/l	140
Sweden: 10 sites: (total of 22 deep boreholes with 305 observations) *Tavinunnanen* *Kamlunge* *Gideå* *Svartboberget* *Forsmark* *Finnsjön* *Fjallveden* *ÅvrÄ,Äspö* *Laxemer* *Klipperås*	Crystalline rock, mostly granitic, but one location with basalt; rock age of 1.5–2.0 × 10^9 y; fracture fillings have been repeatedly activated by hydrothermal and low temperature fluids (45–260 m)								141, 142
	Diluted (shallow) groundwater (Cl^- < 50 mg/l)	6.7	–191	12	50–450	0.3 ± 2.7	Clay, Fe-hydroxide, calcite, quartz, Mn–oxide	Fe-hydroxides, in some cases, can be artifacts resulting from oxidation during sampling	
	Concentrated (deep) groundwater (Cl^- >50 mg/l)	7.4	–149	11	50–450	0.35 ± 1.06	Calcite, troilite, Fe–hydroxide, clay, quartz, Mn-oxide	Calcite precipitation is probably an artifact and increases the reported mean concentration by four-fold	

Table 6 (continued). Nature and Abundance of Colloidal Particles in Various Subsurface Systems

Site Location	Geology (depth from surface/m)	pH	Eh /mv	t /°C	Colloid Diameter (nm)	Colloid Concentration (mg/l)	Colloid Composition	Comments	Refs.
Oklo, Okelobondo Gabon, Africa	Argillaceous rocks, pelites (21–300)								143
Oklo	Diluted groundwater (chloride concentration < 50 mg/l^{-1}) drillhole (21 m)				50–450	1.58 ± 1.57	Fe-hydroxide, pyrophylite	Samples from fractures gave lower values on Al and Si; drilling is therefore believed to increase silica and clay colloids	
Okelobondo	Fractures (300 m)	7.08	350 ± 50	26	50–450	0.14 ± 0.23	Fe-hydroxide, pyrophilite		
Nevada Test Site *[NTS]*, Nevada (US)	Fractured rhyolitic lava (800)	8.4	+	40	4–55	4.5	Quartz, feldspars	Outside nuclear detonation cavity	74
Well 13, Jackass Flats, NV (US)	Rhyolitic welded tuff	7.2		32	100–2000	1.43	Oxides of Ca-Si, Al, Al-Si, Ca-Si-Ti	Sampled from water supply wells (800 l/min); bulk of pore volume in pores ≤ 100 nm, suggesting limited transport paths for colloids	109
Well 20, Nevada test site, NV (US)	Rhyolitic welded tuff	7.7	300	40	15–450	5.08	Layer silicates, plagioclase, carbonate, quartz	Sampled from water supply wells (800 l/min); average pores about 2000 nm (much larger than colloids)	109
Central and Southern Nevada (US): 23	Volcanic, carbonate and alluvial formations were examined; general observa-						Silica (cristobalite, fused silica, or amorphous	Calcite and organic material also identified, but may be sampling artifacts or contaminants.	144

sampling locations in wells and springs	tions for all samples are reported in this row to the right					silica), and possibly small amounts of clay or zeolite	No discernible trends were observed between water chemistry and either concentration or composition	
Pahroe Spring	Volcanic (predominantly ash-flow tuff, ash-fall tuff and rhyolite lava flows)	7.9	16	30–1000 >1000	1.3 0.18	Silica	No discernible trends	144
Peavine Canyon	Volcanic	7.8	11	30–1000 >1000	>0.28 0.38	Silica	No discernible trends	144
Peavine Ranch Well	Volcanic (63–93)	7.7	12	30–1000 >1000	0.51 0.68	Silica	No discernible trends	144
Indian Spring Well	Volcanic (46–131)	8.4	30	30–1000 >1000	1.35 0.18	Silica	No discernible trends	144
Sidehill Spring	Volcanic	7.8	18	30–1000 >1000	0.82 0.18	Silica	No discernible trends	144
Lower Indian Spring	Volcanic	7.8	22	30–1000 >1000	0.66 0.34	Silica	No discernible trends	144
Well 4, NTS	Volcanic (229–424)	8.0	38	30–1000 >1000	0.72 0.30	Silica	No discernible trends	144
Water Well 20, NTS	Volcanic	8.0	27	30–1000 >1000	0.48 0.15	Silica	No discernible trends	144
UE19c, NTS	Volcanic (2587)	8.7	31	30–1000 >1000	0.68 0.13	Silica	No discernible trends	144
Well 8, NTS	Volcanic (381–543)	7.7	27	30–1000 >1000	0.73 0.13	Silica	No discernible trends	144
Cane Springs, NTS	Volcanic	7.5	16	30–1000 >1000	0.36 3.89	Silica	No discernible trends	144
Topopah Spring, NTS	Volcanic	6.9	16	30–1000 >1000	>25.2 2.01	Silica	No discernible trends	144
Crystal Pool	Carbonate aquifer with fractures containing secondary calcite, calcareous clay or, calcareous clay and iron oxide	7.1	33	30–1000 >1000	0.34 0.03	Silica	No discernible trends	144

Table 6 (continued). Nature and Abundance of Colloidal Particles in Various Subsurface Systems

Site Location	Geology (depth from surface/m)	pH	Eh /mv	t /°C	Colloid Diameter (nm)	Colloid Concentration (mg/l)	Colloid Composition	Comments	Refs.
Ash Spring	Carbonate	7.1		31	30–1000 >1000	0.40 0.10	Silica	No discernible trends	144
Cold Creek Spring	Carbonate	8.1		10	30–1000 >1000	0.87 0.10	Silica	No discernible trends	144
Fairbanks Spring	Carbonate			27	30–1000 >1000	0.38 0.28	Silica	No discernible trends	144
Indian Spring	Carbonate	8.0		25	30–1000 >1000	0.59 0.13	Silica	No discernible trends	144
Well C–1, NTS	Carbonate (468–503)	6.6		36	30–1000 >1000	0.55 0.03	Silica	No discernible trends	144
UE16d, NTS	Carbonate (914)	7.6		23	30–1000 >1000	0.35 0.10	Silica	No discernible trends	144
Well A, NTS	Alluvial deposits of sand and gravel cemented by calcium carbonate (490–570)	7.8		27	30–1000 >1000	0.31 0.43	Silica	No discernible trends	144
Beatty Well 2	Alluvial (21–40)	8.2		23	30–1000 >1000	6.48 0.38	Silica	No discernible trends	144
Lathrop Well	Alluvial (125–155)	8.2		26	30–1000 >1000	0.54 0.43	Silica	No discernible trends	144
Gorleben (D)									
Gorleben 8	Sedimentary (115)	7.7		20	2–450	<0.5	Silica/clay	Off-line sampling and analysis; groundwater and sediment samples	103, 133
Gorleben 9	Sedimentary (133)	7.8		20	2–450	<0.5	Silica/clay	Off-line sampling and analysis; groundwater and sediment samples	103, 133

Gorleben 214	Sedimentary			20	10–450	0.005	Silica/NOM	Off-line sampling and analysis ofgroundwater and sediment	125
Well MF12, Morro de Ferro, Poços de Caldas (BZ)	Clay/phonolite (50)	6.0	+200 to +300	20	10–450	0.25	$Fe(OH)_3$ NOM	Analyses of bore holes along flow path provides no evidence of colloid transport through saturated porous rock	115
Maqarin (JD)	Sedimentary rock; 5 sources of basic groundwater	12.5	+	25	10–450	<1	Portlandite	Oxic groundwater containing chromate; highly alkaline due to portlandite produced by natural bitumen combustion	145
Pinal Creek, AZ (US)									47
Well 451	Fine-grained alluvium from a granitic porphyry (< 50)	4.7	250	19		20		Downgradient from inputs of acidic mining wastes; in zone where groundwater pH is rapidly changing	
Well 503	Unconsolidated alluvium from a granitic porphyry (< 50)	5.7	320	19	300–1300	0.1	Gypsum, Fe-oxide, kaolinite	Downgradient from well 451 and plume of acidic waste	
Laughlin, NV (US)	Sand/gravel alluvium below evaporation ponds and ash dump (20–60)	7.1–7.6	100		100–2000	10–100	Silicates	Postulate dissolution of cementing carbonates due to microbially induced increase in groundwater CO_2	64

Table 6 (continued). Nature and Abundance of Colloidal Particles in Various Subsurface Systems

Site Location	Geology (depth from surface/m)	pH	Eh /mv	t /°C	Colloid Diameter (nm)	Colloid Concentration (mg/l)	Colloid Composition	Comments	Refs.
Well WT4, Tel Aviv, Israel	Coastal plain phreatic aquifer; calcareous sandstones (0–16)	6.5–7.0			100–3000	11–33	Calcite and quartz	Postulate detachment of colloids resulting from dissolution of carbonate aquifer matrix; dissolution promoted by long-term irrigation of municipal sewage that resulted in microbiallyinduced increase in pCO_2	146
Cap Cod, MA (US)	Glacial outwash (7)	6.2	−100 to −400 (estimated)	13	100	0.9 to 2	Ferrous phosphate (monodispersed)	Downgradient from infiltration of secondary treated sewage; colloids formed from sewage phosphate and aquifer Fe(II)	13
Coal gasification plant (operated from 1853–1923), CT (US)	Glacial fluvial (6)	6.2	300–780	12	100–3000	≈2 to 3	Silica and clay	Did not observe expected enhancement in transport of polycyclic aromatic hydrocarbons from coal tar deposits	80
McDonalds Branch Watershed, Pine Barrens, NJ (US)									65, 83
"Swamp Deep"	Unconsolidated sandy coastal plain aquifer (8)	4.5	290	12		60	Organic-coated kaolinite, chamosite, some goethite	Anaerobic under swamp; reductive dissolution of Fe-oxide cements postulated to mobilize organic-coated clays	

"Upland Deep"	Unconsolidated sandy coastal plain aquifer (10)	4.8	340	12		<1	Kaolinite	Oxic aquifer in same formation; Fe-oxide cements postulated to bind colloids to aquifer particles	
Harrington, DE (US)									65, 83
Well Lc42–01	Unconsolidated coastal plain aquifer (15)	6.2	100	14		6	Organic-coated muscovite and kaolinite, Fe-oxides	Anoxic aquifer; postulate dissolution of Fe-oxide cements mobilizes colloids	
Well Md22–01	Unconsolidated sandy coastal plain aquifer (5)	4.3	380	18		<1	Muscovite	Oxic aquifer in same formation;Fe-oxide cements postulated to bind colloids to aquifer particles	
Alligator River	Uranium ore body in sandstone (15–25)	6.8–7.2	170–375		18–1000	0.1 to 1.2	Fe and Si species with sorbed U and U daughters	Prefiltration	26
Cigar Lake Saskatchewan (CD)									61
Cigar Lake Hole 75	Upper sandstone (154–159)	6.6	65	7	10–450 >450	0.85 ± 0.26 0.63 ± 0.42	Clay minerals, Fe-Si precipitates, NOM, quartz, carbonate	Well-consolidated sandstone; all Cigar Lake sampling conducted at pumping rates of 30–300 ml/min	61
Cigar Lake Hole 71	Lower sandstone (243–245)	7.5	–304	10	10–450 >450	0.73 ± 0.16 1.76 ± 0.42	Clay minerals, Fe-Si precipitates, NOM, quartz, carbonate	Well-consolidated sandstone	61
Cigar Lake Hole 139	Lower sandstone (437)	6.5	235	5	10–450 >450	0.75 ± 0.2 1.2 ± 0.4	Clay minerals, Fe-Si precipitates, NOM, quartz, carbonate	Well-consolidated sandstone	61

Table 6 (continued). Nature and Abundance of Colloidal Particles in Various Subsurface Systems

Site Location	Geology (depth from surface/m)	pH	Eh /mv	t /°C	Colloid Diameter (nm)	Colloid Concentration (mg/l)	Colloid Composition	Comments	Refs.
Cigar Lake Hole 219	Lower sandstone (414–426)	6.5	124	9	10–450 >450	0.85 ± 0.16 0.85 ± 0.21	Clay minerals, Fe-Si precipitates, NOM, quartz, carbonate	Well-consolidated sandstone	61
Cigar Lake Hole 67	Lower sandstone (346–348)	7.2	–9	5	10–450 >450	1.0 ± 0.2 1.9 ± 0.3	Clay minerals, Fe-Si precipitates, NOM, quartz, carbonate	Formation contains lenses of friable sand, but not near the bore hole	61
Cigar Lake Hole 91	Clay-rich zone (407)	6.1	209	8	10–450 >450	0.6 ± 0.09 4.9 ± 1.5	Clay minerals, Fe-Si precipitates, NOM, quartz, carbonate	Bore hole near altered fracture zone; high hydraulic conductivity	61
Cigar Lake Hole 197	Clay/uranium ore contact (416–421)	6.0	238	10	10–450 >450	0.72 ± 0.16 0.64 ± 0.13	Clay minerals, Fe-Si precipitates, NOM, quartz, carbonate	Highly structured clay	61
Cigar Lake Hole 79	Uranium ore (432)	6.7	80	15	10–450 >450	0.91 ± 0.26 3.4 ± 0.9	Clay minerals, Fe-Si precipitates, NOM, quartz, carbonate		
Cigar Lake Hole 220	Uranium ore (220)	8.0	–243	11	10–450 >450	1.4 ± 0.4 1.9 ± 0.4	Clay minerals, Fe-Si precipitates, NOM, quartz, carbonate	Ore zone contains lenses of friable sand	61
Cigar Lake Hole 199	Altered basement (446 –452)	7.9	109	15	10–450 >450	7.8 ± 4.1 106 ± 3	Clay minerals, Fe-Si precipitates, NOM, quartz, carbonate	Near a major fracture (metapelite rock)	61

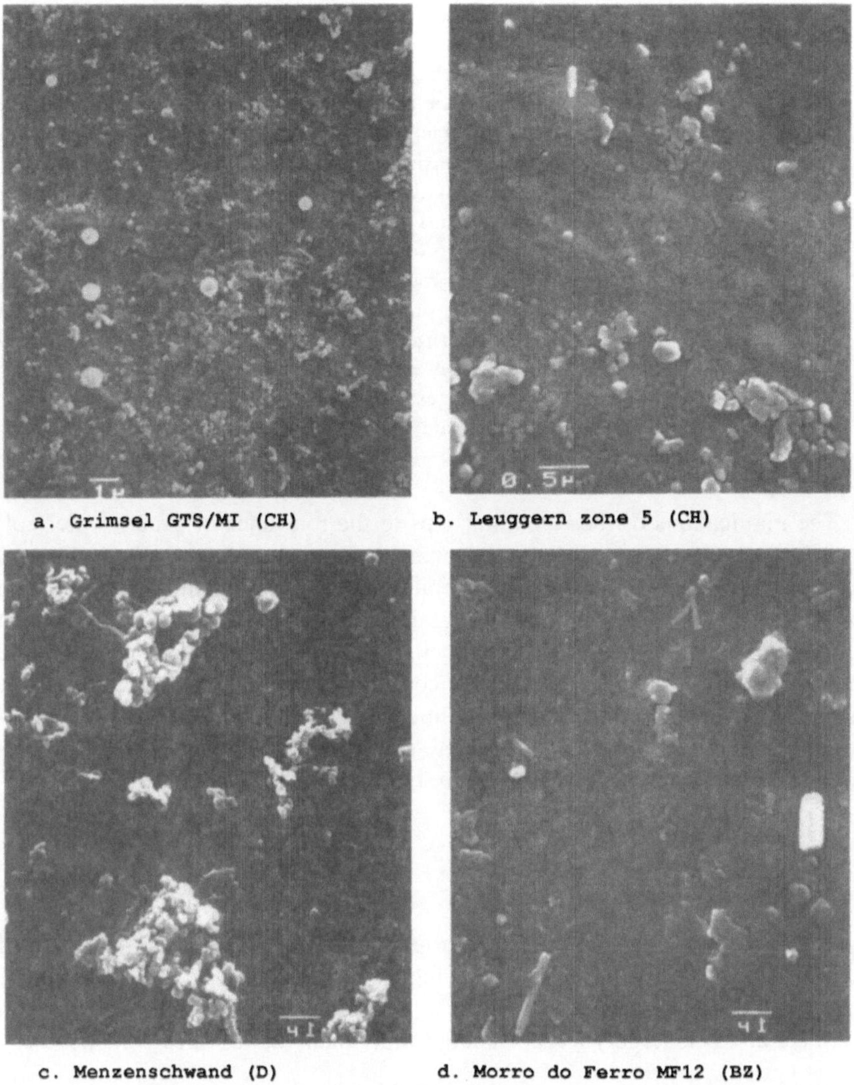

Figure 9. Scanning electron photomicrographs of groundwater colloids and particles from various sites are shown. Details of the sites and the nature and abundance of colloids at these sites are given in Table 6. Locations and the volume of sample filtered through the 1 cm^2 membrane are: (a) Grimsel Test Site (Switzerland), 55 ml filtered;[91] (b) Leuggern (Switzerland), 90 ml filtered;[79] (c) Menzenschwand (Germany), 10 ml filtered;[114] and (D) Morro do Ferro (Brazil), 10 ml filtered.[115]

site (Table 6).[9,109] Sharp gradients in groundwater pH due to infiltration of acidic mining wastes were associated with higher levels of colloids in an unconsolidated alluvium in Arizona.[87] Disturbances occurring on longer geological time scales can also affect colloids. Reducing groundwater resulting from microbial metabolism of

NOM beneath a swamp was associated with increased levels of colloids, compared to oxic zones in adjacent formations.[65]

- In most cases, the composition of the colloids observed in different subsurface systems "makes sense" in terms of the geology of the formation and the nature of the geochemical perturbations. In fractured granitic systems, the observed silica and clay/mica colloids would be consistent with production of primary and secondary minerals from geochemical alteration of the parent rock, and colloids in volcanic tuffs appear to be constituents of the parent rock and alteration products. Particles in a sandstone/uranium ore formation are composed of fracture-filling minerals characteristic of the host rock. Clays postulated to be mobilized by decementation were observed to be present but immobilized within the aquifer matrix at the oxic sites in the same formations.[65] The presence of ferrous phosphate colloids downgradient from sewage infiltration site is consistent with equilibrium solubility calculations suggesting supersaturation with respect to this product within a plume of phosphate (from detergents in the sewage) flowing through an iron-rich aquifer.[13]

The implications of these relationships to the relevance of groundwater colloids in the subsurface transport of contaminants are significant. Waste disposal activities often affect changes in temperature, pH, redox potential and dissolved oxygen levels, ionic strength, and organic carbon concentrations in ways that could promote the formation of stable colloidal suspensions.[68] Assessment of the role of colloids in the dissemination of colloids needs, therefore, to consider the potential generation of colloids in addition to the possible effects of a "background" population of colloids that may exist in uncontaminated environments prior to development of a waste repository.

SUMMARY AND CONCLUSIONS

The motivation for most studies of groundwater colloids has been summarized (i.e., their potential role in migration of highly adsorbed contaminants). Evidence for enhanced transport of contaminants as a result of sorption and cotransport on mobile colloids is largely circumstantial. Individual studies have demonstrated:

1. Colloids are present in groundwater (Table 6), and their abundance is promoted by geohydrochemical perturbations, including those characteristic of waste disposal sites.[68]
2. In a few instances, contaminants associated with colloidal particles have been recovered from monitoring wells much further from the input source than models and laboratory sorption experiments predicted. Unfortunately, there is no clear demonstration that the contaminants migrated by attaching to mobile colloids; solute transport through preferential flow paths and subsequent attachment to particles around the sampling well cannot be unequivocally discounted.
3. There are a few examples that clearly demonstrate transport of (uncontaminated) colloidal particles from a clearly identified source to a monitoring well. Many of these studies involve transport of bacteria or viruses (which have a clearly identi-

fiable signature) from a known source.[50] Some experimental injections of colloids have quantified transport over relatively small spatial scales.[46] Only recently have attempts been made to reconcile field observations of colloid transport with quantitative models derived from filtration theory.[28]

4. Sorption experiments (including a limited number of studies using natural groundwater colloids) have demonstrated that colloids are capable of associating with many metals, radionuclides, and hydrophobic organic contaminants.[4,80,111,112] However, model simulations of colloid-facilitated transport[57,58] suggest that information on the affinity, kinetics, and reversibility of that association is critical to evaluating the significance of colloids to contaminant transport, but this information is largely unavailable for groundwater particles.

While each piece of information described above is integral to a demonstration of the role of colloids in contaminant migration, there are no instances in which all of these elements have been demonstrated within one experiment or set of field observations. However, given the rapid increase of information on groundwater colloids and the increased awareness of the possible role of colloids by the regulatory community, additional data on the involvement of colloids in transport of contaminants may soon be available.

Studies to document the nature and abundance of groundwater colloids or to demonstrate the association of contaminants with field-collected colloidal particles must be reliable and free of confounding artifacts. This chapter has addressed the many issues related to colloid sampling. Studies over the last few years have greatly improved understanding of how colloids must be sampled and analyzed. It is necessary for future investigations to evaluate critically their sampling protocols and to document that their results are not influenced by potential artifacts in sample recovery, preservation, or analysis.

In conclusion, several areas requiring additional research have been identified.[2] The data base of observations of groundwater colloids is growing rapidly, and this trend should be continued. The current data base (Table 6) encompasses a fairly wide range of geochemical environments but still fails to provide the basis for a systematic comparison of hydrological, mineralogical, lithological, geochemical, and microbiological factors that may influence the formation, stability, and transport of colloidal particles. Continued development of innovative *in situ* (down-hole) methods for observing the number, size distribution, and composition of colloids would increase confidence in current descriptions of groundwater colloids.

Information is needed on how far colloids can move in aquifers. Laboratory experimentation and theoretical descriptions of colloid stability and transport must take into account the heterogeneous nature of particles and collector surfaces in aquifers. In many natural systems, control of colloid transport may be dominated by hydrological rather than chemical factors; preferential flow through fractures and secondary pore structure may limit opportunities for collection of colloids on surfaces. Capabilities to describe the physical heterogeneity of formations, such as the size and interconnectedness of pores and fractures, may be a more fundamental limitation to predicting colloid transport than is understanding of geochemical controls of colloid behavior.

The need for more information on the extent of association of different types of colloids with metal, radionuclide, and organic contaminants has been highlighted. Understanding of processes controlling the extent, and especially the kinetics, of association and dissociation must be developed at a fundamental level. Long-term predictions of the role of colloids on contaminant movement must rely on generalized understanding in order to assess the effect of temporal or spatial changes in aqueous chemistry and of aquifer surface properties along a flow path on the transport of contaminants.

Finally, it must be recognized that progress in understanding colloid behavior, as in many areas of environmental chemistry, requires close coordination and interaction between laboratory and field components of research. Progress will depend on effective synthesis of the insights and methodologies of chemists, hydrologists, engineers, and mathematical modelers. Hopefully, integrated research programs in the U.S., Canada, Europe, Australia, and elsewhere will be effective in encouraging this coordination.

ACKNOWLEDGMENTS

We thank L. Shevenell, D. Marsh, and I. McKinley for reviewing this manuscript, L. Liang for many thoughtful suggestions, and C. T. Woodard for secretarial assistance. JFM is supported by the Subsurface Science Program, Environmental Sciences Division, Office of Health and Environmental Research, U.S. Department of Energy (U.S. DOE). The Oak Ridge National Laboratory (ORNL) is managed by Martin Marietta Energy Systems, Inc., under Contract No. DE-AC05-84OR21400 with the U.S. DOE. Publication No. 3939 of the Environmental Sciences Division of ORNL. The colloid research at the Paul Scherrer Institut is carried out in the Laboratory for Material and Nuclear Processes and in the framework of the Waste Management Programme, with partial funding support from NAGRA, Wettingen, Switzerland.

REFERENCES

1. McDowell-Boyer, L.M., J.R. Hunt, and N. Sitar. "Particle Transport through Porous Media," *Water Resour. Res.* 22(13):1901–1921 (1986).
2. McCarthy, J.F. and J.M. Zachara. "Subsurface Transport of Contaminants," *Environ. Sci. Technol.* 23(5):496–503 (1989).
3. Nyhan, J.W., B.J. Drennon, W.V. Abeele, M.L. Wheeler, W.D. Purtymun, G. Trujillo, W.J. Herrera, and J.W. Booth. "Distribution of Plutonium and Americium Beneath a 33-yr-old Liquid Waste Disposal Site," *J. Environ. Qual.* 14(4):501–509 (1985).
4. Penrose, W.R., W.L. Polzer, E.H. Essington, D.M. Nelson, and K.A. Orlandini. "Mobility of Plutonium and Americium Through a Shallow Aquifer in a Semiarid Region," *Environ. Sci. Technol.* 24(2):228–234 (1990).

5. Looney, B.B., M.W. Grant, and C.M. King. "Estimation of Geochemical Parameters for Assessing Subsurface Transport at the Savannah River Plant," Savannah River Laboratory, Aiken, SC, DPST-85-904 (1987a).
6. Looney, B.B., C.M. King, and D.E. Stephenson. "Quality Assurance Program for Environmental Assessment of Savannah River Plant Waste Sites," Report No. DPST-86-725 (1987b).
7. Newman, M.E. "Effects of Alterations in Groundwater Chemistry on the Mobilization and Transport of Colloids," PhD Thesis, Clemson University, Clemson, SC (1990).
8. Coles, D.G. and L.D. Ramspott. "Migration of Ruthenium-106 in a Nevada Test Site Aquifer: Discrepancy Between Field and Laboratory Results," *Science* 215(5): 1235–1237 (1982).
9. Buddemeier, R.W. and J.R. Hunt. "Transport of Colloidal Contaminants in Groundwater: Radionuclide Migration at the Nevada Test Site," *Appl. Geochem.* 3(5):535–548 (1988).
10. Walton, F.B. and W.F. Merritt. "Long-Term Extrapolation of Laboratory Glass Leaching Data for the Prediction of Fission Product Release under Actual Groundwater Conditions," in *Scientific Basis for Nuclear Waste Management, Vol. 2* (New York: Plenum Press, Inc., 1980), pp.155–166.
11. Champ, D.R., J.L. Young, D.E. Robertson, and K.H. Abel. "Chemical Speciation of Long-Lived Radionuclides in a Shallow Groundwater Flow System," *Water Pollut. Res. J. Can.* 19(2):35–54 (1984).
12. Ho, C.H. and N.H. Miller. "Formation of Uranium Oxide Sols in Bicarbonate Solutions," *J. Colloid Interface Sci.* 113(1):232–240 (1986).
13. Gschwend, P.M. and M.D. Reynolds. "Monodisperse Ferrous Phosphate Colloids in an Anoxic Groundwater Flume," *J. Contam. Hydrol.* 1:309–327 (1987).
14. Moriyama, H., M.I. Pratopo, and K. Higashi. "The Solubility and Colloidal Behavior of Neptunium (IV)," *Sci. Total Environ.* 83:227–237 (1989).
15. Maes, A. and A. Cremers. "Highly Selective Ion-Exchange in Clay-Minerals and Zeolites," in *Geochemical Processes of Mineral Surfaces*, J.A. Davis and K.F. Hayes, Eds. (Washington, D.C.: American Chemical Society, 1986), pp. 254–295.
16. Sposito, G. *The Surface Chemistry of Soils* (New York: Oxford University Press, Inc., 1984), p. 234.
17. Morse, J.W. "The Surface Chemistry of Calcium Carbonate Minerals in Natural Waters: An Overview," *Mar. Chem.* 20(1):91–112 (1986).
18. Dalang, F., J. Buffle, and W. Härdy. "Study of the Influence of Fulvic Substances on the Adsorption of Copper(II) Ions at the Kaolinite Surface," *Environ. Sci. Technol.* 18(3):135–141 (1984).
19. Karickhoff, S.W. "Organic Pollutant Sorption in Aquatic Systems," *J. Hydraul. Eng.* 110(6):707–735 (1984).
20. Schwarzenbach, R.P. and J. Westall. "Transport of Nonpolar Organic Compounds from Surface Water to Groundwater. Laboratory Sorption Studies," *Environ. Sci. Technol.* 15(11):1360–1367 (1981).
21. Means, J.C., S.G. Wood, J.J. Hassett, and W.L. Banwart. "Sorption of Amino- and Carboxy-Substituted Polynuclear Aromatic Hydrocarbons by Sediments and Soils," *Environ. Sci. Technol.* 16(2):93–98 (1982).
22. Banerjee, P., M.D. Piwoni, and K. Ebeid. "Sorption of Organic Contaminants to a Low Carbon Subsurface Core," *Chemosphere* 14:1057–1067 (1985).

23. Vinten, A.J.A., B. Yaron, and P.H. Nye. "Vertical Transport of Pesticides into Soil when Adsorbed on Suspended Particles," *J. Agric. Food Chem.* 31(3):662–665 (1983).
24. Vinten, A.J.A. and P.H. Nye. "Transport and Deposition of Dilute Colloidal Suspensions in Soils," *J. Soil Sci.* 36:531–541 (1985).
25. Champ, D.R., W.P. Merritt, and J.L. Young. "Potential for the Rapid Transport of Plutonium in Groundwater as Demonstrated by Core Column Studies," *Scientific Basis for Radioactivity Waste Management* (Materials Research Society, 1982).
26. Short, S.A., R.T. Lowson, and J. Ellis. "$^{234}U/^{238}U$ and $^{230}Th/^{234}U$ Activity Ratios in the Colloidal Phases of Aquifers in Lateritic Weathered Zones," *Geochim. Cosmochim. Acta* 52:2555–2563 (1988).
27. Lyklema, J. "Surface Chemistry of Colloids in Connection with Stability," in *The Scientific Basis of Flocculation*, K.J. Ives, Ed. (Dordrecht, The Netherlands: Sijthoff and Noordhoff, 1978), pp. 3–36.
28. O'Melia, C.R. "Kinetics of Colloid Chemical Processes in Aquatic Systems," in *Aquatic Chemical Kinetics*, W. Stumm, Ed. (New York: John Wiley and Sons, Inc., 1990), pp. 447–474.
29. Khilar, K.C. and H.S. Fogler. "The Existence of a Critical Salt Concentration for Particle Release," *J. Colloid Interface Sci.* 101:214–224 (1984).
30. Cerda, C.M. "Mobilization of Kaolinite Fines in Porous Media," *Colloids Surf.* 27:219–241 (1987).
31. Kia, S.F., H.S. Fogler, and M.G. Reed. "Effect of pH on Colloidally Induced Fines Migration," *J. Colloid Interface Sci.* 118:158–168 (1987).
32. James, R.O. and T.W. Healy. "Adsorption of Hydrolyzable Metal Ions at the Oxide-Water Interface. III. A Thermodynamic Model of Adsorption," *J. Colloid Interface Sci.* 40:65–81 (1972).
33. McDowell-Boyer, L.M. "Migration of Small Particles in Saturated Sand Columns," PhD Thesis, University of California at Berkeley, Berkeley, CA (1989).
34. Tobiason, J.E. and C.R. O'Melia. "Physicochemical Aspects of Particle Removal in Depth Filtration," *J. Am. Water Works Assoc.* 80(12):54–64 (1988).
35. Toran, L. and A.V. Palumbo. "Colloid Transport through Fractured and Unfractured Laboratory Sand Columns," *J. Contam. Hydrol.* 9:289–303 (1992).
36. Puls, R.W. and R.M. Powell. "Transport of Inorganic Colloids through Natural Aquifer Material: Implications for Contaminant Transport," *Environ. Sci. Technol.* 26:614–621 (1992).
37. Liang, L. and J.J. Morgan. "Chemical Aspects of Iron Oxide Coagulation in Water: Laboratory Studies and Implications for Natural Systems," *Aquatic Sci.* 52(1):32–55 (1990).
38. Champlin, J.B.F. and G.G. Eichholz. "Fixation and Remobilization of Trace Contaminants in Simulation Subsurface Aquifers," *Health Phys.* 30:215–219 (1976).
39. Lin, F.-C. "Clay – Coating Reduction of Permeability During Oil – Sand Testing," *Clays Clay Miner.* 33:76–78 (1985).
40. Nightingale, H.I. and W.C. Bianchi. "Ground-Water Turbidity Resulting from Artificial Recharge," *Ground Water* 15(2):146–152 (1977).
41. Muecke, T.W. "Formation Fines and Factors Controlling Their Movement in Porous Media," *J. Pet. Technol.* 144–150 (1979).
42. Sharma, M.M. and Y.C. Yortsos. "Fines Migration in Porous Media," *Am. Inst. Chem. Eng. J.* 33(10):1654–1662 (1987).

43. Goldenberg, L.C., M. Magaritz, and V.S. Mandel. "Experimental Investigation on Irreversible Changes in Hydraulic Conductivity in Seawater-Freshwater Interface in Coastal Aquifers," *Water Resour. Res.* 19:77–85 (1983).
44. Goldenberg, L.C., M. Magaritz, A.J. Amiel, and S. Mandel. "Changes in Hydraulic Conductivity of Laboratory Sand-Clay Mixtures Caused by a Seawater-Freshwater Interface," *J. Hydrol.* 70:329–336 (1984).
45. Hardcastle, J.H. and J.K. Mitchell. "Electrolyte Concentration-Permeability Relationships in Sodium Illite – Silt Mixtures," *Clays Clay Miner.* 22:143–154 (1974).
46. Harvey, R.W., L.H. George, R.L. Smith, and D.R. LeBlanc. "Transport of Fluorescent Microsphere and Indigenous Bacteria Through a Sandy Aquifer: Results of Natural- and Forced-Gradient Tracer Experiments," *Environ. Sci. Technol.* 23:51–56 (1989).
47. Yao, K.-M., M.T. Habibian, and C.R. O'Melia. "Water and Waste Water Filtration: Concepts and Applications," *Environ. Sci. Technol.* 11:1105–1112 (1971).
48. Rajagopalan, R. and C. Tien. "Trajectory Analysis of Deep Bed Filtration with Sphere-in-Cell Porous Media Model," *Am. Inst. Chem. Eng. J.* 22:523–533 (1976).
49. Harvey, R.W. and S.P. Garabedian. "Use of Colloid Filtration Theory in Modeling Movement of Bacteria Through a Contaminated Sandy Aquifer," *Environ. Sci. Technol.* 25(1):178–185 (1991).
50. Keswick, B.H., D.S. Wang, and C.P. Gerba. "The Use of Microorganisms as Groundwater Tracers: A Review," *Groundwater* 20:142–149 (1982).
51. Wood, W.W. and G.G. Ehrlich. "Use of Baker's Yeast to Trace Microbial Movement in Groundwater," *Groundwater* 16:398–403 (1978).
52. Bales, R.C., D.D. Newkirk, and S.B. Hayward. "Chrysotile Asbestos in California Surface Waters—From Upstream Rivers Through Water-Treatment," *J. Am. Water Works Assoc.* 76(5):66–74 (1984).
53. Pilgrim, D.H. and D.D. Huff. "Suspended Sediment in Rapid Subsurface Stormflow on a Large Field Plot," *Earth Surf. Processes Landforms* 8:451–463 (1983).
54. Enfield C.G. and G. Bengtsson. "Macromolecular Transport of Hydrophobic Contaminants in Aqueous Environments," *Groundwater* 26(1):64–70 (1988).
55. Enfield, C.G., G. Bengtsson, and R. Lindqvist. "Influence of Macromolecules on Chemical Transport," *Environ. Sci. Technol.* 23(10):1278–1286 (1989).
56. Smith, M.S., G.W. Thomas, R.E. White, and D. Ritonga. "Transport of Escherichia-Coli Through Intact and Disturbed Soil Columns," *J. Environ. Qual.* 14(1):87–91 (1985).
57. Mills, W.B., S. Liu, and F.K. Fong. "Literature Review and Model (COMET) for Colloid/Metals Transport in Porous Media," *Groundwater* 29:199–208 (1991).
58. Smith, P.A. and C. Degueldre. "Colloid Facilitated Transport of Radionuclides Through Fractured Media," *J. Contamin. Hydrol.* in press (1993).
59. Longworth, G., M. Ivanovich, and M.A. Wilkins. "Uranium Series Disequilibrium Studies at the Broubster Analogue Site," Harwell Laboratory Report, AERER-13609 (1989).
60. Dearlove, J.P.L., G. Longworth, and M. Ivanovich. "Improvement of Colloid Sampling Techniques in Groundwater and Actinide Characterization of the Groundwater System at Gorleben (FRG) and El Berrocal (E)," Harwell Laboratory, United Kingdom Atomic Energy Authority-AEA, D&R 0066 (1990).
61. Vilks, P., J.J. Cramer, D.B. Bachinski, D.C. Doern, and H.G. Miller. "Studies of Colloids and Suspended Particles in the Cigar Lake Uranium Deposit," Migration 91 Conference, Jerez de la Frontera, Spain, October 21–25, 1991.

62. Vilks, P., H.G. Miller, and D.C. Doern. "Natural Colloids and Suspended Particles in the Whiteshell Research Area and Potential Effects on Radiocolloid Formation," *Appl. Geochem.* 6:565–574 (1991).
63. Hamilton, E.I. "Kd Values: An Assessment of Field vs. Laboratory Measurements," in *Application of Distribution Coefficients to Radiological Assessment Models*, T.H. Silbley and C. Myttenaere, Eds. (New York: Elsevier Applied Science, 1986), pp. 35–65.
64. Gschwend, P.M., D. Backhus, J.K. MacFarlane, and A.L. Page. "Mobilization of Colloids in Groundwater Due to Infiltration of Water at a Coal Ash Disposal Site," *J. Contam. Hydrol.* 6:307–320 (1990).
65. Ryan, J.N., Jr. and P.M. Gschwend. "Colloid Mobilization in Two Atlantic Coastal Plain Aquifers," *Water Resour. Res.* 26(2):307–322 (1990).
66. Drever, J.I., Ed. *The Chemistry of Weathering* (Dordrecht, Holland: D. Reidel Publishing Co., 1985), p. 325.
67. Steefel, C.I. and P. Van Cappellen. "A New Kinetic Approach to Modeling Water-Rock Interaction: The Role of Nucleation, Precursors, and Ostwald Ripening," *Geochim. Cosmochim. Acta* 54:2657–2677 (1990).
68. Gschwend, P.M. "Geochemical Factors Influencing Colloids in Waste Environments," in *Concepts in Manipulation of Groundwater Colloids for Environmental Restoration*, J.F. McCarthy and F.J. Wobber, Eds. (Chelsea, MI: Lewis Publishers, Inc., 1993), pp. 41–48.
69. Langmuir, D. "Geochemistry of Iron in a Coastal-Plain Groundwater of the Camden, New Jersey, Area," U.S. Geological Survey Prof. Paper 650-C (1969), pp. C224–C235.
70. Liang, L., J.F. McCarthy, L.W. Jolley, J.A. McNabb, and T.L. Mehlhorn. "Iron Dynamics - Observations of Transformation during Injection of Natural Organic Matter in a Sandy Aquifer," *Geochim. Cosmochim. Acta,* in press (1993).
71. Hunter, K.A. and P.S. Liss. "The Surface Charge of Suspended Particles in Estuarine and Coastal Waters," *Nature* 282:823–825 (1979).
72. Tipping, E. "The Adsorption of Aquatic Humic Substances by Iron Oxides," *Geochim. Cosmochim. Acta* 45:191–199 (1981).
73. Davis, J.A. "Adsorption of Natural Dissolved Organic Matter at the Oxide/Water Interface," *Geochim. Cosmochim. Acta* 46:2381–2393 (1982).
74. Buddemeier, R.W. "Sampling the Subsurface Environment for Colloidal Materials: Issues Problems and Techniques," *Transport of Contaminants in the Subsurface: The Role of Organic and Inorganic Colloidal Particles,* J. F. McCarthy, Ed. DOE/ER-0331, (Washington, D.C.: U.S. DOE, 1986).
75. Degueldre, C., G. Longworth, V. Moulin, P. Vilks, C. Ross, G. Bidoglio, A. Cremers, J. Kim, J. Pieri, J. Ramsay, B. Salbu, and U. Vuorinen. "Grimsel Colloid Exercise: An International Intercomparison Exercise on the Sampling and Characterization of Groundwater Colloids," Paul Scherrer Institut, Internal Report TM-36, Würenlingen and Villigen, Switzerland (1989).
76. Degueldre, C. "Colloid Properties in Granitic Groundwater Systems with Emphasis on the Impact on Safety Assessment of a Radioactive Waste Repository," NAGRA NTB 92-05 and Paul Scherrer Institut, Report No. 39 Würenlingen and Villigen, Switzerland (1992).
77. Barcelona, M.J., J.P. Gibb, and R.A. Miller. "A Guide to the Selection of Materials for Monitoring Well Construction and Ground-Water Sampling," EPA-600/2-84-024 (1984).

78. Barcelona, M.J. "Uncertainties in Groundwater Chemistry and Sampling Procedures," in *Chemical Modeling of Aqueous Systems II, Amer. Chem. Soc. Symp. Ser. 416* (Washington, D.C.: American Chemical Society, 1990), pp. 310–320.
79. Degueldre, C., R. Keil, M. Mohos, B. Van Eygen, and B. Wernli. "Study of the Leuggern Groundwater Colloids," Paul Scherrer Institut, Internal Report TM-43-90-69, Würenlingen and Villigen, Switzerland (1990).
80. Backhus, D.A. "Colloids in Groundwater: Laboratory and Field Studies of their Influence on Hydrophobic Organic Contaminants," PhD Thesis, Massachusetts Institute of Technology, Cambridge, MA (1990).
81. Backhus, D.A., J.N. Ryan, D.M. Groher, J.K. MacFarlane, and P.M. Gschwend. "Sampling Colloids and Colloid-Associated Contaminants in Groundwater," *Groundwater* (in press).
82. Herzig, J.P., D.M. Leclerc, and P. LeGoff. "Flow of Suspensions through Porous Media-Application to Deep Filtration," *Ind. Eng. Chem.* 62:8–35 (1970).
83. Ryan, J.N., Jr. "Groundwater Colloids in Two Atlantic Coastal Plain Aquifers: Colloid Formation and Stability," Master's Thesis, Massachusetts Institute of Technology, Cambridge, MA (1988).
84. Hubbe, M.A. "Detachment of Colloidal Hydrous Oxide Spheres from Flat Solids Exposed to Flow. I. Experimental System," *Colloids Surf.* 16:227–248 (1985).
85. Hubbe, M.A. "Detachment of Colloidal Hydrous Oxide Spheres from Flat Solids Exposed to Flow. II.Mechanism of Release," *Colloids Surf.* 16:249–270 (1985).
86. Hunt, J.R. "Particle Dynamics in Seawater: Implications for Predicting the Fate of Discharged Particles," *Environ. Sci. Technol.* 16:303–309 (1982).
87. Puls, R.W., D.A. Clark, B. Bledsoe, R.M. Powell, and C.J. Powell. "Metals in Groundwater: Sampling Artifacts and Reproducibility," *Haz. Waste Haz. Mater.* 9(2): in press (1992).
88. Barcelona, M.J., J.A. Helfrich, and E.E. Garske. "Sampling Tubing Effects on Groundwater Samples," *Anal. Chem.* 57:460–464 (1985).
89. Holm, T.R., G.K. George, and M.J. Barcelona. "Oxygen Transfer through Flexible Tubing and Its Effects on Groundwater Sampling Results," *Groundwater Monit. Rev.* (1988).
90. Stumm, W. and J.J. Morgan, Eds. *Aquatic Chemistry: An Introduction Emphasizing Chemical Equilibria in Natural Waters* (New York: John Wiley & Sons, Inc., 1981), p. 780.
91. Degueldre, C., B. Baeyens, W. Görlich, J. Riga, J. Verbist, and P. Stadelmann. "Colloids in Water from a Subsurface Fracture in Granitic Rock, Grimsel Test Site, Switzerland," *Geochim. Cosmochim. Acta* 53:603–610 (1989).
92. Lieberman, S.H., G.A. Theriault, S.S. Cooper, P.G. Malone, R.S. Olsen, and P.W. Lurk. "Rapid Subsurface, *In situ* Field Screening of Petroleum Hydrocarbon Contamination Using Laser-Induced Fluorescence Over Optical Fibers," in *Field Screening Methods for Hazardous Wastes and Toxic Chemicals,* L. Williams, Ed. (Washington, D.C.: U.S. EPA, 1991), pp. 57–63.
93. Haas, J.W., T.G. Matthews, and R.B. Gammage. "*In situ* Detection of Toxic Aromatic Compounds in Groundwater Using Fiberoptic UV Spectroscopy," in *Field Screening Methods for Hazardous Wastes and Toxic Chemicals,* L. Williams, Ed. (Washington, D.C.: U.S. EPA, 1991), pp. 677–681.

94. McCarthy, J.F., T.M. Williams, L. Liang, P.M. Jardine, L.W. Jolley, D.L. Taylor, A.V. Palumbo, and L.W. Cooper. "Mobility of Natural Organic Matter in a Sandy Aquifer," *Environ. Sci. Technol.* in press (1993).
95. Palumbo, A.V., P.M. Jardine, J.F. McCarthy, and B.R. Zaidi. "Characterization and Bioavailability of Dissolved Organic Carbon in Deep Subsurfaced Surface Waters," *Proceedings of the First International Symposium on Microbiology of the Deep Subsurface,* C.B. Fliermans and T.C. Hazen, Eds. (Aiken, SC: Westinghouse, Savannah River Company, 1990), pp. 57–68.
96. Walton-Day, K., D.L. Macalady, M.H. Brooks, and V.T. Tate. "Field Methods for Measurement of Groundwater Redox Chemical Parameters," *Groundwater Monit. Rev.* (1990).
97. Brütsch, R., C. Degueldre, and H.J. Ulrich. "Kolloide im Thermalwasser von Bad Zurzach," Paul Scherrer Institut, Internal Report TM-43-91-19 (1991).
98. Bochert, U.K. and W. Dannecker. "On-Line Aerosol Analysis by Atomic Emission Spectroscopy," *J. Aerosol Sci.* 20(8):1525–1528 (1989).
99. Wernli, B. Personal communication. Paul Scherrer Institut, Würenlingen and Villigen, Switzerland, (1991).
100. Danielsson, L. "On the Use of Filters for Distinguishing between Dissolved and Particulate Fractions in Natural Waters," *Water Res.* 16:179–182 (1982).
101. Buffle, J. *Complexation Reaction in Aquatic Systems: An Analytical Approach* (Chichester, England: Ellis Horwood Limited, 1988).
102. Lieser, K.H., S. Sondermeyer, and A. Kliemchen. "Einfluss von Beimengungen auf die Reproduzierbarkeit und Richtigkeit von Analysenergebnissen bei der Bestimmung der Elemente Cd, Cr, Cu, Fe, Mn, and Zn mit der flammenlosen Atomabsorption spectroscopic, Fresenius Z," *Anal. Chem.* 312:520 (1982).
103. Lieser, K.H., B. Gleitsmann, and Th. Steinkopff. "Colloid Formation and Sorption of Radionuclides in Natural Systems," *Radiochim. Acta* 40:39–47 (1986).
104. U.S. Environmental Protection Agency. "Method 200.8 Determination of Trace Elements in Waters and Wastes by Inductively Coupled Plasma-Mass Spectroscopy (Version 4.3)," *ICP Inf. Newsl.* 16(8):460 (1991).
105. Longworth, G., C.A.M. Ross, C. Degueldre, and M. Ivanovich. "Interlaboratory Study of Sampling and Characterization Techniques for Groundwater Colloids," Harwell Laboratory Report AERER-13393 (1989).
106. Waber, U., C. Lienert, and H.P. Von Gunten. "Colloid Related Infiltration of Trace Metal from a River to Shallow Groundwater," *J. Hydrol. Contan.* 6:251–265 (1990).
107. Degueldre, C. and B. Wernli. "Characterization of the Natural Inorganic Colloids from a Reference Granitic Groundwater," *Anal. Chim. Acta* 195:211–223 (1987).
108. Brütsch, R. and C. Degueldre. "Kolloide von Mineralwasser Eglisau III. Probenahme und Charakterisierung Phase 2," Paul Scherrer Institut, Internal Report TM-43-90-38, Würenlingen and Villigen, Switzerland (1990).
109. Buchholtz ten Brink, M., S. Martin, B. Viani, D.K. Smith, and D. Phinney. "Heterogeneities in Radionuclide Transport: Pore Size, Particle Size, and Sorption," *Concepts in Manipulation of Groundwater Colloids for Environmental Restoration*, J. F. McCarthy and F. J. Wobber, Eds. (Chelsea, MI: Lewis Publishers, Inc., 1993), pp. 203–210.

110. Pfeiffer, H.-R., A. Sanchez, and C. Degueldre. "Thermal Springs in Granitic Rocks from the Grimsel Pass (Swiss Alps): The Late Stage of a Hydrothermal System Related to Alpine Orogeny." *Proceedings of the Seventh International Symposium on Rock-Water Interaction.* Y.K. Kharaka and A.S. Maest, Eds. (Rotterdam: A.A. Balkema Publishers, 1992).
111. Higgo, J.J.W. "Review of Sorption Data Applicable to the Geological Environments of Interest for the Deep Disposal of ILW and LLW in the UK," Safety Studies Nirex Radioactive Waste Disposal, British Geological Survey, Keyworth, Nottingham, NSS/R-162 (1988).
112. Vilks, P. and C. Degueldre. "Sorption Behavior of ^{85}Sr, ^{131}I, and ^{137}Cs on Grimsel Colloids," *Appl. Geochem.* 6:553–563 (1991).
113. Driscoll, F.G. "Groundwater and Wells," Johnson Filtration Systems, St. Paul, MN (1986), pp. 268–333.
114. Alexander, W.R., R. Brütsch, C. Degueldre, and B. Hofmann. "Evaluation of Long Distance Transport of Natural Colloids in a Crystalline Groundwater," Paul Scherrer Institut, Internal Report, TM-43-90-20 (1990).
115. Miekeley, N., H. Countinho de Jesus, C. Porto da Silveira, and C. Degueldre. "Chemical and Physical Characterization of Suspended Particles and Colloids in Water from Osamu Utsumi Mine and Morro do Ferro Analog Study Sites, Poços de Caldas, Brazil," NAGRA Technical Report NTB 90-27 Wettingen, Switzerland, and SKB Report TR-90-18 Stockholm, Sweden (1991).
116. DeBlois, R.W. and C.P. Bean. "Counting and Sizing of Submicron Particles by the Resistive Pulse Technique," *Rev. Sci. Instrum.* 41:909 (1970).
117. Baier, R.W., T.L. Cucci, and C.M. Yensch. "Flow Cytometry for Observing the Distribution of Adsorption Sites on Particles," *Limnol. Oceanogr.* 34(5):947–952 (1989).
118. Newman, K.A., F.M. Morel, and K.D. Stolzenbach. "Settling and Coagulation Characteristics of Fluorescent Particles Determined by Flow Cytometry and Fluorometry," *Environ. Sci. Technol.* 24(4):506–519 (1990).
119. Van Dilla, M.A., P.N. Dean, O.D. Laerum, and M.R. Melamed. *Flow Cytometry: Instrumentation and Data Analysis* (New York: Academic Press, Inc., 1985).
120. Pelssers, E., M.S. Cohen, and G.J. Fleer. "Single Particle Optical Sizing," *J. Colloid Interface Sci.* 137:350–361 (1990).
121. Kawaguchi, H., N. Fukawassa, and A. Mizuike. "Investigation of Airborne Particles by Inductively Coupled Plasma Emission Spectroscopy Calibrated with Monodisperse Aerosols," *Spectrochim. Acta* 41B:1277 (1986).
122. Harvey, R.W., R.L. Smith, and L.H. George. "Effect of Organic Contamination upon Microbial Distributions and Heterotrophic Uptake in a Cape Cod, Mass., Aquifer," *Appl. Environ. Microbiol.* 48(6):1197–1202 (1984).
123. Pedersen, K. and S. Ekendahl. "Distribution and Activity of Bacteria in Deep Granitic Groundwater of Southeastern Sweden," *Microb. Ecol.* 20(1):37–52 (1990).
124. Blake, D.F. "Scanning Electron Microscopy," in *Instrumental Surface Analysis of Geologic Materials,* D.L. Perry, Ed. (New York: VCH Publishers, Inc., 1990), pp. 11–43.
125. Degueldre, C., M. Mohos, R. Brütsch, and H. Grimmer. "Characterization of Gorleben Colloid by Ultrafiltration and Scanning Electron Microscopy," Paul Scherrer Institut, Internal Report TM-43-88-05, Würenlingen and Villigen, Switzerland (1988).

126. Emch, R., X. Clivaz, C. Taylor-Denes, P. Vaudaux, and P. Descouts. "Scanning Tunneling Microscopy for Studying the Biomaterial-Biological Tissue Interface," *J. Vac. Sci. Technol.* 8:655 (1990).
127. Emch, R., X. Clivaz, C. Taylor-Denes, P. Vaudaux, D. Lew, and P. Descouts. "Study of the Biocompatibility of Surgical Implant Materials at the Atomic and Molecular Level Using STM," in *Scanning Tunneling Microscopy and Related Methods,* R.J. Behm, Ed. (The Netherlands: Kluwer Academic Publishers, 1990), pp. 349–358.
128. Emch, R., F. Zenhausern, M. Jobin, M. Taborelli, and P. Descouts. "Morphological Difference Between Fibronectin Sprayed on Mica and on PMMA," in Ultramicroscopy 1992, Proceedings of the 6th International Conference on STM and Related Techniques, Switzerland (1991).
129. Zenhausern, F., M. Adrian, R. Emch, M. Taborelli, M. Jobin, and D. Descouts. "Scanning Force Microscopy and Cryo-Electron Microscopy of Tobacco Mosaic Virus as a Test Specimen," *Ultramicroscopy* 42–44:1168–1172 (1992). (Proceedings of the 6th International Conference on Scanning Tunnelling Microscopy and Related Techniques.)
130. Gardner, M.P. "The Measurement of the Molecular Weight of Humic Acid by Ultracentrifugation," United Kingdom Atomic Energy Authority, Harwell Laboratory, Oxfordshire, AERE M 3730 (1989).
131. De Boer, G.B., C. De Weerd, C. Thoenes, and H.W. Goossens. "Laser Diffraction Spectrometry: Frannhofer Diffraction vs. Mie Scattering," *Part. Charact.* 4:14–19 (1987).
132. Buckau, G., R. Stumpe, and J.I. Kim. "Am-Colloid-Generation in Groundwaters and its Speciation by Laser-Induced Photoacoustic Spectroscopy," *J. Less-Common Met.* 122:555–562 (1986).
133. Kim, J.I., G. Buckau, F. Baumgartner, H.C. Moon, and D. Lux. "Colloid Generation and the Actinide Migration in Gorleben Groundwaters," *Mater. Res. Soc. Symp. Proc.* 26:31–40 (1984).
134. Vilks, P., J.J. Cramer, T.A. Shenchuch, and J.P.A. LaRocque. "Colloid and Particulate Matter Studies in the Cigar Lake Natural Analog Study," *Radiochim. Acta* 44–45:305–310 (1988).
135. Kolthoff, I.N., P.J. Eving, and V. Krivam, Eds. "The Nuclear Activation and Radioisotopic Methods of Analysis," in *Treatis on Analytical Chemistry Part I* (New York: John Wiley and Sons, Inc., 1976).
136. Marley, N.A., J.S. Gaffney, K.A. Orlandini, K.C. Picel, and G.R. Choppin. "Chemical Characterization of Size-Fractionated Humic and Fulvic Materials in Aqueous Samples," *Sci. Total Environ.* (in press).
137. Perry, D.L., J.A. Taylor, and C.D. Wagner. "X-Ray-Induced Photoelectron and Auger Spectroscopy," in *Instrumental Surface Analysis of Geologic Material*, D.L. Perry, Ed. (New York: VCH Publishers, Inc., 1990), pp. 45–86.
138. Lieser, K.H., A. Amet, R. Hill, R.N. Stingl, and B. Thybusch. "Colloids in Groundwater and their Influence on Migration of Trace Elements and Radionuclides," *Radiochim. Acta* 49:83–100 (1990).
139. Billion, A., M. Caceci, G. Della Mea, T. Dellis, J.C. Dran, V. Moulin, S. Nicholson, J.C. Petit, J.D.F. Ramsay, P.J. Russell, and M. Theyssier. "The Role of Colloids in the Transport of Radionuclides in Geological Formations," Nuclear Science and Technology, Commission of the European Communities, CEC Report: EUR 13506 EN (1991).

140. Degueldre, C., A. Scholtis, P. Gomez, and A. Laube. "Wellenberg Colloid Exercise, Phase I," Paul Scherrer Institut, Internal Report, TM-43-92, in press (1992).
141. Laaksoharju, M. and C. Degueldre. "Colloids from the Swedish Granitic Groundwater," SKB-Technical Report, in press.
142. Tullborg, E.-L. Geological Information Concerning the 10 Sites in Sweden. Personal communication (1992).
143. Smellie, J., A. Winberg, B. Allant, C. Pettonson, C. Degueldre, M. Laaksoharju, G. MacKenzie, K. Pedersen, E.-L. Tullborg, and M. Wolf. "Background and Status of SKB Participation in the Oklo Project (August 1990-December 1991) and Future Recommendations," in *Oklo Analogue Naturel de Stockage de Déchets Radioactifs,* A.M. Chapuis and P.-L. Blanc, Eds. (France: Commisariat de l'Energie Atomique, 1992), SERG 92–95.
144. Kingston, W.L. "Characterization of Colloids Found in Various Groundwater Environments in Central and Southern Nevada," MS Thesis, University of Nevada, Reno, NV (1989).
145. West, J.M., C. Degueldre, M. Allen, R. Bruesch, S. Gardner, S. Ince, and A.E. Milodowski. "Microbial and Colloidal Populations in the Maqarin Groundwaters," NAGRA Report NTB-91-10, Wettingen, Switzerland (1991).
146. Ronen, D., M. Magaritz, U. Weber, A.J. Amiel, and E. Klein. "Characterization of Suspended Particles Collected in Groundwater under Natural Gradient Flow Conditions," *Water Resour. Res.* 28:1279–1291 (1992).

CHAPTER 7

MANGANESE PARTICLES IN FRESHWATERS

Richard R. De Vitre and William Davison

TABLE OF CONTENTS

0-87371-895-X/93/$0.00+$.50

SOURCES, BEHAVIOR, AND FATE OF Mn PARTICLES

Manganese enters rivers and lakes from a variety of sources. The weathering of rocks and soils directly exposed to surface waters is the largest single natural source. Wet and dry fallout from the atmosphere can be important but may be expected to vary from one location to another, depending on the relative proportion of Mn particles released by natural sources (weathered dust from rocks and soils) and from anthropogenic sources such as the combustion of fossil fuels and the mining and processing of metals. Biogenic sources can also, in certain specific cases, be significant: many higher plants preferentially accumulate Mn in their leaves as compared to other transition metals[1] (Cu, Fe, and Zn, for instance). This contribution may be greatly magnified if an atmospheric pollution source is located nearby.

Manganese Bearing Rocks and Minerals

Although manganese is quite abundant in the earth's crust (0.1 %), there are relatively few Mn rich minerals. Most of the Mn in igneous rocks, resulting from the solidification of magma, is found in ultramafic or basaltic type rocks, including ortho- and clino-pyroxenes, olivines, biotites, amphiboles, and magnetite. Manganese, as well as other transition metals,[2] are minor components (ca. 100 to 2000 mg/kg) of these minerals. Since the manganese in these minerals is present as Mn(II), weathering of these rocks will lead to the release of Mn_{aq}^{2+} which may then undergo complexation, adsorption, oxidation-hydrolysis, or precipitation reactions depending on the prevailing physicochemical conditions.

Mn-specific minerals are quite scarce and, in common with other transition metal minerals, often include two or more formal oxidation states. Among the more common Mn minerals are the oxides and oxyhydroxides such as pyrolusite (MnO_2) and psilomelane ($BaMn(II)Mn(IV)_8O_{16}(OH)_4$), and in near surface deposits manganite (MnOOH), braunite ($(Mn,Si)_2O_3$ and hausmanite (Mn_3O_4). Less common manganese minerals, generally found in more reducing environments, include pyrochroite ($Mn(OH)_2$), manganosite (MnO), and alabanite (MnS). Finally, rhodochrosite ($MnCO_3$) is found in sedimentary rocks, and as an impurity in carbonate rich ores such as calcium carbonates through ionic substitution since the charge and ionic radii of Mn^{2+} and Ca^{2+} are quite similar.

As already mentioned, Mn specific minerals are quite rare. It is therefore unlikely that the concentration in soil pore waters of Mn_{aq}^{2+} due to weathering is controlled simply by the solubility of a given mineral or its rate of solution.

Although these two parameters will play a part, the final concentration will also depend on the prevalent pH, pε (–log electron activity), and pCO_2, as well as

- complexation by dissolved and particulate organic matter
- the activity of dissolved S(-II) species
- surface complexation by colloidal or particulate inorganic phases such as clays, Fe, and Mn oxyhydroxides
- biotic assimilation
- redox interconversion between Mn(II) and Mn(III/IV) by either biotic or abiotic pathways

The redox interconversion is particularly important since the geochemical behavior of Mn in freshwater systems, such as soil pore waters, groundwater, rivers, and lakes (see below), as well as its mobility, is largely controlled by its redox chemistry (see Section 3).

Geochemical Behavior in Freshwater

Reducing environments in which oxygen is absent or at low concentration commonly occur in freshwater. Mn(II), which is soluble and relatively free from complexation, is stable under these conditions, and so it may be present at quite high (10^{-4} mol dm^{-3}) concentrations. Within sediments, for example, the removal of oxygen is usually complete due to its rate of supply by molecular diffusion being slower than the rate of oxidation of labile organic material. Manganese oxyhydroxides, as well as nitrate, iron oxyhydroxides, and sulphate, can act as alternative electron acceptors to oxygen in anaerobic decomposition processes. The sequence of reduction is usually in the approximate thermodynamic order given above,[3] but it may vary due to the local concentrations of each component and the prevalence of particular microorganisms. Oxygen microelectrodes have been used to show that in most reasonably productive aquatic systems, where there is a good supply of labile organic material, oxygen fails to penetrate more than 1 mm into the sediment, and in some very productive systems its penetration cannot be measured (<10 μm).[4] Microelectrodes have illustrated the spatial heterogeneity of sediments attributed to discrete particles rich in organic material.[5,6] In unproductive systems such as oligotrophic lakes or fast flowing rivers oxygen may penetrate to a depth of several cm, and so reduction of manganese may be important at such depths.[7,8]

In productive lakes which thermally stratify, the consumption of oxygen associated with the decomposition of organic material may be sufficient to cause the complete removal of oxygen from the bottom waters, or hypolimnia. This oxygen deficiency may be permanent, as in meromictic systems, or seasonal, as in monomictic or dimictic systems.[9] The concentrations, transformations, and interactions of manganese in such systems have been well studied (see below). There are also several reports[9,23] of lakes where oxygen is depleted or absent from a mid-water zone, but there have been no systematic studies of the manganese chemistry of such systems.

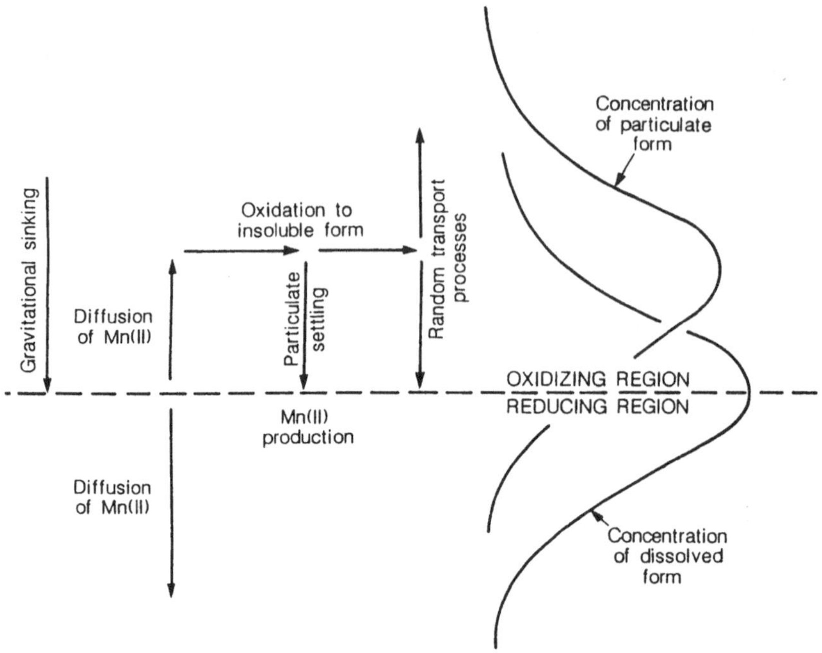

Figure 1. Generation of concentration profiles of dissolved and particulate manganese in the vicinity of a redox boundary. (Adapted from Davison, W. *Chemical Processes in Lakes*, W. Stumm, Ed. (New York: Wiley-Interscience, 1985), pp. 31–53. With permission.)

A generalized conceptual model for the transport of manganese at a redox boundary formed in water columns, sediments, or at the sediment-water interface, has been proposed[10] (Figure 1). It assumes that particulate oxidized forms of Mn are separated from the dissolved reduced form Mn(II) by a well-defined redox boundary. There is a continuous supply of oxidized particulate material to the boundary coming from settling particles in water columns or from the process of accumulation in sediments. This fresh material is instantaneously reduced as it crosses the boundary, providing a localized source of Mn(II). Transport from the boundary is assumed to be random; in water columns, both particles and solution components are subjected to eddy diffusion, and in sediments, solution species are transported by molecular diffusion, which can be augmented by bioirrigation. Unless bioturbation is particularly active, particles within sediments are not transported by random processes.

Mn(II) diffuses upward and downward away from its point source immediately below the redox boundary. The upwardly diffusing Mn(II) will be oxidized at some distance within the oxic zone, which depends on the rate of oxidation. This oxidation creates a local source of particulate material which, depending on the size of the particles, may be subject to gravitational forces causing downward transport. There are also random solution transport processes which encourage both upward and downward movement. If random transport processes operate uniformly throughout the region of the redox boundary, Gaussian-shaped concentration profiles of both Mn(II) and Mn(III/IV) are maintained.

To maintain this steady state, Mn(II) must be removed at depth away from the boundary, and Mn(III/IV) must be removed at some distance above the boundary. $MnCO_3$ has been shown to form within sediments,[8,11] so consuming Mn(II), and particulate material which enters surface waters is removed by flushing. The system is primed by fresh supplies of Mn(III/IV) entering via the inflow.

As lakes and rivers are dynamic systems which are subjected to episodic events such as floods and phytoplankton production, no true steady state is ever achieved. Perhaps the best approximation to such conditions occurs in sediments overlain by permanently well-oxygenated water where Mn(II) has been observed to be at a maximum between 1 and 5 cm below the sediment surface.[8,11] Such profiles may represent a two stage reduction, with a labile manganese fraction being reduced at the sediment surface, the resulting Mn(II) being returned to the water column without affecting the pore-water concentrations.

MANGANESE PARTICLES IN FRESHWATER SYSTEMS

Mn Oxyhydroxide Particles

The production of Mn(III/IV) oxyhydroxide particles within a sediment from the oxidation of upwardly diffusing Mn(II) (Figure 1), can lead to the enrichment of manganese near the surface of the sediment. Surface crusts with high concentrations of iron and manganese have been observed in a number of freshwater environments,[12,13] but the formation of such freshwater manganese nodules has been studied most extensively in Oneida Lake, New York.[14] In this productive but relatively shallow and, thus, well-oxygenated system, nodules form on sediments at intermediate water depths rather than at the deepest locations where the oxidation process may be inhibited. The mechanism illustrated by Figure 1 is an oversimplification, since there are indications that, as well as being supplied by local reduction, Mn(II) may be released from phytoplankton and transported by convection from deeper sites.

Even in oxygenated water columns, the sediment water interface usually has a good supply of labile organic material, and manganese particles which have settled through the water column are captive at this site. Therefore Mn(III/IV) is often reduced at the interface rather than accumulating in the sediment. The Mn(II) which is formed is mainly remobilized into the overlying water, where the rate of vertical eddy diffusion[15] is typically three to five orders of magnitude greater than molecular diffusion. Rapid mixing and dilution in the well-mixed waters of an oxygenated lake usually ensure that the released Mn(II) cannot be measured, but when a lake stratifies prior to deoxygenation, elevated concentrations of manganese above the sediment are immediately observed (Figure 2). Less than 10% of the particulate manganese which reached the sediment of a small

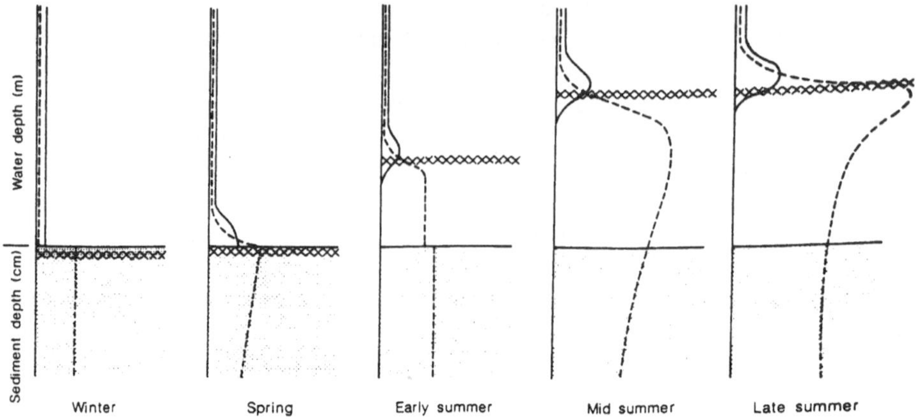

Figure 2. Schematic representation of the oxic-anoxic boundary's seasonal variation in a lake. xxxxx = oxic-anoxic boundary; — particulate Mn concentration profile; ---- dissolved Mn (II) concentration profile.

seasonally anoxic lake during the winter months, when the waters were well oxygenated, was found to accumulate the remaining 90% being re-released into the water column after reduction at the interface.[16] The concentration in the interstitial waters at this time was found to decrease slightly from the sediment surface, in keeping with supply from the interface.[17] The net result of these processes is that most of the manganese entering such a lake is exported via the outflow, and yet it has undergone a redox cycle at the sediment surface. A similar result was found for a large (207 km^2), shallow (mean depth = 6.8 m), well-oxygenated lake where there was strong evidence for pronounced redox cycling and yet 70% of the manganese in the inflow was exported via the outflow.[14] In deep systems, similar cycling at the sediment-water interface occurs, but the released Mn(II) is reoxidized and sinks back to the sediment to form an isolated cycle which allows very little export of manganese.[18-20]

This loss from shallow sediments and retention by deep ones is a general phenomenon which results in the bulk concentration of manganese in sediments increasing with the depth of the overlying water.[10,21-23] Such an observation has paleolimnological applications, as changes in concentrations in long cores have been interpreted in terms of historical changes in water depth.[22] The process of loss from shallow lakes is also consistent with the ratio of sedimentary iron to manganese being 50:1[16] whereas the same ratio in worldwide river water is 3:1.[23]

Exchange Fluxes at the Sediment Water Interface

Quantitative data for rates of supply to sediments and rates of release are available for a range of freshwater lakes (Table 1). Fluxes to and from sediments are remarkably similar for the mainly small, productive systems where measurements have been made, providing support for the notion that most manganese is

Table 1. Examples of Manganese Fluxes Measured in Lakes. For Sediment Trap Data a Mean Provided from a Long Time Period (Usually a Year) and a Range from Shorter Periods of 1–4 Weeks is Given

Type of Flux	Method of Measurement	Flux ($\mu g\ cm^{-2}/d$)	Situation	Ref.
Sediment release	Hypolimnetic accumulation	12.6	Seawater bay, 60–70 m deep	103
Sediment release	Hypolimnetic accumulation	3.4	Productive hard water, 24 m deep	24
Sediment release	Hypolimnetic accumulation	6.6–13.0	Productive hard water, 30 m deep	26
Sediment release	*In situ* benthic chamber	9.4	Reservoir, mean depth 22 m	25
Sediment release	Pore water profiles	0.2	Productive hard water, 24 m deep	24
Sediment release	Pore water profiles	1.3	Reservoir, mean depth 22 m	25
Sediment release	Interfacial mass balance	13	Productive soft water, 16 m deep	17
Sediment release	*In situ* benthic chamber	6.8	Productive hard water, 87 m deep	38
Sediment release	Interfacial mass balance	5.2	Medium productive hard water, 130 m deep	55
Sinking particles	Sediment traps	3.5 (1–35)	Productive soft water, 16 m deep	16
Sinking particles	Sediment traps	5 (2–16)	Medium productive hard water, 130 m deep	55
Sinking particles	Sediment traps	2 (0.5–17)	Small, productive, 8.5 m deep	31
Sinking particles	Sediment traps	0.8 (0.1–4)	Large, mesotrophic hard water, 147 m deep study basin	18
Sinking particles	Sediment traps	15 (2–60)	Productive soft water, 42 m deep basin	19

re-released after contacting the sediment. Where concentration gradients from pore waters to overlying waters were used to assess fluxes, they were 7 to 50 times less than estimates made from the change in concentration in the bottom waters[24] or by benthic chamber.[25] This difference may be due to the hypolimnion accumulation having a component from dissolution of sinking manganese particles, or the pore water gradients may underestimate the flux because 1 cm resolution is insufficient to define the gradient.[17] More manganese is usually collected by traps situated near the bottom of a lake due to the local cycle of release of Mn(II), oxidation, and settling.[16,18,55] In one seasonally anoxic lake, differences in the trapped amounts at different depths in the water column were used to estimate that 56 to 78% of the dissolved manganese accumulating in the hypolimnion comes from dissolution of particles in the water column.[16] Traps provide integrated and therefore averaged information, whereas concentration profiles reflect the instantaneous events of the moment. Analyses of the shape of the concentration profiles from the same lake at the end of the stratification period showed that as much as 80 to 90% of the local supply of manganese could be due to dissolution of sinking particles. Sediment release rates have also been estimated from concentration gradients measured immediately (5 to 50 cm) above the sediment. If release from the sediment is dominant, these fluxes agree with mass balance calculations.[26]

Mn Oxyhydroxide Particles Produced in Water Columns

A pseudo-steady state is approached in the water columns of permanently anoxic basins[27] and towards the end of summer, stratification in a seasonally anoxic system.[24,28,29] The reducing intensity is sufficient to ensure that particulate manganese sinking through the water column is reduced below the redox boundary so that a maximum in the concentration of Mn(II) develops within the water column. A maximum in the concentration of Mn(III/IV) may develop above the Mn(II) maximum as the upwardly diffusing Mn(II) is oxidized.[28-32] Mn(II) usually takes days to months (see Section 3.1) to oxidize in circumneutral lake water, and so the resulting maximum of particulate manganese may occur in relatively shallow water where it is readily dispersed and obscured by mixing processes. Both particulate manganese and Mn(II) are exported from the lake via the outflow, but some of the particulate manganese sinks to complete the cycle.

The conditions which determine the location of the redox boundary, such as the hydrodynamic regime, the supply of readily reducible organic material, and its rate of decomposition are continually changing in most freshwater (Figure 2). Consequently, the simple steady state assumptions appropriate to Figure 1 do not hold. The systematic series of events which occur in seasonally anoxic lakes leads to a well-documented migration of the redox boundary.[10,33] The associated changes in the concentration of soluble and particulate manganese have also been discussed by several authors.[16,17,30,34,35]

In winter the waters of the lake are well mixed and saturated with oxygen, with the oxic/anoxic boundary located a few mm within the sediment. There is a continual small production of Mn(II) at the interface as fresh supplies of manga-

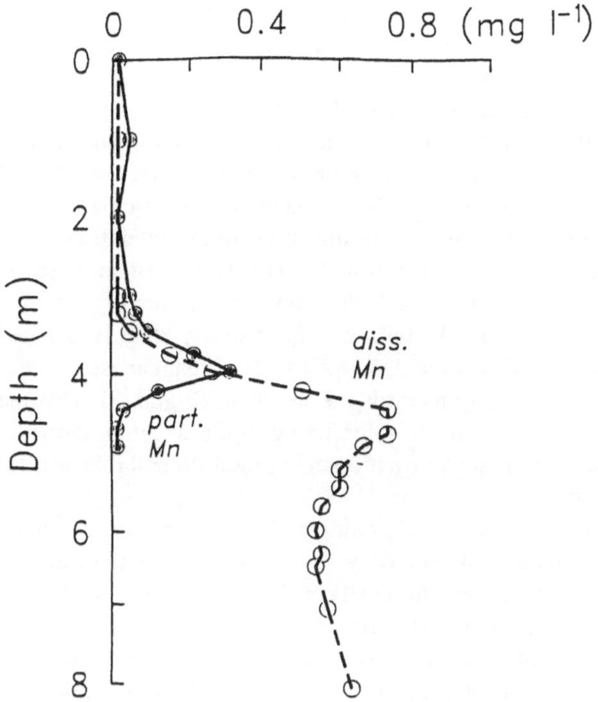

Figure 3. Vertical distribution of dissolved and particulate manganese in Lake Fukami-ike. (Adapted from Yagi, A. *Jpn. J. Limnol.* 49:149–156 (1988). With permission.)

nese particles undergo some reduction, the extent of which being determined by the limited supply of readily reducible organic material which decomposes more slowly at the low winter temperatures. Some of the Mn(II) is released to the overlying water, but the rapid mixing of these waters with consequent dilution prevents measurement. The reducing conditions of the interstitial waters permit quite high concentrations of Mn(II), and there is a slight increase in concentration towards the sediment surface, indicating a small supply from the interface. A bloom of diatoms in early spring provides a fresh supply of organic carbon, and its decomposition, as temperatures rise, consumes oxygen, causing the boundary to migrate so that it is coincident with the sediment-water interface. The additional supply of electrons from the extra organic carbon accelerates the reduction of manganese to Mn(II). There is a noticeable increase in pore water concentrations and a more marked downward gradient. Mn(II) and its oxidation products both appear in the bottom waters which are still well oxygenated, but this is due as much to stratification sufficiently restricting water movements to allow gradients to develop as to the increased supply of Mn(II).

As the bottom waters become devoid of oxygen, the redox boundary migrates up the water column. The nearly constant vertical profile of Mn(II) in early summer is due to the dual supply from the sediment and from reduction of sinking manganese particles occurring within the water column. Consequently, fewer particles are reduced at the interface and the subsurface concentration of Mn(II) declines to give a nearly constant vertical profile within the pore water. Particulate manganese is largely present above the redox boundary. By midsummer the water column reduction of Mn(II) has become more dominant so that a midwater maximum of Mn(II) begins to develop below the redox boundary. Manganese in the bottom waters is then solely supplied by sinking particles, there being a gradient of Mn(II) from the overlying water to the interstitial waters so that the sediments become a sink for Mn(II). The late summer case is typical of a meromictic situation when a well-developed maximum of particulate manganese overlies a maximum of Mn(II) and a near steady state is attained. The profile observed in Lake Fukami-ike[36] shows dissolved manganese concentrations forming a peak and increasing towards the sediment (Figure 3). This may be attributable to very reactive particles solubilizing in the water column below the redox boundary and a less reactive fraction being reduced and released at the sediment-water interface.

There is little doubt that O_2 oxidizes Mn(II) directly, albeit via a microbial process. The reduction of Mn(III/IV) as it sinks through the water column may be brought about by reaction with Fe(II) or S(-II).[29,37] Three attributes of manganese are essential to the above description of events: readily reducible particulate manganese, interstitial Mn(II) easily exhausted, and rate of manganese reduction dependent on supply of organic material. To allow a midwater maximum of Mn(II), manganese particles of both allochthonous and autochthonous origin must be readily reduced. It has been shown that both field collected and laboratory prepared manganese oxides have a reactive and unreactive fraction, so perhaps during deep-water anoxia part of the manganese particles react immediately, the remainder having time to reach the sediment for permanent incorporation or slow subsequent reduction. Compared to the bulk concentration, the interstitial waters represent only a small fraction of the available manganese[38] (~0.01%), so there is a tendency to exhaustion of the Mn(II) supplied from within the sediments, which suggests that the fraction which was unreactive in the water column or at the sediment-water interface remains unreactive even after prolonged residence within the sediment. The largely circumstantial evidence for the dependence of manganese reduction in lakes on the supply of organic material and the temperature is supported by marine studies which have shown a good correlation between manganese fluxes from coastal surface sediments with carbon fluxes and temperature.[39]

Solubility of Mn(II) in the Water Column

The above discussion of manganese in seasonally anoxic systems has focused on the dynamics of reduction and oxidation and has not considered possible thermodynamic controls on Mn(II). Several workers have reported that for rhodochrosite ($MnCO_3$) the ion activity product approaches and sometimes exceeds the solubility product in the bottom waters of seasonally anoxic hard water lakes.[28,38,40-42]

Certainly, $MnCO_3$ has never been identified positively as a major component of the water column particulate material. It is worth noting that in most anoxic basins the alkalinity decreases from the sediment towards the surface waters. This alkalinity gradient would lead to depletion of Mn(II) in the bottom waters if $MnCO_3$ precipitation controlled the concentration of Mn(II), as observed by Sigg et al.[42] Such a downwardly decreasing concentration profile is the type that develops due to redox cycling in the water column (Figure 1).

Whether or not Mn(II) concentrations are controlled by rhodocrosite, manganese is readily released when sediments are acidified, as shown in experiments using artificial enclosures within a lake.[43] In oxygenated acid waters (pH 4.8 to 5.4), Mn(II) was found to be close to saturation with respect to manganite, MnOOH, with typical concentrations of 1 to 2.3 μmol/dm^{-3}.[44] Even in these acidic conditions, however, there was evidence for manganese release from sediments and associated cycling in the water column when concentrations of dissolved oxygen became depleted during stratification (90 μmol/dm^{-3}). Such cycling had little effect on the overall transport through the system. Manganese was a conservative solute, there being no significant difference between the measured annual influx (180 mmol m^{-2} y^{-1}) and outflux (210 mmol m^{-2} y^{-1}). Experimental and in-lake measurements of distribution coefficients between particles and solution, K_D, showed that there was a threshold for manganese adsorption to particles at pH 5. Below this pH, adsorption was represented by a constant, low value of K_D of about 1000 dm^{-3} kg^{-1}, while above pH 5, log K_D increased systematically with pH.

Physicochemical Characteristics of the Particles

The oxidation state of manganese in lake waters is variable, and so particulate manganese is often described as MnOx. When ultrafiltration has been used to investigate the size distribution of particles within a seasonally anoxic lake, most of the manganese present at high concentration in the bottom waters has been found to pass membranes with a molecular weight cut off limit <1000[36] (Figure 4), in keeping with polarographic measurements which have shown that manganese is present as Mn(II).[26,29,45] What is surprising, however, is that there can be appreciable (10 to 50%) proportions of very small colloidal manganese (nominal mol. wt. <50,000),[36] which has again been confirmed by polarography.[29] Polarography provides a measure of the labile species of Mn(II) and generally excludes the measurement of colloidal material with a size larger than ca. 50 nm.[46] Polarographic and atomic absorption spectroscopic measurement of 0.45 μm filtered solutions have shown a difference, ascribable to colloidal manganese.[29] This colloid manganese shows a high degree of stability because its concentration does not decline when sulfide is present.[29] Results obtained using XAD-2 resin indicate that the colloidal manganese might be associated with organic matter.[36]

Immediately above the oxycline, the manganese fraction is dominated by particles (>0.45 mm) rather than colloids.[29,36] In an investigation of temporal and spatial changes in particulate manganese in two seasonally anoxic lakes, Tipping

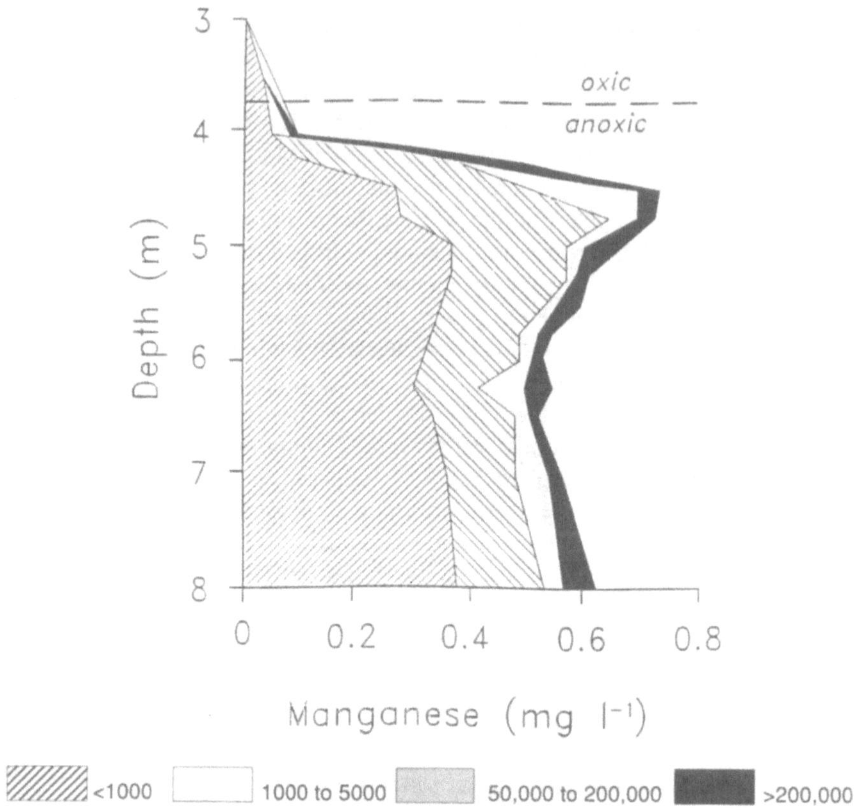

Figure 4. Size fractions, determined by ultrafiltration for manganese in Lake Fukami-ike passing a 0.45 μm filter. (Adapted from Yagi, A. *Jpn. J. Limnol.* 49:149–156. With permission.)

et al.[30] showed that there was no systematic change in the oxidation state with respect to time or depth. Only one bacterial morphotype, Metallogenium, was present in the soft water lake, but in the hard water lake there were three of equal importance: Prosthecomicrobium, Planctomyces, and Metallogenium. Particulate manganese was also present during the summer in the hard water lake as thin crumpled sheets approximately 50 μm across.

In circumneutral lake waters, the manganese particles are thought to be produced entirely by microbial processes, but where the pH is exceptionally high (pH 8 to 10) due to photosynthesis, inorganically oxidized particles may result.[47] Experiments with synthetic preparations of Mn_3O_4 and β-MnOOH indicate that in lake water the negative surface charge is controlled[48] by humic substances and divalent ions such as Ca^{2+}. Studies involving laboratory prepared[49,50] and marine[51] particles have shown that manganese oxides may adsorb trace metals (see Section 4).

Other Mn Particles Formed in Freshwater

Although manganese particles are mainly present in freshwater as oxyhydroxides associated with bacteria, other particles often contain minor amounts of manganese. Typically 0.01% (range 0.001 to 0.1%) of the dry weight of phytoplankton is manganese,[52,53] so for productive waters, with particulate organic carbon of 200 mmol/dm^{-3}, only 5% of the observed particulate manganese of 0.2 mmol/dm^{-3} could be attributed to the organic material.[54] Consequently, even in productive lakes, there is no apparent relationship between the concentrations of particulate organic carbon and particulate manganese,[20,54] nor is manganese in material collected in sediment traps correlated with carbon.[19] This settling flux of particulate Mn has been shown in several studies[16,18,19,55] to be strongly influenced by redox cycling of the oxyhydroxides within the lake. Microprobe analysis of iron oxide particles which formed at a redox boundary in a lake showed them to contain 0.7 to 1.4% manganese,[56] and iron sulfide particles isolated from the bottom waters of a lake were found to be up to 8% manganese.[57] It is not known whether this manganese represents adsorbed Mn(II) or an association of manganese oxyhydroxides with the iron particles. Most minerals, including quartz and alumino silicates, can have a surface coating of oxidized iron and manganese.[58,59]

In anoxic environments, where there are high concentrations of Mn(II) in circumneutral waters, distinct mineral forms of manganese can feasibly be formed by supersaturation of solution components. According to their solubility products[3,60,61] (Equations 1 and 2), rhodochrosite, $MnCO_3$,

$$MnCO_3 = Mn^{2+} + CO_3^{2-} \quad ; \quad pK = 9.7 \tag{1}$$

$$MnS = Mn^{2+} + S^{2-} \quad ; \quad pK = 13.5 \tag{2}$$

and alabandite, MnS, are most likely to occur. Although there have been reports of anoxic lake waters being saturated with respect to $MnCO_3$,[28,40,41] a $MnCO_3$ phase has so far not been identified in the particulate fraction. It has, however, been identified in freshwater sediments by X-ray diffraction, [62] in a relatively pure form,[63] and as a manganosiderite.[64] Robbins and Callender[11] have elegantly modelled the distribution of particulate and interstitial manganese in freshwater sediments in terms of rhodochrosite formation and have obtained excellent fits to the available data. At concentrations of sulfide appropriate to freshwater, MnS is unlikely to form. It has only been suggested as a controlling phase in a lake with marine origins which had an exceptionally high concentration of sulfide of 0.2 g/dm^{-3}, but even here the evidence depended on solubility calculations with no positive identification of a solid phase.[65]

Distinct mineral-forms of manganese (Table 2) have mainly been identified as components of iron and manganese-rich nodules or crusts.[58] Manganese is mainly present as amorphous oxyhydroxides, but todorokite (γ-MnOOH), birnessite (δ-MnO_2), psilomelane ($MnBaMn_8O_{18}.2H_2O$), and rhodochrosite ($MnCO_3$) have all been identified.[13,66] The unique freshwater occurrence of pisilomethane has been attributed to there being insufficient sulphate to form barite, which allows the coprecipitation of barium with manganese.[58] Generally the concentration of barium in manganese nodules is found to be closely correlated with the concentration of manganese.[58] Minor transition metals, including Cu, Co, Ni and Zn, are also associated with the manganese phase.

Manganese has been identified as a component of several phosphate phases including ludlamite, $(Fe, Mn,Mg)_3(PO_4)_2.4H_2O$[67] found in nodules from Lake Erie. It frequently recurs as a minor element of vivianite and apatite, which occur in anoxic sediments.[58]

REDOX INTERCONVERSION OF Mn SPECIES

Oxidative-Hydrolysis of Mn(II)

Mn(II) is thermodynamically unstable in the presence of O_2 in the pH range of natural waters, but the oxidation is kinetically limited. At circumneutral and acidic pH, the homogeneous oxidation of Mn(II) is believed to be insignificant, solutions at pH 8.4 being stable for at least seven years.[68] Homogeneous and heterogeneous catalysts can dramatically accelerate the reaction, and reports of faster homogeneous oxidation rates[69,70] are thought to be due to the presence of catalyzing surfaces.[68] Catalysis by γ-FeOOH[71] and autocatalysis by MnO_2[69] are dependent on pH and oxygen (Equations 3 and 4).

$$-d[Mn(II)]/dt = k_1[Mn(II)][O_2][OH^-]^2[MnO_2]_s \tag{3}$$

$$-d[mn(II)]/dt = k_2[Mn(II)][O_2][OH^-]^2[\gamma\text{-}FeOOH]_s \tag{4}$$

The original work[69] included a homogeneous term in the equations, but as there is no accepted value for this term, which is believed to be negligible, it is not included here. Using values for k_1 and k_2 determined in laboratory experiments,[68,72] it is possible to calculate the half-life for Mn(II) in well-oxygenated waters with typical concentrations of either MnO_2 or FeOOH of 1 μM (Table 3). The data from which these values were derived were largely within the pH range of 8 to 9; extrapolations to lower and higher pH are unproven. Nevertheless, these half-lives show that in neutral waters Mn(II) would be effectively stable if only homogeneous catalysis was operating. Once the pH rises above 9, however, inorganic heterogeneous catalysis can oxidize Mn(II) quite rapidly. So, in productive soft waters, where the pH can exceptionally reach 10 due to CO_2 depletion by photosynthesis, Mn(II) will be short-lived.

Table 2. Characteristics of Manganese Particles Found in Freshwater Environments

Nature of Manganese Compound	Composition	Characterization	Environment	Ref.
Manganous manganite ($MnO_{1.74-1.82}$) or δ-MnO_2	Unknown	XRD[a]	Bog ores (lacustrine Mn crusts ?)	132
Birnessite (Na,Ca,K) $Mn_7O_{14}3H_2O$	(Si,Mn,Fe,Al) > Ca,Mg on quartzite substrate (Mn,Fe) > Ca on gneiss substrate (Si,Al,Mn,Ca,Fe)> (Ti,K,Mg) on basalt substrate	IR spectroscopy, XRD,[a] electron microprobe	Quartzite and gneiss substrates on stream bed, basalt substrate in splash zone	133
Vernadite (δ-MnO_2)	Mn 25–42% Fe 1–13% Ca 1–5% C 7–16%	Tentative from XRD,[a] oxidation state, wet chemistry, crumpled sheet morphology	Anoxic lake water with Mn(II) allowed to oxidize	47
Metallogenium	$MnO_{1.75}$	Optical microscopy and staining, wet chemistry, TEM[b]	Lake water	30,62
Other bacterial morphotypes	$MnO_{1.75}$	Optical microscopy and staining, wet chemistry	Lake water	30,87
Todorokite	Not specified	XRD,[a] treatment with hydroxylamine	Nodules in Green Bay, Lake Michigan	13
Birnessite	Not refined	XRD,[a] treatment with hydroxylamine	Nodules in Green Bay, Lake Michigan	13
Psilomelane	Mn 48% O 30% Ba 8–13% Ca 0.7–1.3% Fe 0.7–1.2%	XRD,[a] treatment with hydroxylamine, electron probe	Nodules in Green Bay, Lake Michigan	13

[a] XRD — X-Ray Diffraction.
[b] TEM — Transmission Electron Microscopy.

In circumneutral natural waters, Mn(II) has been observed to have a half-life of 1 to 100 d,[47,73,74] with residence times appearing to be longer in winter than summer, presumably due to lower temperatures.[35] This rapid removal of Mn(II) when compared to the inorganic rates (Table 3) is due to microbial catalysis. It is difficult to show directly the involvement of microorganisms in oxidations occurring within a lake, but there are numerous strands of circumstantial evidence which have been summarized by Tipping:[75] (a) bacterial cultures oxidize Mn(II), (b) filtered waters oxidize more slowly, (c) bactiocides slow down the reaction, (d) high concentrations of Mn(II) can prevent oxidation — attributed to inhibition of bacterial enzymes, (e) surface catalysis by natural particles cannot account for the high rates, (f) manganese oxidation products in lakes often have "biological" morphologies, and (g) the high oxidation states of precipitates formed in lake water cannot be replicated in simple abiotic solutions.

In a recent review of microbial mediation of manganese oxidation, Nealson et. al.[76] have shown that mechanistic understanding is increasing rapidly. There is a clearly established extracellular mechanism for the freshwater bacteria Leptothix discophera SSI. Cultures of this organism excrete Mn^{2+}-oxidizing proteins which, in association with acidic exopolymers, catalyze the oxidation of Mn(II).[77,78] The oxidation state of the product is close to Mn(III), but aging (days) increases the oxidation state. Manganese oxidation by the surfaces of dormant bacterial spores is believed to be due to proteins binding Mn^{2+}.[78] There is also the possibility that other freshwater organisms, such as Pseudomonas sp., may oxidize manganese through energy-yielding enzymatic reactions.[78,79] Depending on the solution conditions and the organisms present, different oxidation states between Mn(III) and Mn(IV) are produced, but no systematic explanation of the values has yet emerged.

Pronounced diel cycles in the formation rates of particulate manganese have been observed in seawater,[80] with daytime rates being much less than rates measured at night. These observations have been attributed to photoinhibition of the microbially catalyzed Mn(II) oxidation. This effect, along with photodissolution of manganese oxides (see Section 3.1), leads to an increased ratio of Mn(II) to particulate manganese during daylight. Similar processes may be operating in lakes and rivers, but they have yet to be reported.

Structure and Nature of Oxidation Products

There is a bewildering range of oxidation products of Mn(II). Both (III) and (IV) oxidation states are usually involved, although in alkaline solution, Mn(III) oxyhydroxide, MnOOH, is thermodynamically unstable with respect to Mn(IV) oxides and Mn(II).[81,82] In synthetic solutions, the initial form of the oxidation product depends on solution composition and temperature, and a wide range of compounds may result[83-85] (Table 4). Hausmannite (Mn_3O_4) and feitknechtite (β-MnOOH) readily age in the presence of oxygen to form γ-MnOOH which, although metastable, exists for long periods. Disproportionation to the thermodynamically stable products of Mn(II) and Mn(IV) only occurs readily in alkaline

Table 3. Half-Lives ($t_{1/2}$) of MnII in Water at Various pH at Equilibrium with the Atmosphere ($[O_2] = 2.8 \times 10^{-4}$M), Assuming Either Iron or Manganese Oxides have a Concentration of 1 μM. k_1 and k_2 in Equations 1 and 2 are Taken to be 10^{18} M^{-4}/d

pH	$t_{1/2}$
6	70,000 years
7	700 years
8	7 years
9	25 days
10	6 hours

solution. The transformations to γ-MnOOH from Mn_3O_4 and β-MnOOH may be inhibited by complexing ligands such as oxalate, so a similar process may extend the life of these initial metastable products in natural environments.

A range of manganese oxides are found to occur naturally, and there is still some confusion regarding their nomenclature.[85] In manganese nodules, at least two distinct phases occur. One has a diagnostic 97 nm X-ray diffraction line and is known as 100 nm manganite.[59] This γ-MnOOH phase, also referred to as the buserite family or todorokite, can be substituted with cations, such as Ba, Ca, and Mn. The other common phase has a broad band at 244 nm (γ-MnO_2). There is also a less common phase with a band at 71 nm known as 70 nm manganite (δ-MnO_2). It is often called the birnessite family and may be substituted with Na^+ or Ca^{2+}. Structural disorder and degree of oxidation increase from 100 nm manganite through 70 nm manganite to γ-MnO_2.

With the notable exception of their presence in manganese nodules, naturally occurring manganese particles are usually X-ray amorphous, their identification relying heavily on electron microscopic technique. They tend to coexist with and coat many other minerals, biological forms, and organic material. Lind et al.[85] believe that where coatings occur with iron, manganese oxide forms by oxidation of Mn(II) with Fe(III) (Equations 5 to 7), direct oxidation with O_2 being too slow at pH<7 (Table 3).

$$4Fe^{2+} + O_2 + 10H_2O = 4Fe(HO)_3 + 8H^+ \tag{5}$$

$$2Fe(OH)_3 + 2H^+ = 2Fe(OH)_2^+ + 2H_2O \tag{6}$$

$$2Fe(OH)_2^+ + 3Mn^{2+} = Mn_3O_4 + 2Fe^{2+} + 4H^+ \tag{7}$$

As well as amorphous γ-MnO_2, birnessite has been identified in freshwater.[85] Certain phases such as hollandite (α-MnO_2) depend on the presence of other cations — in this case Ba^{2+}.Various manganese minerals, with a range of addi-

Table 4. Oxidation Products of Mn(II) in Synthetic Solutions

Initial Condition and Product			Aged Product
25°C NO_3^-, Cl^-, or SO_4^{2-}	Mn_3O_4 (hausmannite)	protonation → disproportionation →	γMnOOH MnO_2
5°C NO_3^-, Cl^-	βMnOOH (feitknechtite)	rearrangement → disproportionation →	γMnOOH MnO_2
5°C SO_4^{2-}	γMnOOH (manganite)	disproportionation →	MnO_2
>5°C,<25°C NO_3^-, or CL^-	βMnOOH,Mn_3O_4 mixtures		
>5°C,<25°C SO_4^{2-}	γMnOOH,Mn_3O_4 mixtures		

tional substituents, have been identified in soils. The affinity of trace metals for manganese oxides is high. The association may be attributed to simple adsorption, the ion may substitute for Mn^{2+} to form a solid solution (e.g., Zn Mn_2O_4), it may interact to form a mixture of distinct minerals (e.g., Ni, Pb), or it may simply enrich the oxide structure (e.g., Co). γ-MnOOH was the product when Mn(II) in synthetic solutions simulating lake water was allowed to oxidize.[83] When natural, anoxic lake water containing Mn(II) was allowed to oxidize, crumpled sheets of Mn particulate material, about 50 μm across, were formed.[47] Similar sheets were identified in the oxic waters of the same lake, and a similar morphology was observed when Mn(II) in seawater was allowed to oxidize.[86] Some X-ray diffraction structure was observed for the freshwater material, but its interpretation was indefinite. Although Tipping et al.[47] thought that its oxidation state and morphology indicated that it was most likely a disordered vernadite (δ-MnO_2), a number of other oxides, including birnessite, buserite, and hausmannite, would have been consistent with the X-ray diffraction data.

Manganese particles in lake waters have commonly been observed to be linked to bacterial forms, particularly that of Metallogenium, which has been reported in many permanently oxic and seasonally anoxic lakes.[87] The filamentous structures known as Metallogenium have not been proven to be bacteria, although microbiologists generally seem to accept them to be so. They are most commonly found in waters with a low oxygen concentration (< 0.04 mmol/dm^{-3}), where there should be a good supply of Mn(II).[62,83] A systematic study of the temporal and spatial distribution of bacterial morphotypes in a soft and hard water lake showed that there were several morphotypes which contained manganese, including Metallogenium.[30]

Redox titrations using a solution of iodide or an organic reducing agent have been used to determine the oxidation state of manganese particles in lake water. There was little systematic change in the oxidation state with respect to either depth or time in two seasonally anoxic lakes[30] where a mean value of 3.5 (±0.15) was observed. The range in stoichiometries for the two lakes was $MnO_{1.65-1.88}$ (soft water) and $MnO_{1.64-1.99}$ (hard water). Oxidation in the laboratory of Mn(II) in the soft water yielded oxidation states of 3.6 to 3.9, while two experiments involving oxidation of Mn(II) in the hard water resulted in values of 3.1 and 3.9. The oxidation state of Mn associated with Metallogenium in Lake Zurich was 3.2[88] and in a soil at pH 6 it was 3.5 to 3.8.[89] Oxidation of Mn(II) in synthetic solutions (pH 7 to 10) gave values of 2.7 to 3.1; only at pH~12 was a value of 3.8 obtained.[69,90] It appears that at circumneutral pH, the presence of bacteria is required to produce a product of high oxidation state. It is possible that a substantial fraction of the manganese in some particles is simply adsorbed Mn(II).[91] Adsorption of Mn(II) to bacterial particles is considered to be a prerequisite for oxidation, the rate of adsorption, which is faster than the rate of oxidation, determining the rate of removal of Mn(II) from solution.[73,74] When synthetic preparations of Mn_3O_4 and β-MnOOH (oxidation state 3.4) were suspended in lake water, their surface charge was controlled[48] by the adsorption of humic substances and divalent cations, predominantly Ca^{2+}. Manganese particles formed by allowing Mn(II)-rich lake water to oxidize have a low percentage[47] of humic carbon (0 to 2%), but the relatively constant negative charge of such particles in the pH range 4 to 10 of $-(1.4\text{–}1.5) \times 10^{-8}$ $m^2V^{-1}s^{-1}$ is similar to that for synthetic particles exposed to humic substances, suggesting that the surface properties are controlled by these naturally occurring organics.

The elemental composition of manganese oxide material collected directly from a soft water lake and obtained by oxidation of the anoxic lake water was similar but variable. The Ca to Mn ratio varied from 0.01 to 0.68 (usually 0.1) and Mg, Si, P, S, Cl, and Ba were also commonly present.[47] Divalent cations may be associated with the oxide surface by a surface complexation mechanism involving humic material,[47] or they may be a part of the structure.[83] Similar mechanisms may be involved in the known association of some trace metals with manganese particles[51] (see Section 3.1).

Reductive Dissolution of Mn(IV) and Mn(III)

Within the pH range of natural waters, Mn(IV) and Mn(III) form insoluble oxyhydroxides, whereas Mn(II) is soluble. As discussed in Section 3.1, although the Mn(II) oxidation state is thermodynamically unstable in oxygenated surface waters, the rate of oxidation, by chemical homogeneous reactions, is kinetically very slow (see Table 3). Surface catalysis and, in particular, biotically driven oxidation-hydrolysis reactions have been shown to be the important processes.

Dissolution reactions (Reactions 8 and 9) of Mn(III) and Mn(IV) solid phases, under suitable environmental conditions, can greatly increase the mobility of manganese in natural systems (see Table 5).

$$MnOOH_{(s)} + 4H^{+} = 2e^{-} = Mn^{2+} + 2H_2O \tag{8}$$

$$MnO_{2(s)} + 4H^{+} + 2e^{-} = Mn^{2+} + 2H_2O \tag{9}$$

In contrast to oxidation reactions, many reductive dissolution reactions in natural freshwater systems are not kinetically slow and may either be purely chemically driven or may involve direct or indirect biotic reaction pathways (Table 5). As discussed below, photoassisted dissolution reactions[92,93] of manganese oxides are also important in oxic surface waters.

Reaction Pathways

During the past 10 years, considerable progress has been made in our understanding of the reaction pathways involved in reductive dissolution[94,95] of naturally occurring metal oxides such as Mn(III/IV) oxyhydroxides. The overall process may be schematically represented as a number of sequential intermediate reactions (1 through 4) at the particle surface (Figure 5). These steps involve surface complexation, protonation and deprotonation, charge transfer, ligand exchange, and finally steric rearrangement and detachment of the reduced Mn^{II} metal center. Depending on the nature of the reductant, Reaction 1 may lead to the formation of either an outer sphere or an inner sphere type complex, and consequently, both the rate of formation of the surface complex and the rate of charge transfer will be affected. Kinetic rate laws for the reductive dissolution of metal oxides have been discussed extensively by Stone and Morgan[94] in terms of the surface complexation model and the boundary conditions in which each of the reactions in Figure 5 may become rate limiting.

It is well known that hydrous metal oxides have a strong affinity for dissolved metal ions, and the adsorption of Mn(II) by manganese oxides is well documented.[49,69] Consequently, it is reasonable to assume that during the reductive dissolution of manganese oxides, a fraction of the Mn(II) released per unit time will be readsorbed onto the surface of manganese particles. It is therefore important to distinguish between the rate of dissolution d[Mn(II)]/dt and the rate of reduction –d[Mn(IV/III)]/dt.

In general, it is the rate of reduction which is proportional to the surface concentration of the precursor complex, [PC], and the rate law can be expressed in the following way:

$$-d[Mn(IV/III)]/dt = k_{red} \cdot [PC] \tag{10}$$

Table 5. Reductive Dissolution of Manganese Oxides in Natural Aquatic Systems

Reductant	Environmental Conditions	Pertinence	Ref.
Humic type organics	Surface, oxic waters	Increased bioavailability, photoreductive dissolution	92, 93, 141
	Anoxic hypolimnetic waters		42
Fe(II), HS^-	Anoxic hypolimnetic waters, seawater	Mn cycling at a redox interface	29, 37, 104 ,107 132
Phenols	Polluted soils, landfill leachates, waste waters	Waste disposal, soil interactions	134–136
As(III)	Lake sediments	As detoxification	124, 125, 137
Acetate, lactate, succinate	Microorganisms isolated from freshwater and coastal sediments	Biotically mediated anoxic MnO_2 reduction coupled to organic C oxidation	110, 132
Biotic enzymes	Microorganisms isolated from energy production-related waste waters	Anoxic MnO_2 reduction in waste water	138
Bacteria (direct or indirect)	Microorganisms isolated from aquatic media (freshwater, seawater, soils)		37,139,140
		Biotic MnO_x	104, 113–115

1. SURFACE COMPLEXATION

$$\equiv Mn^{IV}OH + Red + H^+ \rightleftharpoons \equiv Mn^{IV}\text{-}Red + H_2O$$

2. 2e CHARGE TRANSFER

$$\equiv Mn^{IV}OMn^{IV}\text{-}Red \rightleftharpoons \equiv Mn^{IV}OMn^{II}\text{-}Red_{(ox)}$$

3. LIGAND EXCHANGE

$$\equiv Mn^{IV}OMn^{II}\text{-}Red_{(ox)} \overset{+H_2O}{\rightleftharpoons} \equiv Mn^{IV}OMn^{II}\text{-}OH + Red_{(ox)} + H^+$$

4. ADSORPTION-DESORPTION

$$\equiv Mn^{IV}OMn^{II}\text{-}OH + 2H^+ \overset{-H_2O}{\rightleftharpoons} \equiv Mn^{IV}OH + Mn^{2+}_{aq}$$

Figure 5. Schematic representation of the surface complexation model applied to the reductive dissolution of the manganese dioxide.

where

k_{red} is an empirical reduction rate constant which has units of inverse time

In the earlier literature, the rate of dissolution, d[Mn(II)]/dt, was generally expressed in the following manner:

$$d[Mn(II)]/dt = k_{red} \cdot [Red] \cdot [MnOx] \qquad (11)$$

where

k_{red} is a second order rate constant with units of $dm^3/mol \cdot s$

Unfortunately, there is at present much confusion in the literature, and Mn(II) adsorption-desorption equilibria are not always taken into account when reduction rate constants are reported. In the following discussion, in order to avoid any further confusion, dissolution rate constants will be referred to as $k_{diss.}$, whereas reduction rate constants with units of $time^{-1}$ will be referred to as k_{red}. Reduction and dissolution rate constants tabulated from the available literature are given in Table 6.

Abiotic Reduction by Organic Reductants

The reductive dissolution of Mn oxides using a wide variety of model organic reductants has been extensively investigated.[96-99] Aromatic model reductants generally reduced Mn oxides, whereas most of the aliphatic compounds tested did not.[97] These studies found that, in general, the observed rates of dissolution could be explained in terms of the surface complexation model. For instance, in a study using hydroquinone,[96] increasing the rate of stirring appeared to have no effect on the rate of dissolution, whereas the adsorption of HPO_4^- and Ca^{2+} inhibited the reaction. This data and other evidence suggest that it is the rate of surface reactions (reactions 2 through 4 in Figure 5) that determine the overall rate and not the rate of formation of the surface complex.

Recently, it has been found that using substituted phenols, the trend followed by the dissolution rate data was a function of the para-, ortho-, and meta-chloro-substitution pattern both for mono- and disubstituted phenols and was consistent with the stability of the phenol oxidation product. This has been interpreted as evidence that it was the charge transfer step that was rate limiting and not the ligand exchange or the detachment step.[99] pH was found to have a very significant effect on the rate of dissolution by substituted phenols, possibly due to either one or a combination of the following two effects: (1) protonation reactions that promote the formation of surface precursor complexes or (2) increased protonation of the precursor complex leading to faster electron transfer kinetics. Finally, the Mn oxide composition and structure were also found to be important, and when different Mn dioxides were compared, the measured rates were found to vary by up to a factor of 3.

Natural organic matter (NOM) such as fulvic and humic acids isolated from both freshwater and seawater has been shown to be able to reduce Mn(III/IV) oxides to dissolved Mn(II).[92,93] The reaction is photocatalyzed: the rate of dissolution in coastal seawater samples was found to increase seven to eight times[93] when exposed to full sunlight as compared to samples in the dark. Photoassisted dissolution has also been reported to be important for iron oxides[100] in the presence of a variety of organic substrates including NOM,[101] and similar photoenhanced mechanisms may be operative for manganese oxides, including:

1. Excitation of the adsorbed ligand, promoting metal charge transfer or internal ligand transitions.
2. A semiconductor type mechanism, where valence band electrons are photo-excited to the conduction band creating at the particle surface a photoelectron/photohole pair which can then interact with the adsorbed ligand.[102]
3. Reduction by radicals or reductants produced by photoionization of NOM. Hydrogen peroxide is known to be produced by this process, and it has been suggested that it could play a role in the reduction of Mn oxides.[93] More recently, however, this suggestion has been questioned.[92]

Waite et al.[92] have recently reported on the photoassisted reductive dissolution of well-characterized manganese oxides by fulvic acids. They found that stirring had no effect (suggesting that the reaction was surface controlled), and there was a linear dependence of the rate of MnO_2 dissolution on light intensity (at 365 nm). There was no change in the rate on addition of catalase or as a function of deoxygenation, indicating that H_2O_2 had no or little effect. pH, in constrast, was a significant parameter. The observed rate of dissolution, $d[Mn(II)_{aq}]/dt$, measured as the release of dissolved Mn(II) to the bulk solution, was much higher at

Table 6. Reductive Dissolution Rate Constants

Reductant/pH	Mn-Oxyhydroxide Phase[a]	Rate Constant[b] k_{red} (min^{-1})	k_{diss} ($dm^3/\mu mol \cdot s$)	Ref.
Fulvic acids pH 4, 25°C	$MnO_{1.94}$	0.55[c] 0.31[d]	— —	92
Fulvic acids pH 7.1, 25°C	$MnO_{1.94}$	1.23[c] 0.53[d]	— —	92
Hydroquinone, pH 7.2	$MnO_{1.64}$	—	$1.4 \cdot 10^{-2}$	96
3-Methoxycatechol pH 7.2	$MnO_{1.64}$	—	0.108	97
Resorcinol pH 7.2	$MnO_{1.64}$	—	1.7615^{-5}	97
HS^-, pH = 8.1	$MnO_{1.94}$	—	0.01–0.132	104
2-Chloro-phenol pH 4.8	$MnO_{1.97}$	0.11	—	99

a See references for a complete characterization of the Mn oxide.
b k_{red} is a surface reduction rate constant defined as -d [Mn(III/IV)]/$_{dt}$ = k_{red}·[Precursor complex] and k_{diss} is a second order dissolution rate constant, i.e., d[Mn(II)]/$_{dt}$ = k_{diss}·[MnO_x].[Red].
c Photoassisted reaction.
d Dark.

pH 4 than at pH 7.1. However, as pointed out by the authors, the rate of reduction should be estimated as the total amount of Mn(II) released per unit time, where [$Mn(II)_T$] is given by Equation 12:

$$\left[Mn(II)_T\right] = \left[Mn(II)_{aq}\right] + \left[Mn(II)_{ads}\right] \tag{12}$$

[Mn(II) $_{ads}$] is the concentration of adsorbed Mn(II) which was estimated from modified Freundlich adsorption isotherms. Assuming the concentration of the surface precursor complex, [PC], reaches a steady state and that only a small fraction of the total concentration of fulvic acids, [HA_T], are adsorbed, the rate of reduction may be written, according to Equation 13, as:

$$-d\left[Mn(IV/III)\right]/dt = d\left[Mn(II)_T\right]/dt = k_{red} \cdot [PC] \tag{13}$$

where

k_{red} is a first order reduction rate constant
[PC] is the surface concentration of the precursor complex,

which can be shown to be given by:

$$[PC] = K^{cond} \cdot [Mn_T] \cdot \frac{[HA_T]}{1 + K^{cond} \cdot [HA_T]} \tag{14}$$

where

K^{cond}, is the conditional formation constant for the surface complex
$[Mn_T]$ is total concentration of manganese oxide surface sites

Using this simplified model of manganese oxide-fulvic acid interaction, and a linear least squares fit to the experimental data, Waite et al.[92] found that the first order reduction rate constants, k_{red}, were greater at pH 7.1 than at pH 4 (see Table 6). As shown in Table 6, this was true both for the photoassisted reaction and the dark reaction. In constrast, as already mentioned, they found that the observed rate of dissolution, $dMn(II)_{aq}/dt$, measured as the release of dissolved Mn(II) to the bulk solution was much higher at pH 4 than at pH 7.1. This apparant anomaly can be readily understood since a lower pH will:

1. favor the release of Mn(II) to the solution through adsorption-desorption equilibria
2. favor the adsorption of fulvic acid and therefore increase the surface concentration of the precursor complex and therefore the rate of charge transfer

Abiotic Reduction by Inorganic Reductants

Under the pH and pε conditions found in freshwater, MnO_2 reduction by naturally occurring inorganic reductants such as sulfide, nitrite, ammonia, and ferrous iron is thermodynamically favorable.[3] Burdige and Nealson[104] have reported that neither nitrite nor ammonia reduced δ-MnO_2, while the reaction with sulfide was rapid and complete within 5 to 10 min. The observed end products were S_8 and possibly polysulfides; the average stoichiometry of the reaction was found to be 0.95 (± 0.07) in agreement with the following reaction:

$$\delta - MnO_2 + HS^- + 3H^+ = Mn^{2+} + 2H_2O + S^o \tag{15}$$

However, the above authors also reported that only 44 to 56% of the sulfide removed could be accounted for as precipitated elemental sulfur. Possible explanations for this discrepancy were the formation of polysulfides or colloidal S^O

neither of which were analytically determined in the procedures used by Burdige and Nealson.[104] More recently, Aller and Rude[105] have reported that solid phase sulfides and elemental sulfur are completly oxidized to sulfate by manganese oxides in marine sediments, suggesting that the missing "elemental sulfur"in Burdige and Nealson's experiments may have been further oxidized to sulfate according to:

$$3MnO_2 + S^o + 4H^+ = 3Mn^{2+} + 2H_2O + SO_4^{2-} \quad (16)$$

The oxidation of sulfide by MnO_2 has also been discussed by Luther[106] in terms of molecular orbital theory, and is thought to involve a two electron transfer from the HS^- p_z orbital to the Mn(IV) d_z orbital. This type of σ - σ charge transfer would consequently imply an inner sphere type mechanism.

In anoxic freshwater, in addition to sulfide, ferrous iron can also act as an electron donor and reduce Mn(IV/III) solid phases to soluble Mn(II). In a field study, De Vitre et al.[29] found that natural MnO_x particles formed at a redox boundary in a eutrophic lake were reduced by both reductants. At pH 7.2, in the presence of a 30-fold excess (relative to MnO_x) of either Fe(II) or S(-II), typical half-reaction times were found of 10 to 20 and 20 to 200 min, respectively. Thus, under natural conditions, the reaction with Fe(II) appears to be even faster than the reaction with sulfide. The reduction of synthetic MnO_2 by ferrous iron has also been investigated in the laboratory by Myers and Nealson,[107,108] who also found that the reaction was fast and that, although both sulfide and ferrous iron can reduce MnO_2, in natural systems biotic mediation was important for both reactions.

Biotic Reduction Pathways

Reduction of Mn oxides by microbial catalysis has been widely reported in lakes[16,109-111] as well as in soils and groundwaters[111,112] and has been reviewed by Ehrlich[113] and by Ghiorse.[114] It is often assumed that when Mn oxides act as the terminal electron acceptors for bacterial respiration, the reduction may only occur under anoxic conditions.[115] However, as pointed out by Ehrlich,[113] this may not always be the case, and a number of bacteria appear to be able to reduce Mn oxides aerobically.[116,117] Furthermore, for some microorganisms, Mn reduction may not represent a form of respiration but possibly a detoxification process[118] or a means of making Mn^{2+} available in a Mn deficient environment.[93] De Vrind et al.[119] have also suggested that for the vegetative cells of Bacillus SG1, manganese oxide reduction was a means of making Mn^{2+} available for sporulation.

Since, as already discussed, Mn oxides may readily be reduced chemically by a number of organic and inorganic species present in freshwater, the importance of direct bacterial reduction of Mn oxides remains an open question. It is, however, obvious that in freshwater, indirect reduction by bacterial exopolymers and ectoenzymes[120] as well as by inorganic bacterial metabolites such as Fe(II), S(-II), etc. is widespread[104,107,108,110] although its study is still in its infancy.

OUTLOOK AND CONCLUSIONS

Our understanding of the cycling of manganese and the behavior and the fate of Mn particles in freshwater systems (soils, rivers, lakes, and in the atmosphere — rain and fog waters) has certainly greatly increased over the last 10 years. The redox-driven interconversion between dissolved Mn and particulate Mn species is now better understood, but the exact roles and interdependencies of thermodynamic, kinetic, and biotic controls on the geochemical behavior of Mn is still unclear. It should, furthermore, be borne in mind that Mn cycling at a redox interface or even within a given body of water is intimately linked to other elemental cycles. For instance, the microbial reduction of manganese oxides is biotically and abiotically coupled to iron and sulfur species[107,108] and is ultimately determined by the roles of carbon and oxygen.

The role of manganese particles in the cycling of trace components in lakes is only just beginning to emerge. Generally, the water column concentrations of Cu, Zn, Pb, and Cd are not appreciably influenced by dramatic changes in the concentration of dissolved and particulate manganese associated with redox cycling[121,122] although sediment traps have revealed that fluxes of Cu and Zn may be associated with manganese.[19,55] Co, Ni, and to some extent Cr are more strongly associated with the Mn oxide phases and are released as it is dissolved in the water column.[122] Within the sediment, redistribution of trace metals originally associated with sedimenting particles may occur. Belzile et al.[123] have, for instance, shown that diagenetic Mn oxides were enriched with Zn and Ni, whereas As and Cu were mostly associated with Fe oxyhydroxides.

It is also becoming increasingly clear that our knowledge of the behavior and fate of Mn particles in aquatic systems will require us to understand surface interactions (ligand adsorption, heterogeneous charge transfer, protonation-deprotonation reactions, etc.) and to measure surface concentrations of precursor or successor complexes. For instance, reaction kinetics between Mn(IV/III) oxyhydroxide particles and toxic xenobiotic organic or inorganic species must be understood at the molecular level in order to be able to model their behavior in real systems. The role of Mn oxides as a detoxifier in freshwater systems is only beginning to be investigated. For instance, in lakewater, arsenite is oxidized by both diagenetic manganese and iron oxyhydroxide particles[124,125] to less toxic arsenate. Xenobiotic organics such as substituted phenols (chloro-phenols, nitrophenols, and others) may be degraded by oxidation by Mn oxides[98,99] under the conditions found in freshwater. These reactions are strongly pH dependent, and it has been suggested that they may be particularly significant in fog and rain waters where low pH values are common.

Spectroscopic techniques are likely to play an increasing role in the further appreciation of the nature of the particles and their surface interactions. McBride[126,127] has clearly demonstrated the usefulness of electron spin resonance spectroscopy for the unambiguous determination of the oxidation state of manganese in both solids and solution and the potential of Fourier transform infrared spectroscopy for providing information about the bonding of organic components. As yet, such applications are largely restricted to studies of synthetic compounds in the laboratory, there being few applications to natural conditions.[128] X-ray adsorption spectroscopy techniques are proving to be powerful tools for the elucidation of the structures of hydrous oxides and their modification by adsorption.[129] Use of such techniques to investigate structural differences in particles dynamically interacting within water columns should prove very fruitful.

Although the colloidal nature of a considerable fraction of manganese particles in lakes has been clearly demonstrated, the origin, fate, and dynamics of this fraction remains obscure. Understanding the origin requires a better appreciation of microbial oxidation processes and the role of the bacteria in the nucleation and growth of particles. Progress here will probably depend on the use of electron microscopic techniques allied to systematic experiments involving cultures under near natural, but manipulated conditions. To appreciate the dynamics, rates of aggregation will need to be measured under natural conditions. Particle sizing procedures will only be useful if they can be used in natural conditions. Other chapters in this volume show that field flow fractionation[130] and light scattering methods are likely candidates.[131] We do not know whether the different-sized colloids have different reactivities with respect to reductive dissolution or adsorption of other components, such as phosphate and trace metals. Collection into different size fractions perhaps using field flow fractionation or ultrafiltration will be a necessary prerequisite of any investigation.

Most work on photoreduction and oxidation of manganese has been done on seawater.[80] Similar systematic studies of freshwater are now required. Here there are differences due to the generally higher concentrations of both manganese and organics. The composition of freshwater is much more diverse than that of seawater, however, and an appreciation of these processes for a diverse range of pH and light regimes is required.

REFERENCES

1. Gosz, J.R., Likens G.E., and F.H. Borman. "Nutrient Budgets for Undisturbed Ecosystems (Watersheds)" *Ecol. Monogr.* 43:173–182 (1973).
2. Wedepohl, K.H. "Geochemical Behavior of Manganese," in *Geology and Geochemistry of Manganese, Vol. 1, General Problems*, J.M. Varentsov and G. Grassely, Eds. (Budapest: Akad. Kiado, 1980).
3. Stumm, W. and J.J. Morgan. *Aquatic Chemistry*, 2nd ed. (New York: Wiley-Interscience, 1981).
4. Revsbech, N.P. and B.B. Jorgensen. "Microelectrodes: Their Use in Microbial Ecology," in *Advances in Microbial Ecology, Vol. 9*, K.C. Marshall, Ed. (New York: Plenum Press, Inc., 1986).

5. Gundersen, J.K. and B.B. Jorgensen. "Microstructure of Diffusive Boundary Layers and the Oxygen Uptake of The Sea Floor," *Nature* 345:604–607 (1990).
6. Reimers, C.E., S. Kalhorn, S.R. Emerson, and K.H. Nealson. "Oxygen Consumption Rates in Pelagic Sediments from the Central Pacific: First Estimates from Microelectrode Profiles," *Geochim. Cosmochim. Acta* 48:903–910 (1984).
7. Jones, J.G., M.J.L.G. Orlandi, and B. Simon. "A Microbiological Study of Sediments from the Cumbrian Lakes." *J. Gen. Microbiol.* 115:37–48 (1979).
8. Nembrini, G., J.A. Capobianco, J. Garcia, and J.M. Jaquet. "Interaction between Interstitial Water and Sediment in Two Cores of Lac Léman, Switzerland," *Hydrobiologia* 92:363–375 (1982).
9. Hutchinson, G.E. *A Treatise on Limnology; Geography, Physics and Chemistry, Vol. 1* (New York: John Wiley & Sons, Inc., 1957).
10. Davison, W. "Conceptual Models for Transport at a Redox Boundary," in *Chemical Processes in Lakes*, W. Stumm, Ed. (New York: Wiley-Interscience, 1985), pp. 31–53.
11. Robbins, J.A. and E. Callender. "Diagenesis of Manganese in Lake Michigan Sediments," *Am. J. Sci.* 275:512–533 (1975).
12. Gorham, E. and D.J. Swaine. "The Influence of Oxidizing and Reducing Conditions upon the Distribution of Some Elements in Lake Sediments," *Limnol. Oceanogr.* 10:268–279 (1965).
13. Callender, E. and C.J. Bowser. "Freshwater Ferromanganese Deposits," in *Handbook of Strata-Bound and Stratiform Ore Deposits, Vol. 7*, K.H. Wolf, Ed. (New York: Elsevier North Holland, Inc., 1976).
14. Dean, W.E., W.S. Moore, and K.H. Nealson. "Manganese Cycles and the Origin of Manganese Nodules, Oneida Lake, New York, USA," *Chem. Geol.* 34:53–64 (1981).
15. Imboden, D.M. and R.P. Schwarzenback. "Spatial and Temporal Distribution Of Chemical Substances In Lakes: Modelling Concepts," in *Chemical Processes in Lakes*, W. Stumm, Ed. (New York: Wiley-Interscience, 1985), pp. 1–30.
16. Davison, W., C. Woof, and E. Rigg. "The Dynamics of Iron and Manganese in a Seasonally Anoxic Lake; Direct Measurement of Fluxes using Sediment Traps," *Limnol. Oceanogr.* 27:987–1003 (1982).
17. Hamilton-Taylor, J. and E.B. Morris. "The Dynamics of Iron and Manganese in the Surface Sediments of a Seasonally Anoxic Lake," *Arch. Hydrobiol.* 72:135–165 (1985).
18. Stabel, H.H. and J. Kleiner. "Endogenic Flux of Manganese to the Bottom of Lake Constance," *Arch. Hydrobiol.* 98:307–316 (1983).
19. Hamilton-Taylor, J., M. Willis, and C.S. Reynolds. "Deposition Fluxes of Metals and Phytoplankton in Windermere as Measured by Sediment Traps," *Limnol. Oceanogr.* 29:695–710 (1984).
20. Sholkovitz, E.R. and D. Copland. "The Major-Element Chemistry of Suspended Particles in the North Basin of Windermere," *Geochim. Cosmochim. Acta* 46:1921–1930 (1982a).
21. Delfino, J.J., G.C. Bortlesen, and G.F. Lee. "Distribution of Mn, Fe, P, Mg, K, Na and Ca in the Surface Sediments of Lake Mendota, Wisconsin," *Env. Sci. Technol.* 11:1189–1192 (1969).
22. Takamatsu, T., M. Kawashima, and M. Kogama. "Manganese Concentration in the Sediment as an Indicator of Water Depth, Paleo-water Depth During the Last Few Million Years," *Kokuritsu Kogai Kenkyusho Kenkyo Hokoku* 75:63–67 (1985).

23. Wetzel, R.G. *Limnology*, 2nd ed. (New York: CBS College Publishing, 1983).
24. Stauffer, R.E. "Cycling of Manganese and Iron in Lake Mendota, Wisconsin," *Env. Sci. Technol.* 20:449–457 (1986).
25. Sakata, M. "Diagenetic Remobilization of Manganese, Iron, Copper and Lead in Anoxic Sediment of a Freshwater Pond," *Wat. Res.* 19:1033–1038 (1985).
26. Davison, W. and C. Woof. "A Study Of The Cycling Of Manganese And Other Elements in a Seasonally Anoxic Lake, Rostherne Mere, UK," *Wat. Res.* 18:727–734 (1984).
27. Kjensmo, J. "The Development and Some Main Features of 'Iron-Meromictic' Soft Water Lakes," *Arch. Hydrobiol. Suppl.* 32:137–312 (1967).
28. Mayer, L.M., F.P. Liotta, and S.A. Norton. "Hypolimnetic Redox and Phosphorus Cycling in Hypereutrophic Lake Sebasticook, Maine," *Wat. Res.* 16:1189–1196 (1982).
29. De Vitre, R.R., J. Buffle, D. Perret, and R. Baudat. "A Study of Iron and Manganese Transformations at the O2/S(-II) Transition Layer in a Eutrophic Lake (Lake Bret, Switzerland): A Multimethod Approach," *Geochim. Cosmochim. Acta* 52:1601–1613 (1988).
30. Tipping, E., J.G. Jones, and C. Woof. "Lacustrine Manganese Oxides: Mn Oxidation States and Relationships to Mn Depositing Bacteria," *Arch. Hydrobiol.* 105:161–175 (1985).
31. Yagi, A. and I. Shimodaira. "Seasonal Changes of Iron and Manganese in Lake Fukami-ike. Occurrence of Tubid Manganese Layer," *Jpn. J. Limnol.* 47:279–289 (1986).
32. Balistrieri, L.S., J.W. Murray, and B. Paul. "The Cycling of Iron and Manganese in Lake Sammamish, WA," *Limnol. Oceanogr.* in press (1992a).
33. Buffle, J., R.R. De Vitre, D. Perret, and G.G. Leppard. "Physico-chemical Characteristics of a Colloidal Iron Phosphate Species Formed at the Oxic-Anoxic Interface of a Eutrophic Lake," *Geochim. Cosmochim. Acta* 53:399–408 (1989).
34. Davison, W. "Supply of Iron and Manganese to an Anoxic Lake Basin," *Nature* 290:241–243 (1981).
35. Davison, W. "Iron and Manganese in Lakes," *Earth-Science Rev.* in press (1993).
36. Yagi, A. "Dissolved Organic Manganese in the Anoxic Hypolimnion of Lake Fukami-ike," *Jpn J. Limnol.* 49:149–156 (1988).
37. De Vitre, R.R. "Multimethod Characterization of the Forms of Iron, Manganese, and Sulfur in a Eutrophic Lake (Bret, Vaud, Switzerland)," PhD Thesis, University of Geneva, Switzerland (1986).
38. Meyer, J.S. and R. Gachter. "Manganese Release from Sediment in a Eutrophic Lake," *Limnol. Oceanogr.* in press.
39. Hunt, C.D. "Variability in the Benthic Mn^{2+} Flux in Coastal Marine Ecosystems Resulting from Temperature and Primary Production," *Limnol. Oceanogr.* 28:913–923 (1983).
40. Delfino, J.J. and G.F. Lee. "Chemistry of Manganese in Lake Mendota, Wisconsin," *Environ. Sci. Technol.* 2:1094–1100 (1968).
41. Verdouw, H. and E.M.J. Dekkers. "Iron and Manganese in Lake Vechten: Dynamics and Role in the Cycle of Reducing Power," *Arch. Hydrobiol.* 89:509–532 (1980).
42. Sigg, L., C.A. Johnson, and A. Kuhn. "Redox Conditions and Alkalinity Generation in a Seasonally Anoxic Lake (Lake Griefen)," *Mar. Chem.* 36:9–26 (1991).

43. Schindler, D.W., R.H. Hesslein, R. Wagemann, and W.S. Broecker. "Effects of Acidification on Mobilization of Heavy Metals and Radionuclides from the Sediments of a Freshwater Lake," *Can. J. Fish. Aquat. Sci.* 37:373–377 (1980).
44. White, J.R. and C.T. Driscoll. "Manganese Cycling in an Acidic Adirondack Lake," *Biogeochem.* 3:87–103 (1987).
45. Davison, W., J. Buffle, and R.R. De Vitre. "Direct Polarographic Determination of O_2, Fe(II), Mn(II), S(-II) and Related Species in Anoxic Waters," *Pure Appl. Chem.* 60:1535–1548 (1988).
46. Van Leeuwen, H., R. Cleven, and J. Buffle. "Voltammetric Techniques for Complexation Measurements in Natural Aquatic Media. Role of the Size of Macromolecular Ligands and Dissociation Kinetics of Complexes," *Pure Appl. Chem.* 61:255–274 (1989).
47. Tipping, E., D.W. Thompson, and W. Davison. "Oxidation Products of Mn(II) in Lake Waters," *Chem. Geol.* 44:359–383 (1984).
48. Tipping, E. and M.J. Heaton. "The Adsorption of Aquatic Humic Substances by Two Oxides of Manganese," *Geochim. Cosmochim. Acta* 47:1393–1397 (1983).
49. Murray, J.W. "The Interactions of Metal Ions at the Manganese Dioxide-Solution Interface," *Geochim. Cosmochim. Acta* 39:505–519 (1975).
50. Gadde, R.R. and H.A. Laitinen. "Studies of Heavy Metal Adsorption by Hydrous Iron and Manganese Oxides," *Anal. Chem.* 46:2022–2026 (1974).
51. Balistrieri, L.S. and J.W. Murray. "The Surface Chemistry of Sediments from the Panama Basin: the Influence of Mn Oxides on Metal Adsorption," *Geochim. Cosmochim. Acta* 50:2235–2243 (1986).
52. Gerloff, G.C. and F. Skoog. "Availability of Iron and Manganese in Southern Wisconsin Lakes for the Growth of Microcystis Aeruginosa," *Ecology* 38:551–556 (1957).
53. Udel'nova, T.M., E.N. Kondral'eva, and E.A. Boichenko. "Iron and Manganese Content in Various Photosynthesizing Microorganisms," *Mikrobiologiya* 37:197–200 (1968).
54. Sholkovitz, E.R. and D. Copland. "The Chemistry of Suspended Matter in Esthwaite Water, a Biologically Productive Lake with Seasonally Anoxic Hypolimnion," *Geochim. Cosmochim. Acta* 46:393–410 (1982b).
55. Sigg, L., M. Sturm, and D. Kistler. "Vertical Transport of Heavy Metals by Settling Particles in Lake Zurich," *Limnol. Oceanogr.* 32:112–130 (1987).
56. Tipping, E., C. Woof, and D. Cooke. "Iron Oxide from a Seasonally Anoxic Lake," *Geochim. Cosmochim. Acta* 45:1411–1419 (1981).
57. Davison, W. and D.P.E. Dickson. "Mössbauer Spectroscopic and Chemical Studies of Particulate Iron Material from a Seasonally Anoxic Lake," *Chem. Geol.* 42:177–187 (1984).
58. Jones, B.F. and C.J. Bowser. "The Mineralogy and Related Chemistry of Lake Sediments," in *Lakes: Chemistry, Geology, Physics*, A. Lerman, Ed. (New York: Springer-Verlag, 1978).
59. Salomons, W. and U. Förstner. *Metals in the Hydrocycle* (Berlin: Springer-Verlag, 1984).
60. Johnson, K.S. "Solubility of Rhodochrosite ($MnCO_3$) in Water and Seawater," *Geochim. Cosmochim. Acta* 46:1805–1809 (1982).
61. Emerson, S., L. Jacobs, and B. Tebo. "The Behavior of Trace Metals in Marine

Anoxic Waters:Solubilities at the Oxygen-Hydrogen Sulfide Interface," in *Trace Metals in Seawater*, C.S. Wong, E. Boyle, K.W. Bruland, J.D. Burton, and E.D. Goldberg, Eds. (New York: Plenum Press, Inc., 1983).

62. Jaquet, J.M., G. Nembrini, J. Garcia, and J.P. Vernet. "The Manganese Cycle in Lac Léman, Switzerland: the Role of Metallogenium," *Hydrobiol.* 91:323–340 (1982).
63. Callender, E., C.J. Bowser, and R. Rossman. "Geochemistry of Ferromanganese and Manganese Carbonate Crusts from Green Bay, Lake Michigan," *Trans. Am. Geophys. Union* 54:340–356 (1974).
64. Degens, E.T., G. Kulbricki. "Hydrothermal Origin of Metals in Some East African Rift Lakes," *Mineral. Deposita* 8:388–404 (1973).
65. Shigematsu, T. and M. Tabushi, Y. Nishikowa, T. Muroga, and Y. Matsunaga. "Geochemical Study on Lakes Mikata," *Ball. Inst. Chem. Res. Kyoto Univ.* 39:43–56 (1961).
66. Bowser, C.J., E. Callender, and R. Rossman. "Electron-probe and X-ray Studies of Freshwater Ferromanganese Nodules from Wisconsin and Michigan," *Geol. Soc. Am. Abs.* 2:500–501 (1970).
67. Nriagu, J.O. and C.I. Dell. "Diagenetic Formation of Iron Phosphates in Recent Lake Sediments," *Amer Mineral.* 59:934–946 (1974).
68. Diem, D. and W. Stumm. "Is Dissolved Mn^{2+} Being Oxidized by O_2 in Absence of Mn-Bacteria or Surface Catalysts?" *Geochim. Cosmochim. Acta* 48:1571–1573 (1984).
69. Morgan, J.J. "Chemical Equilibria and Kinetic Properties of Manganese in Natural Waters," in *Applications of Water Chemistry*, S.D. Faust and J.V. Hunter, Eds. (New York: John Wiley & Sons, Inc., 1967).
70. Caughlin, R.W. and I. Matsui. "Catalytic Oxidation of Aqueous Mn(II)," *J. Catal.* 41:108–123 (1976).
71. Davies S.H.R. and J.J. Morgan, "Manganese(II) Oxidation Kinetics on Metal Oxide Surfaces," *J. Colloid Interface Sci.* 129:1, 63–77 (1989).
72. Morel, F.M.M. *Principles of Aquatic Chemistry* (New York: John Wiley & Sons, Inc., 1983).
73. Chapnick, S.D., W.S. Moore, and K.H. Nealson. "Microbially Mediated Manganese Oxidation in a Freshwater Lake," *Limnol. Oceanogr.* 27:1004–1014 (1982).
74. Kawashima, M., T. Takamatsu, and M. Koyama. "Mechanisms of Precipitation of Manganese II in Lake Biwa, a Freshwater Lake," *Wat. Res.* 22:613–618 (1988).
75. Tipping, E. "Temperature Dependence of Mn(II) Oxidation in Lakewaters: a Test of Biological Involvement," *Geochim. Cosmochim. Acta* 48:1353–1356 (1984).
76. Nealson, K.H., R.A. Rosson, and C.R. Myers. "Mechanisms of Oxidation and Reduction of Manganese," in *Metal Ions and Bacteria*, T.J. Beveridge and R.J. Doyle, Eds. (New York: John Wiley & Sons, Inc., 1989).
77. Adams, L.F. and W.C. Ghiorse. "Characterization of Extracellular Mn^{2+}-Oxidizing Activity and Isolation of an Mn^{2+}-Oxidizing Protein from Leptothrix discophera ss–1," *J. Bacteriol.* 169:1279–1285 (1987).
78. Adams, L.F. and W.C. Ghiorse. "Oxidation State of Mn in the Mn Oxide Produced by Leptothrix discophera ss–1," *Geochim. Cosmochim. Acta* 52:2073–2076 (1988).
79. Kepkay, P.E. and K.H. Nealson. "Growth of Manganese Oxidizing Pseudomonas sp. in Continuous Culture," *Arch. Microbiol.* 148:63–67 (1987).
80. Sunda, W.G. and S.A. Huntsman, "Diel Cycles in Microbial Manganese Oxidation and Manganese Redox Speciation in Coastal Waters of the Bahama Islands," *Limnol. Oceanogr.* 35:325–338 (1990).

81. Chiswell, B. and M.B. Mokhtar. "The Speciation of Manganese in Freshwaters," *Talanta* 33:669–677 (1986).
82. Hem, J.D. "Redox Processes at Surfaces of Manganese Oxide and their Effects on Aqueous Metal Ions," *Chem. Geol.* 21:199–218 (1978).
83. Stumm, W., D. Diem, and R. Giovanoli. "Chemistry of Manganese in Natural Waters," *Thal. Jugoslavica* 16:177–180 (1980).
84. Stumm, W. and R. Giovanoli. "On the Nature of Particulate Manganese in Simulated Lakewaters," *Chimia 30*:423–425(1976).
85. Lind, C.J., J.D. Hem, and C.E. Roberson. "Reaction Products of Manganese-Bearing Waters," in *Chemical Quality of Water and the Hydrological Cycle*, R.C. Averett and D.M. McKnight, Eds. (Chelsea, MI: Lewis Publishers, Inc., 1987).
86. Hastings, D. and S. Emerson. "Oxidation of Manganese by Spores of a Marine Bacillus: Kinetics and Thermodynamic Considerations," *Geochim. Cosmochim. Acta* 50:1819–1824 (1986).
87. Klaveness, D. "Morphology, Distribution and Significance of the Manganese-Accumulating Microorganism Metallogenium in Lakes," *Hydrobiol.* 56:25–33 (1977).
88. Giovanoli, R., R. Brätsch, D. Diem, G. Osman-Sigg, and L. Sigg. "The Composition of Settling Particles in Lake Zurich," *Schweiz. Z. Hydrol.* 42:89–100 (1980).
89. Bromfield, S.M. "The Properties of a Biologically Formed Manganese Oxide, Its Availability to Oats and Its Solution by Root Washings," *Plant Soil* 9:325–337 (1958).
90. Hem, J.D. "Rates of Manganese Oxidation in Aqueous Systems," *Geochim. Cosmochim. Acta* 45:1369–1374 (1981).
91. Emerson, S., S. Kalhom, S. Jacobs, B.M. Tebo, K. Nealson and R.A. Rosson. "Environmental Oxidation Rate of Manganese (II): Bacterial Catalysis," *Geochim. Cosmochim. Acta* 46:1073–1079 (1982).
92. Waite, T.D., I.C. Wrigley, and R. Szymczak. "Photoassisted Dissolution of a Colloidal Manganese Oxide in the Presence of Fulvic Acid," *Environ. Sci. Technol.* 22/7:778–785 (1988).
93. Sunda, W.G., S.A. Huntsman, and G.R. Harvey. "Photoreduction of Manganese Oxides in Seawater and Its Geochemical and Biological Implications," *Nature* 301:234–236 (1983).
94. Stone, A.T. and J.J. Morgan. "Kinetics of Chemical Transformations in the Environment," in *Aquatic Chemical Kinetics: Reaction Rates of Processes in Natural Waters*, W. Stumm, Ed. (New York: Wiley-Interscience, 1990).
95. Stumm, W. and E. Wieland. "Dissolution of Oxide and Silicate Minerals: Rates Depend on Surface Speciation," in *Aquatic Chemical Kinetics: Reaction Rates of Processes in Natural Waters*, W. Stumm, Ed. (New York: Wiley-Interscience, 1990).
96. Stone, A.T. and J.J. Morgan. "Reduction and Dissolution of Manganese(III) and Manganese(IV) Oxides by Organics. I. Reaction with Hydroquinone," *Environ. Sci. Technol.* 18:450–456 (1984a).
97. Stone, A.T. and J.J. Morgan. "Reduction and Dissolution of Manganese(III) and Manganese(IV) Oxides by Organics. II. Survey of the Reactivity of Organics," *Environ. Sci. Technol.* 18:617–624 (1984b).
98. Stone, A.T. "Reductive Dissolution of Manganese(III/IV) Oxides by Substituted Phenols," *Environ. Sci. Technol.* 21:979–988 (1987a).
99. Ulrich, H.-J. and A.T. Stone. "Oxidation of Chloro-Phenols Adsorbed to Manganese Oxide Surfaces," *Environ. Sci. Technol.* 23:421–428 (1989).

100. Sulzberger, B., "Photo Redox Reactions at Hydrous Metal Oxide Surface: A Surface Coordination Chemistry Approach," in *Chemical Kinetics: Reaction Rates of Processes in Natural Waters*, W. Stumm, Ed. (New York: Wiley-Interscience, 1990).
101. Waite, T.D. and F.M.M. Morel. "Photoreductive Dissolution of Colloidal Iron Oxides in Natural Waters," *Environ. Sci. Technol.* 18:860–868 (1984).
102. Pichat, P. and M.A. Fox. "Photocatalysis on Semiconductors," in *Photoinduced Catalysis*, Part D, M.A. Fox and M. Chanon, Eds. (Amsterdam: Elsevier North-Holland, Inc., 1988).
103. Kawana, K., T. Shiozawa, A. Hoshika, T. Tanimotor, and O. Takimura. "Diffusion of Manganese from Bottom Sediments in Beppu Bay," *Bull. Soc. Franer-Japonaise Oceanogr.* 18:131–137 (1980).
104. Burdige, D.J. and K.H. Nealson, "Chemical and Microbiological Studies of Sulfide-Mediated Manganese Reduction." *Geomicrobiol. J.* 4:361–387 (1986).
105. Aller, R.C. and P.D. Rude. "Complete Oxidation of Solid Phase Sulfides by Manganese and Bacteria in Anoxic Marine Sediments," *Geochim. Cosmochim. Acta* 52:751–765 (1988).
106. Luther, G.W., III. "The Frontier-Molecular-Orbital Theory Approach in Geochemical Processes," in *Aquatic Chemical Kinetics:Reaction rates of Processes in Natural Waters*, W. Stumm, Ed. (New York: Wiley-Interscience, 1990).
107. Myers, C.R. and K. Nealson. "Microbial Reduction of Manganese Oxides:Interactions with Iron and Sulfur," *Geochim. Cosmochim. Acta* 52:2727–2732 (1988a).
108. Myers, C.R. and K. Nealson. "Bacterial Manganese Reduction and Growth with Manganese Oxide as the Sole Electron Acceptor," *Science* 240:1319–1321 (1988b).
109. Jones, J.G., S. Gardener, and B.M. Simon. "Reduction of Ferric Iron by Eutrophic Bacteria in Lake Sediments," *J. Gen. Microbiol.* 130:45–51 (1984).
110. Lovely, D.R. and E.J.P. Phillips, "Novel Mode of Microbial Energy Metabolism: Organic Carbon Oxidation Coupled to Dissimilatory Reduction of Iron and Manganese," *Appl. Environ. Microbiol.* 54:1472–1480 (1988).
111. Gottfreund, J., G. Schmitt, and R. Schweisfurth. "Wertigkeitswechsel von Manganspecies durch Bakterien in Naehrloesungen und in Lochergestein," *Landwirtsch. Forschung* 38:80–86 (1985).
112. Gottfreund, J. and R. Schweisfurth. "Mikrobiologische Oxidation und Reduktion von Manganspecies," *Fresenius Z. Anal. Chem.* 316:634–638 (1983).
113. Ehrlich, H.L. "Manganese Oxide Reduction as a Form of Anaerobic Respiration," *Geomicrobiol. J.* 5 (3/4):423–431 (1987).
114. Ghiorse, W.C. "Microbial Reduction of Manganese and Iron," in *Biology of Anaerobic Microorganisms*, A.J.B. Zehnder, Ed. (New York: John Wiley & Sons, Inc., 1988).
115. Nealson, K.H. "The Microbial Manganese Cycle," in *Microbial Biochemistry*, W. Krumbein, Ed. (Oxford: Blackwell Scientific, 1983), pp. 191–222.
116. Ehrlich, H.L. "Bacterial Leaching of Manganese Ores," in *Biogeochemistry of Ancient and Modern Environments*, P. A. Trudinger, Ed. (Berlin: Springer-Verlag, 1980).
117. Trimble, R.B. and H.L. Ehrlich. "Bacteriology of Manganese Nodules. III. Reduction of MnO_2 by Two Strains of Nodule Bacteria," *Applied Microbiol.* 16:695–702 (1968).
118. Dubinina, G.A. "Functional Role of Bivalent Iron and Manganese Oxidation in Leptothrix Ochracea," *Mikrobiologiya* 47:783–789 (1978), in Russian.

119. de Vrind, J.P.M., F.C. Boogerd, and E.W. de Vrind-de Jong. "Manganese Reduction by a Marine Bacillus Species," *J. Bacteriol.* 167(1):30–34 (1986).
120. Price, N.M. and F.M.M. Morel. "Role of Extracellular Enzymatic Reactions in Natural Waters," in *Aquatic Chemical Kinetics: Reaction Rates of Processes in Natural Waters*, W. Stumm, Ed. (New York: Wiley-Interscience, 1990).
121. Morfett, K., W. Davison, and J. Hamilton-Taylor. "Trace Metal Dynamics in a Seasonally Anoxic Lake," *Environ. Geol. Water Sci.* 11:107–114 (1988).
122. Balistrieri, L.S., J.W. Murray, and B.P. Paul. "The Biogeochemical Cycling of Trace Metals in the Water Column of Lake Sammamish, WA: Response to Seasonally Anoxic Conditions," *Limnol. Oceanogr.* in press (1992b).
123. Belzile, N., R. R. De Vitre, and A. Tessier. "In Situ Collection of Diagenetic Iron and Manganese Oxyhydroxides from Natural Sediments," *Nature* 340:376–377 (1989).
124. Oscarson, D.W., P.M. Huang, W.K. Liaw, and U.T. Hammer. "Kinetics of Oxidation of Arsenite by Various Manganese Dioxides," *Soil Sci. Soc. Amer.* 47:644–648 (1983).
125. De Vitre, R.R., N. Belzile, and A. Tessier. "Speciation and Adsorption of As on Diagenetic Iron Oxyhydroxides," *Limnol. Oceanogr.* 36(7):1480–1485 (1991).
126. McBride, M.B. "Oxidation of 1,2- and 1,4-dihydroxybenzene by Birnessite in Acidic Aqueous Suspension," *Clays Clay Miner.* 37:479–486 (1989).
127. McBride, M.B. "Adsorption and Oxidation of Phenolic Compounds by Iron and Manganese Oxides," *Soil Sci. Soc. Amer. J.* 51:1466–1472 (1987).
128. Carpenter, R. "Quantitative Electron Spin Resonance (ESR) Determinations of Forms and Total Amount of Mn in Aqueous Environmental Samples," *Geochim. Cosmochim. Acta* 47:875–885 (1983).
129. Charlet, L. and A. Manceau. "Structure, Formation and Reactivity of Hydrous Oxide Particles: Insights from X-ray Absorption Spectroscopy," in *Environmental Particles, Vol. 2*, H.P. van Leeuwen and J. Buffle, Eds. (Chelsea, MI: Lewis Publishers, Inc., 1993).
130. Beckett, R. and B.T.Hart. " Use of Field Flow Fractionation Techniques to Characterize Aquatic Particles," in *Environmental Particles, Vol. 2*, H.P. van Leeuwen and J. Buffle, Eds. (Chelsea, MI: Lewis Publishers, Inc., 1993).
131. Schurtenberger, P. and M.E. Newman. "Characterization of Biological and Environmental Particles Using Static and Dynamic Light Scattering," in *Environmental Particles, Vol. 2*, H.P. van Leeuwen and J. Buffle, Eds. (Chelsea, MI: Lewis Publishers, Inc., 1993).
132. Burdige, D.J. and K.H. Nealson. "Microbial Manganese Reduction by Enrichment Cultures From Coastal Marine Sediments," *Appl. Environ. Microbiol.* 50(2):491–497 (1985).
133. Ljunggren, P. "Differential Thermal Analysis and X-ray Examination of Fe and Mn Bog Ores," *Geol. For. Stockh. Forh.* 77:135–147 (1955).
134. Potter, R.M. and G.R. Rossman. "Mineralogy of Manganese Dendrites and Coatings," *Am. Mineral.* 64:1219–1226 (1979).
135. Reinhard, M., N.L. Goodman, and J.F. Barker. "Occurrence and Distribution of Organic Chemicals in Two Landfill Leachates Plumes. *Environ. Sci. Technol.* 18:953–961 (1984).
136. Wegman, R.C.C. and A.W.M. Hofstee. "Chlorophenols in Surface Waters of the Netherlands," *Water Res.* 13:651–657 (1979).
137. Leuenberger, C., R. Caney, J.W. Graydon, E. Molnar-Kubica, and W Giger. "Per-

sistent Organics in Pulp Mill Effluents: OccurRence and Behavior in a Biological Treatment plant," *Chimia* 37:345–354 (1983).

138. Huang, P.M., D.W. Oscarson, W.K. Liaw, and U.T. Hammer, "Dynamics and Mechanisms of Arsenite Oxidation by Freshwater Sediments," *Hydrobiologia* 91/92:315–322 (1982).
139. Francis, A.J. and C.J. Dodge. "Anaerobic Microbial Dissolution of Transition and Heavy Metal Oxides," *Appl. Environ. Microbiol.* 54(4):1009–1014 (1988).
140. Ghiorse, W.C. and H.L. Ehrlich. "Effects of Seawater Cations and Temperature on Manganese Dioxide Reductase Activity in a Marine Bacillus," *Appl. Microbiol.* 28(5):785–792 (1974).
141. Mopper, K. and B.F. Taylor. "Biochemical Cycling of Sulfur: Thiols in Coastal Marine Sediments," in *Organic Marine Chemistry*, M.L. Sohn, Ed. *Am. Chem. Soc. Symp. Ser.* 305:324–329 (1986).
142. Sunda, W.G. and S.A. Huntsman, "Effect of Sunlight on Redox Cycles of Manganese in the South-western Sargasso Sea," *Deep Sea Res.* 35:1297–1317 (1988).

CHAPTER 8

PHYSICOCHEMICAL AGGREGATION AND DEPOSITION IN AQUATIC ENVIRONMENTS

Charles R. O'Melia and Christine L. Tiller

TABLE OF CONTENTS

0-87371-895-X/93/$0.00+$.50

INTRODUCTION

Particles in water may be in motion, suspended in an aqueous phase such as a river or an estuary, or fixed in space, components of a porous medium such as a lake sediment or an aquifer. Moving particles can come into contact with each other; when they attach to each other, the process is termed aggregation or coagulation. Suspended particles may also come into contact with stationary ones or with other fixed solid boundaries; when they attach, the process is termed deposition or filtration.

It has been convenient to consider the processes of physicochemical aggregation and deposition as comprised of two separate and sequential steps, a transport step followed by an attachment step.[1] Particle transport depends upon hydrodynamics and external forces such as gravity; it is primarily a physical process. The attachment of two suspended particles or of a suspended particle to a stationary surface is often dominated by the chemistry of the aqueous phase and by the surface chemical properties of the two solids involved. This distinction between transport and attachment, or physics and chemistry, is not perfectly sharp. Some colloidal forces (van der Waals forces) must be present to allow transport of a particle to the surface of another solid. Similarly, surface chemistry does not always control attachment; physical forces are sufficient to describe "favorable" attachment when chemical particle-particle interactions are small.

This chapter is written with two objectives: (1) to describe certain physical and chemical phenomena that affect aggregation and deposition in aquatic systems, focussing on similarities between these two environmental processes, and (2) to recommend applications for these concepts in the study and management of aquatic systems.

Aggregation and deposition are primarily kinetic phenomena. For example, the rate of aggregation of a suspension can be written as follows:

$$\frac{dn}{dt} = -k_a n^2 \tag{1}$$

where

n is the number concentration of particles in suspension at time t
k_a is a second order rate constant that is a function of physical and chemical properties of the system

In a similar manner, the deposition of particles from a suspension to the surface of a porous medium can be described as

$$\frac{dn}{dL} = -k_d n \tag{2}$$

where

L is the distance along the length of the porous medium
k_d is a pseudo-first order rate constant that also depends upon physical and chemical properties of the system

Deposition or filtration involves collisions between suspended particles (n) and stationary collectors in the porous medium; the number of these stationary collectors is included in the distance L. This chapter is focused on k_a and k_d and, specifically, on the chemical and physical contributions to them.

For aggregation or coagulation we can write

$$k_a = \alpha_a \beta \tag{3}$$

where

β is a mass transport coefficient with dimensions such as m^3/s^1; it is largely physical and often evaluated theoretically
α_a is a dimensionless sticking coefficient that is primarily chemical and often determined experimentally

For deposition or filtration we write

$$k_d \propto \alpha_d \eta \tag{4}$$

where

η is a dimensionless mass transport coefficient that, like β, is primarily physical and frequently determined theoretically
α_a, α_d is a dimensionless sticking coefficient or attachment probability that is typically chemical and measured experimentally

TRANSPORT

In this section the mass transport coefficients for aggregation (betas) and deposition (etas) are described. Betas in aggregation are considered first, followed by etas in deposition. Similarities and differences between these coefficients are then discussed.

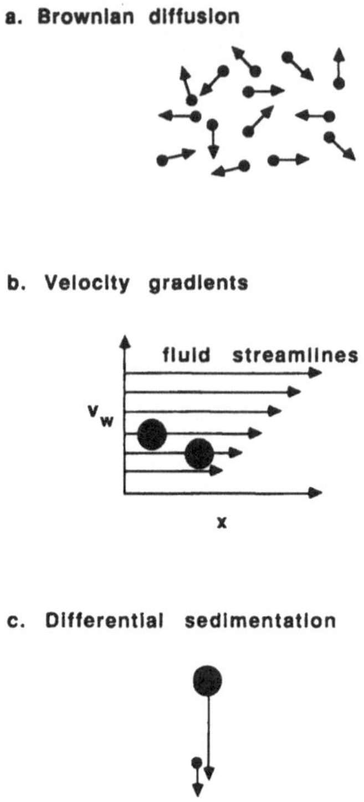

Figure 1. Schematic representation of rectilinear models for the transport of suspended particles in aquatic systems: (a) Brownian diffusion or perikinetic flocculation, (b) velocity gradients or orthokinetic flocculation, and (c) differential sedimentation.

Three physical processes of mass transport are considered. The first is Brownian, or molecular diffusion, by which random motion of small particles is brought about by thermal effects. The driving force for this transport is a function of kT, the product of Boltzmann's constant (k, 1.38×10^{-23} J/K) and the absolute temperature (T, K). The resistance force is provided by viscous drag on the diffusing particle. Considering these forces, Einstein[2] derived the following expression for the diffusion coefficient of a particle:

$$D = \frac{kT}{3\pi\mu d_p} \tag{5}$$

where

D is the molecular or Brownian diffusion coefficient of an uncharged spherical particle of diameter d_p (m)
μ is the absolute viscosity of the fluid ($N.s\,m^{-2}$)

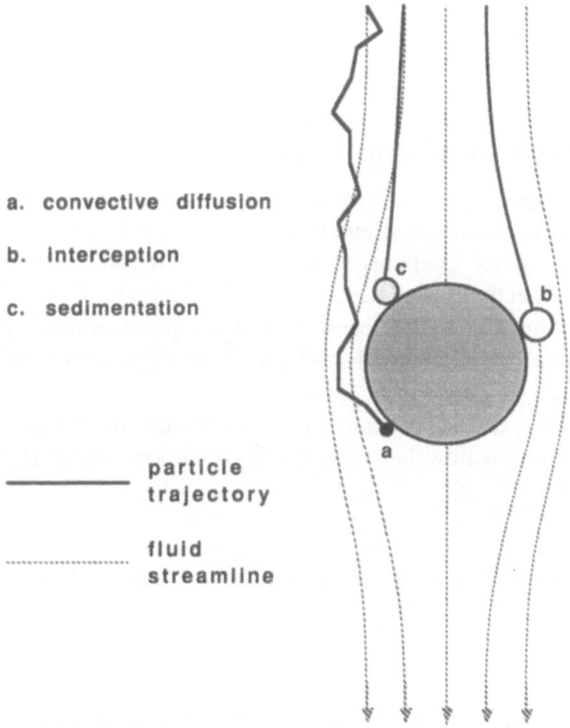

Figure 2. Schematic representation of transport mechanisms for deposition in porous media in aquatic systems: (a) convective Brownian diffusion, (b) interception or velocity gradients, and (c) sedimentation.

Transport of particles in suspension by Brownian diffusion (Figure 1) depends on thermal and viscous effects only and is not dependent on such factors as gravity. In deposition or filtration phenomena, Brownian diffusion is usually coupled with fluid flow in a process called convective diffusion (Figure 2).

The second process affecting particle transport in aquatic systems is fluid flow, either laminar or turbulent. Velocity differences or gradients occur in all real flowing fluids. Particles that follow the motion of the suspending fluid will then travel at different velocities. These fluid and particle velocity differences or gradients can produce interparticle contacts among particles suspended in the fluid (Figure 1). Particle transport in this case depends upon the mean velocity gradient in the fluid, G (s^{-1}). When one of the particles is stationary, as in a porous medium, and the other is transported by a flowing fluid, contact is said to be by interception (Figure 2).

The third process considered here is gravity, which produces vertical transport of particles and which depends upon the buoyant weight of the particles. Large, dense suspended particles can contact smaller or less dense ones in a process termed differential sedimentation (Figure 1). If one of the particles is stationary, as in a packed bed filter, contacts of suspended particles with the fixed particle can be said to occur by convective sedimentation (Figure 2).

Aggregation

Smoluchowski: A Rectilinear Model

Particle transport by Brownian diffusion, velocity gradients, and differential sedimentation is illustrated schematically in Figure 1. Transport by Brownian diffusion and velocity gradients, frequently termed perikinetic and orthokinetic flocculation, respectively, was described by Smoluchowski[3] in 1917. Transport by differential sedimentation has been considered subsequently by others.[4,5] This approach considers that all particles move in straight lines until contacts occur between them; it is a rectilinear model.

For a heterodisperse suspension, i.e., a suspension containing particles of more than one size, Smoluchowski's description of the kinetics of aggregation can be written as follows:[3]

$$\frac{dn_k}{dt} = \frac{1}{2}\sum_{i+j=k} \alpha_a \beta(i,j) n_i n_j - n_k \sum_{all\, i} \alpha_a \beta(i,k) n_i \qquad (6)$$

where

the subscripts i, j, and k refer to discrete particle sizes or size classes
n_i and n_j are the number concentrations (m^{-3}) of particles of sizes i and j, respectively
$\beta(i,j)$ and $\beta(i,k)$ are mass transport coefficients or collision frequency functions between particles of the size classes indicated, commonly taken as the sum of the transport provided by the three physical mechanisms shown in Figure 1

The first term on the right side of Equation 6 describes the formation of particles of size k by aggregation involving two smaller particles whose total volume is that of size k; i.e., the condition i + j = k under the summation states that v(i) + v(j) = v(k) where v is the volume of a particle (m^3). This is termed the coalesced sphere assumption; particle volume is conserved during the aggregation process. This means, for example, that aggregate pore volume is not included in this description of the process. The second term on the right side describes the loss of particles of size k by aggregation with particles of any other size to form larger aggregates. The term on the left side depicts the change in the concentration of particles of size k with time as aggregation proceeds.

Mass transport coefficients for the three transport processes are as follows:[3-5]

Brownian diffusion

$$\beta_{bd}(i,j) = \frac{2kT}{3\mu} \frac{\left(d_i + d_j\right)^2}{d_i d_j} \qquad (7)$$

Velocity gradients

$$\beta_{vg}(i,j)=\frac{1}{6}\left(d_i+d_j\right)^3 G \tag{8}$$

Differential sedimentation

$$\beta_{ds}(i,j)=\frac{\pi g}{72\mu}\left(\rho_p-\rho\right)\left(d_i+d_j\right)\left|d_i-d_j\right| \tag{9}$$

where

ρ_p and ρ are the densities of the particles and the fluid, respectively (kg/m^{-3}) and g is the gravity acceleration (9.80 m/s^{-2})

As stated previously, it is often assumed that the three transport processes are additive, i.e.,

$$\beta(i,j)=\beta_{bd}(i,j)+\beta_{vg}(i,j)+\beta_{ds}(i,j) \tag{10}$$

Important assumptions in the Smoluchowski approach to aggregation rates are (1) the rectilinear model and (2) that aggregates are coalesced spheres. The assumption of rectilinear motion neglects hydrodynamic interactions and short range forces between approaching particles. The hydrodynamic interactions are of two types: (1) changes around each suspended particle in the undisturbed flow fields that result from the basic transport processes such as fluid shear or gravity and (2) hydrodynamic drag between the particles as they come into close proximity. At very close separating distances, these short range hydrodynamic interactions become important. It becomes increasingly difficult to drain fluid from the narrowing gap between the particles; velocity gradients at the solid-water boundaries in the gap become very large, resulting in large drag forces on the particles. In fact, drag forces become infinite as the separating distance goes to zero, so that interparticle contacts cannot occur unless other short range attractive forces can overcome this hydrodynamic interaction. Attractive van der Waals forces provide this necessary interaction, so that contacts can occur. This short range physical or hydrodynamic effect is termed hydrodynamic retardation or the lubrication effect. Each of these two hydrodynamic interactions reduces the frequency of interparticle contacts from rates predicted with Smoluchowski's approach (Equations 7 through 10).

The coalesced sphere assumption, in which particle volume is conserved, underestimates the actual collision rates that occur. Fluid is incorporated into pores within aggregates as coagulation proceeds; a result is that the effective target volume for interparticle contacts by fluid motion and by gravity is larger than predicted with the Smoluchowski approach. The two assumptions in Smolu-

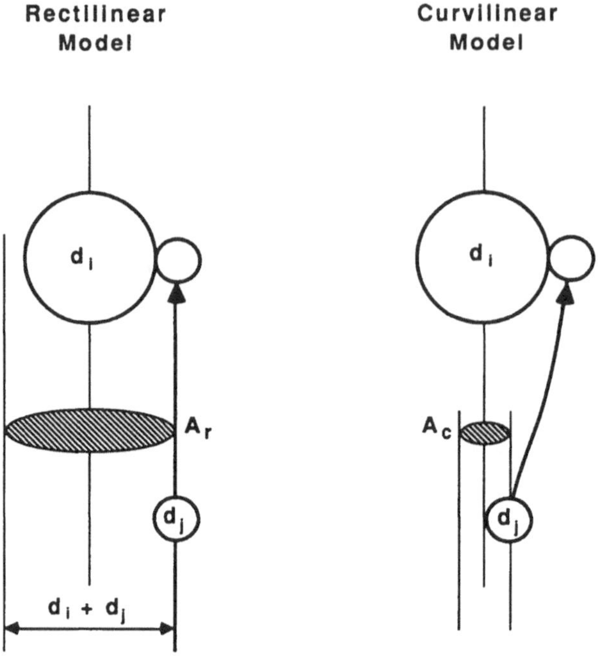

Figure 3. Illustration of rectilinear and curvilinear trajectories in particle transport by differential sedimentation. (From Han, H. and D. F. Lawler. *J. Am. Water Works Assoc.* 84:461–474 (1981). With permission.)

chowski's approach considered here, i.e., rectilinear motion and coalesced spheres, produce opposite effects on aggregation rates predicted by Equation 6. The rectilinear assumption overpredicts aggregation rates while the coalesced sphere assumption underpredicts them. These effects are discussed subsequently in this chapter.

Curvilinear Aggregation Models

Analytical solutions for the collision rates between suspended particles do not exist for cases where hydrodynamic interactions between two approaching particles are taken into account, but the use of computers has provided numerical solutions which can then be presented in graphical form or in fitted equations. Results are available for Brownian diffusion,[6-8] fluid shear,[8-10] and differential sedimentation.[8,11]

Differences between the rectilinear and curvilinear approaches to particle transport processes in aggregation are illustrated schematically for transport by differential sedimentation in Figure 3 adapted from Han and Lawler.[11] In each case, the upper, larger particle is settling by gravity toward the lower, smaller particle. In the rectilinear or Smoluchowski model, all small particles of size d_j that reside below the larger one and within the area described by the shaded area

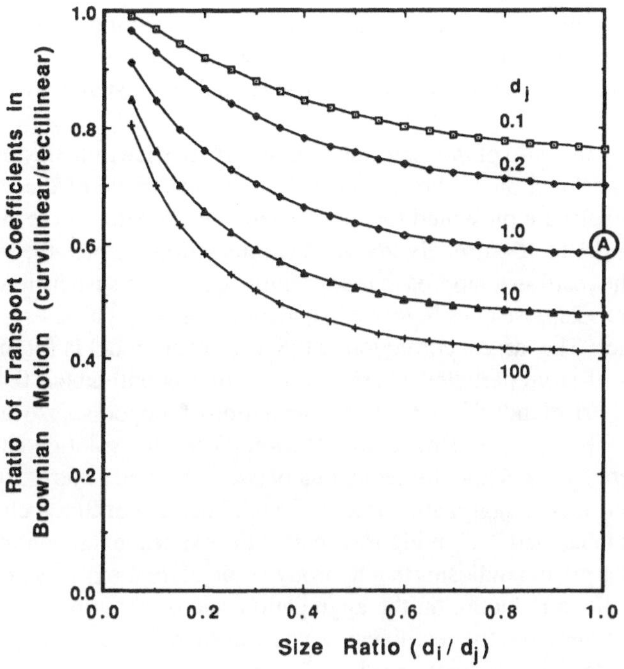

Figure 4. Effect of particle size on mass transport coefficients in Brownian motion. The ratio of rectilinear (Smoluchowski) to curvilinear transport coefficients is plotted as a function of the ratio of the size of a small particle (d_i) to a larger one (d_j); the Hamaker constant (A) is assumed as 10 kT. The circled A refers to a case described in the text. (From Han, H. and D. F. Lawler. *J. Am. Water Works Assoc.* 84:461–474 (1981). With permission.)

denoted by A_r with diameter ($d_i + d_j$) can come into contact with the larger particle; $\beta_{ds}(i,j)$ is described by Equation 9. In the curvilinear case where all hydrodynamic interactions between the two particles are considered, only those small particles in the shaded area denoted as A_c can come into contact with the larger particle. The area A_c can be determined numerically. The result is that the actual collision rate is less than the rectilinear rate and the actual mass transport coefficient is equal to $[(A_c/A_r)\ \beta_{ds}(i,j)]$. Han and Lawler[11] calculated reductions in the rectilinear transport rate by differential sedimentation ranging from about 0.3 to 0.001, so the effects of hydrodynamic interactions on this transport process can be substantial in many cases.

The effects of hydrodynamic interactions on particle transport by Brownian diffusion are described in Figure 4. These results are taken from the work of Han and Lawler;[8] similar studies have been done by Spielman,[6] Valioulis and List,[7] and others. The results in Figure 4 are derived from calculations in which attractive van der Waals interactions are included as the separating distance between the particles becomes very small. Even here, two choices are available, "retarded" or "unretarded" interactions. For retarded van der Waals forces, the time lag in the travel between the two particles of the electric field resulting from the oscillating dipoles that produce the interaction is considered. The results in Figure 4 were obtained using this retarded interaction. The Hamaker constant was assumed to be 10 kT.

In Figure 4 the ratio of the curvilinear transport coefficient to the rectilinear one is plotted as a function of the size ratio (smaller to larger) of the two colliding particles. Results are presented for cases where the diameter of the larger particle ranges from 0.1 to 100 μm. Hydrodynamic interactions reduce the transport rate relative to the rectilinear approach more for monodisperse suspensions ($d_i/d_j \rightarrow 1$) than for heterodisperse ones ($d_i/d_j << 1$) and more for large particles (large d_j) than for small ones. The largest correction to the rectilinear model is for monodisperse suspensions of large particles where the correction is only about 0.4.

Swift and Friedlander[12] studied the coagulation of monodisperse latex particles (diameter = 0.871 μm) by Brownian diffusion. When the solution chemistry was such that repulsive chemical interactions between particles were expected to be absent, experimental aggregation rates were 62.5 percent of Smoluchowski kinetics. The circle labeled "A" in Figure 4 marks the experimental conditions (d_i/d_j = 1; d_j = 0.871 μm) and indicates that hydrodynamic corrections to Smoluchowski's analysis predict a reduction in the aggregation rate of about 40%. The observed reduction in rates from the rectilinear model was therefore primarily physical and consistent with the curvilinear model.

Results obtained by Han and Lawler[8] for the effects of hydrodynamic interactions on particle transport by fluid shear are presented in Figure 5. These results are based on the work of Adler.[9,10] The effects of particle size, velocity gradient, and van der Waals interaction are characterized by a dimensionless group, H_A, defined as follows:

$$H_A = \frac{A}{18\pi\mu d_j^3 G} \tag{11}$$

where

A is the Hamaker constant (J)
d_j is the diameter of the larger of two interacting particles

Retarded van der Waals interactions are used; *A* is selected as 10 kT. The ratio of the curvilinear transport coefficient to the rectilinear case is plotted as a function of the size ratio, as in Figure 4.

Hydrodynamic interactions in aggregation by fluid shear are relatively small for monodisperse suspensions and become increasingly important as the systems become more heterodisperse, i.e., as the size ratio decreases. The hydrodynamic effects become more important as H_A becomes smaller; increasing the velocity gradient or the size of the larger particle increases the interaction and the deviation from the rectilinear model.

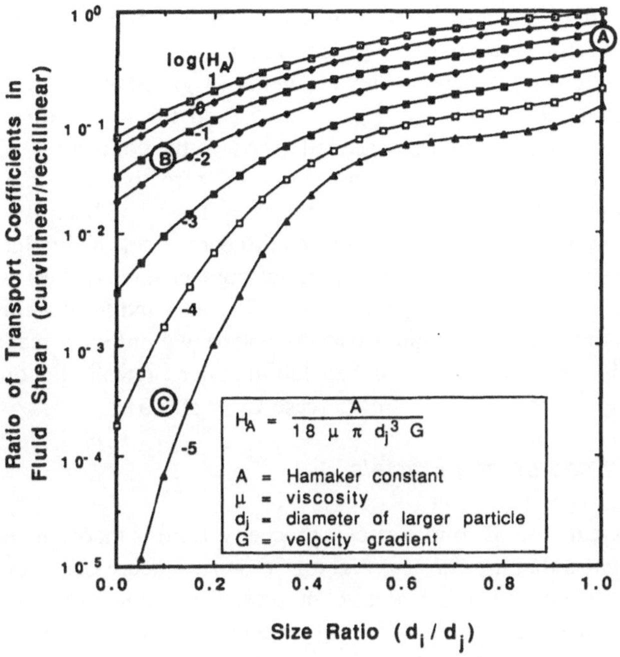

Figure 5. Effect of particle size on mass transport coefficients in fluid shear. The ratio of rectilinear (Smoluchowski) to curvilinear transport coefficients is plotted as a function of the logarithm of the dimensionless group, H_A, defined in the Figure. The circled A, B, and C refer to cases discussed in the text. (From Han, H. and D. F. Lawler. *J. Am. Water Works Assoc.* 84:461–474 (1981). With permission.)

Three cases are illustrated in Figure 5, marked by the circles labeled "A", "B", and "C." Case A refers to experiments by Swift and Friedlander[12] on the coagulation of monodisperse latex particles (diameter = 0.871 μm) in shear flow and in the absence of repulsive chemical interactions. Assuming a velocity gradient of 20 s^{-1}, H_A is 0.0535, log H_A is –1.27, and d_i/d_j is 1.0 for these experimental conditions. The circle labeled "A" in Figure 5 marks these conditions and indicates that the hydrodynamic corrections to Smoluchowski's model predict a reduction in the aggregation rate by fluid shear of about 40%. The experimental measurements showed a reduction of 64%. As with Brownian diffusion, this reduction was largely hydrodynamic and consistent with the curvilinear model.

Case B in Figure 4 is a representation of shear coagulation in a surface water such as a lake or the ocean. Considering the interactions between a bacterium (d_i = 1 μm) and an algal cell (d_j = 10 μm) in a shear field with G = 0.1 s^{-1} at 20°C, H_A is 0.007, log H_A is –2.15, and the size ratio is 0.1. Under these circumstances, the curvilinear approach predicts contact rates by fluid shear that are in the order of 3% of the Smoluchowski model. Case C is a representation of aggregation in a water or wastewater treatment plant, again considering contacts between a small particle (1 μm) and a larger one (10 μm); in this case fluid mixing is provided to achieve a velocity gradient of 20 s^{-1}. For this situation H_A is 3.6×10^{-5}, log H_A is −4.45, and the size ratio is again 0.1. Hydrodynamic interactions are predicted to reduce the contact rate to only about 0.03% of the rectilinear case, indicating that the effectiveness of fluid shear in producing interparticle contacts in treatment systems has been overrated.

These three examples show that the rectilinear model of Smoluchowski can provide a useful representation of the initial contact rates in laboratory studies using monodisperse suspensions (cases A in Figures 4 and 5), that this model can overpredict contacts in lakes and oceans by one to two orders of magnitude (case B in Figure 5), and that it overestimates initial contact rates in treatment systems by three orders of magnitude or more (case C, Figure 5).

Coalesced Spheres and Fractals

As the coagulation of solid particles proceeds, fluid is incorporated into pores in the aggregates that are formed. Aggregate density decreases, and total aggregate volume increases as the process continues. The result is that the target cross sections or collision diameters of the aggregates increase, thereby increasing the rates of interparticle contacts brought about by Brownian diffusion, fluid shear, and differential sedimentation. Observations of natural and technological systems[13,14] indicate that the aggregates in these systems are fractals and have fractal dimensions less than three. An excellent review of theoretical and experimental aspects of fractal aggregates in natural processes is provided by Meakin.[15] Jiang and Logan[16] have applied fractal mathematics to aquatic systems. In this volume, Jackson and Lochmann[17] discuss the coalesced sphere assumption and the use of fractals in modeling coagulation in marine ecosystems.

The relationship between the mass and the length of a fractal aggregate can be written as follows:

$$l \propto m^{1/D_f} \tag{12}$$

where

- l is a characteristic length dimension of the aggregate
- m is the mass of the aggregate
- D_f is the fractal dimension

The coalesced sphere assumption corresponds to $D_f = 3$. The fraction of fluid incorporated (aggregate porosity) increases with the size of the aggregate when $D_f < 3$, and a given mass of solid material in such an aggregate will have a larger collision diameter than the coalesced sphere diameter. Interparticle contact rates will be more rapid than described by the rectilinear model.

Deposition

The Clean Bed Approach

Particle transport from a flowing fluid to a single stationary spherical grain in a porous medium is illustrated in Figure 2. Three transport processes are presented: convective Brownian diffusion, interception (analogous to fluid shear), and sedimentation. The approach, adapted from the work of Friedlander[18] in air filtration, was introduced to water filtration by O'Melia and co-workers.[1,19] Substantial additions and improvements have been made by Spielman[20,21] and Tien[22,23] and their co-workers.

Particle deposition is often related to the depth or length of the porous medium (filter) within which suspended particles are removed from the flowing fluid. For a monodisperse suspension, the rate of removal of particles with respect to the depth of a clean filter bed is written as follows:[19]

$$\frac{dn}{dL} = -\frac{3}{2}\alpha_d \eta(p,c)\frac{(1-\varepsilon)}{d_c}n \tag{13}$$

where

L is the distance along the length or depth of the filter medium (m)
ε is the porosity of the filter (dimensionless)
$\eta(p,c)$ is the particle transport or collision frequency function (dimensionless) for contacts between suspended particles with diameter d_p and media grains or collectors with diameter d_c

The collision frequency function is commonly taken as the sum of the three physical processes shown in Figure 2.

Transport coefficients for the three transport processes are as follows:

Convective Brownian diffusion

$$\begin{aligned}\eta_{cbd}(p,c) &= 4.0A_s^{1/3}Pe^{-2/3} \\ &= 0.897A_s^{1/3}\left(\frac{kT}{\mu d_p d_c U}\right)^{2/3}\end{aligned} \tag{14}$$

Interception (fluid motion)

$$\eta_i(p,c) = \frac{3}{2}A_s\left(\frac{d_p}{d_c}\right)^2 \tag{15}$$

Sedimentation

$$\eta_s(p,c) = A_s \frac{v_s}{U} = \frac{(\rho_p - \rho)g}{18\mu U} A_s d_p^2 \qquad (16)$$

where

Pe is the dimensionless Peclet number, the ratio of transport by fluid flow (advection) to transport by molecular or Brownian diffusion, given by $2d_c U/D$
d_p and d_c are the diameters (m) of the suspended particle and the stationary collector, respectively
U is the undisturbed or approach velocity (m/s)
v_s is the settling velocity of the suspended particle (m/s^1)
A_s is a dimensionless parameter that depends on the porosity of the media

It is introduced to account for the effects of neighboring grains or stationary collectors on the flow field around a single collector, such as illustrated by the streamlines in Figure 2. Using a model for flow in porous media developed by Happel,[24] A_s is given as follows:

$$A_s = \frac{(1-\gamma^5)}{\left(1 - \frac{3}{2}\gamma + \frac{3}{2}\gamma^5 - \gamma^6\right)} \qquad (17)$$

in which $\gamma = (1 - \varepsilon)^{1/3}$. As in aggregation, it is often assumed that the three transport processes are additive:

$$\eta(p,c) = \eta_{cbd}(p,c) + \eta_i(p,c) + \eta_s(p,c) \qquad (18)$$

Two important assumptions in this approach to deposition are (1) neglect of hydrodynamic retardation and (2) clean collectors. Unlike the Smoluchowski approach to aggregation, this model for deposition is curvilinear. However, it is not complete. It does include the effects of a single collector on the flow of the fluid (Figure 2) and, with Happel's model, the effect of neighboring collectors on the flow around a single media grain. However, the results depicted by Equations 14 through 16 do not include short range hydrodynamic forces that slow the drainage of fluid from the narrowing gap between the approaching suspended particle and the stationary grain of filter media. This lubrication effect can reduce contact rates below those predicted with Equations 14 through 16.

As deposition proceeds in a porous bed, the particles that are removed from suspension can act as new collectors. Transport by convective Brownian diffusion and by interception increases as the size of the collector is decreased (Equations 14 and 15), so that small retained particles can be very effective filter media. The result is that the mass transport rate increases with time as the original clean media become "dirty." The assumption of clean collectors in deposition, like the coalesced sphere assumption in aggregation, underestimates the transport rate after the start of the process. The incomplete curvilinear approach to deposition overpredicts deposition rates, while the clean bed assumption underpredicts them after the onset of the process. These effects are discussed subsequently in this chapter.

Hydrodynamic Retardation

The clean bed approach to particle deposition in porous media described by Equations 13 through 17 is a curvilinear model, but it is a somewhat simplified one. Hydrodynamic retardation, the difficulty in draining fluid from the gap between the suspended particle and the stationary collector as the gap between them becomes very small, is not included in these relationships. An excellent summary of this aspect of deposition is provided by Rajagopalan and Tien.[22] Analytical solutions describing the process are not available; these authors have performed numerical analyses and summarized their results in the following correlating equation:

$$\eta(p,c) = 4A_s^{1/3}Pe^{-2/3} + A_s Lo^{1/8}R^{15/8} + 3.38 \times 10^{-3} A_s Gr^{1.2} R^{-0.4} \tag{19}$$

where

Lo, R, and Gr are dimensionless numbers
Lo is a van der Waals force number given as $4A/(9\pi\mu d_p^2 U)$
R is the size ratio written as d_p/d_c
Gr is the ratio of the Stokes' settling velocity of the suspended particle to the undisturbed flow velocity and is equal to $(\rho_p - \rho) g d_p^2/(18\ \mu U)$

Considering the right side of Equation 19, the first term is the transport coefficient for convective Brownian diffusion. It is identical to the right side of Equation 14; hydrodynamic retardation does not appear to be significant in particle deposition by convective diffusion. The second and third terms describe interception and sedimentation. These processes are affected by hydrodynamic retardation; as suspended particle size increases, contacts become less than those predicted by Equations 15 and 16. Laboratory experiments[25,26] of the filtration of monodisperse suspensions under solution conditions that eliminate the effects of repulsive interparticle chemical forces have been successfully described by mass transport theory as represented by Equation 19.

Ripening

When particle deposition occurs in a bed of porous media, the retained particles can serve as filter media for subsequent suspended particles passing through the system. Since the number of single collectors or filter media actually increases as deposition occurs, the rate of deposition and efficiency of the process can increase as time goes on. This is commonly observed.

When suspended particles become retained filter media, their small size makes them effective sites for deposition (Equations 14, 15, and 19; d_p becomes d_c as deposition proceeds). The view that deposited particles become filter media is based on experimental observations. First, the removal of suspended particles by fibers in air filtration has been shown photographically to produce chains, dendrites, or "trees" growing from the fiber surface as deposition proceeds.[27] These dendrites appear to be fractals, and the ripening of filters has analogies to the formation of aggregates in suspension. Second, in technological filtration processes, filtrate quality continues to improve over considerable time periods when chemical conditions are favorable, so that suspended particles can adhere to previously retained particles.[28] Third, the effects of suspended particle size on the energy required to maintain flow through a porous medium as deposition proceeds are dramatic and surprising.[28,29] For a given mass of material removed, small particles produce much larger energy losses than large ones. These results collectively indicate that retained particles can serve as efficient collectors (serve as filter media for other suspended particles) and produce substantial energy losses (exert drag on the flowing suspension).

While deposition can become more rapid or effective as the process proceeds, models to describe this enhancement have had limited success. Difficulties in describing deposit morphology in filtration resemble those in describing aggregate morphology in coagulation. Similarly, just as the formation of fractal aggregates leads to interparticle contact rates that are greater than those predicted by curvilinear models for particle transport in suspended systems (e.g., Figures 4 and 5), the formation of fractal deposits in filtration leads to particle contact rates that are greater than the clean bed case (Equation 19).

Transport Summary

When considering *initial* aggregation kinetics, Smoluchowski's equations can work well in determining mass transport coefficients (betas) for monodisperse suspensions in Brownian diffusion and fluid shear. This is the case in many laboratory experiments. For heterodisperse suspensions, the rectilinear models overestimate initial interparticle contact rates when transport is by fluid shear or by differential sedimentation, and this overestimation can be very substantial in some cases. For transport by Brownian diffusion, the rectilinear model does provide a good estimate of initial contact rates.

When considering *initial* deposition rates, curvilinear mass transport models that include hydrodynamic retardation provide good estimates of mass transport coefficients (etas) and deposition rates. Hydrodynamic retardation has little effect on mass transport of submicron particles by convective diffusion; it does reduce the deposition of larger particles, and this effect can be described.

As aggregation *proceeds*, aggregate porosity and structure can lead to substantial increases in aggregation rates, especially under chemical conditions favorable for attachment and after some aggregation has occurred. Models for this process are under development. For transport by fluid shear or differential sedimentation, refinements in the rectilinear model to account for hydrodynamic interactions indicate reduced contact rates, while considerations of aggregate structure indicate increased rates. Hill[30,31] has suggested approaches for aggregation in marine systems. Some compensation must occur, and the use of the rectilinear model to describe coagulation rates as aggregation proceeds, as suggested by Jackson and Lochmann,[17] has merit and useful simplicity.

Analyses of particle transport as deposition *proceeds* are not well developed. An early model is too simple.[29] Recent experiments[32] are instructive. A Monte Carlo approach may be helpful.

ATTACHMENT

In this section, the attachment of particles in aggregation and deposition are discussed. The origins of colloidal stability in natural aquatic systems are considered. Experimental determinations of attachment probabilities, alphas (α_a and α_d) are presented. Alphas in aggregation are considered first, followed by alphas in deposition.

Natural Organic Matter

There is extensive evidence from fresh waters, estuaries, and the oceans that the surface properties and the colloidal stability of particles in natural waters are affected by naturally occurring organic substances dissolved in these waters. These effects of natural organic matter (NOM) in establishing colloidal stability are also expected to occur in subsurface environments and to affect the passage/retention of pollutants in groundwater aquifers.

The predominant fraction of NOM in natural waters is comprised of humic substances. These are anionic polyelectrolytes of low to moderate molecular weight; they have both aromatic and aliphatic components and can be surface active; they are refractive, and some can persist in aquatic environments; they are produced by biological processes and are present to some extent in all natural waters. They adsorb on most surfaces and impart stability to particles in aquatic systems. Some investigators have observed that major divalent metal ions in natural waters (in particular, Ca^{2+}) can exert destabilizing effects on natural and synthetic particles at low to moderate metal ion concentrations. A consequence of

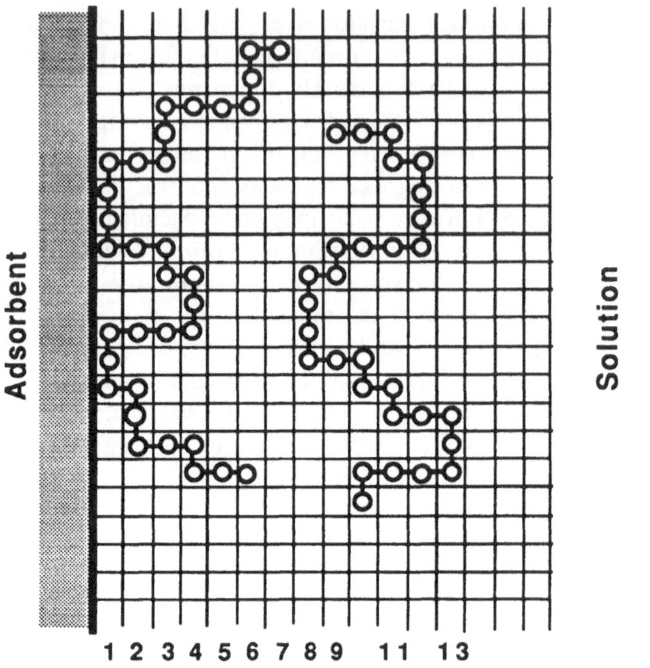

Figure 6. Schematic representation of uncharged macromolecules in solution and adsorbed at an interface, showing trains, loops, and tails of an adsorbed macromolecule. (From Lyklema, J. *Flocculation, Sedimentation, and Consolidation, Proc. of the Engineering Foundation Conference,* K. J. Ives, Ed. (The Netherlands: Sitjhoff and Noordhoff, 1978), pp. 3–36. With permission.)

these observations is that particles and associated particle-reactive pollutants may be expected to remain stable in soft waters that are high in NOM; in such waters they will remain in suspension in lakes and travel long distances in aquifers. When particles are present in hard waters that are low in NOM, they may be expected to be unstable and to be deposited in lake sediments and aquifers, retaining particle-reactive pollutants in bottom sediments and clogging aquifers. The consequences of colloidal stability on environmental quality can be diverse and substantial.

The effects of humic substances and divalent metal ions on the colloidal stability of natural particles are common observations without clear origins. Early studies[33,34] showed that particles carry a net negative charge in the presence of natural organic matter, regardless of the composition of the solid phase. This was taken as indirect evidence that humic substances stabilize particles electrostatically; calcium ions, in turn, might contribute electrostatically to particle destabilization. More recently,[35] there has been speculation that nonelectrostatic (steric) effects are involved in the stabilization of particles by humic materials and that specific chemical interactions (complex formation) are involved in colloid destabilization by divalent metal ions so that colloidal stability is very sensitive to divalent metal ions, even at low metal concentrations. There are, however, no conclusive evidence and no consistent theory for these hypotheses. Some speculations follow.

Colloidal Stability

A consideration of the interaction between two suspended particles (e.g., in a lake) or between a suspended particle and a stationary collector (e.g., in a groundwater aquifer) can begin with a determination of the structure of a single solid-solution interface. With this established, the interaction between two solid bodies (as characterized by force, free energy, or disjoining pressure) can be addressed.

The magnitudes of the interaction forces that determine stability in particle-particle interactions depend on the properties of each solid-solution interface. Electrostatic forces are almost universal in water since few particles are uncharged in aquatic environments, at least over a substantial pH range. Electrolytes and pH strongly affect the magnitude of these electrostatic forces. Specific chemical interactions between solutes and surfaces can be decisive in establishing surface properties and structure.[36] Steric effects are potentially most significant in systems containing uncharged particles with adsorbed nonionic macromolecules. In this case, the amount of adsorbed polymer and solvent quality are important factors. The relative importance of electrostatic and steric effects is not clear in systems containing charged surfaces and adsorbed macromolecules, especially if the macromolecules are ionizable.

There are two potential contributions to the structure of a solid/solution interface: (1) an electric double layer (termed EDL) and (2) a macromolecular adsorbed layer (denoted here as MAL). Separately these two structures are understood fairly well, at least in simplified cases. However, less is known about the effects of charge in the EDL on the conformation of adsorbed macromolecules in the MAL and, reciprocally, about the effects of charged and uncharged macromolecules on EDL properties.

These two distinct contributions result in two distinct mechanisms for colloidal stability, termed electrostatic and steric stabilization. Electrostatic stabilization results from energetically unfavorable overlap of the diffuse ion atmospheres surrounding all charged particles in water. The well known Derjaguin-Landau-Verwey-Overbeek (DLVO) theory[37] of the electrical double layer and colloidal stability provides a model for evaluation of the combined effects of electrostatic repulsion and van der Waals attraction between surfaces in the absence of adsorbed macromolecules and when all chemicals can be treated as point charges. When electrostatic repulsion dominates over van der Waals attraction, the results are termed "slow" coagulation or aggregation and "unfavorable" deposition or filtration. A good didactic summary has been given by Lyklema.[38]

Steric stabilization can result when polymers are adsorbed at solid-water interfaces. Large polymers can adsorb with some segments on a solid surface and with loops and tails extending into solution (Figure 6).[39] Steric stabilization results from energetically unfavorable intermolecular interactions and entropically unfavorable compression in overlapping macromolecular adsorbed layers. Steric effects are likely to be important when the thickness of the macromolecular adsorbed layer is about the same as or larger than the diffuse layer thickness. In the absence of charge effects, the energy of interaction between the adsorbed layers on two particles can be calculated by summing the various contributions to the free energy of the system as a function of the distance separating the surfaces. Under conditions of unfavorable intermolecular interactions, a repulsive interaction energy develops when the particles are close enough so that dangling tails (Figure 6) of adsorbed macromolecules interact. On close approach, conformational entropy losses can give rise to strong repulsion if desorption of the macromolecules does not occur. When surface coverage is incomplete so that open sites remain available for adsorption, dangling tails can adsorb to the opposite surface, leading to aggregation.

In considering the conformations of macromolecules at interfaces, a segment can be operationally defined as a small portion of a macromolecule that can be considered as a single unit in terms of both functionality and flexibility (e.g., a monomeric unit in a polymer). The most important structural characteristics of the MAL are the segment density distribution near the surface and the conformations of the adsorbed molecules. The conformation of a flexible linear molecule adsorbed at an interface is described in terms of adsorbed trains, extended loops, and up to two extended tails (Figure 6). The equilibrium adsorbed state is determined by a balance of conformational entropy losses, intermolecular interactions, and surface-solute interactions. The surface/solvent/macromolecule interactions may be specific, hydrophobic, or electrostatic in origin. Specific interactions include van der Waals forces, hydrogen bond formation, and chemical reactions such as complexation and ligand exchange.

Polymer adsorption has been the subject of extensive experimental and theoretical investigations. The first stochastic model of polymer adsorption was developed in 1953 by Simha and co-workers.[40] Currently, several advanced models of polymer and polyelectrolyte adsorption are based on the work of Scheutjens and Fleer,[41,42] termed here the SF theory. Reviews of polymer adsorption include publications by Vincent and Whittington[43] and Fleer and Lyklema.[44] The adsorption of strong polyelectrolytes is discussed by Hesselink.[45] SF theory has been extended to polyelectrolytes.[46,47] Recent work has addressed weak polyelectrolytes.[48-50] An excellent recent review of the adsorption of polyelectrolytes has been given by Cohen Stuart et al.[51]

SF theory is a statistical thermodynamic model in which chain conformations are formulated as step-weighted random walks on a lattice. The simplified problem involves the adsorption of flexible linear monodisperse polymers comprised of identical uncharged segments at a uniform planar surface. Interactions among surface, adsorbate and solvent are quantified using adjustable composite parameters; mechanistic details are neglected. The equilibrium MAL is determined by minimizing the total free energy of the system. That is, the canonical partition function is maximized with respect to the number of chains in each possible

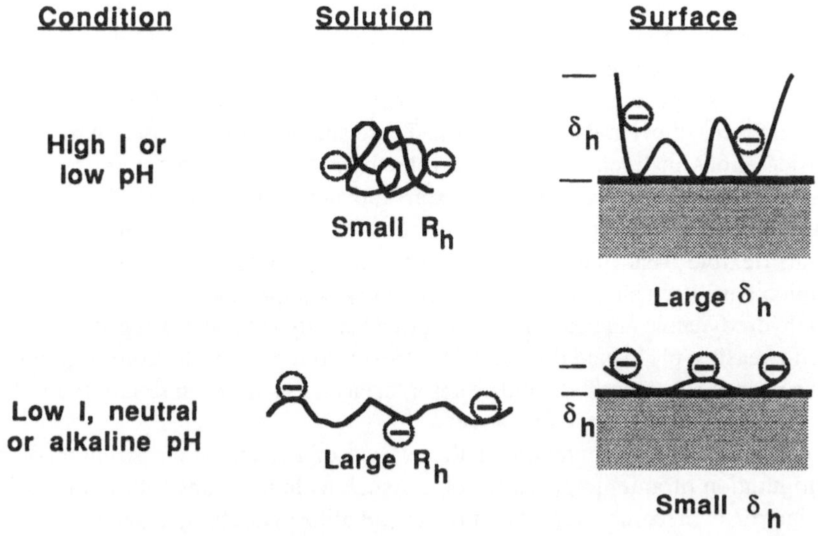

Figure 7. Schematic description of the effects of ionic strength (I) and pH on the conformations of a humic molecule in solution and at a surface. R_h denotes the hydrodynamic radius of the molecule in solution, and δ_h denotes the hydrodynamic thickness of the adsorbed natural anionic polyelectrolyte. (From Yokoyama, A., K.R. Srinivasan, and H.S. Fogler. *Langmuir* 5:534–538 (1989). With permission.)

configuration, allowing specification of the distribution of trains, loops, and tails in the interfacial region.

In general, SF theory for uncharged macromolecules predicts an exponential decay in the segment density distribution close to a surface. At greater distances, tails dominate the segment density, and the decay with distance is much less. For moderately sized polymers with molecular weights in the order of a few hundred thousand, a typical root-mean-square adsorbed layer thickness is 5 to 10 nm. Tails typically extend as far as 20 nm and can have significant influences in particle-particle interactions. The extent of adsorption increases with the molecular weight and concentration of the polymer, with surface affinity, and with decreasing solvent quality.

Because of intermolecular electrostatic repulsion, polyelectrolytes adsorb in flat conformations. With increasing electrolyte concentration, polyelectrolytes behave more like uncharged polymers, and adsorption can increase with increasing ionic strength. Because macromolecule and surface charge are usually dependent on pH, adsorption is a strong function of pH.

Natural Aquatic Environments

As stated previously, the colloidal stability of particles in aquatic systems has been observed to depend very substantially on natural organic matter and divalent metal ions such as Ca^{2+} and Mg^{2+}. Effects of pH and ionic strength have not been studied as extensively. Very few experiments to determine colloidal stability in aquatic systems have been made. Theories for the origins of colloidal stability in aquatic systems are few and qualitative; none has been tested experimentally in such systems. Speculations are presented here, with emphasis placed on NOM and the influences of pH and ionic strength. These suggestions are based in part on the considerations of linear, flexible polyelectrolytes at interfaces described previously. Humic substances are not linear and not completely flexible; they are branched and complex; they are also, however, weak polyelectrolytes. As with linear, flexible weak polyelectrolytes used in modeling, their configurations in solution have been shown to vary significantly with pH and ionic strength,[52] with their hydrodynamic diameters increasing substantially with decreasing ionic strength and increasing pH. Since the use of SF theory to represent the configurations of these natural substances at solid-water interfaces has not been tested, the following discussion is in part speculative.

A schematic representation of the effects of pH and ionic strength on the configuration of anionic polyelectrolytes such as humic substances is presented in Figure 7.[53] In fresh waters at neutral and alkaline pHs, charged macromolecules assume extended shapes (large hydrodynamic radius, R_h) as a result of intramolecular electrostatic repulsive interactions. When adsorbed at interfaces under these conditions, they assume flat configurations (small hydrodynamic thickness, δ_h). At high ionic strength or at low pH, the polyelectrolytes have a coiled configuration in solution (small R_h) and extend further from the solid surface when adsorbed (large δ_h).

The ionic strength of aquatic systems from mountain springs to the ocean varies from about 10^{-4} to 0.7 mol/dm^{-3}, with corresponding diffuse layer thicknesses in the EDL ranging from about 30 to 0.3 nm. The pH in aquatic environments may vary from 3 to perhaps 9, with most waters in the range from pH 6 to 8.3. Aquatic humic substances range in molecular weight from 500 to perhaps 50,000 or more. Because of these wide variations in pH, ionic strength, and molecular weight, the thicknesses of the macromolecular adsorbed layer (δ_{MAL}) and the electrical double layer (δ_{EDL}) at solid-solution interfaces can be expected to vary widely in aquatic systems. Following Warszynski,[54] the dimensionless ratio ($\delta_{MAL}/\delta_{EDL}$) will be used as a partial description of the relative effects of macromolecular and simple electrostatic effects on colloidal stability.

In a mountain stream, a neutral fresh water lake, an activated sludge tank, or the core of a reverse osmosis (RO) membrane treating a surface water supply, the thickness of the EDL (Debye length, δ_{EDL}) is large, in the order of 10 to 30 nm. Under the same solution conditions, the thickness of an adsorbed layer of anionic natural organic matter will be small, perhaps less than 5 nm. Under these conditions, $\delta_{MAL}/\delta_{EDL}$ will be $<< 1$. Colloidal particles in such systems may have negative charges that are due primarily to adsorbed organic matter, but the interactions among such particles are primarily electrostatic ones between similar diffuse layers. In contrast, in the ocean, an acid lake, a sludge digester, or the concentration boundary layer at the surface of an RO membrane, δ_{MAL} is increased substantially due to ionic strength or pH effects (Figure 7), while δ_{EDL} can be quite small. In these cases $\delta_{MAL}/\delta_{EDL} > 1$. The stability of colloidal particles in such systems may be determined by interactions between the organic adsorbed layers of colliding particles. Estuarine environments comprise a very important group of intermediate cases where the effects of the EDL and the MAL may be comparable.

Alphas in Aggregation

It is customary and convenient to describe the colloidal stability of particles in an aqueous suspension by a sticking or stability factor, α_a, defined as follows:

$$\alpha_a = \frac{\textit{rate at which particles attach}}{\textit{rate at which particles collide}} \tag{20}$$

This sticking or stability factor is thus the ratio of the collision rate producing aggregates to the total rate at which collisions occur by Brownian diffusion, fluid shear, and differential sedimentation. When repulsive interactions are such that a suspension is completely stable, all collisions are unsuccessful and $\alpha_a = 0$. For a completely unstable or sticky suspension in which repulsive interactions are absent or insignificant, all collisions produce aggregation and $\alpha_a = 1$. The sticking factor, α_a, is the reciprocal of another measure of colloidal stability, the Fuchs stability factor, W ($\alpha_a = W^{-1}$). In natural environments, $0 < \alpha_a \leq 1$ and $\infty > W \geq 1$.

Theories of colloidal stability are helpful in understanding why particles are stable and in identifying important chemical properties of solutions and solids that affect stability, but they are not able to provide accurate quantitative predictions of aggregation rates when chemical repulsive interactions produce low attachment probabilities (low alphas). Experimental determinations of alphas are possible in many cases and provide means for quantifying the effects of chemical parameters on aggregation rates.

Experiments can be performed in which the rates at which particles aggregate or coagulate are measured; these aggregation rates are related to attachment rates, the numerator in Equation 20. Calculations of experimental alphas involve these aggregation rate measurements and also determinations of collision rates, the denominator in Equation 20. In one approach, chemical conditions are used in which repulsive particle interactions are expected to be negligible, and the aggregation rate observed under these conditions is assumed to be the collision rate in the denominator of Equation 20. Values of α_a are then calculated for any experimental aggregation rate by dividing it by the fastest rate observed. When this is done, the fastest aggregation rate should be compared with theories of particle transport such as those described previously to validate the experimental procedures. In another approach, particle collision rates are calculated from mass transport theories described previously. Since there are a number of theories, there can be a number of possible values of α_a for the same experimental aggregation rate. For monodisperse suspensions this is not a significant error; for heterodisperse suspensions the differences among the models depend on the transport process and experimental conditions and can be substantial.

Two types of experiments have been used by several investigators; each is based on a different transport process. For particles larger than a micron or so in size, Brownian diffusion is not important; experiments conducted in systems with substantial, controlled, and known velocity gradients are used to produce interparticle contact rates that are describable and reproducible. As shown in Figure 5, differences between the curvilinear and rectilinear models can be substantial for high shear rates and heterodisperse suspensions; the circled C denotes a case in which actual transport is only 0.03% as fast as the Smoluchowski model. For submicron particles, Brownian diffusion dominates interparticle contacts, and Smoluchowski's theory of perikinetic flocculation or a suitable modification is used to describe interparticle contacts for both monodisperse and heterodisperse suspensions.

When fluid shear dominates particle transport, a combination of Equations 1, 3, and 8 leads to the following expression:

$$ln\frac{n_t}{n_o} = -\frac{4}{\pi}\alpha_a\phi Gt \tag{21}$$

where

n_t and n_o are the number concentrations of particles at time t and at the beginning of the experiment, respectively
ϕ is the volume fraction of solid material in the suspension, assumed constant in this analysis

Rearranging,

$$\alpha_a = \frac{\pi}{4\phi Gt}ln\frac{n_o}{n_t} \tag{22}$$

All terms on the right side of Equation 22 are constants or are experimentally accessible. Aggregation is measured by determining n_t and n_o; particles should be larger than about a micron in size so that transport by Brownian diffusion is not important. Smoluchowski's rectilinear model and the coalesced sphere assumption are used to describe the interparticle transport rate; the initial particle volume concentration and the average shear rate must be measured. The coalesced sphere assumption (constant ϕ) is a good one when initial aggregation rates are used. The rectilinear model (Equation 8) is useful for monodisperse suspensions; as stated previously, it becomes inaccurate as the size heterogeneity of suspensions increases. As coagulation proceeds, the effects of these two assumptions tend to

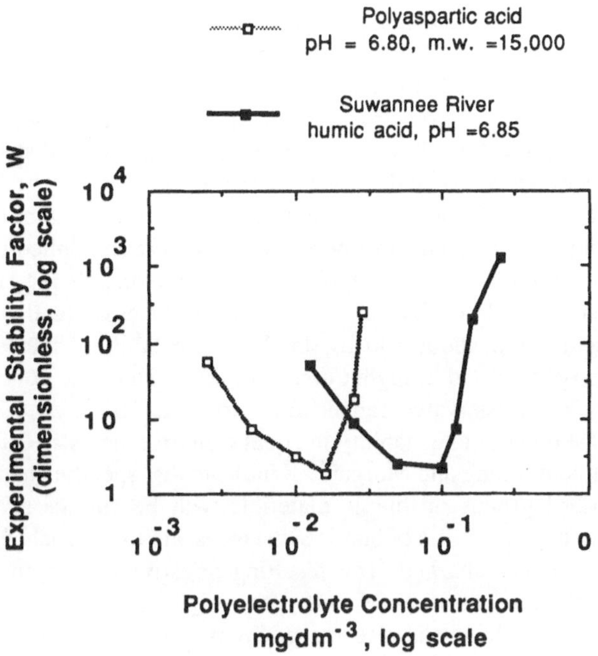

Figure 8. Experimental values of the stability ratio (*W*) as a function of the concentrations of two polyelectrolytes (mg/dm^{-3}) for a hematite suspension (17 mg/dm^{-3}) in 1 millimolar NaCl. (From Liang, L. "Effects of Surface Chemistry on Kinetics of Coagulation of Submicron Iron Oxide Particles (a-Fe_2O_3) in Water," PhD Thesis, California Institute of Technology, Pasadena (1988). With permisison.)

compensate each other. This approach has been used as described here or with some modifications to determine alphas in laboratory studies (e.g., Swift and Friedlander[12]) and in natural waters.[55,56]

Experimentally based particle stability coefficients (alphas) for fresh surface waters range over two orders of magnitude, from about 0.001 to more than 0.1, and support the view that solution chemistry can control the colloidal stability of natural particles in aquatic environments. These estimates are based on initial aggregation rates of somewhat heterodisperse suspensions. Consequently, interparticle contact rates are probably somewhat overestimated, so that the actual attachment probabilities are somewhat greater than these values.

For submicron particles, Brownian diffusion controls particle transport, and hydrodynamic interactions are less significant than for larger particles. Light scattering techniques can be used to determine aggregation rates. Some recent results by Liang[57] are presented in Figure 8. Results are reported in terms of the Fuchs stability factor, and range from slightly more than 1 up to 1000. Corresponding values of α_a are slightly less than 1 to 0.001. These experiments were conducted at near neutral pH, where hematite is positively charged in the absence of specific adsorption of soluble species. Two polyelectrolytes were used, polyaspartic acid with a molecular weight of about 15,000 and NOM, a humic acid isolated from the Suwannee River. Both polyelectrolytes contain carboxyl groups and are negatively charged at neutral pH.

Two points are made here with the results in Figure 8. First, organic polyelectrolytes including NOM can have substantial effects on colloidal stability. At low concentrations of added polyelectrolytes, hematite is kinetically stable as a positively charged colloid. Polyaspartic acid provides essentially complete destabilization (attachment rate equals transport rate, W = 1) at an applied concentration of only about 20 $\mu g/dm^{-3}$; higher applications reverse the charge and produce a negatively charged colloid. The humic acid also produces destabilization at a low applied concentration, about 100 $\mu g/dm^{-3}$, and restabilizes the hematite as a negatively charged colloid at higher concentrations. Since the concentrations of humic materials in fresh waters range from 1 to about 50 mg/dm^{-3}, the potential effects of NOM on colloidal stability in aquatic environments are illustrated well by this experiment. Liang and Morgan[58,59] indicate that specific chemical interactions of carboxyl groups on humic materials with Fe surface sites contribute significantly to the adsorption of humic substances and to reversal of charge when hematite is positively charged. The resulting negatively charged particles are kinetically stable but, as stated previously, the origins of this stability are not yet clear. A second conclusion is also made. Simply put, attachment probabilities (alphas) in aggregation phenomena are experimentally accessible parameters. This point will be discussed subsequently.

Alphas in Deposition

The sticking coefficient, stability factor, or attachment probability, α_d, of particles depositing in porous media is defined as follows:

$$\alpha_d = \frac{\textit{rate at which particles attach to porous media}}{\textit{rate at which particles collide with porous media}} \tag{23}$$

This sticking factor is the ratio of the collision rate producing deposited particles to the total rate at which suspended particles contact the stationary collectors in the porous medium of interest. These collisions occur in aquatic systems by convective Brownian diffusion, fluid motion (interception), and convective sedimentation. When repulsive interactions are such that all collisions between suspended particles and stationary media are unsuccessful in producing attachment, $\alpha_d = 0$; for a completely sticky interaction in which all contacts produce deposition, $\alpha_d = 1$.

Theories of colloidal stability provide understanding about why particles do not deposit, i.e., why deposition is unfavorable. These theories can also identify important chemical properties of solutions, suspended particles, and stationary collectors that affect particle deposition. Unfortunately, they are not able to provide accurate descriptions of deposition rates, even in simple inorganic solutions, when chemical repulsive interactions produce low attachment probabilities

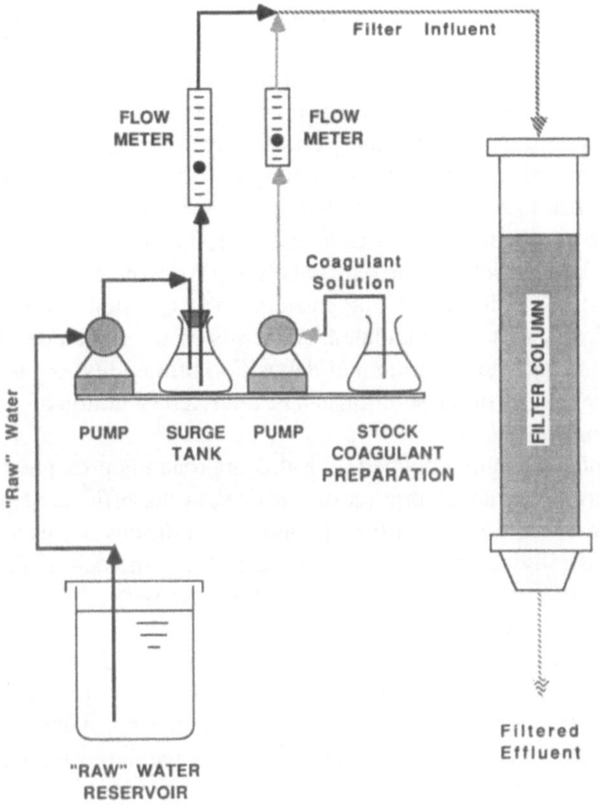

Figure 9. Schematic diagram of filtration apparatus for determining experimental attachment probabilities (α_d).

(low alphas).[25,26,60] Experimental determinations of alphas are possible in many cases and provide means for quantifying the effects of chemical parameters on deposition rates.

Integration of Equation 13 yields the following:

$$ln\frac{n_L}{n_o} = -\frac{3}{2}(1-\varepsilon)\alpha_d\eta(p,c)\left(\frac{L}{d_c}\right) \tag{24}$$

in which n_L and n_o are the number concentrations of particles leaving the bed with length L and at the inlet to the bed, respectively. Rearranging,

$$\alpha_d = \frac{2d_c}{3(1-\varepsilon)\eta(p,c)} ln\frac{n_o}{n_L} \tag{25}$$

With the exception of η(p,c), all terms on the right side of Equation 25 are constants or can be determined experimentally. The mass transport coefficient,

$\eta(p,c)$, can be calculated using Equation 19. This calculation assumes a clean deposit surface and requires knowledge of the flow velocity, particle size, and fluid temperature. The Hamaker constant for the particle-collector interaction in water must be calculated or assumed; it often is not known with accuracy, but calculations of $\eta(p,c)$ are not sensitive to this parameter.

An experimental apparatus for evaluating α_d is illustrated in Figure 9. A sample of the suspension to be tested is applied at a known flow rate and temperature to a bed of known media composition and size, length, and bed porosity. The concentrations of particles applied to and eluting from the porous medium (n_o and n_L) are measured. If the size of the particles in suspension is known, the single collector efficiency can be calculated. All terms on the right side of Equation 25 are then known and the experimental deposition probability, α_d, is calculated. If desired, the chemistry of the solution can be altered by addition of selected solutes including coagulants.

An example of results obtained with this approach is presented in Figure 10. Measurements of the concentration of particles in the effluents from laboratory filters are presented for five different solution conditions corresponding to KCl concentrations ranging from 10^{-3} to 10^{-1} mol/dm^{-3}. In each of the five experiments, mass transport was unchanged; latex particles with a diameter of 0.753 μm were filtered at a velocity of 1.36×10^{-3} m/s through beds containing spherical glass beads of diameter 0.2 mm. The filter beds were 20 cm long and had a porosity of 0.40. Differences in removal among the five experiments result from differences in solution chemistry, i.e., KCl concentration. After an initial period during which clean water is displaced from the bed, the removal efficiency of each clean bed reaches a constant value taken as n_L/n_o for use in calculating experimental values of α_d with Equation 22.

The results of these and other calculations are presented in Figure 11. The experimental sticking coefficients vary from about 0.5 when high salt concentrations eliminate repulsive chemical interactions to 0.004 in more dilute KCl when interacting diffuse layers hamper attachment. Two points are made here from the results in Figures 10 and 11. First, solution chemistry can control particle deposition from flowing suspensions to solid surfaces. Second, alphas in deposition phenomena are experimentally accessible parameters in many cases. This point will be discussed subsequently.

Attachment Summary

Surface and solution chemistry controls attachment in aggregation and deposition phenomena in aquatic environments. Natural organic matter is present in all aquatic systems, and its ability to adsorb at solid-water interfaces results in NOM dominating the chemical aspects of colloidal stability in most natural systems.

Theories for predicting chemical effects in aggregation and deposition rates are not accurate. Experimental determinations of α_a and α_d are necessary in all cases and also feasible in many cases.

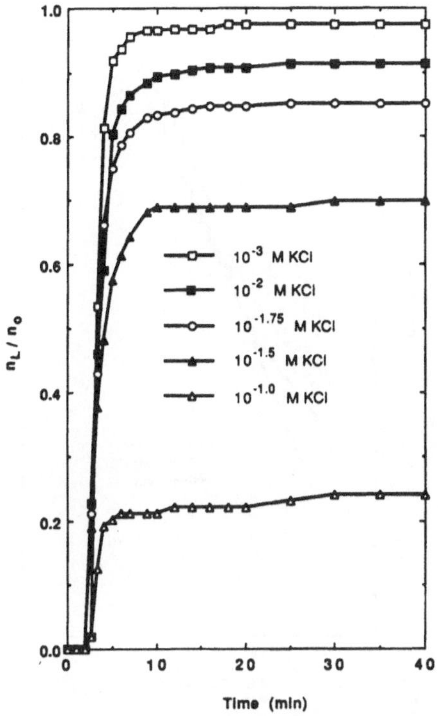

Figure 10. Results of filtration experiments to determine the effect of solution conditions on particle deposition in porous media. Suspensions of latex particles (0.753 μm in diameter) containing different concentrations of indifferent electrolyte (KCl) were filtered through beds of glass beads. (From Elimelech, M. "The Effect of Particle Sile on the Kinetics of Deposition of Brownian Particles in Porous Media," PhD Thesis, The John Hopkins University, Baltimore, MD (1989). With permission.)

Mass transport theories (betas and etas) are needed in the experimental determination of attachment probabilities (alphas). These theories are best developed for the initial aggregation of monodisperse suspensions and the initial deposition of monodisperse suspensions on clean surfaces. Experimental alphas and theoretical betas and etas are unavoidably coupled.

CONCLUDING REMARKS

Until the 1950s, research and applications in environmental protection were directed at microbial pathogens and at managing oxygen resources in natural waters by oxidizing degradable organic matter in biological treatment plants. In the 1960s, attention was directed toward soluble nutrients such as phosphorus and nitrogen and their impact on cultural eutrophication. The present focus in protecting the environment is on substances that, in general, attempt to leave the aqueous phase

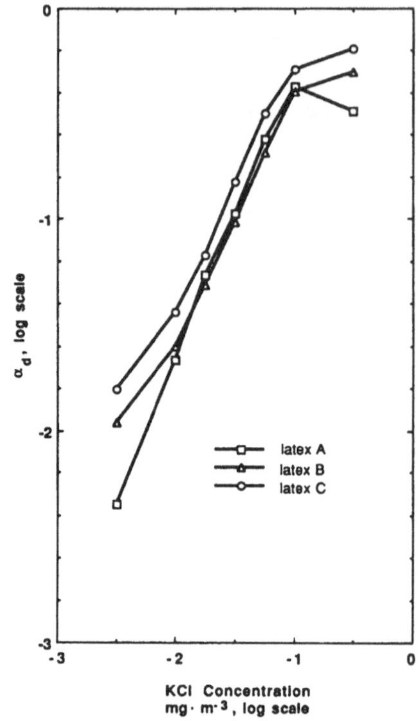

Figure 11. Experimental sticking coefficients (α_d) as functions of solution chemistry (KCl concentration, mol/dm^{-3}) for three different latex suspensions; A = 0.046 µm; B = 0.378 µm; C = 0.753 µm. (From Elimelech, M. "The Effect of Particle Sile on the Kinetics of Deposition of Brownian Particles in Porous Media," PhD Thesis, The Johns Hopkins University, Baltimore, MD (1989). With permission.)

by partitioning into the atmosphere or to solid phases. PCBs in the Great Lakes are volatile, partition into the tissues of fish, and deposit in lake sediments. Carbon dioxide is injected into the atmosphere by fossil fuel combustion and is eventually transported to ocean sediments primarily as particulate $CaCO_3$. Solvents buried in hazardous waste sites are prevented from volatilizing to the atmosphere and so will partition there into organic particles or be transported in reluctant solution in groundwater aquifers. Some of the synthetic organic compounds released to the Rhine River by the industrial accident in Basel, Switzerland were particle-reactive, accumulated in sediments of the Rhine, and had great impact on benthic organisms. Radionuclides released by the Chernobyl accident were transported atmospherically, collected by rainfall and fog, and accumulated as dry deposition on plant surfaces. This event as well as any other exemplifies the coupling of the atmospheric, aquatic, and terrestrial conveyor belts which exists in the global environment, and the significance of particles and interfaces in environmental processes.

Modeling the fate of particles and particle-reactive pollutants in aquatic environments requires ability to predict aggregation and deposition rates. These rates can be considered conveniently as two steps, one primarily physical and the other chemical. While physical theories of mass transport in aggregation and deposition phenomena can profit from new research, some important uncertainties produce compensating effects. These theories are probably underutilized in modeling environmental processes. Chemical theories are much less developed, but laboratory experiments can be made that characterize the effects of chemistry on attachment probabilities, and these results can be applied to models of field situations. In doing so, the unavoidable coupling of theoretical betas and etas with laboratory alphas should be recognized as these results are used in practice.

It is possible to use models to predict the aggregation of small particles having dimensions of nanometers to form large aggregates with sizes of centimeters or larger. While such predictions may be instructive, they may not be accurate. As aggregates grow in size, transport mechanisms change, aggregate morphology is altered, and breakup by hydrodynamic forces can occur. In particular, the ability of the rectilinear modeling approach, with its compensating assumptions of rectilinear motion and coalesced spherical aggregates, to predict extended aggregation accurately requires testing and, if necessary, modification.

The application of alphas determined experimentally in laboratory measurements to field situations needs additional testing. For aggregation studies, field samples can be used directly in laboratory experiments. For deposition or filtration in porous media, selection of an appropriate test system is less clear. To the extent that solution chemistry and, in particular, dissolved natural organic matter establish colloidal stability in aquatic systems, the use of natural water samples in laboratory experiments with model media such as glass beads and model colloidal particles such as hematite or latex is convenient. The use of natural particles and media, while experimentally more difficult, may be preferable.

ACKNOWLEDGMENTS

The work described herein was supported in part by the U.S. National Science Foundation under grant BCS-9112766. The authors gratefully acknowledge the comments of James J. Morgan and the editors of this volume.

REFERENCES

1. O'Melia, C.R. and W. Stumm. "Theory of Water Filtration," *J. Am. Water Works Assoc.* 59(11):1393–1412 (1967).
2. Einstein, A. *Investigations on the Theory of the Brownian Movement*, R. Furth, Ed. (New York: Dover Publications, Inc., 1956).

3. Smoluchowski, M. "Versuch einer mathematischen Theorie der Koagulationskinetic kolloider Lösungen," *Z. Phys. Chemie* 92:129–168 (1917).
4. Findheisen, W. "Zur Frage der Regentropfenbildung in reinem Wasserwolken," *Meteorol. Z.* 56:365–368 (1939).
5. Camp, T.R. and P.C. Stein "Velocity Gradients and Internal Work in Fluid Motion," *J. Boston Soc. Civ. Eng.* 30:219–237 (1943).
6. Spielman, L.A. "Viscous Interactions in Brownian Coagulation," *J. Colloid Interface Sci.* 33:562–571 (1970).
7. Valioulis, I.A. and E.J. List. "Collision Efficiencies of Diffusing Spherical Particles: Hydrodynamic, van der Waals, and Electrostatic Forces," *Adv. Colloid Interface Sci.* 20:1–20 (1984).
8. Han, M. and D.F. Lawler. "The (Relative) Insignificance of G in Flocculation," *J. Am. Water Works Assoc.* submitted.
9. Adler, P.M. "Heterocoagulation in Shear Flow," *J. Colloid Interface Sci.* 83:106–115 (1981).
10. Adler, P.M. "Interaction of Unequal Spheres 1. Hydrodynamic Interaction: Colloidal Forces," *J. Colloid Interface Sci.* 84:461–474 (1981).
11. Han, H. and D.F. Lawler. "Interactions of Two Settling Spheres: Settling Rates and Collision Efficiency," *J. Hydraul. Eng.* 117:1269–1289 (1991).
12. Swift, D.L. and S.K. Friedlander. "The Coagulation of Hydrosols by Brownian Motion and Laminar Shear Flow," *J. Colloid Sci.* 19:621–647 (1964).
13. Logan, B.E. and D.B. Wilkinson. "Fractal Geometry of Marine Snow and Other Biological Aggregates," *Limnol. Oceanogr.* 35:130–136 (1990).
14. Logan, B.E. and D.B. Wilkinson. "Fractal Dimensions and Porosities of *Zoogloea ranigera* and *Saccharomyces cerevisae* Aggregates," *Biotechnol. Bioeng.* 38:389–396 (1991).
15. Meakin, P. "Fractal Aggregates in Geophysics," *Rev. Geophys.* 29: 317–354 (1991).
16. Jiang, Q. and B.L. Logan. "Fractal Dimensions of Aggregates Determined from Steady-State Size Distributions," *Environ. Sci. Technol.* 25:2031–2038 (1991).
17. Jackson, G.A. and S. Lochmann. "Modeling Coagulation of Algae in Marine Ecosystems," in *Sampling and Characterization of Environmental Particles II*, J. Buffle and H.P. van Leeuwen, Eds. (Chelsea, MI: Lewis Publishers, Inc., 1993).
18. Friedlander, S.K. "Theory of Aerosol Filtration," *Ind. Eng. Chem.* 50:1161–1164 (1958).
19. Yao, K.-M., M.T. Habibian, and C.R. O'Melia. "Water and Wastewater Filtration: Concepts and Applications," *Environ. Sci. Technol.* 5:1105–1112 (1971).
20. Fitzpatrick, J.A. and L.A. Spielman. "Filtration of Aqueous Latex Suspensions Through Beds of Glass Spheres," *J. Colloid Interface Sci.* 43:350–369 (1973).
21. Spielman, L.A. "Particle Capture from Low-Speed Laminar Flows," *Ann. Rev. Fluid Mech.* 9:297–319 (1977).
22. Rajagopalan, R. and C. Tien. "Trajectory Analysis of Deep-Bed Filtration with the Sphere-in-Cell Porous Media Model," *Am. Inst. Chem. Eng. J.* 22:523–533 (1976).
23. Tien, C. *Granular Filtration of Aerosols and Hydrosols* (Boston, MA: Butterworth Publishers, 1989).
24. Happel, J. "Viscous Flow in Multiparticle Systems: Slow Motion of Fluids Relative to Beds of Spherical Particles," *Am. Inst. Chem. Eng. J.* 4:197–201 (1958).
25. Tobiason, J.E. and C.R. O'Melia. "Physicochemical Aspects of Particle Removal in Depth Filtration," *J. Am. Water Works Assoc.* 80(12):54–64 (1988).

26. Elimelech, M. and C.R. O'Melia. "Kinetics of Deposition of Colloidal Particles in Porous Media," *Environ. Sci. Technol.* 24:1528–1536 (1990).
27. Billings, C.E. "Effects of Particle Accumulation in Aerosol Filtration," W. M. Keck Laboratory, California Institute of Technology, Pasadena (1966).
28. Habibian, M.T. and C.R. O'Melia. "Particles, Polymers, and Performance in Filtration," *J. Environ. Eng. Div. ASCE* 101:567–583 (1975).
29. O'Melia, C.R. and W. Ali. "The Role of Retained Particles in Deep Bed Filtration," *Prog. Water Technol.* 10:167–182 (1973).
30. Hill, P.S. and A.R.M. Nowell. "The Potential Role of Large, Fast-Sinking Particles in Clearing Nepheloid Layers," *Philos. Trans. R. Soc. London* A331:103–117 (1990).
31. Hill, P.S. "Reconciling Aggregation Theory with Observed Vertical Fluxes Following Phytoplankton Blooms," *J. Geophys. Res.* in press.
32. Darby, J.L., D.F. Lawler, and T.P. Wilshusen. "Depth Filtration of Wastewater: Particle Size and Ripening," *Res. J. WPCF* 63:228–238 (1991).
33. Niehof, R.A. and G.I. Loeb "The Surface Charge of Particulate Matter in Seawater," *Limnol. Oceangor.* 17:7–16 (1972).
34. Hunter, K.A. and P.S. Liss "The Surface Charge of Suspended Particles in Estuarine and Coastal Waters," *Nature* 282:823–825 (1979).
35. Tipping, E. and D.C. Higgins. "The Effect of Adsorbed Humic Substances on the Colloid Stability of Haematite Particles," *Colloids Surf.* 5:85–92 (1982).
36. Stumm, W. *Chemistry of the Solid-Water Interface* (New York: Wiley-Interscience, in press).
37. Verwey, E.J.W. and J.Th.G. Overbeek. *Theory of the Stability of Lyophobic Colloids* (Amsterdam: Elsevier-North Holland, 1948).
38. Lyklema, J. "Surface Chemistry of Colloids in Connection with Stability," in *The Scientific Basis of Flocculation*, K.J. Ives, Ed. (The Netherlands: Sitjhoff and Noordhoff, 1978), pp. 3–36.
39. Lyklema, J. "How Polymers Adsorb and Affect Colloid Stability," in *Flocculation, Sedimentation, and Consolidation, Proceedings of the Enginering Foundation Conference,* B.M. Moudgil and P. Somasundaran, Eds. (New York: Engineering Foundation, 1985), pp. 3–21.
40. Simha, R., H.L. Frisch, and F.R. Eirich. "The Adsorption of Flexible Molecules," *J. Phys. Chem.* 57:584–589 (1953).
41. Scheutjens, J.M.H.M. and G.J. Fleer. "Statistical Theory of the Adsorption of Interacting Chain Molecules. I. Partition Function, Segment Density Distribution, and Adsorption Isotherms," *J. Phys. Chem.* 83:1619–1635 (1979).
42. Scheutjens, J.M.H.M. and G.J. Fleer."Statistical Theory of the Adsorption of Interacting Chain Molecules. II. Train, Loop, and Tail Size Distribution," *J. Phys. Chem.* 84:178–190 (1980).
43. Vincent, B. and S.G. Whittington. "Polymers at Interfaces and in Disperse Systems," *Colloids Surf.* 12:1–117 (1982).
44. Fleer, G.J. and J. Lyklema. "Adsorption of Polymers," in *Adsorption from Solution at the Solid/Liquid Interface*, G.D. Parfitt and C.H. Rochester, Eds. (London: Academic Press, Inc., 1983).
45. Hesselink, F.Th. "Adsorption of Polyelectrolytes from Dilute Solutions," in *Adsorption from Solution at the Solid/Liquid Interface,* G.D. Parfitt and C.H. Rochester, Eds. (London, Academic Press, Inc., 1983).

46. van der Schee, H.A. and J. Lyklema. "A Lattice Theory of Polyelectrolyte Adsorption," *J. Phys. Chem.* 88:6661–6667 (1984).
47. Papenhuijzen, J., H.A. van der Schee, and G.J. Fleer. "Polyelectrolyte Adsorption. I. A New Lattice Theory," *J. Colloid Interface Sci.* 104:540–552 (1985).
48. Evers, O.A., G.J. Fleer, J.M.H.M. Scheutjens, and J. Lyklema. "Adsorption of Weak Polyelectrolytes from Solution," *J. Colloid Interface Sci.* 111:446–454 (1986).
49. Blaakmeer, J., M.R. Böhmer, M.A. Cohen Stuart, and G.J. Fleer. "Adsorption of Weak Polyelectrolytes on Highly Charged Surfaces. Poly(acrylic acid) on Polystytrene Latex with Strong Cationic Groups," *Macromolecules* 23:2301–2309 (1990).
50. Böhmer, M.R., O.F. Evers, and J.M.H.M. Scheutjens. "Weak Polyelectrolytes Between Two Surfaces: Adsorption and Stabilization," *Macromolecules* 23:2288–2301 (1990).
51. Cohen Stuart, M.A., G.J. Fleer, J. Lyklema, W. Norde, and J.M.H.M. Scheutjens. "Adsorption of Ions, Polyelectrolytes, and Proteins," *Adv. Colloid Interface Sci.* 34:477– 535 (1991).
52. Cornell, P.H., R.S. Sommers, and P.V. Roberts. "Diffusion of Humic Acid in Dilute Aqueous Solution," *J. Colloid Interface Sci.* 110:149–164 (1986).
53. Yokoyama, A., K.R. Srinivasan, and H.S. Fogler. "Stabilization Mechanism by Acidic Polysaccharides. Effects of Electrostatic Interactions on Stability and Stabilization," *Langmuir* 5:534–538 (1989).
54. Warszynski, P. "The Influence of Polymer Adsorption on Deposition Kinetics of Colloid Particles I. Theory," *Colloids Surf.* 39:79–92 (1989).
55. Ali, W., C.R. O'Melia, and J.K. Edzwald. "Colloidal Stability of Particles in Lakes: Measurement and Significance," *Water Sci. Technol.* 17:701–712 (1985).
56. Weilenmann, U., C.R. O'Melia, and W. Stumm. "Particle Transport in Lakes: Models and Measurements," *Limnol. Oceanogr.* 34:1–18 (1989).
57. Liang, L. "Effects of Surface Chemistry on Kinetics of Coagulation of Submicron Iron Oxide Particles (α–Fe_2O_3) in Water," PhD Thesis, California Institute of Technology, Pasadena (1988).
58. Liang, L. and J.J. Morgan. "Chemical Aspects of Iron Oxide Coagulation in Water: Laboratory Studies and Implications for Natural Systems," *Aquatic Sci.* 52:32–55 (1990).
59. Liang, L. and J.J. Morgan. "Coagulation of Iron Oxide Particles in the Presence of Organic Molecules," in *Chemical Modeling of Aqueous Systems II*, D.C. Melchior and R.L. Bassett, Eds. ACS Symposium Series 416 (Washington, D.C.: American Chemical Society, 1990).
60. Elimelech, M. and C.R. O'Melia."Effect of Particle Size on Collision Efficiency in the Deposition of Brownian Particles with Electrostatic Energy Barriers," *Langmuir* 6:1153–1163 (1990).
61. Elimelech, M. "The Effect of Particle Size on the Kinetics of Deposition of Brownian Particles in Porous Media," PhD Thesis, The Johns Hopkins University, Baltimore, MD (1989).

CHAPTER 9

MODELING COAGULATION OF ALGAE IN MARINE ECOSYSTEMS

George A. Jackson and Steve Lochmann

TABLE OF CONTENTS

0-87371-895-X/93/$0.00+$.50

INTRODUCTION

Aggregates more than 1 cm in length have been observed by scientific divers in the surface waters of the ocean.[1-4] Such aggregates are composed of a variety of biologically-derived organic material, including algal cells, fecal pellets, and animal feeding structures, such as mucus nets from pteropods and feeding structures "houses" from larvaceans.[5] The component particles of these aggregates are brought together by classical coagulation processes as well as by animal feeding. By virtue of their large size, they sink faster than smaller particles and can be an important means for the removal of particulate matter from the ocean's surface. However, the variety of sources of organic particles and the mechanisms for their aggregation make this a difficult system to study systematically.

Algal blooms offer a relatively simple system with which to start the analysis of the role of aggregation in marine ecosystems. Typical algal blooms occur when nutrient and light conditions allow the rapid growth of a phytoplankton population near the ocean surface. A bloom usually ends when the algae deplete their nutrient supply and slow their growth while they continue to sediment out of surface waters or to be consumed by zooplankton. The fallout of such material is responsible for the rapid movement of algal material from the surface to the bottom of the North Atlantic, more than 4500 m below.[6-8] Because blooms occur when there is insufficient grazing pressure to keep phytoplankton concentrations down, other processes such as coagulation have the potential to determine the fate of the algae. Coagulation has been invoked as the primary removal process of algal material in the North Sea[9,10] and in the Barents Sea.[11]

An algal bloom is a particularly good system in which to study coagulation dynamics because the high particle concentrations should promote high coagulation rates. Furthermore, the relative unimportance of the grazers makes this a biologically simpler system, with only one source of particles. Finally, aggregation has been observed and inferred after algal blooms to be important to the dynamics of the system.

Coagulation theory has been applied to understanding phytoplankton in freshwater lakes where it was used more to explain mass fluxes from surface waters than to describe plankton dynamics.[12,13] Models of the coagulation of algal blooms have focused on the effect of coagulation processes on the development of the algal bloom[14,15] and on the fate of the particles when the bloom is over.[16]

Coagulation theory has been applied to other marine particle systems. In the earlier volume of this series, Honeyman and Santschi[17] have discussed the roles that colloidal particles have in scavenging surface-active radioisotopes and that coagulation has in accelerating their sedimentation. Recent observations show the colloidal system to be very dynamic.[18]

In this paper, we will discuss the effect that coagulation can have on the development of a bloom. As part of this, we will suggest changes that have to be made in the coagulation kernels to accommodate them to a biological oceanography context.

Coagulation requires the collision of particles and their sticking together. This article focuses on the mechanisms and rates of collisions, using a simplistic representation of sticking. A companion article in this volume by C. R. O'Melia discusses the chemical nature of interactions between particles.[19]

OVERVIEW OF CLASSIC COAGULATION THEORY

The Kinetic Coagulation Equations

Coagulation models can be formulated assuming a monodisperse system, in which all of the particles are initially the same, or assuming a continuous size distribution.[20,21] For the monodisperse case, an aggregate is composed of an integral number of monomers. The size class of a particle is denoted by a subscript which equals the number of monomers that it contains. Thus, r_i denotes the radius of a particle containing *i* monomers. In the continuous version, the radius could be an explicit function of the mass m: r(m). Some formulations use the particle volume *v* as the master variable rather than mass. This approach is useful when volume is conserved and, hence, particle density is constant.

Classic coagulation models emphasize three mechanisms for particle-particle contact: Brownian diffusion, laminar and turbulent shear, and differential sedimentation. For each of these, the rate R_{ij} of forming new particles of mass m_i + mj from the collision of two smaller particles of masses m_i and m_j is given in terms of the number concentrations of the two colliding particles n_i and n_j, a rate constant called the coagulation kernel β_{ij}, and an efficiency factor E_{ij}:

$$R_{ij} = E_{ij} n_i n_j \beta_{ij} \quad (1)$$

In the absence of any nonaggregation processes, the rate of change in the concentration of a given particle size equals the sum of all rates from reactions that create that size particle, less the sum from all those reactions that consume it:[20,21]

$$\frac{dn_i}{dt} = 0.5 \sum_{j=1}^{i} E_{j,i-j} \beta_{j,i-j} n_j n_{i-j} - n_i \sum_{j=1}^{\infty} E_{i,j} \beta_{i,j} n_j \left(1 + \delta_{ij}\right) \quad (2)$$

where

$\delta_{ij} = 0$ when $i \neq j$, $=1$ when $i = j$ (the Kronecker delta)

For the case of continuous particle size distributions,

$$\frac{dn(m)}{dt} = 0.5\int_0^m E(m-m',m')\beta(m-m',m')n(m-m')n(m')dm' \\ -n(m)\int_0^\infty E(m,m')\beta(m,m')n(m')dm' \tag{3}$$

The three different mechanisms for particle-particle contact, discussed above, are usually assumed to operate independently, allowing the total rate of formation of new particles to be expressed as the sum of the individual rates. Using the subscripts *Br*, sh, and ds to denote the different mechanisms, this leads to[20]

$$R_{ij} = n_i n_j\left(E_{Br,ij}\beta_{Br,ij} + E_{sh,ij}\beta_{sh,ij} + E_{ds,ij}\beta_{ds,ij}\right) \tag{4}$$

There are at least two different approaches to using E_{ij} and β_{ij}. The first uses a simple formulation to describe the hydrodynamics of contact in β_{ij} and uses E_{ij} to add the higher order correction terms; the second approach incorporates all of the hydrodynamic correction terms in β_{ij} and uses E_{ij} to describe the probability that any contact between two particles will result in their sticking together. In the second approach, E_{ij} is frequently assumed to be the same for all mechanisms of contact and for all particle sizes. This is used in this paper. E_{ij} equals the sticking efficiency α.

Early coagulation models assumed that all particles were spheres of constant density. This is equivalent to assuming that volume, as well as mass, is conserved when two particles aggregate. Recent work, both theoretical and experimental, has shown that the particles formed by aggregation processes are often fractals, having fractal dimensions <3.[22,23] Thus, the fraction of fluid medium incorporated within the particle structure increases with the size of the particle. The resulting relationship between mass and length l of the aggregate can be expressed using the *fractal dimension* d_{fr}:

$$l \propto m^{1/d_{fr}} \tag{5}$$

For aggregates of constant density, d_{fr} is 3 and Equation 5 could express the relationship between radius and volume of a sphere. However, aggregates may have d_{fr} values as low as 1.75.[23] As a result, a given mass will have a much larger radius than would a solid sphere with the same mass. The standard way to measure the length of a fractal is to use the root mean-square radius, called the radius of gyration r_g. It is given by[24]

$$r_g = \left(M^{-1}\sum_{k=1}^{i} m_k r_k^2\right)^{0.5} \tag{6}$$

where

M is the total aggregate mass
m_k and r_k are the mass and distance from the center of mass of the k-th subparticle in the aggregate
i is the total number of such subparticles

If all of the subparticles have the same mass, then Equation 6 reduces to

$$r_g = \left(i^{-1} \sum_{k=1}^{i} r_k^2 \right)^{0.5} \tag{7}$$

An aggregate composed of algal cells is not, strictly speaking, a fractal because Equation 5 is not satisfied for particles smaller than an algal cell. Some authors highlight this distinction by speaking of such aggregates as having a "cluster fractal dimension."[25] Furthermore, aggregates that are composed of more than one size particle, such as those composed of algae and zooplankton fecal pellets, may have different values of d_{fr}. The important property of an aggregate, as far as this chapter is concerned, is that the mass of an aggregate can be related to its size by Equation 5.

This different radius to mass relationship can affect the estimates for aggregation rates in several ways. Because the coagulation kernels are functions of the radius of the interacting particles, the values of β change from their values when all particles are solid because particle settling rates are functions of particle radius and density. Using a fractal density thus affects the differential sedimentation kernel. The open structure of fractal aggregates allows some water to flow through it, changing the capture cross section for both differential sedimentation and shear. Lastly, the fractal radius to mass relationship can also affect the particle dynamics by changing loss rates caused by sedimentation from the system.

Brownian Motion

The diffusion coefficient for Brownian motion, D, of a spherical particle of radius r is given by[20]

$$D = kT(6\pi\eta r)^{-1} \tag{8}$$

where

T is the absolute temperature
k is the Boltzmann constant
η is the viscosity

For $r = 0.5$ µm, $T = 300$ K, and $\eta = 1.$ mPa s^{-1}, D = 4.4×10^{-13} m^2 s^{-1}.

The coagulation kernel for Brownian motion of two particles of radius r_i and r_j and Brownian diffusivities D_i and D_j is given by:[20]

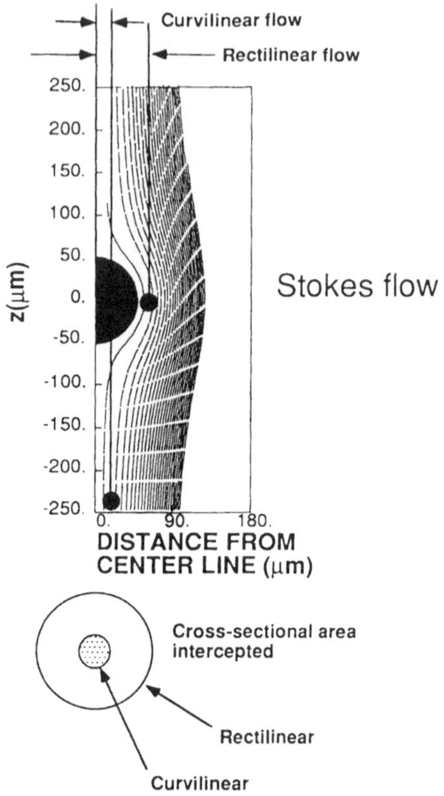

Figure 1. Capture cross sections used to calculate the coagulation kernels for differential sedimentation. Shown are the cross sections for the rectilinear model and the curvilinear models. The rectilinear model assumes that the capture cross section is determined by the sum of the radii of the two particles; the curvilinear model modifies this approach by using the flow field of the larger particle to determine the capture cross section. The two concentric circles at the bottom of the figure represent the cross sections for the two cases. The smaller particle has radius r_i; the larger has radius r_j.

$$\beta_{Br,ij} = 4\pi\left(D_i + D_j\right)\left(r_i + r_j\right) \tag{9}$$

The fractal nature of aggregates formed by diffusive processes was recognized early in the development of fractal theory.[26-28] The effect on the coagulation kernel has been addressed by arguing that the r_i and r_j used in Equations 8 and 9 can be calculated using the radius of gyration calculated in Equation 5.[26,29-31]

Differential Sedimentation

The coagulation kernel for differential sedimentation has been calculated at several levels of accuracy. The first level, known as the rectilinear formulation,

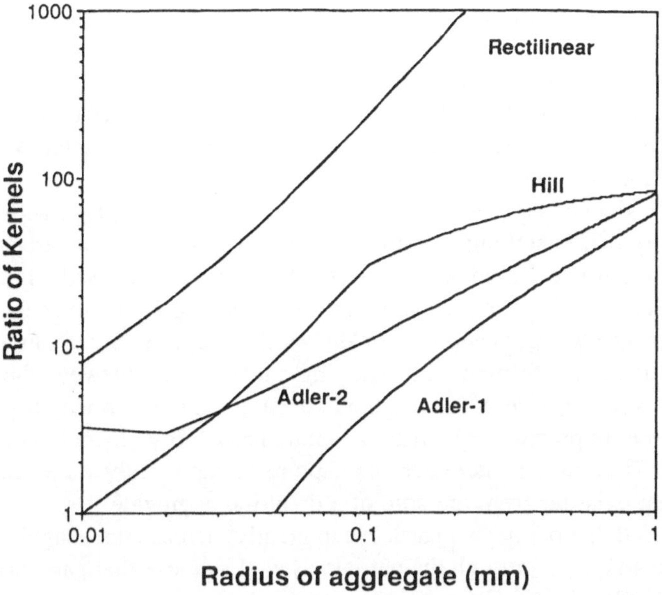

Figure 2. Coagulation kernels for different approximations to differential sedimentation, relative to that calculated for the curvilinear approximation (Equation 11) as a function of aggregate size. The smaller particle is assumed to be 10 μm. Adler-1: for the interception of any smaller particle, whose center intersects the larger, porous particle; Adler-2: for the interception of any smaller particle, whose outer edge touches the larger, porous particle; Rectilinear: the rectilinear case; Hill: the interception calculated using Hill's modification of the shear kernel with his adjustable parameter equal to 10. Particle density calculated as in Jackson and Lochman[15]; porosity as in Valioulis and List[36]; streamlines as in Adler.[34]

assumes that any particle with radius r_i that is within a distance $r_i + r_j$ of the centerline far upstream of the larger, faster settling particle with radius r_j will collide (Figure 1). If w_i and w_j are their fall velocities, then the rate of collision with the larger particle will be $\pi (r_i + r_j)^2 |w_j - w_i| n_i$. The rate of all collisions is n_j times this. Thus, the coagulation kernel for the rectilinear case will be[20]

$$\beta_{ds,ij} = \pi\left(r_i + r_j\right)^2 \left|w_j - w_i\right| \tag{10}$$

The next level of accuracy, known as the curvilinear formulation, incorporates the flow field of the larger particle, which is assumed to be a sphere falling at a low Reynolds number with a flow field described by Stokes flow. As a particle falls, it pushes water away from its fall line. This effectively decreases the capture cross section. The smaller particle whose center is at $r_i + r_j$ at closest approach was initially closer to the centerline of the larger particle (Figure 1). As a result, the coagulation kernel is smaller:[20]

$$\beta_{ds,ij} = 0.5\pi r_i^2 \left|w_i - w_j\right| \tag{11}$$

The effect of making this correction is greatest for small particles impacting large ones (Figure 2).

Further corrections have been made which include calculating the flow field as determined by both (spherical) particles and including the effect of attraction or repulsion between surfaces.[32] The hydrodynamic correction decreases the contact rate while the chemical forces can increase it.[33]

Adler[34] calculated the flow field in and around a porous sphere with a constant permeability that is settling at a constant rate. Any particles on streamlines that intercept the particle should be captured by the aggregate. He used this to calculate the cross section of water that would intersect the aggregate. In order to use his results, one needs a permeability for the settling particle. Sutherland and Tan[35] simulated aggregate formation, developing a relationship between the aggregate permeability and its size. Valioulis and List[36] applied this permeability to Adler's result in order to predict the increased capture probability of small particles by an aggregate. They further increased the capture radius by 20% to account for the tendrils that exist far from the core of a dendritic aggregate.

Flow through a permeable particle can greatly increase the coagulation kernel of particles that pass through the particle as well as those that pass close enough to touch (Figure 2).[34,35] For a 10 μm solitary alga and an aggregate ten times as large, these corrections increase the capture cross section, and hence the coagulation kernel for differential sedimentation, by over a factor of 10. This is still small relative to the coagulation kernel calculated by the rectilinear approximation.

Because an aggregate with a fractal mass distribution does not have a uniform porosity, it may have greater internal flow and a greater capture cross section than a particle with uniform porosity. As a result, the appropriate coagulation kernel would be between those for the permeable model and for the rectilinear formulation. More work is needed to determine differential sedimentation kernel more accurately by incorporating fractal aggregates.

Simulations have shown that the fall velocity can be related to the radius of gyration of an aggregate.[37,38] The hydraulic radius r_h is the radius of a sphere which contains the particle mass of an aggregate as well as accompanying fluid and which has the same fall velocity as the aggregate would have using Stokes law. Simulations show that the hydraulic radius is proportional to the radius of gyration with a proportionality constant which increases with decreasing fractal dimension and which can increase with aggregate size.[38] For a large aggregate, r_h/r_g is about 2 with $d_{fr} = 1.49$ and about 1 to 1.4 with $d_{fr} = 1.79$. Because the radius of gyration is related to the aggregate mass by the fractal dimension, one can compensate for the fractal nature when calculating the fall velocity if one knows d_{fr}.

Hill[16] tested the sensitivity of a model determining the fate of oceanic particles to changes in the differential sedimentation kernel as a way to estimate the potential impact of porous aggregates on coagulation rates. The model results were very sensitive to changes in the kernel, with the kernels closer to the

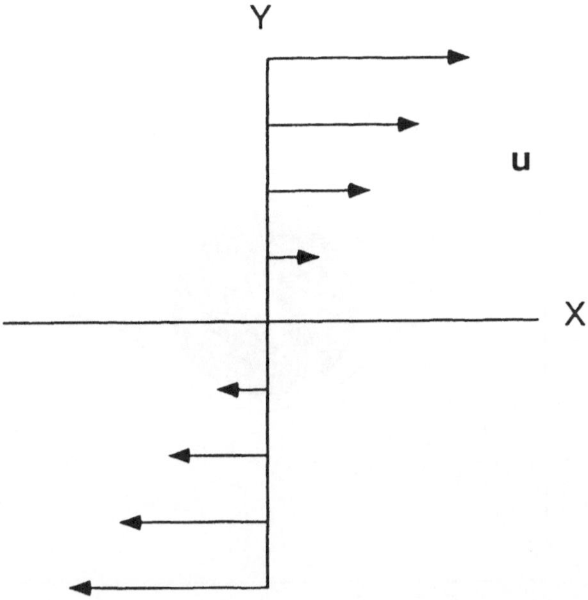

Figure 3. The velocity flow in laminar shear in the absence of any particles. The velocity u is in the x direction with a strength that is proportional to y. There is no change in the z direction.

rectilinear kernel giving the greatest coagulation and particle sedimentation rates.

Shear Coagulation

Shear flow is fluid flow in which the water velocity changes with position. Simple shear is described by equations of the form

$$u_x = \gamma y, u_y = u_z = 0$$

where

u_x, u_y, and u_z are the fluid velocity components in the x, y, and z directions
γ is the shear gradient

(Figure 3). A freely-suspended sphere in the presence of simple shear will rotate, carrying a mass of water with it.[43,44] Motion of this trapped water is characterized by closed paths, or streamlines. The minimum thickness of this layer is 0.156 r at x = z = 0, where r is the sphere's radius (Figure 4).[43]

The coagulation kernel for laminar shear is given by[20]

$$\beta_{sh,ij} = \frac{4}{3}\gamma\left(r_i + r_j\right)^3 \quad (12)$$

This is the rectilinear version of the laminar shear kernel.

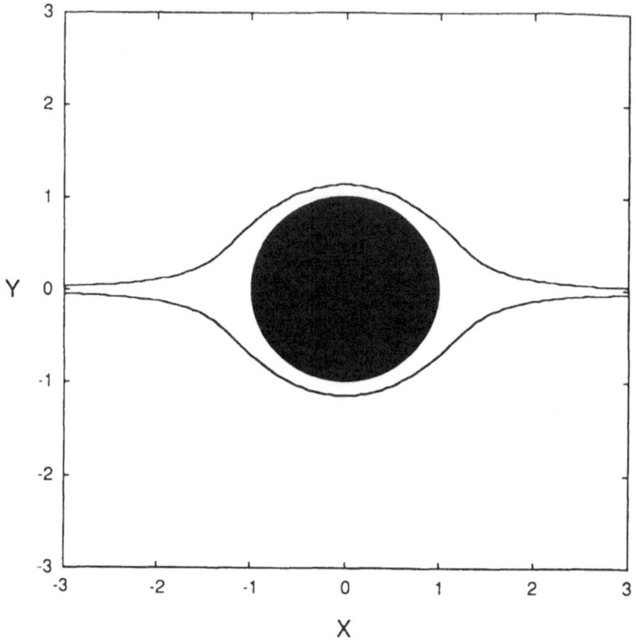

Figure 4. The thickness of the nonexchanging layer of water around a particle in simple shear flow. Flow is from the left for y > 0, from the right for y < 0. The minimum size for a small particle to touch the larger is the minimum thickness of the layer of nonexchanging water, 0.156a.

The flow around a single particle in shear flow has been solved and applied to calculating the curvilinear version of β_{sh}.[32,43,44,46] Values for β_{sh} have also been calculated for the flow between two interacting spheres[32,43,46] and for the particle trajectories when molecular forces are included.[32,45] Because the results are too complicated to present analytically, they are usually presented in graphical or tabular form. One conclusion is that the water entrained by a particle acts as a significant barrier to contact between it and a smaller particle. Although this can be partially overcome by the attraction of van der Waals forces between the particles, there is still a substantial decrease in the contact rate when the radius of the smaller particle is less than 10% of the radius of the larger.

Adler[34] has determined the streamlines in and around a porous particle in shear flow. Valioulis and List[36] used this relationship to calculate the capture cross section and coagulation kernel for shear. They noted the inadequacy of this approach when calculating the coagulation of two porous particles.

Much of the interest in shear flow is the result of it being the manifestation of turbulent flow in the smallest, viscous, size range. One approach to describing such flow is to break the laminar flow into purely rotational and purely straining flows. Arguing that the rotational component does not affect transport, Saffman and Turner[39] and Hill[16] focused on the purely straining component. For turbulent shear, the rectilinear coagulation kernel becomes[39]

$$\beta_{sh,ij} = 1.3\left(\frac{\varepsilon}{\nu}\right)^{0.5}\left(r_i + r_j\right)^3 \tag{13}$$

where

ε is the turbulent energy dissipation rate
ν is the kinematic viscosity of the fluid

If γ is estimated as $(\varepsilon/\nu)^{0.5}$, the two equations are the same. The curvilinear version of this is[16]

$$\beta_{sh,ij} = 10\left(\frac{\varepsilon}{\nu}\right)^{0.5}\left(r_i + r_j\right)^2 r_i \tag{14}$$

where

$r_i << r_j$

While this is reasonable in the absence of a rotating particle, it neglects the motion of water trapped next to the particle. As previously noted, this can have a large effect on mass transfer from solution, including the particle contact rates given by the coagulation kernel.[40,41] As a result, these results overestimate the collision frequency between large and small particles.

OCEANOGRAPHIC CONSIDERATIONS

The following sections discuss ways that the marine environment and its organisms, particularly phytoplankton, affect the way that one must consider coagulation. In some cases, the modifications are illustrated by their effects on the nutrient-driven model of an algal bloom.

Models of Bloom Dynamics Incorporating Coagulation

The simplest phytoplankton growth/coagulation model assumed that single cells formed the basis for the system, that they were dividing at a constant specific growth rate μ, and that the ultimate fate for any particles was to sediment out of the mixed layer:[14]

$$\begin{aligned}\frac{dn_1}{dt} &= \mu n_1 - \alpha n_1 \sum_{i=1}^{\infty} n_i \beta_{1i}\left(1 + \delta_{1i}\right) - n_1 w_1 Z^{-1} \\ &= +\ growth - coagulation\ \ loss - sedimentation\end{aligned} \tag{15}$$

$$\frac{dn_i}{dt} = 0.5\alpha \sum_{j=1}^{i-1} n_j n_{i-j} \beta_{j,i-j} - \alpha n_i \sum_{j=1}^{\infty} \left(1 + \delta_{ij}\right) n_j \beta_{ij} - n_i w_i Z^{-1} \quad for\ i \neq 1 \tag{16}$$

$$+ \ coagulation \ \ gain - coagulation \ \ loss - sedimentation$$

where

Z is the depth of the surface mixed layer
$E_{ij} = \alpha$

Results showed that solitary algal cells could have a maximum concentration. This maximum was determined by a balance between the formation of new cells by division and the loss of single cells to aggregates and, ultimately, to settling out of the mixed layer. The maximum concentration could be estimated by assuming that it happened at steady state and that all of the single cell aggregation was caused by collision with other single cells. This simplification was the result of an assumption that collisions with aggregates were negligible, an assumption which simulations showed to be true. It also neglected any nonalgal particles which could be important in natural situations.[16] The resulting estimate of the maximum concentration n_{cr}, was:

$$n_{cr} = 0.048\left(\mu - w_1 / Z\right)\left(\alpha\gamma\right)^{-1} r_1^{-3} \tag{17}$$

n_{cr} is a function only of shear coagulation because Brownian coagulation was assumed to be insignificant for particles >1 μm and because there are no collisions between two identical cells falling at the same speed.

A more sophisticated model included dependence of algal growth on environmental factors by making the growth rate a function of nutrient concentration and light intensity.[15] Expressed in the continuous particle size distribution form (see Equation 3), the growth equations become:

$$\begin{aligned} \frac{dn(m,t)}{dt} &= \mu n - nwZ^{-1} \\ &+ \frac{1}{2}\int_0^m \alpha\left(m_1, m - m_1\right)\beta\left(m_1, m - m_1\right) n\left(m_1, t\right) n\left(m - m_1, t\right) dm_1 \\ &- n(m,t)\int_0^\infty \alpha\left(m, m_1\right)\beta\left(m, m_1\right) n\left(m_1, t\right) dm_1 \end{aligned} \tag{18}$$

where

m is the particle mass

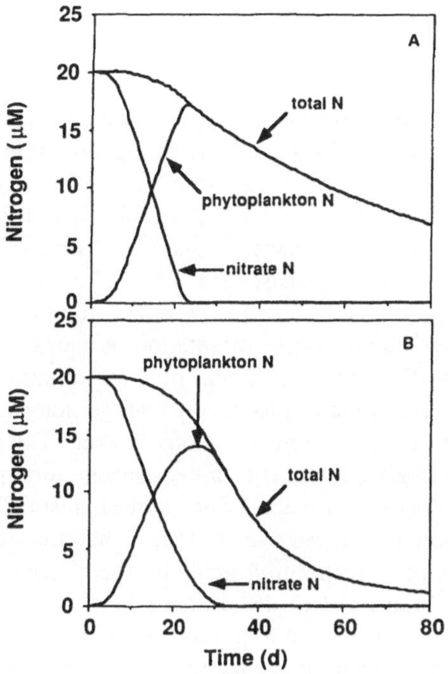

Figure 5. Effect of coagulation on the model predictions for growth of an algal population. (A) in the absence of coagulation; (B) in the presence of coagulation. The algal population is seeded into a 50 m mixed layer in which the nitrate concentration is 20 μm. Growth of the algae decreases the nitrate concentration and increases the particulate nitrogen concentration (phytoplankton). Phytoplankton and total nitrogen concentrations are decreased by particles falling from the mixed layer. This particle flux increases as the phytoplankton concentration increases and as coagulation forms larger, faster settling particles.

This can be reformulated in sections, each section spanning a range with its upper limit twice its lower limit.[15,42] While not required, the ratio of the upper to lower masses in a section being two makes the computation of sectional coagulation coefficients simpler.[42] These equations can be numerically integrated in conjunction with equations determining the light attenuation, nitrate concentration, and their relationship to μ for different initial values of nitrate, mixed layer depth, algal size, and shear rates. Solitary algae are no longer monodisperse but are distributed over the sectional mass range, with the mass of the smallest alga being half that of the largest.

This second model for algal growth coagulation has shown two dominant effects (Figure 5). As in the first model, losses of single cells to aggregates can occur at a rate great enough to balance the formation of new cells by algal division. The parameter n_{cr} remains a useful estimate for this situation. Second, even for situations where the coagulation rate does not limit the maximum algal concentration, coagulation can enhance the loss of algal material from the surface layer.

Collision Kernels/Mechanisms

Cell Swimming

Excluding the diatoms, most marine algae as well as many of the bacteria are flagellated and can swim.[47-49] Measured swimming rates of marine dinoflagellates range between 8 and 500 μm/s.[47,50] In addition, several dinoflagellate species have been shown to move to regions of more desirable chemical concentrations.[51,52] The chemotactic responses in bacteria, such as *Escherichia coli*, have been intensively studied.[53,54] They consist of a series of runs in which an organism moves at constant speed and direction followed by tumbles in which it does not move but does change its directional orientation. As a result, the subsequent run is in a new, random direction. An increasing concentration of a desirable substance during a run causes the run to be, on average, longer. The average run is shorter if the chemical signal decreases along the run. The result of many such runs and tumbles is to give a net drift up the concentration gradient in what has been described as a "biased diffusion." Such a mechanism allows an organism to respond to concentration changes over distances that are much larger than their lengths. For a constant concentration gradient, the concentration differences at two locations are proportional to the distance between them. In the case of a bacterium which is less than 1 μm long but has an average run length of 10 μm, this could increase the concentration change during a run by an order of magnitude over that along its length.

In the absence of any chemical cues, this motion is a random walk, which can be described in terms of a diffusion coefficient D_r:[53]

$$D_r = \frac{1}{3} v_c^2 \tau_0 \tag{19}$$

where

v_c is the swimming speed of the organism during a run
τ_0 is the length of time of the average run

This must be modified if the tumble direction is biased by the previous run direction. For bacterium swimming at 12.5 μm/s for an average of 0.67 s,[53] $D_r = 3.5 \times 10^{-11}$ m^2/s; for a dinoflagellate swimming at 100 μm/s for an average of 1 s, $D_r = 3.3 \times 10^{-9}$ m^2/s.

An equivalent coagulation kernel can be formulated for the case of organisms which use this biased random walk to move:

$$\beta_{Rij} = 4\pi\left(D_{C,i} + D_{C,j}\right)\left(r_i + r_j\right) \tag{20}$$

Such an approach can be used to estimate the diffusivity of single swimming cells. Aggregates of swimming organisms would not be expected to move, as the cells'

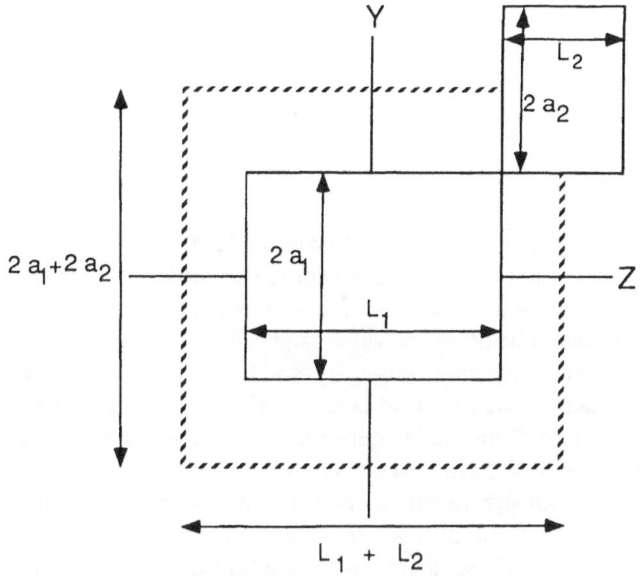

Figure 6. Interactions between two helical diatom chains caused by shear. Diatom chains are modeled as cylinders of length L and radius a. The coordinate system is the same as that of Figure 3.

combined motions would cancel each other. Thus, the enhancement to the coagulation kernel by algal swimming only occurs for single cells. It could be used to create a new estimate for the critical concentration (Equation 16)

$$n_{cr} = \left(\mu - w_1 Z^{-1}\right)\beta_{Rij}^{-1} \tag{21}$$

$$\approx \left(\mu - w_1 Z^{-1}\right)\left(16\pi\alpha D_{r1}\right)^{-1} \tag{22}$$

For a dinoflagellate with $r_1 = 50$ μm, $D_r = 3.3 \times 10^{-9}$ m²/s, $\alpha = 0.1$, $\mu = 1$, $d = 1.16 \times 10^{-5}$ s^{-1}, and $w_1 = 0$, this implies n_{cr} of 14 cells/ml. If a shear coagulation with $\gamma = 0.1$ s^{-1} controlled the population, then $n_{cr} = 440$ cells/ml by Equation 16.

Dinoflagellates are frequently found in dense blooms known as "red tides." For example, *Gymnodinium flavum* has been observed in concentrations of 6000 cells/ml, implying that n_{cr} must be greater than this concentration.[55] The cells had an $r \approx 15$ μm. Reported swimming speeds for members of the genus *Gymnodinum* range from 0.72 to 1 m/h (0.2 to 0.3 mm/s). Assuming that the higher swim speed describes *G. flavum*, then $D_r = 3. \times 10^{-8}$/m²/s. If 6000 cells/ml represents a lower bound on n_{cr} that is determined by the stickiness of the cells, and the other parameters are as given above, then α is at most 8.5×10^{-5}. This is substantially less than the typical values of 10^{-3} measured for diatoms[68] and suggests that low stickiness may be a necessary adaptation for dinoflagellates to allow them to achieve such high concentrations without aggregating.

Algal Chains

Diatoms form one of the dominant groups of algae in marine systems.[56] One of their characteristics is the silica shell composed of two large pieces, the valves, and smaller connecting structures. When a diatom divides, each of the two daughter cells keeps one of the old valves and makes a new one. After the division, the two daughters may stay united by small connective structures. Multiple divisions can lead to long chains of cells, which form a partially flexible string of cells. The ability to form chains depends on the species and on physiological factors. For example, the blue-green alga *Anabaena flos-aquae* can form helices in which the individual cells are about 5 μm in diameter, the radius of a helical coil about 15 μm, and the distance between coils about 15 μm.[57] Other species form shorter chains whose length is a function of cellular nutritional status. Several diatom species form chains with as few as 10 cells per chain when grown under low nutrient conditions.[58-60]

In the presence of a shear, a flexible string will rotate and bend. For sufficiently long and flexible chains, the string takes the form of a helix, with its long axis perpendicular to both the flow and the shear.[61]

Consider a helical chain as a cylinder with its long axis oriented perpendicular to the flow. Two such chains of lengths L_1 and L_2, radii a_1 and a_2, and concentrations n_1 and n_2 will interact whenever their centers are within $a_1 + a_2$ of each other in the the y direction and within $0.5\ (L_1 + L_2)$ on either side of the z direction (Figure 6). If γ is the shear gradient, the flux of L_2-particles relative to a particle at $x = 0$ is $n_2\gamma y$. The transport J of L_2-particles hitting a L_1-particle is

$$J = 2\int_0^{a_1+a_2} \left(L_1 + L_2\right) n_2 \gamma y\, dy \tag{23}$$

$$= \left(L_1 + L_2\right)\gamma n_2 \left(a_1 + a_2\right)^2 \tag{24}$$

The total rate of such collisions is then $n_1 J$ or $(L_1 + L_2)\gamma\ (a_1 + a_2)^2 n_1 n_2$. Thus the coagulation kernel for shear becomes

$$\beta_{sh,ij} = \gamma\left(L_1 + L_2\right)\left(a_1 + a_2\right)^2 \tag{25}$$

One might expect the values of a_1 and a_2 to be set by the nature of the diatoms, regardless of number of cells. The values of L_1 and L_2 are then proportional to the

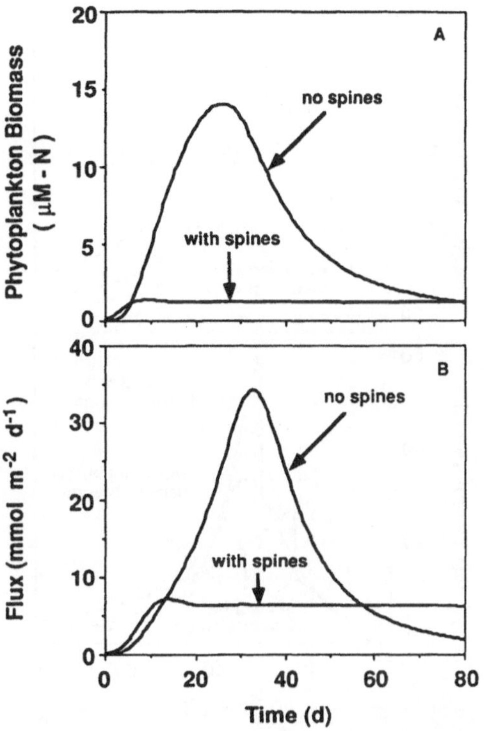

Figure 7. Effect of spines on the growth and flux of an algal bloom. (A) phytoplankton biomass in the presence and absence of spines; (B) particulate flux from the surface mixed layer in the presence and absence of spines. The spines increase the coagulation kernels so that they drastically decrease n_{cr}. Because of the low concentrations, total algal growth, as opposed to algal growth rate, stays low and lowers the export of organic matter. The case for no spines is that of Figure 5B. Phytoplankton biomass is measured by its nitrogen content. The algal cell radius is 10 µm, the spine 50 µm. The computational model used is the same as in Figure 5.

number of cells. Thus the cross section depends linearly on the mass, rather than on the cube root of the mass, as when the radius is the important variable.

Spines

Algal cells, particularly diatom cells, can have spines of silica or chitin extending away from the cell for distances larger than the cell diameter. For example, diatoms of the species *Thalassiosira fluviatilis* raised in culture had diameters of 12 to 15 µm and chitin spines as long as 70 µm.[62] Such spines can increase the interaction cross section of an alga. In the case of an alga in shear, the alga will rotate, causing the spines to sweep out a region larger than the cross section of the algal cell itself.[61] Additionally, one algal cell overtaking a second will have its spines spinning in a direction opposite that of the second cell by virtue of one being on top and the other beneath.

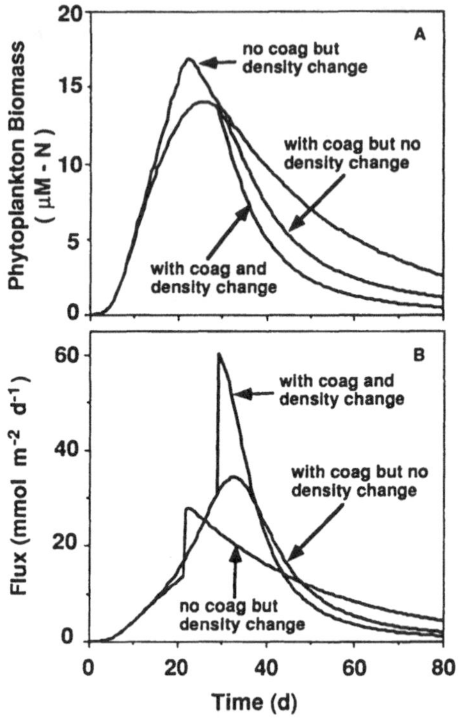

Figure 8. Effect of variable particle settling velocities on particle densities and settling velocities. (A) particulate concentrations associated with the standard coagulation situation (Figure 5B), with a change of particle density but no coagulation, and with a change in particle concentration in the presence of coagulation; (B) fluxes associated with the above cases. Change in density involves a doubling of all particle settling velocities when the algal specific growth rate is lower than 0.05/d. The maximum phytoplankton concentration is almost the same in the two coagulation cases, although there is faster removal rate of the algal cells when the settling rate increases.

Valioulis and List[36] confronted the same problem in the case of sewage particles by assuming that the effective radius increased by 20%. Similarly, a fixed length equal to the length of the spines can be added to the contact radius of all particles to describe the effect of spines. Thus, if δ_{sp} is the length of the spines, coagulation kernels for shear and differential sedimentation become

$$\beta_{Br,ij} = 4\pi\left(D_i + D_j\left(r_i + r_j\right)\right) \tag{26}$$

$$\beta_{ds,ij} = 0.5\pi\left(r_i + 2\delta_{sp}\right)^2\left|w_i - w_j\right| \tag{27}$$

$$\beta_{sh,ij} = 1.3\gamma\left(r_i + r_j + 2\delta_{sp}\right)^3 \tag{28}$$

These kernels are based on the assumption that spines effectively make a particle a sphere with a radius greater by δ_{sp}. While this may overestimate the actual collision rates because the spines are not solid, it does allow an estimate of the general effect of spines.

The impact of the spines on algal concentrations is to greatly increase the rate of coagulation (Figure 7). The algae achieve a much lower maximum concentration because of the intense control on their populations. Nutrient concentrations take a long time to decrease, and vertical fluxes are small.

Variable Densities/Fall Velocities

Marine diatoms can vary the rate at which they sink in response to nutrient stress.[58-60,63,64] Under conditions of nutrient stress, they tend to increase the rates at which they sink. For most species, the highest and lowest sinking velocities are within a factor of two of each other. Some species, such as *Thalassiosira rotula*[58] and *Coscinodiscus wailesii*,[63] have sinking rates that are a factor of 10 greater under times of nutrient stress than the sinking rates when they are nutrient replete. These changes in sinking rate are accompanied by relatively small changes in cell size, implying that much of the sinking rate change is related to changes in cellular densities.

Such changes in settling velocity should change the differential sedimentation kernel, increasing the contact rates. They should also increase the rate at which cells leave the mixed layer. If changes in settling velocity are the result of changes in algal cell density, and if all particles, cells and aggregates, have the same density, then a doubling of algal settling velocity will double the settling velocities of all particles and the coagulation kernels for differential sedimentation.

Coagulation rates caused by changes in settling velocity at the time of nutrient depletion for a case in which settling velocity is doubled when the specific growth rate fell to less than 0.05/d are substantially greater than those for a case with no change in the settling velocity. Settling velocity changes which occur when nutrients are depleted do not affect the maximum biomass of a phytoplankton bloom (Figure 8A). The rate at which material is removed from the system thereafter increases proportionately to the doubling of the settling velocity.The flux jumps from 30 to 60 mmol/m^2/d when settling velocity is doubled (Figure 8B). Flux increases may be due to more rapid settling of all the particles in the system and to enhanced coagulation through changes in the differential sedimentation kernel. A third case, in which the settling velocity is doubled as before but the effect of coagulation is removed, can separate the two effects. The maximum flux with doubling of settling velocity but without coagulation is 28 mmol/m^2/d, whereas the maximum flux with settling velocity doubling and coagulation effects is 60 mmol/m^{-2}/d. The increase in flux observed after settling velocity doubling is due in part to enhanced coagulation and is not simply an artifact of more rapid settling of preexisting particles.

Solitary algal cells in any one environment have a range of settling velocities. As a result, there can be a nonzero coagulation kernel for differential sedimentation which can be significant in initiating rapid coagulation.[65] The sectional model that has been used here incorporates cell contacts from differential sedimentation because single cells are allowed to vary in mass by a factor of two and thus have a range of settling rates. In fact, the coagulation kernel for this interaction is small compared to that of shear coagulation.

Variable Stickiness

An important parameter determining the coagulation rate of a collection of particles is the probability that they will stick upon contact, a. Kiørboe et al.[68] have observed changes in a for cultures of three species of diatoms. They grew the algae in batch cultures, in which algal cells were added to nutrient rich solutions and allowed to grow until they had depleted the solution of nutrients. Kiørboe et al. measured a during different algal nutritional states. One algal species, *Skeletonema costatum*, had an essentially constant value for a of almost 0.1 for most of its growth cycle. A different species, *Thalassiosira pseudonana*, had a value of a as small as 10^{-4} during the nutrient rich, high growth phase but a much larger value, almost 0.1, when the nutrients were depleted and growth stopped. This effect was reversed when the culture was renewed with fresh nutrients, reversing the slowdown of algal growth.

The simulations here increase a for nutrient-deplete algae in the algal growth and coagulation model by switching the stickness from 0.001 to 0.25 when the specific growth rate decreases to less than 0.05/d because of nutrient depletion. The results show that this can have a dramatic effect on the maximum algal concentration, the rate of coagulation, and the export of particulate matter from the surface waters (Figure 9). The low initial value of a allows the algal concentration to accumulate to a concentration far beyond the maximum it could achieve at the high value of a. When the stickiness changes, the subsequent coagulation occurs much faster than it would otherwise because of the high particle concentration. This is made particularly rapid by the essentially squared relationship between the rate of coagulation and the particle concentration. As a result, the particulate matter leaves the surface layer in a particularly short period of time, here about 5 d. While the transition from small to large a may not be as abrupt as assumed here, the fact that the particulate matter can increase to relatively large values before coagulation starts can have a significant effect on coagulation rates.

Fractal Nature of the Aggregate

Coagulation by Brownian motion is generally considered to be unimportant for aquatic particles larger than 1 mm.[66] Thus, while Brownian motion is important for the colloids believed to be involved with the removal of chemically-reactive elements such as thorium,[67] it will not be important in determining the fate of the larger biological particles comprising diatom blooms and found in marine snow. It is unfortunate for our understanding of coagulation of large particles that the theory of aggregation of fractal particles is best developed for Brownian motion rather than for the shear and differential sedimentation mechanisms that dominate for larger particles.

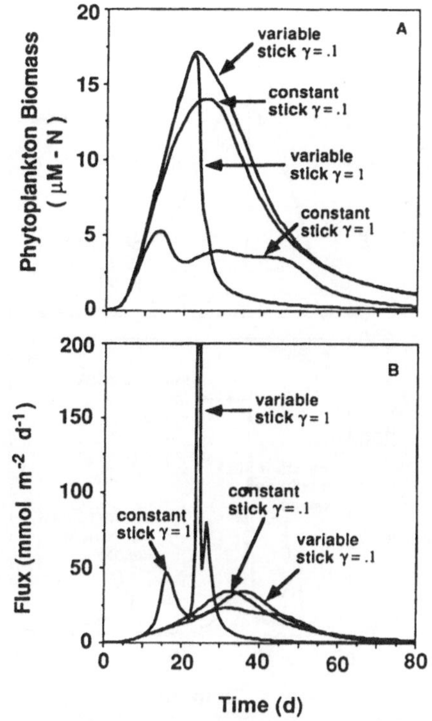

Figure 9. Effect of variable stickiness on the growth and flux of an algal bloom. (A) particulate concentrations; (B) particulate fluxes from the euphotic zone. Algal stickiness is 0.001 until the nutrient depletion slows the algal growth rate to 0.05/d, when α increases to 0.25. The effect is to rapidly increase the formation of aggregates and the flux from the surface mixed layer. The effect is most pronounced at shear, $\gamma = 1$/s. The standard case in previous figures corresponds to the $\gamma = 0.1$/s constant case.

Logan and Wilkinson[69] have used the measured relationship between aggregate size and fall velocity[3,70] to calculate a fractal dimension for marine snow. The fractal dimensions for aggregates composed of diatoms and of general material were within one standard deviation of each other. Values ranged from 1.26 to 2.14.

The use of the rectilinear kernels for shear and differential sedimentation can place a bound on the effect that flow through an aggregate can have on the development of an algal bloom. Their use in the equations describing the development of a bloom does, in fact, have a large effect (Figure 10).

Hill[16] simulated the effect of varying the differential sedimentation kernel on the removal of the products of a bloom in a similar way and also found that the form of the kernel could vastly increase the rate of removal of material from surface waters.

Heterogeneous Nature of Particulate Material

Marine snow is a mixture of materials of different origins, sizes, and densities.[5,71,72] These include millimeter-sized gelatinous appendicularean "houses,"

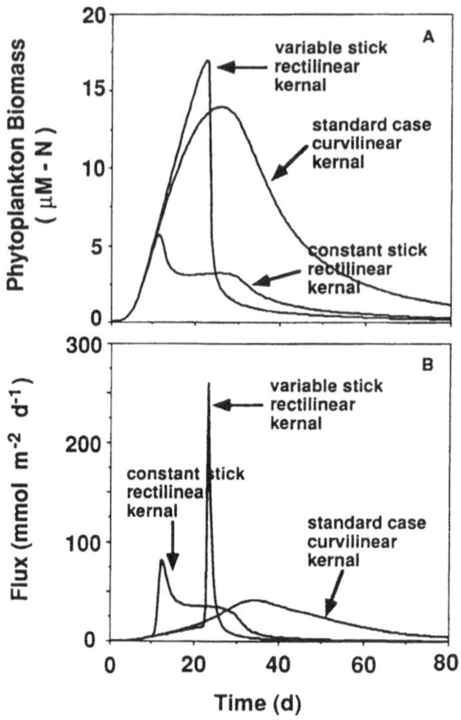

Figure 10. Effect of using the rectilinear and curvilinear kernels for differential sedimentation on the development of an algal bloom. (A) particulate concentrations; (B) particulate fluxes from the euphotic zone. The rectilinear case is used as a surrogate for a coagulation kernel, which adequately accounts for the porous nature of an aggregate. The larger coagulation kernel associated with the rectilinear kernel limits the particle concentration at a lower concentration and causes rapid particle transport earlier than with the standard curvilinear case. The combination of variable stickiness (Figure 9) and rectilinear kernels causes a very sharp increase in coagulation into large particles, resulting in the fastest particle export.

pteropod mucus feeding webs, polychaetes, and fecal matter, as well as phytoplankton. Each of these has its own characteristic size and density. Incorporating this mixture into the mathematical theory of coagulation which has been most extensively developed for monodisperse particles will require extensive modification of traditional coagulation models.

DISCUSSION

McCave[66] made an exhaustive review of expected coagulation rates in the ocean, emphasizing coagulation of inorganic particles in the benthic and midwater regions. He believed that biological feeding processes would make coagulation of algae unimportant, stating that "the larger particles, particularly 'snow,' are not produced by aggregation of small ones; they start out with a large organic substrate." Subsequent field observations and more detailed modeling indicate otherwise. Riebesell[9,10] followed a diatom bloom in the North Sea as the algae grew and, ultimately, coagulated. His work is the best documented set of observations on the development of an algal bloom. Algae were not in aggregates until late in the bloom, when algal abundances were high. The movement to aggregates occurred rapidly once it had started, only when the bloom was near its peak, as bloom models have predicted.[14,15]

Riebesell[10] calculated a value for $n_{cr} = 1.30 \times 10^4$ cells/ml, compared to the observed maximum concentration of 1020 cells/ml. He suggested that the formula for n_{cr} was inaccurate for diatoms in chains; that their collision frequencies should be greater than those for spheres. Because his value for n_{cr} was calculated using the spherical-equivalent radius for a chain of eight diatoms, it actually represents the concentration of diatom chains. The value for cell n_{cr} should be eight times this value, 1.04×10^5 cells/ml. This higher value further accentuates the difference between the maximum observed cell concentration and n_{cr}, with the two differing by a factor of 100.

This algal bloom was dominated at its peak by the diatom *Asterionella glacialis*. Electron micrographs showed that it formed chains and had spines about 15 μm long. The coagulation kernel for a chain (Equation 24) could be used to calculate n_{cr}, but an eight diatom chain is too short to develop a helical structure. Using Equation 27 for the shear kernel of a diatom with spines to develop an expression for the critical concentration yields an equation for n_{cr} in which the spine length plus the algal radius substitutes for the radius. *Asterionella* has an effective radius of 20 μm, four times the actual radius of 5 μm; the new n_{cr} is $1/4^3 = 1/64$ times the old. The resulting ratio of n_{cr}, to observed maximum concentration of 1.6 is much closer than that calculated using the n_{cr} which does not account for spines. Thus, the modified coagulation kernels developed here help the understanding of natural systems.

Phaeocystis pouchetii is a marine alga which forms blooms, many of which are not consumed by zooplankton grazers. Aggregation-enhanced sedimentation may be important for *Phaeocystis* removal from the euphotic zone after an algal bloom.[11] *Phaeocystis* is a prymensiophyte alga which can exist as either a swimming flagellated individual 3 to 10 μm in size or as a member of a spherical colony which can be as large as a few millimeters in diameter.[76] It forms blooms in high latitude regions, including the Bering and Barents Seas and the waters of Antarctica. Copepods avoid consuming *Phaeocystis* colonies when the algae are healthy but will consume them at other times.[77] Colonies release organic matter at rates that are inversely related to the nutrient concentrations in the surrounding water and that are associated with the colonies becoming sticky.[74] The dominant mechanism of *Phaeocystis* loss from the water column after a bloom appears seems to be settling rather than zooplankton grazing.[11,73]

Wassman et al.[11] have observed *Phaeocystis* colony settlement rates of about 0.03 to 0.06/d, which they suggested might result from aggregation of the colonies. If the colonies, which were found at abundances of 13.5 colonies/ml, had diameters of about 500 μm,[77] $\alpha = 0.1$, $\mu = 0.1/d$, and $\gamma = 0.1/s$, then the expected $n_{cr} = 3.6$ colonies/ml. The fact that this is below the observed colony concentration suggests that the conditions would, indeed, have been favorable for coagulation to occur. The removal rate observed by Wassman is similar to the rates calculated using the algal bloom-coagulation model (Figure 5).

Hill[16] simulated the fate of particles after an algal bloom when they were subject to coagulation, finding that such coagulation was enhanced by particles present as part of the background particle spectrum. In order to match observed vertical fluxes of organic material, he had to increase the particle contact rates, arguing that there is an enhancement caused by nonspherical particle shape and porosity. Furthermore, he required values of α of 0.1 or larger. This examination of the effects of modifying the coagulation kernels on particle export emphasizes the importance of developing kernels that better reflect the solid sphere nature of marine particles.

Coagulation models that emphasize algal blooms are useful tools for studying the interaction between organisms and coagulation, but they lack many important features of real ecosystems. As such, these models represent only the first step. The presence of other particles, such as fecal pellets and detritus, can affect coagulation rates. Hill's[16] inclusion of a background particle spectrum shows this. However, a fully-developed model algal coagulation model that includes a background particle spectrum will need to include a mechanism for generating and maintaining it. Such a model may be useful to explain the dynamics of marine colloids as well. Furthermore, coagulation is not the only process that removes algal particles. A more complete model of coagulation must incorporate zooplankton feeding as a competitive process for the algae.

CONCLUSIONS

Oceanic ecosystems are systems where coagulation is an important if not fully understood process. Algal blooms represent a simple case to explain and to test the coagulation models. It is clear that the use of coagulation kernels will have to be modified to model real world particles accurately. The theoretical emphasis on understanding all of the nuances of spherical particles has been important for us to understand aspects of coagulation, but it is not adequate to understand coagulation in marine systems. Future advances will depend more on a fuller understanding of how the porous, fractal nature of aggregates affects the coagulation kernels, particularly those involving shear and dif ferential sedimentation. An important part of this understanding will involve the incorporation of the fluid flow through the aggregates.

ACKNOWLEDGMENTS

We are grateful to Jeff Haney for his help with programming. Paul Hill and Meredith Newman provided helpful critiques. This work was supported by Office of Naval Research Contract N00014 87-K0005 and U. S. Department of Energy grant DE-FG05-85-ER60341.

REFERENCES

1. Trent, J.D., A.L. Shanks, and M.W. Silver. "In Situ and Laboratory Measurements on Macroscopic Aggregates in Monterey Bay, California," *Limnol. Oceanogr.* 23: 626–635 (1978).
2. Alldredge, A.L. and C.C. Gotschalk. "Direct Observations of the Mass Flocculation of Diatom Blooms: Characteristics, Settling Velocities and Formation of Diatom Aggregates," *Deep-Sea Res.* 36:159–171 (1988).
3. Alldredge, A.L. and C. Gotschalk. "In Situ Settling Behavior of Marine Snow," *Limnol. Oceanogr.* 33:339–351 (1988).
4. Kranck, K. and T. Milligan. "Macroflocs from Diatoms: In Situ Photography of Particles in Bedford Basin, Nova Scotia," *Mar. Ecol. Prog. Ser.* 44:183–189 (1990).
5. Alldredge, A.L. and M.W. Silver. "Characteristics, Dynamics and Significance of Marine Snow," *Prog. Oceanogr.* 20:41–82 (1988).
6. Billet, D.S.M., R.S. Lampitt, A.L. Rice, and R.F.C. Mantoura. "Seasonal Sedimentation of Phytoplankton to the Deep-Sea Benthos," *Nature* 302:520–522 (1983).
7. Lochte, K. and C.M. Turley. "Bacteria and Cyanobacteria Associated with Phytodetritus in the Deep Sea," *Nature* 333:67–69 (1988).
8. Thiel, H., O. Pfannkuche, G. Schriever, K. Lochte, A.J. Gooday, V. Hemleben, R.F.G. Mantoura, C.M. Turley, J.W. Patching, and F. Riemann. "Phytodetritus on the Deep-Sea Floor in a Central Oceanic Region of the Northeast Atlantic," *Biol. Oceanogr.* 6:203–239 (1990).
9. Riebesell, U. "Particle Aggregation during a Diatom Bloom. I. Physical Aspects," *Mar. Ecol. Prog. Ser.* 69:273–280 (1991).
10. Riebesell, U. "Particle Aggregation during a Diatom Bloom. II. Biological Aspects," *Mar. Ecol. Prog. Ser.* 69:281–291 (1991).
11. Wassmann, P., M. Vernet, B.G. Mitchell, and F. Rey. "Mass Sedimentation of *Phaeocystis pouchetii* in the Barents Sea," *Mar. Ecol. Prog. Ser.* 66:183–195 (1990).
12. O'Melia, C.R. and K.S. Bowman. "Origins and Effects of Coagulation in Lakes," *Schweiz. Z. Hydrol.* 46:64–85 (1984).
13. Weilenmann, U., C.R. O'Melia, and W. Stumm. "Particle Transport in Lakes: Models and Measurements," *Limnol. Oceanogr.* 341–418 (1989).
14. Jackson, G.A. "A Model of the Formation of Marine Algal Flocs by Physical Coagulation Processes," *Deep-Sea Res.* 37:1197–1211 (1990).
15. Jackson, G.A. and S. Lochmann. "Effect of Coagulation on Nutrient and Light Limitation of an Algal Bloom," *Limnol. Oceanogr.* 37:77–89 (1992).
16. Hill, P.S. "Reconciling Aggregation Theory with Observed Vertical Fluxes Following Phytoplankton Blooms," *J. Geophys. Res.* 97:2295–2308 (1992).
17. Honeyman, B.D. and P.H. Santschi. "The Role of Particles and Colloids in the Transport of Radionuclides and Trace Metals in the Oceans," *Environmental Particles, Vol. 1*, J. Buffle and H.P. van Leeuwen, Eds. (Chelsea, MI: Lewis Publishers, Inc., 1992).
18. Wells, M.L. and E.D. Goldberg. "Occurrence of Small Colloids in Sea Water," *Nature* 353:342–344 (1991).

19. O'Melia, C.R. "Physicochemical Aggregation and Deposition in Aquatic Systems," in *Environmental Particles, Vol. 2*, H. P. van Leeuwen and J. Buffle, Eds. IUPAC Environmental Analytical and Physical Chemistry Series (Chelsea, MI: Lewis Publishers, 1993).
20. Pruppacher, H.R. and J.D. Klett. *Microphysics of Clouds and Precipitation* (Boston D. Riedel Publishing Co., 1980).
21. O'Melia, C.R. "An Approach to Modeling of Lakes," *Schweiz. Z. Hydrol.* 34:1–33 (1972).
22. Witten, T.A. and M.E. Cates. "Tenuous Structures Disorderly Growth Processes," *Science* 232:1607–1612 (1986).
23. Meakin, P. "Fractal Aggregates in Geophysics," *Rev. Geophys.* 29:317–354 (1991).
24. Tanford, C. *Physical Chemistry of Macromolecules* (New York: John Wiley & Sons, Inc., 1961).
25. Feder, J. *Fractals* (New York: Plenum Press, Inc., 1988).
26. Sutherland, D.N. "A Theoretical Model of Floc Structure," *J. Colloid Interface Sci.* 25:373–380 (1967).
27. Sutherland, D.N. and I. Goodarz-Nia. "Floc Simulation: the Effect of Collision Sequence," *Chem. Eng. Sci.* 26:2071–2085 (1971).
28. Witten, T.A. and L.M. Sander. "Diffusion-Limited Aggregation, a Kinetic Critical Phenomenon," *Phys. Rev. Lett.* 47:1400–1403 (1981).
29. Feder, J. and T. Jøssang. "A Reversible Reaction Limiting Step in Irreversible Immunoglobulin Aggregation," in *Scaling Phenomena in Disordered Systems*, R. Pynn and A. Skjeltorp, Eds. (New York: Plenum Press, Inc., 1985), pp. 99–149.
30. Pratsinis, S.E., S.K. Friedlander, and A.J. Pearlstein. "Aerosol Reactor Theory: Stability and Dynamics of a Continuous Stirred Tank Aerosol Reactor," *AIChE J* 32:177–185 (1986).
31. Koch, W. and S.K. Friedlander. "The Effect of Particle Coalescence on the Surface Area of a Coagulating Aerosol," *J. Colloid Interface Sci.* 140:419–427 (1990).
32. Adler, P.M. "Interaction of Unequal Spheres. I. Hydrodynamic Interaction: Colloidal Forces," *J. Colloid Interface Sci.* 84:461–474 (1981).
33. Han, M. and D.F. Lawler. "Interactions of Two Settling Spheres: Settling Rates and Collision Efficiency," *J. Hydraul. Eng.* 117:1269–1289 (1991).
34. Adler, P.M. "Streamlines in and around Porous Particles," *J. Colloid Interface Sci.* 81:531–535 (1981).
35. Sutherland, D.N. and C.T. Tan. "Sedimentation of a Porous Sphere," *Chem. Eng. Sci.* 25:1948–1950 (1970).
36. Valioulis, I.A. and E.J. List. "Numerical Simulation of a Sedimentation Basin. I. Model Development," *Environ. Sci. Technol.* 18:242–247 (1984).
37. Meakin, P., Z.-Y. Chen, and J.M. Deutch. "The Transitional Friction Coefficient and Time Dependent Cluster Size Distribution of Three Dimensional Cluster-Cluster Aggregation," *J. Chem. Phys.* 82:3786–3789 (1985).
38. Rogak, S.N. and R.C. Flagan. "Stokes Drag on Self-Similar Clusters of Spheres," *J. Colloid Interface Sci.* 134:206–218 (1990).
39. Saffman, P.G. and J.S. Turner. "On the Collision of Drops in Turbulent Clouds," *J. Fluid Mech.* 1:16–30 (1956).
40. Batchelor, G.K. "Mass Transfer from a Particle Suspended in Fluid with a Steady Linear Ambient Velocity Distribution," *J. Fluid Mech.* 95:369–400 (1979).

41. Batchelor, G.K. "Mass Transfer from Small Particles Suspended in Turbulent Fluid," *J. Fluid Mech.* 98:609–623 (1980).
42. Gelbard, F., Y. Tambour, and J.H. Seinfeld. "Sectional Representations for Simulating Aerosol Dynamics," *J. Colloid Interface Sci.* 76:541–556 (1980).
43. Kao, W.V., R.G. Cox, and S.G. Mason. "Streamlines around Single Spheres and Trajectories of Pairs of Spheres in Two-Dimensional Creeping Flows," *Chem. Eng. Sci.* 32:1505–1515 (1977).
44. Poe, G.G. and A. Acrivos. "Closed-Streamline Flows Past RotatingSingle Cylinders and Spheres: Inertia Effects," *J. Fluid Mech.* 72:605–623 (1975).
45. Han, M. "Mathematical Modeling of Heterogeneous Flocculent Sedimentation," PhD Thesis, University of Texas, Austin (1989).
46. Batchelor, G.K. and J.T. Green. "The Hydrodynamic Interaction of Two Small Freely-Moving Spheres in a Linear Flow Field," *J. Fluid Mech.* 56:375–400 (1972).
47. Sournia, A. "Form and Function in Marine Phytoplankton," *Biol. Rev.* 57:347–394 (1982).
48. Varon, M. and M. Shilo. "Ecology of Aquatic Bdellovibrios," *Adv. Aquat. Microbiol.* 2:1–48 (1980).
49. Walsby, A.E. and C.S. Reynolds. "Sinking and Floating," in *The Physiological Ecology of Phytoplankton*, I. Morris, Ed. (Oxford: Blackwell Scientific Publications, 1980).
50. Levandowsky, M. and P.J. Kaneta. "Behaviour in Dinoflagellates," in *The Biology of Dinoflagellates*, F.J.R. Taylor, Ed. (Oxford: Blackwell Scientific Publications, 1987), pp. 360–397.
51. Spero, H.J. "Chemosensory Capabilities in the Phagotrophic Dinoflagellate *Gymnodinium fungiforme*," *J. Phycol.* 21:181–184 (1985).
52. Fitt, W.K. "Chemosensory Responses of the Symbiotic Dinoflagellate *Symbiodinium microadriatica* (Dinophycae)," *J. Phycol.* 21:62–67 (1985).
53. Berg, H.C. *Random Walks in Biology* (Princeton, NJ: Princeton University Press, 1983).
54. Berg, H.C. and D.A. Brown. "Chemotaxis in *Escherichia coli* Analyzed by Three-Dimensional Tracking," *Nature* 239:500–504 (1972).
55. Cullen, J.J., S.G. Horrigan, M.E. Huntley, and F.M.H. Reid. "Yellow Water in La Jolla Bay, California, July 1980. I. A Bloom of the Dinoflagellate, *Gymnodinium flavum* Kofoid and Swezy," *J. Exp. Mar. Biol. Ecol.* 63:67–80 (1982).
56. Round, F.E., R.M. Crawford, and D.G. Mann. *The Diatoms* (Cambridge: Cambridge University Press,1990).
57. Booker, M.J. and A.E. Walsby. "The Relative Form Resistance of Straight and Helical Blue-Green Algal Filaments," *Br. Phycol.* 14:141–150 (1979).
58. Smayda, T.R. "Experimental Observations on the Flotation of Marine Diatoms. I. *Thalassiosira cf. nana, Thalassiosira rotula,* and *Nitzchia seriata*," *Limnol. Oceanogr.* 10:499–509 (1965).
59. Smayda, T.R. "Experimental Observations on the Flotation of Marine Diatoms. II. *Skeletonema costatum* and *Rhizosolenia setigera*," *Limnol. Oceanogr.* 11:18–34 (1966).
60. Smayda, T.R. "Experimental Observations on the Flotation of Marine Diatoms. III. *Bacteriastrum hyalinum* and *Chaetoceros lauderi*," *Limnol. Oceanogr.* 11:35–43 (1966).
61. Goldsmith, H.L. and S.G. Mason. "The Microrheology of Dispersions," in *Rheology — Theory and Application, Vol. 2*, F.R. Eirich, Ed. (New York: Academic Press, Inc., 1967), pp. 85–250.

62. McLauchlan, J., A.G. McInnes, and M. Falk. "Studies on the Chitan (Chitin: poly-N-acetylglucosamine) Fibers of the Diatom *Thalassiosira fluviatilis* Hystedt. I. Production and Isolation of Chitan Fibres," *Can. J. Bot.* 43:707–713 (1965).
63. Bienfang, P.K., P.J. Harrison, and L.M. Quarmby. "Sinking Rate Response to Depletion of Nitrate, Phosphate, and Silicate in Four Marine Diatoms," *Mar. Biol.* 67:295–302 (1982).
64. Bienfang, P.K. and P.J. Harrison. "Sinking-Rate Response of Natural Assemblages of Temperate and Subtropical Phytoplankton to Nutrient Depletion," *Mar. Biol.* 83:293–300 (1984).
65. Passarelli, R.E. and R.C. Srivastava. "A New Aspect of Snowflake Aggregation Theory," *J. Atmos. Sci.* 36:484–493 (1979).
66. McCave, I.N. "Size Spectra and Aggregation of Suspended Particles in the Deep Ocean," *Deep-Sea Res.* 31:329–352 (1984).
67. Honeyman, B.D. and P.H. Santschi. "A Brownian-Pumping Model for Oceanic Trace Metal Scavenging: Evidence from TH Isotope," *J. Mar. Res.* 47:951–992 (1989).
68. Kiørboe, T., K.P. Andersen, and H.G. Dam. "Coagulation Efficiency and Aggregate Formation in Marine Phytoplankton," *Mar. Biol.* 107:235–245 (1990).
69. Logan, B.E. and D.B. Wilkinson. "Fractal Geometry of Marine Snow and Other Biological Aggregates," *Limnol. Oceanogr.* 35 :130–136 (1990).
70. Logan, B.E. and Alldredge, A.L. "The Increased Potential for Nutrient Uptake by Flocculating Diatoms," *Mar. Biol.* 101:443–450 (1989).
71. Silver, M.W. and M.W. Gowing. "The 'Particle' Flux: Origins and Biological Components," *Prog. Oceanogr.* 26:75–113 (1991).
72. Shanks, A.L. and E.W. Edmondson. "The Vertical Flux of Metazoans (Holoplankton, Meiofauna, and Larval Invertebrates) Due to Their Association with Marine Snow," *Limnol. Oceanogr.* 35:455–463 (1990).
73. Smith, W.O., L.A. Codispoti, D.M. Nelson, T. Manley, E.J. Buskey, H.J. Niebauer, and G.F. Cota. "Importance of *Phaeocystis* Blooms in the High-Latitude Ocean Carbon Cycle," *Nature* 352:514–516 (1991).
74. Lancelot, C. "Factors Affecting Phytoplankton Extracellular Release in the Southern Bight of the North Sea," *Mar. Ecol. Prog. Ser.* 12:115–121 (1983).
75. Lancelot, C. "Metabolic Changes in *Phaeocystis pouchetii* (Hariot)Lagerheim during the Spring Bloom in Belgian Coastal Waters," *Estuar. Coast. Shelf Sci.* 18:593–600 (1984).
76. Lancelot, C., G. Billen, A. Sournia, T. Weisse, F. Colijn, M.J.W. Veldhuis, A. Davies, and P. Wassman. "*Phaeocystis* Blooms and Nutrient Enrichment in the Continental Coastal Zones of the North Sea," *Ambio* 16:38–46 (1987).
77. Estep, K.W., J.Ch. Nejstgaard, H.R. Skjoldal, and F. Rey. "Predation by Copepods upon Natural Populations of *Phaeocystis pouchetii* as a Function of the Physiological State of the Prey," *Mar. Ecol. Prog. Ser.* 67:235–249 (1990).

INDEX

A

B

C

D

I

K

L

M

N

O

P